**Neuroimaging in Developmental
Clinical Neuroscience**

Neuroimaging in Developmental Clinical Neuroscience

Edited by

Judith M. Rumsey

Neurodevelopmental Disorders Branch
Division of Developmental Translational Research
National Institute of Mental Health, Rockville, MD, USA

Monique Ernst

Neurodevelopment of Reward Systems, Mood and Anxiety
Disorders Program
Emotional Development and Affective Neuroscience Branch
National Institute of Mental Health, Bethesda, MD, USA

With a Foreword by

Husseini K. Manji

Chief, Laboratory of Molecular Pathophysiology
Director for the Mood and Anxiety Disorders Program
National Institute of Mental Health, Bethesda, MD, USA

CAMBRIDGE
UNIVERSITY PRESS

CAMBRIDGE UNIVERSITY PRESS
Cambridge, New York, Melbourne, Madrid, Cape Town, Singapore, São Paulo,
New Delhi

Cambridge University Press
The Edinburgh Building, Cambridge CB2 8RU, UK

Published in the United States of America by
Cambridge University Press, New York

www.cambridge.org
Information on this title: www.cambridge.org/9780521883573

First published 2009

Printed in the United Kingdom at the University Press, Cambridge

A catalog record for this publication is available from the British Library

Library of Congress Cataloging-in-Publication Data
Neuroimaging in developmental clinical neuroscience / edited by Judith M.
Rumsey, Monique Ernst; with a foreword by Husseini K. Manji.
    p. ; cm.
   Rev. ed. of: Functional neuroimaging in child psychiatry / edited by Monique
Ernst and Judith M. Rumsey. 2000.
   Includes bibliographical references and index.
   ISBN 978-0-521-88357-3 (hardback)
1. Pediatric neuropsychiatry.   2. Brain–Imaging.   3. Developmental
neurophysiology.   I. Rumsey, Judith M.   II. Ernst, Monique, 1953–
III. Functional neuroimaging in child psychiatry.   IV. Title.
   [DNLM: 1. Mental Disorders–diagnosis. 2. Brain Diseases–diagnosis.
3. Child Development. 4. Child. 5. Diagnostic Imaging–methods.
6. Neuropsychology–methods. WS 350 N4943 2009]
   RJ486.5.N485 2009
   618.92′890754–dc22
                                                                          2008044125

ISBN 978-0-521-88357-3 hardback

Contents

List of Contributors   *page* vii
Foreword   xiii
Preface   xvii

Section 1 Normal developmental processes 1

Introduction to Section 1   3

1 Morphological development of the brain: what has imaging told us?   5
Lisa H. Lu and Elizabeth R. Sowell

2 Neural correlates of the development of cognitive control   22
Silvia A. Bunge and Eveline A. Crone

3 Neurobiology of emotion regulation in children and adults   38
Ian H. Gotlib and Jutta Joormann

4 Goal-directed behavior: evolution and ontogeny   53
Monique Ernst and Michael G. Hardin

5 Charting brain mechanisms for the development of social cognition   73
Kevin A. Pelphrey and Susan B. Perlman

6 Language and the developing brain: insights from neuroimaging   91
Kristin McNealy, Mirella Dapretto and Susan Bookheimer

Section 2 Atypical processes in developmental neuropsychiatric disorders 109

Introduction to Section 2   111

7   A pathophysiology of attention deficit/
      hyperactivity disorder: clues from
      neuroimaging   113
      Jeffery N. Epstein

8   Brain imaging of autism spectrum
      disorders   130
      Monica Zilbovicius, Nathalie Boddaert
      and Nadia Chabane

9   Neuroimaging of schizophrenia and its
      development   147
      Matcheri S. Keshavan, Vaibhav A. Diwadkar,
      Konasale Prasad and Jeffrey A. Stanley

10   Cortico-limbic development in bipolar
      disorder: a neuroimaging view   161
      Jessica H. Kalmar, Maulik P. Shah and Hilary
      P. Blumberg

11   Anxiety and depressive disorders   183
      Daniel S. Pine

12   Disturbances of fronto-striatal circuits
      in Tourette syndrome and obsessive-
      compulsive disorder   199
      Rachel Marsh, Daniel A. Gorman, Jason Royal
      and Bradley S. Peterson

13   From genes to brain to behavior:
      the case of fragile X syndrome   217
      Susan M. Rivera and Allan L. Reiss

14   Alcohol exposure and the developing
      human brain   229
      James M. Bjork

15   Neuroimaging as a tool for unlocking
      developmental pathophysiology in
      anorexia and bulimia nervosa   245
      Guido K. W. Frank and Walter H. Kaye

Section 3   Ethical issues   259

      Introduction to Section 3   261

16   Legal and ethical considerations in
      pediatric neuroimaging research   263
      Clinton D. Hermes

Section 4   Techniques and integration
              with other research approaches   277

      Introduction to Section 4   279

17   BOLD fMRI: an update with emphasis
      on pediatric applications   281
      Liat Levita, Rebecca M. Jones and B. J. Casey

18   Magnetic resonance spectroscopy:
      methods and applications in developmental
      clinical neuroscience   296
      Marisa M. Silveri, Deborah Yurgelun-Todd
      and Perry Renshaw

19   Diffusion tensor imaging in developmental
      clinical neuroscience   314
      Dae-Shik Kim

20   Arterial spin labeling perfusion magnetic
      resonance imaging in developmental
      neuroscience   326
      Jiong-Jiong Wang, Hengyi Rao
      and John A. Detre

21   Neuroimaging of treatment effects in
      developmental neuropsychiatric disorders   344
      Steven R. Pliszka and David C. Glahn

22   Functional alleles, neuroimaging and
      intermediate phenotypes in the
      deconstruction of complex behavioral
      variation   365
      David Goldman, Beata Buzas and Ke Xu

Section 5   Progress and future directions   383

      Introduction to Section 5   385

23   Neuroimaging in developmental clinical
      neuroscience today   387
      Judith M. Rumsey and Monique Ernst

Appendix A: Functional MRI educational
resources and software   401
Appendix B: Neuroinformatics and
neuroethics resources   403
Glossary   407
Index   437

Contributors

James M. Bjork, Ph.D.
Laboratory of Clinical and Translational Studies
National Institute of Alcohol Abuse and Alcoholism
National Institutes of Health
Bethesda, MD, USA

Hilary P. Blumberg, M.D.
Associate Professor of Psychiatry and Diagnostic
Radiology
Director, Mood Disorders Research Program
Yale University School of Medicine
New Haven, CT, USA

Nathalie Boddaert, M.D.
Necker Hospital
Pariv V University
Paris, France

Susan Bookheimer, Ph.D.
Professor, Psychiatry and Biobehavioral Sciences
UCLA School of Medicine
Ahmanson-Lovelace Brain Mapping Center
University of California, Los Angeles
Los Angeles, CA, USA

Silvia A. Bunge, Ph.D.
Assistant Professor
Psychology Department
Center for Mind and Brain
University of California
Davis, CA, USA

Beata Buzas, Ph.D.
National Institute of Alcohol Abuse and Alcoholism
Rockville, MD, USA

B. J. Casey, Ph.D.
Director of the Sackler Institute for Developmental
Psychobiology
Sackler Professor of Developmental Psychobiology
Departments of Psychiatry, Neurology and
Neuroscience
Weill Cornell Medical College
New York, NY, USA

Nadia Chabane, M.D., Ph.D.
Robert Debre Hospital
Paris VII University
Paris, France

Eveline A. Crone, Ph.D.
Brain and Development Laboratory
Department of Developmental Psychology
Leiden University
Leiden, The Netherlands

Mirella Dapretto, Ph.D.
Associate Professor
Department of Psychiatry and Biobehavioral
Sciences
Ahmanson-Lovelace Brain Mapping Center
University of California, Los Angeles
Los Angeles, CA, USA

John A. Detre, M.D.
Associate Professor
Departments of Neurology and Radiology
University of Pennsylvania
Philadelphia, PA, USA

Vaibhav A. Diwadkar, Ph.D.
Department of Psychiatry and Behavioral
Neuroscience
Wayne State University School of Medicine
Detroit, MI, USA

Jeffery N. Epstein, Ph.D.
Associate Professor
Department of Pediatrics
Cincinnati Children's Hospital Medical Center
Cincinnati, OH, USA

Monique Ernst, M.D., Ph.D.
Head, Neurodevelopment of Reward Systems
Emotional Development and Affective
Neuroscience Branch
Mood and Anxiety Disorders Program
National Institute of Mental Health
Bethesda, MD, USA

Guido K. W. Frank, M.D.
University of Colorado at Denver and Health
Sciences Center
Department of Psychiatry
The Children's Hospital (Pavilion)
Denver, CO, USA

David C. Glahn, Ph.D.
Director of Neuroimaging Core in Psychiatry
Associate Professor
Department of Psychiatry and Research
Imaging Center
University of Texas Health Science Center
at San Antonio
San Antonio, TX, USA

David Goldman, M.D.
Chief, Laboratory of Neurogenetics
National Institute on Alcohol Abuse and Alcoholism
Bethesda, MD, USA

Daniel A. Gorman
Department of Psychiatry
Hospital for Sick Children and
University of Toronto
Toronto, Ontario, Canada

Ian H. Gotlib, Ph.D.
Psychology Department
Stanford University
Stanford, CA, USA

Michael G. Hardin, M.S.
Graduate Research Fellow
Emotional Development and Affective Neuroscience
Branch
Mood and Anxiety Program
National Institute of Mental Health
Bethesda, MD, USA

Clinton D. Hermes, J.D.
Senior Vice President and General Counsel
St. Jude Children's Research Hospital
Memphis, TN, USA

Rebecca M. Jones, M.Phil.
Sackler Institute for Developmental Psychobiology
Weill Medical College of Cornell University
New York, NY, USA

Jutta Joormann, Ph.D.
Assistant Professor
Department of Psychology
University of Miami
Coral Gables, FL, USA

Jessica H. Kalmar, Ph.D.
Associate Research Scientist
Yale University School of Medicine
New Haven, CT, USA

Walter H. Kaye, M.D.
Professor of Psychiatry
University of Pittsburgh Medical Center
Western Psychiatric Institute and Clinic
Pittsburgh, PA, USA

Matcheri S. Keshavan, M.D.
Professor and Associate Chair
Department of Psychiatry and Behavioral
Neuroscience
Wayne State University School of Medicine
Detroit, MI, USA

Dae-Shik Kim, Ph.D.
Associate Professor
Boston University School of Medicine
Boston, MA, USA

Liat Levita, Ph.D.
Instructor of Psychology in Psychiatry
Sackler Institute for Developmental Psychobiology
Weill Medical College of Cornell University
New York, NY, USA

Lisa H. Lu, Ph.D.
Postdoctoral Fellow
Laboratory of Neuro Imaging
David Geffen School of Medicine
University of California
Los Angeles, CA, USA

Rachel Marsh, Ph.D.
Division of Child Psychiatry
Department of Psychiatry
Columbia University
New York State Psychiatric Institute
New York, NY, USA

Kristin McNealy
Doctoral Student in Neuroscience
Neuroscience Interdepartmental Program
Center for Culture, Brain and Development
Ahmanson-Lovelace Brain Mapping Center
University of California, Los Angeles
Los Angeles, CA, USA

Kevin A. Pelphrey, Ph.D.
Associate Professor
Department of Psychology
Carnegie Mellon University
Pittsburgh, PA, USA

Susan B. Perlman, M.A.
Doctoral Student
Department of Psychology
Carnegie Mellon University
Pittsburgh, PA, USA

Bradley S. Peterson, M.D.
Professor
Department of Psychiatry
Columbia College of Physicians and Surgeons
Somers, NY, USA

Daniel S. Pine, M.D.
Chief, Emotional Development and Affective
Neuroscience Branch
National Institute of Mental Health
Bethesda, MD, USA

Steven R. Pliszka, M.D.
Professor and Vice Chair
Chief, Division of Child and Adolescent Psychiatry
Department of Psychiatry
University of Texas Health Sciences Center at
San Antonio
San Antonio, TX, USA

Konasale Prasad, M.D.
Department of Psychiatry
University of Pittsburgh School of Medicine
Pittsburgh, PA, USA

Hengyi Rao, Ph.D.
Department of Psychiatry and Neurology
Center for Functional Neuroimaging
University of Pennsylvania
Philadelphia, PA, USA

Allan L. Reiss, M.D.
Howard C. Robbins Professor
Department of Psychiatry and Behavioral Sciences
Director of the Center for Interdisciplinary Brain
Sciences Research
Stanford University School of Medicine
Stanford, CA, USA

Perry Renshaw, M.D., Ph.D.
Department of Psychiatry
McLean Brain Imaging Center
McLean Hospital
Harvard University Medical School
Belmont, MA, USA

Susan M. Rivera, Ph.D.
Assistant Professor
Department of Psychology
M.I.N.D. Institute

Center for Mind and Brain
University of California, Davis
Davis, CA, USA

Jason Royal, Ph.D.
Division of Child and Adolescent Psychiatry
Columbia University
New York State Psychiatric Institute
New York, NY, USA

Judith M. Rumsey, Ph.D.
Chief, Neurodevelopment and Neuroimaging
Program and
Executive Function and Attention Deficit
Hyperactivity Disorder Program
Neurodevelopmental Disorders Branch
Division of Developmental Translational Research
National Institute of Mental Health
Bethesda, MD, USA

Maulik P. Shah, B.A.
Medical student
Mood Disorders Research Program
Yale University School of Medicine
New Haven, CT, USA

Marisa M. Silveri, Ph.D.
Department of Psychiatry
Harvard Medical School
Cognitive Neuroimaging and Neuropsychology
Laboratory
Brain Imaging Center
McLean Hospital
Belmont, MA, USA

Elizabeth R. Sowell, Ph.D.
Associate Professor of Neurology
David Geffen School of Medicine
Laboratory of Neuro Imaging
University of California, Los Angeles
Los Angeles, CA, USA

Jeffrey A. Stanley, Ph.D.
Department of Psychiatry and Behavioral
Neuroscience

Wayne State University School of Medicine
Detroit, MI, USA

Henning U. Voss, Ph.D.
Assistant Professor of Physics in Radiology
Citigroup Biomedical Imaging Center
Weill Cornell Medical College
New York, NY, USA

Jiong-Jiong Wang, Ph.D.
Research Assistant Professor
Departments of Radiology and Neurology
Center for Functional Neuroimaging
University of Pennsylvania
Philadelphia, PA, USA

Ke Xu, M.D., Ph.D.
Department of Psychiatry
School of Medicine

Yale University
New Haven, CT, USA

Deborah Yurgelun-Todd, Ph.D.
Cognitive Neuroimaging/Neuropsychology
Laboratory
Brain Imaging Center
McLean Hospital
Harvard Medical School
Belmont, MA, USA

Monica Zilbovicius, M.D.
Unité INSERM-CEA
Service Hospitalier Frederic Joliot
Orsay, France

Foreword

Neuroimaging has in many ways revolutionized the field of psychiatry, and its ability to help clinicians and researchers understand the mechanisms of childhood psychiatric disorders should not be underestimated. This volume reviews recent developments in neuroimaging techniques and their implications for child psychiatry. It is a unique book in that it focuses on children and integrates brain mapping with genetics and behavioral testing. What is truly astounding is the rapid evolution of this field since the seminal first edition of this book appeared in 2000.

In the past decade, one of the enormous shifts in our thinking about psychiatry has been the growing appreciation that many, if not all, major psychiatric disorders have their antecedents in childhood. It is now clear that the major psychiatric disorders are serious, debilitating, life-shortening illnesses that affect millions of people worldwide. The major psychiatric disorders are clearly "chronic illnesses of the young," characterized by multiple episodes of symptom exacerbation, residual symptoms between episodes, and functional impairment. These illnesses arise from the complex, developmentally determined interaction of multiple genes and environmental factors, and the phenotypic expression of the disease includes not only affective disturbance, but also a constellation of cognitive, motor, autonomic, endocrine, and sleep/wake abnormalities. Research on the biological underpinnings of the major psychiatric disorders has therefore begun to focus less on absolute changes in individual neurotransmitters, and more on the role of neural circuits and synapses, and the processes controlling their function.

Revolutionary techniques in neuroimaging and genetics, combined with epidemiological and longitudinal research, have now firmly established that psychiatric disorders are developmental disorders. They emerge while the brain is still developing and evolve into the more easily recognized chronic and disabling course. The field of childhood neuroimaging has played a key role in redefining this thinking. As Section 2 of this book so thoroughly describes, neuroimaging has helped identify atypical brain developmental processes associated with a variety of disorders. These include: attention deficit/hyperactivity disorder (ADHD) (Chapter 7); autism and autism spectrum disorders (Chapter 8); schizophrenia (Chapter 9); bipolar disorder (Chapter 10); anxiety and depressive disorders (Chapter 11); Tourette syndrome and obsessive-compulsive disorder (OCD) (Chapter 12); fragile X syndrome (Chapter 13); alcohol-related disorders (Chapter 14); and eating disorders (Chapter 15). The work outlined in these chapters clarifies the notion that studying these disorders early in life is most likely to eventually allow us to make a major impact on their course. The neuroimaging of normally developing children provides investigators with the unparalleled opportunity to assess human structural brain development and the key domains of normal developmental processes underlying cognitive and regulatory aspects of human behavior. With regards to psychopathology, neuroimaging provides investigators with the opportunity to study and identify risk factors that may appear before the symptoms of any illness and which are therefore unconfounded by treatment. Among its many valuable advantages, neuroimaging also provides investigators with the tools to track the longitudinal course of an illness, to determine correlates of symptom reduction, and to directly assess the effects of medications or other interventions.

The chapters included in Sections 3 and 4 of this book describe many of the most recent advances in neuroimaging with respect to ethical considerations, and technology, experimental design, and data analysis. They also address which neuroimaging techniques can safely be used to study children – an issue of enormous importance. The most striking thing about these sections of the book is how much the field has expanded – both in scope and volume – since 2000. Because neuroimaging affords us the possibility of directly studying the organ of interest, it has significantly altered how we conceptualize, diagnose, understand, and treat psychiatric disorders. Our expanded understanding, in turn, has led us to refine and improve neuroimaging techniques. There have been advances not only in neuroimaging technology (and particularly in temporal and spatial resolution), but also the adaptation and development of novel cognitive neuroscience techniques. This integrated circle of growth has allowed investigators to conduct innovative translational research directly with children.

The National Institutes of Health (NIH) broadly defines translational research as "the process of applying ideas, insights and discoveries generated through basic scientific inquiry to the treatment or prevention of human disease." True translational research takes this idea a step further and encompasses the bidirectional interplay between clinical and laboratory research – from bench to bedside and back again. Toward that end, neuroimaging has allowed investigators to use methods in children derived directly from cognitive neuroscience approaches used in various settings. These approaches create a basis for improving our understanding, assessment, and treatment of psychiatric disorders in children and adults; these approaches also aid the investigation of mechanisms of treatment response and provide imaging phenotypes for understanding genetic effects on brain and behavior. Ultimately, there is potential for the interventional application of the techniques themselves.

Much has been written lately about the future of predictive and, ultimately, preemptive tools in psychiatry. Part of the rationale for such thinking is the knowledge gleaned from neuroimaging and other studies showing that structural and other changes precede the development of psychiatric disorders in asymptomatic children and adolescents.

This knowledge has worked synergistically with that derived from the growing field of early intervention in psychiatric disorders to change our thinking from palliative to predictive to preventive care in psychiatry. There is thus a clear need to refine these multifactorial diseases into mechanism-based subcategories so that particular target-based therapies can be matched to particular markers in subgroups of patients. Biological markers or biomarkers are quantitative measurements that provide information about biological processes, a disease state, or response to treatment. The neuroimaging biomarkers discussed in this volume thus hold the potential to provide a better understanding of the etiology and pathophysiology of the complex and heterogeneous psychiatric disorders. Moreover, biomarkers hold considerable potential to identify patients who are likely to respond to a particular treatment modality. Indeed, the Food and Drug Administration has recently changed its definitions and requirements of biomarkers to include broader categories, and to encourage submissions of putative biomarkers.

Since the first edition of this book was published in 2000, an extraordinary number of articles have been published regarding the ability of early intervention to delay the onset or positively influence the course of many psychopathological conditions. Neuroimaging – both of children and adults – has been an integral part of this evolving field. However, more research is certainly needed to further elucidate this topic, to identify and refine the prodromal manifestations of each psychopathological condition, to introduce and use appropriate measuring instruments, and to apply validated interventions, particularly in children. Neuroimaging is one of the most powerful tools in our armament to predict psychopathology. Finally, and as Chapter 16 of this book so excellently points out, there are multiple ethical issues concerning neuroimaging in children, and these must be examined very carefully.

Given the varied nature of presentation and variability in course of many of the major psychiatric illnesses, separating and appropriately treating at-risk individuals constitutes a serious challenge. Careful selection of the subjects who might benefit from an early intervention and appropriate study designs to correctly evaluate outcome, though ongoing, are still needed. Prospective large-scale studies from high-risk populations with appropriate biological markers, likely neuroimaging in nature, could help identify "real" high-risk subjects and develop new treatment algorithms. Moreover, the primary outcome measure of randomized controlled trials should be not only syndromic resolution, but also functional and psychosocial recovery, along with cognitive improvement. The results from studies conducted with high-risk subjects will eventually help identify clinical prodromes and biological markers in subjects from the general population as well.

As the outstanding chapters in this volume have highlighted, the future of early intervention depends on our ability to identify individuals at risk for developing major psychiatric illnesses, and the capacity to provide targeted treatment that specifically prevents onset or recurrence of episodes. Though such work will not easily be put into practice, we now have significant clues about not only disease onset and progression, but also about the tools that are needed to help implement successful early intervention.

The back cover of the previous 2000 edition of this book noted that the book's focus on children and on integrating brain mapping with genetics and behavioral testing was "an interface *that is likely to become* fundamental to functional neuroimaging" (italics mine). Eight years later, these prescient words reflect how quickly extraordinary innovations can be adopted in medicine when they truly advance our understanding. The excellent chapters contained in this book highlight how far the field of neuroimaging in childhood has brought us but, more importantly, how far it can yet take us.

Husseini K. Manji, M.D.

Preface

Neuroimaging in Developmental Clinical Neuro-science provides a broad review of the use of neuroimaging in the emerging field of developmental clinical neuroscience. This volume updates and expands the earlier *Functional Neuroimaging in Child Psychiatry*, edited by Monique Ernst and Judith Rumsey and published in 2000. Since the publication of the earlier volume, there has been an exponential surge in the number of neuroimaging publications addressing the development of the pediatric brain. Paralleling this growing literature is a widening of the scope and diversity of the studies, with respect to both scientific focus and the methods applied.

A heightened focus has been directed onto normative development in a variety of domains, including cognitive control, emotion regulation, goal-directed behavior, social cognition, and language. Recent advances in imaging technologies, such as blood-oxygen-level-dependent (BOLD) functional magnetic resonance imaging (fMRI), diffusion tensor imaging (DTI), perfusion imaging using arterial spin labeling (ASL), and magnetic resonance spectroscopy (MRS) now provide a rich armamentarium of techniques whose integration with one another and with other neuroscience approaches offers unprecedented opportunities for elucidating human neurodevelopment. Structural magnetic resonance imaging, one of the oldest of the techniques included in this volume, has likewise advanced, as new analytic techniques have enhanced our ability to derive meaningful information from anatomical images.

These approaches are increasingly being applied to delineate neurocircuitry involved in developmental neuropsychiatric disorders and have begun to provide a basis for advancing assessment and treatment. Beyond this, these approaches are being expanded to investigate mechanisms of treatment response and to provide imaging phenotypes for understanding genetic effects on brain and behavior. And finally, there are hints of potential for interventional applications of these techniques themselves.

This book is organized into five sections. Section 1 is dedicated to the understanding of normative pediatric brain-behavior development. This section describes our knowledge of human structural brain development from a neuroimaging perspective. Key domains of normal developmental processes underlying cognitive and regulatory aspects of human behavior are addressed in six chapters that focus primarily on functional brain measures, including fMRI and electrophysiology. Section 2 considers atypical developmental processes associated with a variety of disorders, including attention deficit/hyperactivity disorder (ADHD), autism spectrum disorders, schizophrenia, bipolar disorder, anxiety and depressive disorders, Tourette syndrome and obsessive-compulsive disorder (OCD), fragile X syndrome, alcohol-related disorders, and

eating disorders. Section 3 will provide the reader with essential knowledge of legal, regulatory, and ethical guidelines and considerations governing research involving children. Section 4 addresses neuroimaging techniques that can safely be used to study children. Also included are two special topics that integrate neuroimaging with other neuroscience approaches – the study of responses to pharmacological and other interventions and imaging genetics, an emerging approach that shows promise for deconstructing complex behavioral phenotypes. Finally, in Section 5 we highlight the value of a developmental perspective and the progress achieved in understanding human brain maturation and developmental neuropsychiatric disorders using these approaches. Finally, we outline some critical needs and challenges and consider directions for future research.

The authors wish to thank David Shore, M.D., Associate Director for Clinical Research, National Institute of Mental Health (NIMH) and Benedetto Vitiello, M.D., Chief, Child and Adolescent Treatment and Preventive Intervention Research Branch, NIMH for their helpful review of selected material. We also wish to thank our families for their patience and support while we dedicated so much of our energy toward this unique volume.

Normal developmental processes

Introduction to Section 1

The advent of magnetic resonance imaging (MRI) in the 1980s and functional MRI in 1991 was revolutionary for developmental clinical neuroscience. In addition to improved resolution relative to older imaging techniques, the lack of ionizing radiation and safety made it ethically possible to study healthy, typically developing children and to do so repeatedly. The rapid development of a variety of techniques, all using conventional MRI scanners, quickly expanded research to allow researchers to study, not only structural brain development, but functional and metabolic development as well.

To provide an understanding of normal brain development, the six chapters in this section describe our current knowledge of and approaches to studying structural and functional brain development. Lu and Sowell describe and richly illustrate the morphological development of the human brain (Chapter 1). Bunge and Crone address the processes, neural correlates of, and developmental changes in, cognitive control (Chapter 2). Gotlib and Joormann discuss emotion regulation and stress reactivity, including both neuroimaging and neuroendocrine aspects (Chapter 3). In Chapter 4, Ernst and Hardin describe a heuristic model of decision making and preliminary developmental findings. In Chapter 5, Pelphrey and Perlman describe key constructs and early findings in the emerging field of developmental social cognitive neuroscience. Finally, McNealy, Dapretto, and Bookheimer provide a comprehensive look at the neural correlates of the multifaceted aspects of language development (Chapter 6).

Not only is the literature described key to understanding healthy, typical, or normal brain development, but it also lays a foundation for understanding deviations in development associated with neuropsychiatric disorders, particularly those beginning in childhood and adolescence. The domains and processes covered are critical to self-regulation and healthy behavioral functioning and are those most often impaired in developmental neuropsychiatric disorders. Much of the work to date has involved cross-sectional comparisons of relatively small samples of different age groups. Going forward, longitudinal designs, increasingly being adopted in structural imaging studies, hold promise for mapping developmental trajectories with improved sensitivity to maturational changes.

Morphological development of the brain: what has imaging told us?

Lisa H. Lu and Elizabeth R. Sowell

Introduction

Rapid advances in imaging technology have yielded a significant wealth of knowledge about maturational trajectories of human brain development. Magnetic resonance imaging (MRI) findings are based on signal intensity variations which differentiate tissue types, and regional developmental changes in tissue distribution are thought to reflect cellular changes known from postmortem studies. Imaging studies render longitudinal evaluations possible, and examination of within-subject development across time increases our ability to detect maturational changes embedded within the anatomical variability between individuals. We describe normal morphological maturation findings from three basic approaches for analyzing T_1- and T_2-weighted data: volumetric, voxel-based morphometry (VBM), and cortical pattern-matching (CPM). Researchers are beginning to link morphological changes to cognitive development, and these efforts represent the next wave of fruitful investigations.

Normal morphological maturation

Volumetric image analysis

First attempts to quantify structural maturation in vivo used stereotaxic region definition schemes because image resolution was relatively low (i.e.,

4- to 5-mm slice thickness) compared to more recent studies (i.e., 1- to 1.5-mm slice thickness) (Giedd et al., 1996a; Jernigan et al., 1991; Reiss et al., 1996). Some of these studies assessed whole brain tissue volumes (Caviness et al., 1996; Courchesne et al., 2000), while others used manual region definition on a slice-by-slice basis using anatomical landmarks (Giedd et al., 1996b; Lange et al., 1997; Sowell and Jernigan 1998; Sowell et al., 2002c). Algorithms have also been used to warp standardized lobar measures to individual brains and thus define lobar regions automatically (Giedd et al., 1999).

These early studies found decreasing cortical gray matter volume with age while white matter volume increased with age (Jernigan and Tallal, 1990; Jernigan et al., 1991; Pfefferbaum et al., 1994; Reiss et al., 1996) after controlling for differences in overall brain volume (Pfefferbaum et al., 1994; Reiss et al., 1996). Within lobar regions, frontal and parietal lobes increased in gray matter during childhood years (Giedd et al., 1999, 2006; Sowell et al., 2002c) before decreasing during adolescent years (Giedd et al., 1999, 2006). These changes in lobar gray matter volume occurred concomitantly with increasing white matter volume in the corresponding lobes (Giedd et al., 1999, 2006; Sowell et al., 2002c). The most notable changes during childhood and adolescence occurred in more dorsal cortices (Jernigan et al., 1991). More ventral cortices of the temporal lobes changed less dramatically between childhood and

Neuroimaging in Developmental Clinical Neuroscience, eds. Judith M. Rumsey and Monique Ernst. Published by Cambridge University Press. © Cambridge University Press 2009.

adolescence (Giedd et al., 1996b; Sowell et al., 2002c). Most studies showed decreasing subcortical gray matter volume with age with overall brain volume controlled (Jernigan et al., 1991; Reiss et al., 1996; Sowell et al., 2002c), although one study found decreasing striatal volume only for boys (Giedd et al., 1996a) and another reported increasing hippocampal volume for girls and increasing amygdalar volume for boys (Giedd et al., 1996b).

Voxel-based morphometry

Volumetric studies can address gross lobar structural changes but are limited in addressing maturational changes with finer spatial resolution. Voxel-based morphometry (VBM) (Ashburner and Friston, 2000), which was initially used to evaluate functional imaging data, has been used to assess structural effects during normal development on a voxel-by-voxel basis (Paus et al., 1999; Sowell et al., 1999a, 1999b). Essentially, VBM entails automated spatial normalization of volumes into a standard coordinate space and scaling of images so that each voxel coordinate is anatomically comparable across subjects. Tissue segmentation and spatial smoothing is then used to assess localized differences in gray matter and/or white matter.

We used VBM to localize age-related gray matter density reductions between childhood and adolescence in 18 normally developing individuals between 7 and 16 years of age (Sowell et al., 1999a). Gray matter density refers to the proportion of signal intensity that is segmented as gray matter by automated segmentation software within a smoothing kernel. We found that gray matter volume reduction observed in frontal and parietal lobes in the volumetric studies was mainly driven by gray matter density reduction in the dorsal region of these cortices. The parietal cortex changed the most in both the volumetric and VBM assessments of gray matter, and relatively little change occurred in the more ventral cortices of the temporal and occipital lobes (Figure 1.1). The prominent finding in the parietal lobes, relative to the frontal lobes, was not expected given the known posterior to anterior

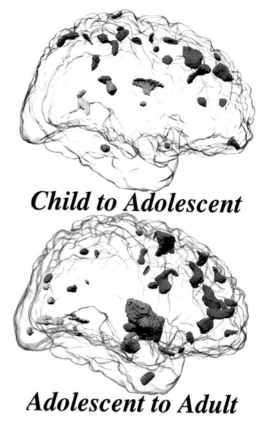

Figure 1.1. Age-related reductions in gray matter density. *Top*: Statistical map representing gray matter density reduction between childhood and adolescence. *Bottom*: Gray matter density reduction between adolescence and adulthood. These maps are three-dimensional renderings of traditional statistical maps shown inside the transparent cortical surface of one representative subject's brain. Lobes and the subcortical region were defined anatomically on the same subject's brain. Color coding is applied to each cluster based on its location within the representative brain. Clusters are shown in the frontal lobes (*purple*), parietal lobes (*red*), occipital lobes (*yellow*), temporal lobes (*blue*), and subcortical region (*green*). Reproduced with permission from Sowell et al. (2004b).

progression of maturational cellular events. We tested the hypothesis that frontal lobe changes occur later in adolescence by conducting a VBM study focusing on the adolescent to adult age range (Sowell et al., 1999b). Whereas cortical changes were

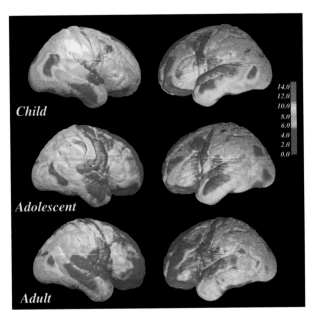

Figure 1.2. Age-related differences in cortical surface variability. Maps showing cortical surface variability in the average child ($n = 14$), the average adolescent ($n = 11$), and the average young adult ($n = 10$). The color bar indicates variability within each group as the root mean square magnitude (in mm) of displacement vectors required to map each individual onto the group average surface mesh. Note that this map is representative of residual brain shape variability after affine transformation (translation, rotation, scaling, and skew/shear) into the standard space created by the International Consortium for Brain Mapping from 305 normal adult brains. Higher variability is observed in posterior temporal regions and in the postcentral gyrus in all three age groups, with relatively less variability in precentral and anterior temporal gyri. Adapted with permission from Sowell et al. (2002b).

diffusely distributed in dorsal frontal and parietal regions between childhood and adolescence, cortical maturation between adolescence and adulthood was localized to large regions of dorsal, mesial, and orbital frontal cortex with relatively little gray matter density reduction in other brain regions (Sowell et al., 1999b), as shown in Figure 1.1. These results are consistent with studies showing that the frontal lobes are essential for such functions as response inhibition, emotion regulation, planning, and organization (Fuster, 2001), which may not be fully developed in adolescents.

Figure 1.1 also shows gray matter density reductions in subcortical structures. Gray matter density loss in the lentiform nuclei was more evident between adolescence and adulthood (Sowell et al., 1999b). In a study that examined white matter change using VBM, Paus and colleagues found prominent white matter density increases in the arcuate fasciculus in the temporo-parietal region and in the posterior limb of the internal capsule in subjects 4–17 years of age (Paus et al., 1999).

Cortical pattern-matching

A limitation with VBM is that brain volumes are typically spatially normalized across subjects using automated image registration. Current VBM techniques cannot control for variability in sulcal patterns across individuals. Variability in sulcal patterns differs by cortical region and is more pronounced the further the region is from the center of the brain. As shown in Figure 1.2, sulcal pattern can vary up to 14 mm even after spatial normalization, particularly in the

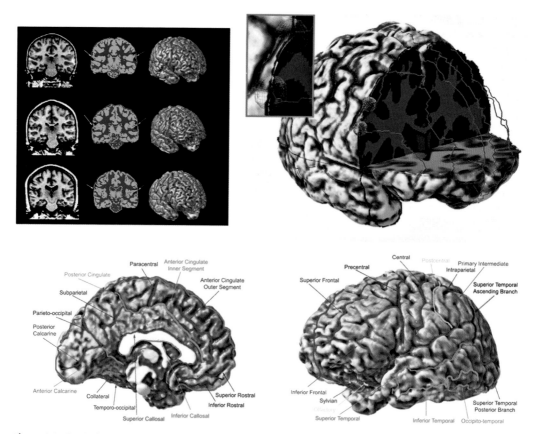

Figure 1.3. Cortical pattern-matching (CPM) methods. *Top left*: Three representative brain image data sets with the original MRI, tissue-segmented images, and surface renderings, with sulcal contours shown in pink. *Top right*: Surface rendering of one representative subject with cutout showing tissue-segmented coronal slice and axial slice superimposed within the surface. Sulcal lines are shown where they would lie on the surface in the cutout region. Note the sample spheres over the right hemisphere inferior frontal sulcus (lower sphere) and on the middle region of the precentral sulcus (upper sphere) that illustrate varying degrees of gray matter density (GMD). In the blown-up panel, note that the upper sphere has a higher GMD than does the lower sphere as it contains only blue pixels (gray matter) within the brain. The lower sphere also contains green pixels (white matter) that would lower the GMD within it. In the actual analysis, GMD was measured within 15-mm spheres centered across every point over the cortical surface. *Bottom*: Sucal anatomical delineations are defined according to color. These are the contours drawn on each individual's surface rendering according to a reliable, written protocol (Sowell et al., 2002b). Reproduced with permission from Sowell et al. (2004b).

posterior temporal regions (Sowell et al., 2002b). Without taking this variability into account, VBM techniques are not likely to match cortical surface regions well across subjects. Cortical pattern-matching (CPM) methods (Thompson et al., 2004) allow us to account for interindividual differences in cortical patterns and achieve better matching of

cortical anatomy across subjects, which may improve statistical power to localize age-related changes. Figure 1.3 depicts some major steps involved in using CPM methods. Briefly, CPM involves creating a three-dimensional geometric model of the cortical surface extracted from the MRI volume (MacDonald et al., 1994) and then flattening it to a two-dimensional planar format (Thompson and Toga, 1997; Thompson and Toga, 2002). A warping transformation is then applied that aligns the sulcal anatomy of each subject with an average sulcal pattern derived for the group. To improve sulcal alignment across subjects, all sulci that occur consistently can be manually defined on the surface rendering and used as anchors to restrict this transformation. Once cortical pattern and shape have been matched across subjects, morphological measures such as gray matter thickness or distance of the surface from the center of the brain (reflecting localized brain growth) can then be compared with greater anatomical accuracy. This method of measuring gray matter thickness has been validated against histological data (Sowell et al., 2004a). We have measured both gray matter thickness (i.e., distance from the white matter/gray matter interface to the gray matter/cerebrospinal fluid border) and gray matter density. Because both methods of quantifying gray matter yield very similar results, we will use the term gray matter density (GMD) for the remainder of the chapter. Cortical pattern-matching techniques can also be used to study cortical expansion, brain surface asymmetry, and subcortical growth. Combined, these studies have highlighted regional patterns of cortical change with age that have not been appreciated with techniques such as volumetric image analysis or VBM.

Gray matter density

Gray matter density studies using CPM have largely concurred with findings from VBM studies. In a study of 35 individuals between 7 and 30 years of age, we found local GMD loss distributed over the dorsal frontal and parietal lobes between childhood and adolescence (Sowell et al., 2001b). Between adolescence and adulthood, a dramatic increase in local GMD loss was observed in the frontal lobes, parietal GMD loss was reduced relative to earlier years, and a relatively small, circumscribed region of local GMD increase was observed in the left peri-Sylvian region. We tested for non-linear age effects on GMD change in a larger sample of 176 normal

individuals ranging in age from 7 to 87 years (Sowell et al., 2003). As shown in Figure 1.4, non-linear age effects were significant over most dorsal aspects of frontal and parietal regions and in the orbitofrontal cortex. There was a dramatic decline in GMD between 7 and 60 years of age. This pattern in attenuated form was evident in most lateral and medial brain surface regions (Figure 1.5) except for the peri-Sylvian region in the inferior parietal and posterior temporal cortices (Figure 1.4). The non-linear age effects in bilateral peri-Sylvian regions were inverted, showing a subtle increase in GMD in the first three decades of life, then a decline in later decades (Figure 1.4).

We were able to corroborate cortical thickening in the peri-Sylvian region using a separate longitudinal data set of children between 5 and 11 years of age (Sowell et al., 2004a). Again, bilateral peri-Sylvian thickening was found, but in the left hemisphere, thickening was extended more anteriorly to encompass the inferior frontal gyrus. In contrast, Gogtay and colleagues did not find increasing GMD in the peri-Sylvian region of 13 subjects (age ranged from 4 to 21 years) studied longitudinally (Gogtay et al., 2004). They found a flat GMD growth curve in superior temporal gyrus and inferior frontal gyrus. However, this growth pattern did contrast with the growth pattern in remaining regions of the brain, where GMD loss was seen, similar to our findings. These findings from different data sets across independent research groups corroborate that the growth pattern of the peri-Sylvian region differs from that of remaining regions of the cortex. Gogtay and colleagues also corroborate the pattern of GMD loss we observed between childhood and adulthood (Sowell et al., 2001b). They found evidence of GMD loss in dorsal parietal cortices early in development, then spreading rostrally over the frontal cortex and caudally and laterally over the parietal, occipital, and temporal cortices during adolescence (Gogtay et al., 2004). Within frontal and parietal subregions, GMD loss progressed from primary cortices (precentral and postcentral gyri) to association and tertiary cortices. Frontal, occipital, and temporal poles matured early, whereas

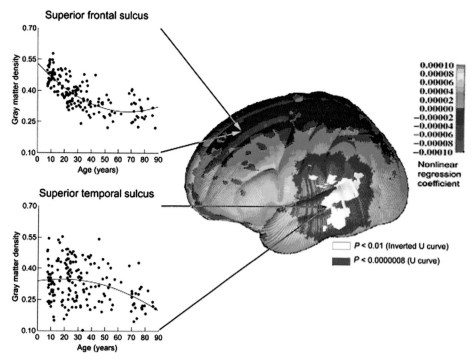

Figure 1.4. Statistical map of nonlinear age effects of gray matter density (GMD) on the lateral brain surface. This left frontal view shows age effects of GMD on the lateral surface of the brain between childhood and old age. Regions shown in red correspond to regression coefficients that have statistically significant positive non-linear age effects at a threshold of $p = 0.000\,000\,8$ (U-shaped curves with respect to age), and regions shown in white correspond to statistically significant negative non-linear age effects at a threshold of $p = 0.01$ (inverted U-shaped curves). Shades of green to yellow represent positive partial regression coefficients for the quadratic term, and shades of blue, purple and pink represent negative partial coefficients. The pattern of non-linear age effects was similar in the left and right (not shown) hemispheres, except that none of the negative non-linear age effects in the right posterior temporal lobe reached a threshold of $p = 0.01$. Scatterplots of age effects with the best-fitting quadratic regression curve are shown for sample surface points in the superior frontal sulcus (top) and the superior temporal sulcus (bottom). Gray matter density within the 15-mm sphere centered on the sample surface point (matched across subjects) is shown on the y-axis. Reproduced with permission from Sowell et al. (2003).

other parts of the temporal lobe matured last (Gogtay et al., 2004).

Cortical expansion

The precise relationship between signal change as detected on MR images and actual physiological changes taking place at the cellular level is not yet clear. It is known from postmortem studies that synaptic pruning and loss of dendritic processes occur during childhood and adolescence, with synaptic elimination ending earlier in primary sensory cortices (around age 12) than in the prefrontal cortex (around age 16) (Huttenlocher, 1979; Huttenlocher and Dabholkar, 1997; Huttenlocher and De Courten, 1987). Increase in GMD may be affected by the number, size, and packing density of neurons, which, in turn, are affected by genetics, hormones, growth factors, nutrients, and other environmental variables (e.g., toxins, trauma, stress,

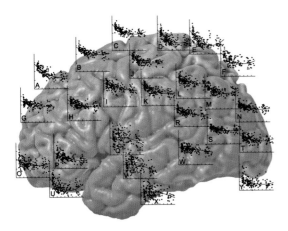

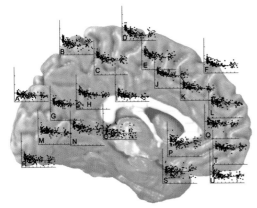

Figure 1.5. Scatterplots of gray matter density (GMD) on lateral (top) and medial (bottom) left hemisphere surfaces. Shown are scatterplots for GMD at various points over the brain surface where measurements were taken. The axes for every graph are identical to those in Figure 1.4. Age in years is represented on the *x*-axis (range 0–90) and GMD on the *y*-axis (range 0.1–0.7). *Top*: Anatomical location associated with each graph on the lateral surface: (A) superior frontal gyrus, (B) superior frontal sulcus (SFS), middle region, (C) SFS, posterior region, (D) precentral gyrus, (E) postcentral gyrus, (F) superior parietal gyrus, (G) SFS, anterior region, (H) inferior frontal sulcus (IFS), middle region, (I) IFS, posterior region, (J) precentral sulcus, (K) central sulcus, (L) postcentral sulcus, (M) intraparietal sulcus, (N) secondary intermediate sulcus, (O) olfactory gyrus, (P) IFS, anterior region, (Q) Sylvian fissure, (R) primary intermediate sulcus, (S) superior temporal sulcus (STS), ascending branch, (T) transverse occipital sulcus, (U) olfactory sulcus, (V) STS, main body,

enrichment level of the environment) (Giedd et al., 1996a). Concomitant with these neuronal changes is the proliferation of myelination, which begins near the end of the second trimester of fetal development and is protracted into the second decade of life and beyond (Yakovlev and Lecours, 1967). Giedd and colleagues have hypothesized that myelination is primarily responsible for changes in the size of the brain and its components, as it is estimated that a total loss of synaptic boutons would account for only a 1%–2% decrease in volume (Giedd et al., 1996a).

Examining localized brain growth together with changes in GMD may provide clues regarding cellular mechanisms underlying GMD change. Some have speculated that cortical thinning could in part be due to increased proliferation of myelin into the periphery of the cortical neuropil. In other words, tissue with gray matter signal in the younger subjects may actually be unmyelinated peripheral fibers, so the MR signal value that is "gray matter" in the younger subjects could change to white matter in the older subjects (for a discussion see Sowell et al., 2003, 2004b). Data from our laboratory and others using volumetric methods suggest that gray matter is replaced by white, given that white matter volumes increase and gray matter volumes decline (Courchesne et al., 2000; Giedd et al., 1999; Jernigan et al., 1991). Our CPM studies of localized brain growth described below suggest that "gray matter" thinning may not be the best term to

Caption for Figure 1.5. (*cont.*)

(W) STS, posterior branch, (X) inferior temporal sulcus, (Y) inferior occipital gyrus. *Bottom*: Anatomical location associated with each graph on the medial surface: (A) cuneus, (B) pre-cuneus, (C) posterior cingulate, (D) precentral gyrus, (E) paracentral sulcus, (F) superior frontal gyrus, (G) parieto-occipital sulcus, (H) subparietal sulcus, (I) isthmus region, (J, K, L and Q) anterior cingulate, (M) posterior calcarine sulcus, (N) anterior calcarine sulcus, (O) callosal sulcus, (P) genu, (R) retrocalcarine sulcus, (S) gyrus rectus, (T) superior rostral sulcus, (U) frontomarginal gyrus. Adapted with permission from Sowell et al. (2003).

Figure 1.6. Statistical map of the correlation between gray matter density (GMD) and distance from center (DFC) across 35 individuals from 7 to 30 years of age. Anatomically, the central sulcus and Sylvian fissure are highlighted. Highlighted in red are regions where the negative relationship (i.e., less GMD associated with more extensive DFC) is highly statistically significant ($p = 0.000\,001$). Shown in shades of green to yellow are regions where the negative correlation between GMD and DFC does not reach threshold of $p = 0.05$. Shown in shades of blue, purple, and pink are regions where the correlation is positive. Note that none of the positive correlations between GMD and DFC is significant, even when $p = 0.01$ was used as a threshold. © 2001 by the Society for Neuroscience Sowell et al. (2001b).

describe morphological changes we observe with MRI (Sowell et al., 2001b, 2004a).

We quantified localized brain growth by measuring the distance from the midline decussation of the anterior commissure ($x = 0$, $y = 0$, $z = 0$) to each of the 65 536 matched brain surface points. Differences in the length of the distance from center (DFC) line at each brain surface point between different age groups (e.g., children and adolescents) suggest local growth in that location, and statistical analysis at each point can be conducted. In a group of 35 individuals between 7 and 30 years of age, we found brain growth in bilateral dorsal frontal and parietal regions that correlated significantly with concomitant GMD loss (Figure 1.6) (Sowell et al., 2001b). In these regions, thinner gray matter was associated with more brain expansion. Myelination would presumably only result in volume increase, given that myelin consists of space-occupying glial cells (Friede, 1989), whereas neuronal factors underlying maturation could be associated with either volume decrease (due to synaptic pruning) or volume increase (increase in soma size, expansion of dendritic arborization, etc.). Gray matter density loss simultaneous with brain expansion in the dorsal frontal and parietal regions likely resulted from a combination of increased myelination and neuronal factors, as it is known that expansion of myelin into

the cortical neuropil could result in a reduction of brain tissue that segments as gray matter on MRI, and non-myelinated fibers do not have typical white matter value signals on T1-weighted images (Barkovich et al., 1988). Interestingly, GMD gain in the inferior parietal regions did not correlate significantly with localized brain expansion, nor did brain expansion in the temporo-occipital junction correlate with GMD gain (Sowell et al., 2001b).

Brain surface asymmetry

Asymmetries in sulcal patterns are of considerable interest, particularly in the peri-Sylvian region given the functional lateralization of language in this region (reviewed in Geschwind and Galaburda, 1985). Postmortem studies have shown that, in adults, the Sylvian fissure is longer in the left hemisphere compared to the right (Galaburda et al., 1978; Ide et al., 1996), and in vivo vascular imaging studies have shown that the Sylvian fissure angles up more dramatically at its posterior end in the right hemisphere compared to the left (LeMay and Culebras, 1972). Greater left planum temporale length in the left hemisphere compared to the right has been observed in postmortem studies of infants, indicating that these asymmetry patterns may be independent of maturational change and

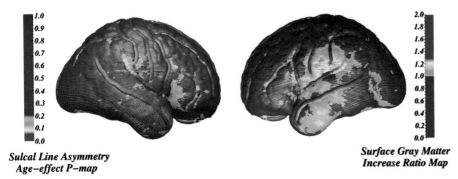

**Sulcal Line Asymmetry
Age–effect P–map**

**Surface Gray Matter
Increase Ratio Map**

Figure 1.7. Difference between childhood and young adulthood in Sylvian fissure asymmetry and gray matter density (GMD). Cortical GMD increase, expressed as a ratio of GMD in the child group to GMD in the adult group at each surface point, is shown according to the color bar on the right. Warmer colors represent regions where GMD is greater in adults than in children, with, at maximum, a ~25% increase in gray matter in the peri-Sylvian region. The probability map of age effects in sulcal line asymmetry is laid over the cortical surface map of GMD increase. Probability values corresponding to the age-related increase in asymmetry are color coded using the color bar on the left. Significant age effects in sulcal line asymmetry indicate greater asymmetry between the left and right Sylvian fissures in adulthood than in children. Note the spatial correspondence between regions of increased asymmetry and increased gray matter between childhood and young adulthood in the left hemisphere. Reproduced with permission from Sowell et al. (2002b).

the acquisition of language abilities (Witelson and Pallie, 1973).

Using CPM methods, we were able to measure displacement between left and right hemisphere sulci and compare asymmetry maps in 35 individuals between 7 and 30 years of age. We found that asymmetries in peri-Sylvian regions continued to develop between childhood and young adulthood (Sowell et al., 2002b). In fact, the longer left Sylvian fissure (compared to the right) and more sloped right Sylvian fissure (compared to the slope of the left) were more pronounced in young adults than in children. Figure 1.7 shows the statistical map of the differences between children and young adults in sulcal asymmetry, overlaid on the surface map of GMD change in the same sample of subjects. Age effects in sulcal asymmetry in this map indicate greater asymmetry between left and right Sylvian fissures in adulthood than in childhood. Overlaying this age effect map of sulcal asymmetry on the map of GMD change shows that the localized GMD increase in the left hemisphere corresponds with the increase in Sylvian fissure asymmetry from childhood to adulthood.

In another report, we assessed total brain surface asymmetry by mapping the distance from matched surface points in the left hemisphere to the same location in the reflected right hemisphere (Sowell et al., 2002a). Figure 1.8a shows the magnitude and direction of asymmetry at each brain surface point. Peak asymmetry was observed in the posterior peri-Sylvian region, where the distance between anatomically homologous surface points in the left and sulcally matched right brain surface was between 6 and 12 mm. Vector maps showed that the direction of displacement between hemispheres was primarily in the anterior–posterior axis, with the left being more posterior than the right. Prominent within-individual asymmetry in the posterior peri-Sylvian region and prominent interindividual variability in the posterior temporal region of the right hemisphere (Blanton et al., 2001; Sowell et al., 2002b; Figure 1.2) suggest that right posterior temporo-parietal regions may undergo more dynamic morphological change than corresponding left hemisphere regions.

We also mapped GMD asymmetry across the brain surface to explore whether sulcal asymmetry

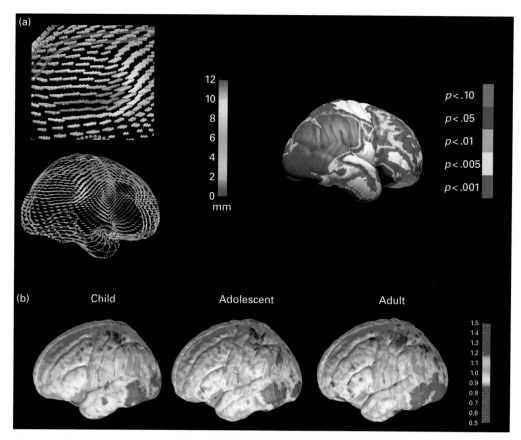

Figure 1.8a, b. Brain surface asymmetry. (a) The arrows in these maps show the direction and displacement between analogous surface points in the left and right hemispheres for a group of 62 normal controls between 7 and 30 years old. The base of each arrow represents the left hemisphere surface point location, and the tip of the arrow represents the analogous surface point location in the right hemisphere (a flipped and reflected version). Group differences (in millimeters) are mapped in color according to the corresponding color bar. Note that maximal asymmetry, up to 12 mm difference, is found in the peri-Sylvian region, shown enlarged to enhance detail. Statistical maps are shown on the right documenting the significance of displacement between the left and right hemispheres according to the corresponding color bar (note white regions have $p \geq 0.10$). (b) Ratio maps of left hemisphere GMD (at each surface point) to right GMD (at corresponding surface point). According to the color bar, 1 (green shades) represents symmetry. Warmer colors (less than 1) represent regions where there is more gray matter in the right hemisphere than in the left. Cooler colors (greater than 1) represent regions where there is more gray matter in the left than in the right. Adapted with permission from Sowell et al. (2002a).

could be related to localized gray matter change. Figure 1.8b shows greater GMD in the right posterior temporal lobe than in homologous left hemisphere regions (Sowell et al., 2002a). This finding is generally consistent with right-greater-than-left gray matter asymmetry in the posterior temporal region reported in a volumetric study of children and adolescents (Giedd et al., 1996b) and in VBM and CPM studies of young adults (Luders et al., 2006; Watkins et al., 2001). Left-greater-than-right white matter asymmetry in this general region has been reported in imaging (Penhune et al., 1996) and postmortem studies (Anderson et al., 1999). Given that changes in brain tissue that segment as gray matter in MR images may be due to myelination or neuronal factors, the reverse asymmetry of white matter (leftward) in the posterior temporal region may be consistent with rightward gray matter asymmetry in this region. Age effects in rightward gray matter asymmetry between the children and adolescent groups and between the adolescent and young adult groups were not significant, suggesting that the pattern of gray matter asymmetry is established early in development (Sowell et al., 2002a). Since gray matter asymmetry does not appear to change with age, whereas Sylvian fissure asymmetry becomes more pronounced with age, factors other than GMD change are likely to contribute to Sylvian fissure morphology during development.

Subcortical growth

Cortical pattern-matching methods could be applied to subcortical structures after extracting a three-dimensional surface model of the structure from two-dimensional MR images and identifying anatomical landmarks by which to anchor warping transformations. Currently, efforts have been limited to surface extractions (Gogtay et al., 2006; Koikkalainen et al., 2007; Thompson et al., 2000) due to challenges in identifying landmarks reliably within each subcortical structure. Koikkalainen and colleagues examined striatal shape and volume in a group of healthy adults between ages 19 and 55 (Koikkalainen et al., 2007). They found inward expansion primarily in the anterior caudate and putamen regions, which presumably contributed to decreasing striatal volume with age (normalized). This suggests a heterogeneous pattern of basal ganglia volume decrease in adults and elicits questions of whether regional differences in subcortical gray matter change may be present in children and adolescents, in whom previous findings of normalized striatal volume decrease have been reported (Jernigan et al., 1991; Reiss et al., 1996; Sowell et al., 2002c). Heterogeneous growth of the hippocampus has been documented in both children and adults using

a longitudinal approach and stereotaxic scaling to allow for intrinsic growth to be represented accurately while adjusting for baseline differences in brain volume (Gogtay et al., 2006). In this sample of 31 subjects, hippocampal volume remained unchanged with development. However, the posterior hippocampus enlarged over time, while the head of the hippocampus showed volume loss with age. The meaning of anterior–posterior differences in hippocampal volume change is not yet clear.

An issue related to studies using surface extraction, which usually test statistical significance at tens of thousands of surface points, is controlling for false positives. We have dealt with this issue by permuting data within a-priori-defined regions of interest (ROI) (see Nichols and Holmes, 2004 for an explanation of permutation test). If the observed number of surface points reaching statistical significance has less than a 5% probability of occurring by chance within an ROI, we accept that there is a significant finding within that ROI. With the cortical mantle, we have typically used lobar ROIs. With subcortical structures, the challenge is to establish ROIs a priori so that regional differences in structural morphology can be evaluated with some method of controlling for false positives. Until this issue is addressed, it is difficult to distinguish chance findings from real findings in subcortical morphology.

Summary of morphological maturation from childhood to young adulthood

It is important to keep in mind that the pattern of morphological change identified will depend on the age range studied. For example, there may appear to be a discrepancy between volumetric studies showing increasing frontal and parietal gray matter volume during childhood (Giedd et al., 1999, 2006; Sowell et al., 2002c) and CPM studies showing decreasing GMD in these same regions (Sowell et al., 2001b, 2003, 2004a). These volumetric studies sampled children as young as age 4 whereas the youngest age in the CPM studies was 7 years. Pfefferbaum and colleagues' volumetric study of ages spanning infancy to adulthood showed peak cortical gray matter volume at 4 years of age (Pfefferbaum et al., 1994). It may be that inclusion of children at the younger age range is necessary to detect increasing trends in gray matter. Another age-related issue pertains to the extent of the age range studied. In a large cross-sectional sample encompassing individuals from age 7 to 87, we found curvilinear relationships between GMD and age for most cortical brain regions (Sowell et al., 2003). Curvilinear patterns of gray matter maturation across many cortical regions were corroborated by findings from longitudinal studies (Giedd et al., 1999; Gogtay et al., 2004). Cross-sectional studies with a truncated age range may not be able to detect such curvilinear trajectories and instead report linear age effects (Sowell et al., 2002c). The accruing evidence for curvilinear maturation patterns points to the need to test for non-linear age effects in cross-sectional studies and to obtain more than two time points in longitudinal studies so that growth curves can be examined.

Across volumetric, VBM, and CPM methods, available data suggest an increase in gray matter during early childhood (Pfefferbaum et al., 1994). At some point gray matter peaks, after which gray matter loss occurs through most regions of the cortex (Gogtay et al., 2004; Sowell et al., 2001b, 2003, 2004a). The trajectory of gray matter loss is heterogeneous across the cortical surface. Dramatic changes proceed from posterior to anterior dorsal regions such that parietal gray matter loss is most prominent from childhood to adolescence and frontal gray matter decrease is most salient from adolescence to adulthood (Gogtay et al., 2004; Sowell et al., 2001b, 2004b). Frontal gray matter thinning occurs concurrently with localized brain expansion (Sowell et al., 2001b), suggesting that an increase in space-occupying myelin is a likely physiological factor contributing to gray matter thinning, along with other neuronal factors such as synaptic pruning or somal size change (Giedd et al., 1996a; Sowell et al., 2004b). The peri-Sylvian region appears to have a different growth trajectory than remaining regions of the brain; gray matter increases rather than decreases (Sowell et al., 2003, 2004a). In the left hemisphere, gray matter thickening in the inferior parietal region corresponds with the region of the Sylvian fissure where the right and left fissures become more asymmetrical with age (Sowell et al., 2002b). Deformation maps quantifying the degree of cortical surface variability between individuals show the inferior parietal and posterior temporal region to be most variable across individuals (Sowell et al., 2002a, 2002b).

Much less is known about the maturational pattern of subcortical structures. Available evidence suggests that normalized striatal volume decreases with age (Giedd et al., 1996a; Jernigan et al., 1991, Reiss et al., 1996; Sowell et al., 2002c; Thompson et al., 2000), although comparison of findings on subcortical gray is complicated by the inclusion of mesial temporal structures in some studies, while others distinguish them from the striatum and thalamus. Absolute hippocampal volume does not appear to change as much with age (Giedd et al., 1996b; Gogtay et al., 2006), but one study showed decreasing normalized mesial temporal volume between childhood and adolescence (Sowell et al., 2002c).

Relationship between morphology and cognition

The earliest study to examine the correlation between in vivo cerebral volume and intelligence quotient (IQ) found that larger total gray matter volume predicted higher IQ scores in children between 5 and 17 years, after removing effects of gender and age (Reiss et al., 1996). Gray matter volume of the prefrontal cortex was more highly correlated with IQ than other regional cortical volumes, accounting for 20% of the variance in IQ (Reiss et al., 1996).

These results contrast with our results using both volumetric methods and CPM to obtain GMD at each cortical surface point. We found frontal gray matter thinning to be related to improvement in verbal learning in children between 7 and 16 years of age after age and brain volume were controlled (Sowell et al., 2001a). In an independent longitudinal sample of children studied between ages 5 and 11, we examined changes in raw scores of vocabulary performance (Sowell et al., 2004a). Vocabulary skill is known to correlate highly with full-scale IQ (Sattler, 2001) and is often used to estimate IQ when there are time constraints. We found that gray matter thinning across diverse regions of the left hemisphere was associated with greater vocabulary improvement scores (Sowell et al., 2004a). An explanation for the discrepancy between our longitudinal findings (cortical thinning with improvement) and Reiss and colleagues' results (increased gray matter volume with greater IQ) is that their finding represents morphological variation among individuals with different levels of cognitive ability, whereas our finding represents within-individual morphological maturation associated with improving cognitive skill. Because our subjects were normally developing children without neurological or psychiatric conditions, their IQ at time 1 and at time 2 theoretically should have remained stable. Our finding does not reflect morphological variation associated with IQ; rather it reflects how learning is associated with morphological maturation.

Another potential explanation for the discrepancy between these studies is offered by Shaw and colleagues' report of morphological variation between individuals of different IQ ranges (Shaw et al., 2006). They tested over 300 children between 7 and 19 years of age and analyzed cortical thickness trajectories of three IQ groups: average, high, and superior intelligence. The superior intelligence group started out with a relatively thinner cortex at age 7, but then showed a marked increase in cortical thickness, peaking at approximately 11 years before thinning, such that by age 19, there was no difference in cortical thickness between the superior intelligence and other groups. The average intelligence group showed a steady decline in cortical thickness between 7 and 19 years of age, and the high average group followed an intermediate pattern. While overall thickness decline was seen in all groups, trajectory of maturation varied by intelligence, and the directionality (thickening or thinning) and rate of growth varied depending on the age range of focus. This study showed that the anatomical maturation pattern of gifted individuals differs from that of the general population. Neither the Reiss et al. (1996) study nor the Sowell et al. (2004a) study reported the IQ range of their subjects. Shaw et al.'s (2006) finding points out the need to control for IQ level in studies aiming to correlate anatomical maturation with cognitive skill development.

The finding of cortical thickening in the peri-Sylvian region within a context of gray matter thinning in surrounding regions (Sowell et al., 2003, 2004a) prompted us to speculate that peri-Sylvian thickening may be associated with improving language skills. Alternatively, peri Sylvian thickening was seen bilaterally and may reflect a general maturational pattern that is non-specific to language development. Using a longitudinal design, we correlated thickness change with improvement in language skill and improvement in motor skill, both of which are known to mature from 5 to 11 years, the age range of subjects in this study (Lu et al., 2007). Language skill was defined by performance on two standardized measures of phonological awareness, and motor skill was defined by performance on standardized tests of hand strength and fine motor skill. We found that cortical thickening in the left inferior frontal gyrus was related to improvements in phonological processing skill (Figure 1.9a). Crucially, this relationship did not appear to be an effect of general maturation because thickness change in this region did not correlate with motor skill improvement. Improving motor skill was related to cortical thinning in more dorsal regions of left frontal cortex (Figure 1.9b), where correlation with phonological processing was not found. This double dissociation suggests that gray matter thickening in the left inferior frontal gyrus is specific to maturation of phonological processing skills. This study represents the first effort to associate maturation of a specific cognitive skill with localized morphological change measured in vivo.

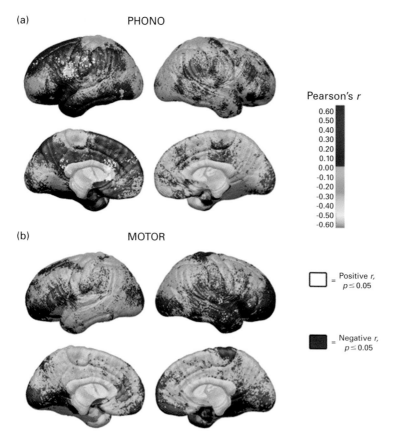

Figure 1.9. Double dissociation of gray matter thickness change with skill improvement from ages 5 to 11. Statistical map of Pearson's *r* correlation coefficients is rendered using the color bar on the right. Regions shown in white correspond to Pearson's *r* correlation coefficients that have statistically significant positive effects at a threshold of $p \leq 0.05$. Regions shown in red correspond to Pearson's *r* correlation coefficients that have statistically significant negative effects at a threshold of $p \leq 0.05$. On the medial surface, corpus callosum and brainstem regions are masked out with gray. (a) Correlation between change in gray matter thickness and phonological processing skill improvement over time. Significant positive relationship in the left inferior frontal gyrus indicates gray matter thickening in those with more improvement in phonological processing skill. No significant relationship between thickness change and skill change is seen for the right peri-Sylvian region. (b) Correlation between thickness change and motor skill improvement. No significant relationship is seen in the left inferior frontal gyrus. Significant negative correlation is evident in more dorsal regions of the left frontal lobe, indicating gray matter thinning in those with more improved motor skills. Adapted with permission from (Lu et al., 2007).

Conclusion

Technological advances are now in place to study morphological changes that occur as cognitive skills develop. Cross-sectional studies allow comparison of morphology among individuals of different age, skill, and intelligence levels. However, it is very difficult to disentangle effects of learning that lead to differences in skill level on brain morphological development from effects associated with individual differences in IQ. Longitudinal designs enable the elucidation of relationships between skill acquisition and changing brain morphology in individuals of different IQ levels who serve as their own controls using within-subject designs. Furthermore, it is possible to model age along with longitudinal changes in skill level to disentangle age effects on morphological change from effects unique to experience and skill acquisition. Such studies would lead us toward an understanding of the effects of nature versus nurture on brain morphology, but require that the field invests in studies using longitudinal designs.

This chapter has focused on morphological changes between childhood, adolescence, and adulthood. It should be pointed out that all studies reviewed here defined adolescence by vague age ranges, with the lower end between 10 and 12 years and the higher end between 16 and 18 years. We know of no studies that evaluated hormonal levels with human morphological development. As estrogen and androgen are known to affect structural plasticity (Baron-Cohen et al., 2005; De Lacalle, 2006), it will be important to evaluate whether brain morphological maturation during adolescence is linked to changes in hormonal levels. Because hormonal changes during puberty differ for boys and girls, sex differences in brain morphology may shed light on the neural bases of differences in cognitive strengths between males and females.

Correlations between morphological and skill maturation, although instructive, reveal only associations and cannot elucidate causality. Neuroscience must still rely on animal studies using controlled experimental designs to learn whether morphological maturation enables the acquisition of skills or if skill acquisition drives morphological change. Despite the many challenges and limitations imaging technology still faces, the ability to study the human brain in vivo has significantly increased our knowledge of morphological development of the human brain.

REFERENCES

Anderson, B., Southern, B. D., and Powers, R. E. (1999) Anatomic asymmetries of the posterior superior temporal lobes: a postmortem study. *Neuropsychiatry Neuropsychol Behav Neurol*, **12**(4), 247–54.

Ashburner, J. and Friston, K. J. (2000) Voxel-based morphometry – the methods. *NeuroImage*, **11**(6 Pt 1), 805–21.

Barkovich, A. J., Kjos, B. O., Jackson, D. E., Jr., et al. (1988) Normal maturation of the neonatal and infant brain: MR imaging at 1.5 T. *Radiology*, **166**(1 Pt 1), 173–80.

Baron-Cohen, S., Knickmeyer, R. C., and Belmonte, M. K. (2005) Sex differences in the brain: implications for explaining autism. *Science*, **310**(5749), 819–23.

Blanton, R. E., Levitt, J. G., Thompson, P. M., et al. (2001) Mapping cortical asymmetry and complexity patterns in normal children. *Psychiatry Res*, **107**(1), 29–43.

Caviness, V. S., Jr., Kennedy, D. N., Richelme, C., et al. (1996) The human brain age 7–11 years: a volumetric analysis based on magnetic resonance images. *Cereb Cortex*, **6**(5), 726–36.

Courchesne, E., Chisum, H. J., Townsend, J., et al. (2000) Normal brain development and aging: quantitative analysis at in vivo MR imaging in healthy volunteers. *Radiology*, **216**(3), 672–82.

De Lacalle, S. (2006) Estrogen effects on neuronal morphology. *Endocrine*, **29**(2), 185–90.

Friede, R. L. (1989) Gross and microscopic development of the central nervous system. In: Friede, R. L. (ed.) *Developmental Neuropathology*, 2nd edn. Berlin: Springer, pp. 2–20.

Fuster, J. M. (2001) The prefrontal cortex – an update: time is of the essence. *Neuron*, **30**(2), 319–33.

Galaburda, A. M., Sanides, F., and Geschwind, N. (1978) Human brain. Cytoarchitectonic left-right asymmetries in the temporal speech region. *Arch Neurol*, **35**(12), 812–17.

Geschwind, N. and Galaburda, A. M. (1985) Cerebral lateralization. Biological mechanisms, associations, and pathology: I. A hypothesis and a program for research. *Arch Neurol*, **42**(5), 428–59.

Giedd, J. N., Blumenthal, J., Jeffries, N. O., et al. (1999) Brain development during childhood and adolescence: a longitudinal MRI study. *Nat Neurosci*, **2**(10), 861–3.

Giedd, J. N., Clasen, L. S., Lenroot, R., et al. (2006) Puberty-related influences on brain development. *Mol Cell Endocrinol*, **254–255**, 154–62.

Giedd, J. N., Snell, J. W., Lange, N., et al. (1996a) Quantitative magnetic resonance imaging of human brain development: ages 4–18. *Cereb Cortex*, **6**(4), 551–60.

Giedd, J. N., Vaituzis, A. C., Hamburger, S. D., et al. (1996b) Quantitative MRI of the temporal lobe, amygdala, and hippocampus in normal human development: ages 4–18 years. *J Comp Neurol*, **366**(2), 223–30.

Gogtay, N., Giedd, J. N., Lusk, L., et al. (2004) Dynamic mapping of human cortical development during childhood through early adulthood. *Proc Natl Acad Sci U S A*, **101**(21), 8174–9.

Gogtay, N., Nugent, T. F., 3rd, Herman, D. H., et al. (2006) Dynamic mapping of normal human hippocampal development. *Hippocampus*, **16**(8), 664–72.

Huttenlocher, P. R. (1979) Synaptic density in human frontal cortex – developmental changes and effects of aging. *Brain Res*, **163**(2), 195–205.

Huttenlocher, P. R. and Dabholkar, A. S. (1997) Regional differences in synaptogenesis in human cerebral cortex. *J Comp Neurol*, **387**(2), 167–78.

Huttenlocher, P. R. and De Courten, C. (1987) The development of synapses in striate cortex of man. *Hum Neurobiol*, **6**(1), 1–9.

Ide, A., Rodriguez, E., Zaidel, E., et al. (1996) Bifurcation patterns in the human sylvian fissure: hemispheric and sex differences. *Cereb Cortex*, **6**(5), 717–25.

Jernigan, T. L. and Tallal, P. (1990) Late childhood changes in brain morphology observable with MRI. *Dev Med Child Neurol*, **32**(5), 379–85.

Jernigan, T. L., Trauner, D. A., Hesselink, J. R., et al. (1991) Maturation of human cerebrum observed in vivo during adolescence. *Brain*, **114**(Pt 5), 2037–49.

Koikkalainen, J., Hirvonen, J., Nyman, M., et al. (2007) Shape variability of the human striatum – effects of age and gender. *NeuroImage*, **34**(1), 85–93.

Lange, N., Giedd, J. N., Castellanos, F. X., et al. (1997) Variability of human brain structure size: ages 4–20 years. *Psychiatry Res*, **74**(1), 1–12.

LeMay, M. and Culebras, A. (1972) Human brain – morphologic differences in the hemispheres demonstrable by carotid arteriography. *N Engl J Med*, **287**(4), 168–70.

Lu, L. H., Leonard, C. M., Thompson, P. M., et al. (2007) Normal developmental changes in inferior frontal gray matter are associated with improvement in phonological processing: a longitudinal MRI analysis. *Cereb Cortex*, **17**, 1092–99.

Luders, E., Narr, K. L., Thompson, P. M., et al. (2006) Hemispheric asymmetries in cortical thickness. *Cereb Cortex*, **16**(8), 1232–8.

Macdonald, D., Avis, D., and Evans, A. (1994) Multiple surface identification and matching in magnetic resonance images. *Proc Vis Biomed Comput*, **2359**, 160–169.

Nichols, T. and Holmes, A. (2004) Nonparametric permutation tests for functional neuroimaging. In: Frackowiak, R. S. J., Friston, K. J., Frith, C. D. et al. (eds.) *Human Brain Function*, 2nd edn. New York: Elsevier, pp. 887–910.

Paus, T., Zijdenbos, A., Worsley, K., et al. (1999) Structural maturation of neural pathways in children and adolescents: in vivo study. *Science*, **283**(5409), 1908–11.

Penhune, V. B., Zatorre, R. J., Macdonald, J. D., et al. (1996) Interhemispheric anatomical differences in human primary auditory cortex: probabilistic mapping and volume measurement from magnetic resonance scans. *Cereb Cortex*, **6**(5), 661–72.

Pfefferbaum, A., Mathalon, D. H., Sullivan, E. V., et al. (1994) A quantitative magnetic resonance imaging study of changes in brain morphology from infancy to late adulthood. *Arch Neurol*, **51**(9), 874–87.

Reiss, A. L., Abrams, M. T., Singer, H. S., et al. (1996) Brain development, gender and IQ in children. A volumetric imaging study. *Brain*, **119**(Pt 5), 1763–74.

Sattler, J. M. (2001) *Assessment of Children: Cognitive Applications*, 4th edn. San Diego, CA: Jerome M. Sattler, Publisher, Inc.

Shaw, P., Greenstein, D., Lerch, J., et al. (2006) Intellectual ability and cortical development in children and adolescents. *Nature*, **440**(7084), 676–9.

Sowell, E. R. and Jernigan, T. L. (1998) Further MRI evidence of late brain maturation: limbic volume increases and changing asymmetries during childhood and adolescence. *Dev Neuropsychol*, **14**, 599–617.

Sowell, E. R., Delis, D., Stiles, J., et al. (2001a) Improved memory functioning and frontal lobe maturation between childhood and adolescence: a structural MRI study. *J Int Neuropsychol Soc*, **7**(3), 312–322.

Sowell, E. R., Peterson, B. S., Thompson, P. M., et al. (2003) Mapping cortical change across the human life span. *Nat Neurosci*, **6**(3), 309–15.

Sowell, E. R., Thompson, P. M., Holmes, C. J., et al. (1999a) Localizing age-related changes in brain structure between childhood and adolescence using statistical parametric mapping. *NeuroImage*, **9**(6 Pt 1), 587–97.

Sowell, E. R., Thompson, P. M., Holmes, C. J., et al. (1999b) In vivo evidence for post-adolescent brain maturation in frontal and striatal regions. *Nat Neurosci*, **2**(10), 859–61.

Sowell, E. R., Thompson, P. M., Leonard, C. M., et al. (2004a) Longitudinal mapping of cortical thickness and brain growth in normal children. *J Neurosci*, **24**(38), 8223–31.

Sowell, E. R., Thompson, P. M., Peterson, B. S., et al. (2002a) Mapping cortical gray matter asymmetry patterns in adolescents with heavy prenatal alcohol exposure. *NeuroImage*, **17**(4), 1807–19.

Sowell, E. R., Thompson, P. M., Rex, D., et al. (2002b)
Mapping sulcal pattern asymmetry and local
cortical surface gray matter distribution in vivo:
maturation in perisylvian cortices. *Cereb Cortex*,
12(1), 17–26.

Sowell, E. R., Thompson, P. M., Tessner, K. D., et al. (2001b)
Mapping continued brain growth and gray matter
density reduction in dorsal frontal cortex: inverse
relationships during postadolescent brain maturation.
J Neurosci, **21**(22), 8819–29.

Sowell, E. R., Thompson, P. M., and Toga, A. W. (2004b)
Mapping changes in the human cortex throughout
the span of life. *Neuroscientist*, **10**(4), 372–92.

Sowell, E. R., Trauner, D. A., Gamst, A., et al. (2002c)
Development of cortical and subcortical brain
structures in childhood and adolescence: a structural
MRI study. *Dev Med Child Neurol*, **44**(1), 4–16.

Thompson, P. M. and Toga, A. W. (1997) Detection,
visualization and animation of abnormal anatomic
structure with a deformable probabilistic brain atlas
based on random vector field transformations.
Med Image Anal, **1**(4), 271–94.

Thompson, P. M. and Toga, A. W. (2002) A framework for
computational anatomy. *Comput Vis Sci*, **5**, 1–12.

Thompson, P. M., Giedd, J. N., Woods, R. P., et al. (2000)
Growth patterns in the developing brain detected by
using continuum mechanical tensor maps. *Nature*,
404(6774), 190–3.

Thompson, P. M., Hayashi, K. M., Sowell, E. R., et al.
(2004) Mapping cortical change in Alzheimer's disease,
brain development, and schizophrenia. *NeuroImage*,
23 (Suppl 1), S2–18.

Watkins, K. E., Paus, T., Lerch, J. P., et al. (2001) Structural
asymmetries in the human brain: a voxel-based
statistical analysis of 142 MRI scans. *Cereb Cortex*,
11(9), 868–77.

Witelson, S. F. and Pallie, W. (1973) Left hemisphere
specialization for language in the newborn.
Neuroanatomical evidence of asymmetry. *Brain*,
96(3), 641–6.

Yakovlev, P. I. and Lecours, A. R. (1967) The myelogenetic
cycles of regional maturation of the brain. In:
Minkowski, A. (ed.) *Regional Development of the Brain
in Early Life*. Oxford: Blackwell Scientific, pp. 3–70.

2

Neural correlates of the development of cognitive control

Silvia A. Bunge and Eveline A. Crone

Introduction

One of the most salient ways in which our behavior changes during childhood and adolescence is that we get better at working towards long-term goals, ignoring irrelevant information that could distract us from these goals, and controlling our impulses – in other words, we exhibit improvements in *cognitive control* (Casey et al., 2005a; Diamond, 2002; Zelazo, 2004). Cognitive control relies on *working memory*, or the ability to keep relevant information in mind as needed to carry out an immediate goal. What precisely is changing in a child's brain over time, enabling him or her to better control his/her thoughts and behavior? To what extent do these neural changes result from experience and practice, and to what extent do they result from predictable developmental changes in brain structure? What are the elemental control processes that develop during childhood?

A number of brain imaging studies have been conducted in recent years in an effort to tackle these and other difficult questions about the developing brain (Casey et al., 2005b; Luna and Sweeney, 2004). In comparison to what is known about changes in brain structure over development (see Chapter 1 by Lu and Sowell), far less is known about the resulting changes in brain function. Below, we focus on recent event-related functional magnetic resonance imaging (fMRI) studies examining age-related changes in cognitive control.

Cognitive control processes

The terms *executive function* and *cognitive control* refer to cognitive processes associated with the control of thought and action. Putative control functions include the ability to: (1) selectively attend to relevant information while filtering out distracting information (*selective attention and interference suppression*); (2) work with information that is currently being held in working memory (*manipulation*); (3) flexibly switch between tasks (*task-switching*); (4) inhibit inappropriate response tendencies (*response inhibition*); and (5) represent contextual information that determines whether a thought is relevant or whether an action is appropriate (e.g. *task-set representation*).

Developmental studies suggest that some of these abilities may mature at different rates during childhood and adolescent development (Welsh et al., 1991). For example, the ability to inhibit a motoric response matures earlier than the ability to inhibit a response when the task additionally requires selective attention (Van den Wildenberg and Van der Molen, 2004). Likewise, the ability to switch between task rules develops earlier than the ability to keep a difficult rule online (Crone et al., 2004b).

One of the key questions in developmental research has been whether the ontogenic changes in cognitive control are associated with the development of a single mechanism, such as the capacity to store or

Neuroimaging in Developmental Clinical Neuroscience, eds. Judith M. Rumsey and Monique Ernst. Published by Cambridge University Press. © Cambridge University Press 2009.

process information (Case, 1995; Dempster, 1993), or with a set of mechanisms. Recent advances using structural equation modeling indicate that at least three cognitive control functions – working memory, task-switching, and response inhibition – are separable latent constructs with distinct developmental trajectories (Brocki and Bohlin, 2004; Huizinga et al., 2006). Thus, the trends observed in developmental research suggest that different cognitive control functions have separable developmental trajectories. As discussed below, fMRI, event-related potentials (ERPs) and other brain imaging techniques can be used to probe the neural changes underlying improvements in different cognitive control functions.

To effectively control behavior, it is necessary to update our responses to certain situations on the basis of the outcome of these responses; this capacity is referred to as *feedback-based learning*. An important component of learning based on feedback is the ability to know when one has made an error, a process referred to as *error monitoring*. Another important component is the ability to update one's assessment of the probable outcome of an action, based on past experience. The ability to process positive and negative feedback and to update one's behavior accordingly is the topic of several recent developmental neuroimaging studies, as discussed below. These feedback-based learning mechanisms are assumed to be tightly interconnected with working memory, task-switching, response inhibition, and other cognitive control functions.

Neural substrates of cognitive control

Prefrontal cortex (PFC) has long been implicated in executive function. Indeed, in 1895, the Italian physiologist Bianchi observed that experimental ablations of the frontal lobes in monkeys destroyed the ability to synthesize incoming percepts together and to integrate these with outgoing motor commands. Since this early work over a century ago, countless experiments have been conducted with a variety of techniques to better understand how PFC exerts control over other brain regions.

It is now clear that PFC should not be considered a monolithic structure; indeed, this expansive region constitutes roughly one-third of the human brain and consists of a number of different subregions. Each of these subregions is thought to provide a distinct contribution to cognition, via different cellular characteristics and anatomical connectivity. As such, the use of a technique with high spatial resolution, such as fMRI, is essential for determining whether two cognitive control tasks rely on the same or different PFC subregions – and whether developmental improvements in two cognitive control tasks rely on changes in similar or distinct neural circuits.

During development, gray matter loss occurs at different rates in different subregions of the brain and is considered an index of the time course of maturation of a region (Sowell et al., 2003; see Chapter 1). The dynamics of gray matter increases and decreases, particularly in the PFC, are associated with differences in intellectual ability (Shaw et al., 2006). Within PFC, gray matter reduction is completed earliest in orbitofrontal cortex (OFC), followed by ventrolateral PFC (VLPFC) and then by dorsolateral PFC (DLPFC) (Gogtay et al., 2004). We have argued that differences in maturational time course between PFC subregions partially account for differences in the rate of development of distinct cognitive control processes (e.g., Bunge and Zelazo, 2006).

Despite the nearly exclusive focus on PFC in the cognitive control literature thus far, this structure does not operate in a vacuum. Interactions between PFC and parietal cortex are important for working memory and for the control of action (Fuster, 1997). Interestingly, research on changes in brain structure over childhood indicates that portions of the parietal cortex develop slowly, in concert with mid-lateral PFC (Gogtay et al., 2004). As discussed in the section on working memory below, increased functional connectivity between PFC and parietal cortex is observed over childhood; the strengthening of prefrontal–parietal connections is likely to be a critical component of developmental improvements in cognitive control.

Neurodevelopmental changes in working memory and cognitive control

Below, we first discuss brain imaging studies focusing on developmental changes in working memory and several cognitive control functions: interference suppression, manipulation, response inhibition, and response selection. We then discuss recent studies that investigate the ability to control behavior by learning about the outcomes associated with specific actions. This set of studies focuses on the ability to process errors and update behavior accordingly. Finally, we will discuss research focusing on reasoning ability. The findings featured here are summarized in Table 2.1.

Working memory

Since the first fMRI study of working memory in children over a decade ago (Casey et al., 1995), most such studies have focused on pure maintenance, and, specifically, visuospatial working memory (VSWM) (Klingberg et al., 2002; Kwon et al., 2002; Luna et al., 2001; Scherf et al., 2006). Event-related fMRI studies have shown that regions that have been strongly implicated in VSWM in adults – the superior frontal sulcus (SFS) and the intraparietal sulcus (IPS) – are increasingly engaged over childhood (Klingberg et al., 2002; Kwon et al., 2002). Moreover, a comparison of fMRI data and diffusion tensor imaging data reveals that increased fractional anisotropy in fronto-parietal white matter – suggestive of increased strength of anatomical connectivity between these regions – is positively correlated with blood-oxygen-level-dependent (BOLD) activation in the SFS and IPS and with VSWM capacity (Olesen et al., 2003). These and other data indicate that increased interaction between the SFS and IPS over development is important for improvements in VSWM (Edin et al., 2007).

While SFS and IPS are increasingly engaged during spatial working memory over childhood and adolescence, other regions are engaged to a lesser extent over development. For instance, Scherf, Luna and colleagues found that children recruit only limited activation from core working memory regions (DLPFC and parietal regions) and instead rely primarily on ventromedial regions (caudate nucleus and anterior insula) (Scherf et al., 2006). In adolescence, by contrast, they observed refinements of the specialized network found in adults (Casey et al., 2008; Durston et al., 2006; Konrad et al., 2005; Scherf et al., 2006). These results suggest that the maturation of adult-level cognition involves first a shift from an immature network to an adult network including more mature performance-enhancing regions, and next an increase in the efficiency within those necessary regions.

Although the majority of developmental fMRI studies of working memory have focused on VSWM, we have conducted a study on non-spatial working memory development in which participants had to remember a series of three nameable objects (Crone et al., 2006c). Children aged 8–12 made more errors than 13- to 17-year-olds and 18- to 25-year-olds, but engaged similar brain regions during task performance. Positive correlations between accuracy and activation across the entire group were observed in all regions of interest: VLPFC, DLPFC, and superior parietal cortex. These correlations remained significant after controlling for age, indicating that the level of engagement of these regions in and of itself impacts performance. Taken together with the VSWM studies, these studies suggest that the basic working memory circuitry is in place by middle childhood, but they also provide evidence for the strengthening of working memory circuitry during middle childhood.

Interference suppression

A VSWM study by Olesen, Klingberg, and colleagues included a period of distraction, during which participants were asked to ignore stimuli appearing in various locations on the screen (Olesen et al., 2007).

Table 2.1 Developmental studies of working memory and cognitive control functions

Publication	Year	Technique	Paradigm	Sample sizes and ages	Brain activation changes with increasing age
Working memory					
Kwon et al.	2002	fMRI	2-back VSWM task	8 (7–12), 8 (13–17), 7 (18–22)	Increased LPFC, PPC (DLPFC, VLPFC, PMC, angular gyri, IPS, SPL)
Klingberg et al.	2002	fMRI	VSWM task	13 (9–18)	Increased superior frontal cortex, intraparietal cortex
Olesen et al.	2003	fMRI, DTI	VSWM task	23 (8–18)	Increased superior frontal cortex, intraparietal cortex
Tsujimoto et al.	2004	Optical	VSWM task	16 (5–6), 7 (22–28)	No significant difference in bilateral LPFC
Scherf et al.	2006	fMRI	ODR task	30 (8–47)	Different neural circuit for children
Crone	2006	fMRI	Object working memory task	14 (8–12), 12 (13–17), 18 (18–25)	Increased accuracy associated with greater VLPFC, DLPFC, and superior parietal activation, even when accounting for age
Interference suppression/selective attention/task-switching					
Booth et al.	2003	fMRI	Conjunction search	12 (9–12), 12 (20–30)	Decreased ACC, thalamus associated with attention
Konrad et al.	2005	fMRI	Attention network task	16 (8–12), 16 (20–34)	Increased right MFG associated with alerting; increased right temporo-parietal junction, decreased SFG association with reorienting; increased DLPFC, decreased STG associated with selective attention
Olesen et al.	2007	fMRI	VSWM task with distractor task	13 (13), 11 (19–25)	During delay, increased MFG, IPS, SPL associated with maintenance and decreased frontal cortex associated with interference
Crone et al.	2006	fMRI	Set shift task	17 (8–12), 17 (13–17), 20 (18–25)	Increased pre-SMA/SMA by adolescence
Manipulation					
Kwon et al.	2002	fMRI	2-back VSWM task	8 (7–12), 8 (13–17), 7 (18–22)	Increased LPFC, PPC (DLPFC, VLPFC, PMC, angular gyri, IPS, SPL)
Crone et al.	2006	fMRI	Object working memory task	14 (8–12), 12 (13–17), 18 (18–25)	Recruitment of right DLPFC, bilateral superior parietal cortex during manipulation delay period
Response inhibition/selection					
Luna et al.	2001	fMRI	Antisaccade task	11 (8–13), 15 (14–17), 10 (18–30)	Decreased SMG and increased BG, IPS, SC, FEF, lateral cerebellum with DLPFC peaking in adolescence
Tamm et al.	2002	fMRI	Go/no-go task	19 (8–20)	Increased left IFG, decreased left SFG/MFG, increased focalization of activation

Table 2.1 (*cont.*)

Publication	Year	Technique	Paradigm	Sample sizes and ages	Brain activation changes with increasing age
Booth et al.	2003	fMRI	Go/no-go task	12 (9–12), 12 (20–30)	Decreased fronto-striatal network (MCC, MFG, medial bilateral SFG, left CN) associated with inhibition
Durston et al.	2006	fMRI	Go/no-go task	14 (9–12)	Increased VLPFC and decreased other prefrontal areas
Rubia and Smith	in press	fMRI	Stop signal task	21 (10–17), 26 (20–42)	Increased right inferior PFC
Adleman et al.	2002	fMRI	Stroop task	8 (7–11), 11 (12–16), 11 (18–22)	Increased parietal cortex until adolescence and then increased left MFG until adulthood
Schroeter et al.	2004	Optical	Stroop task	23 (7–13), 14 (19–29)	Increased DLPFC with performance and decreased left LPFC
Marsh et al.	2006	fMRI	Stroop task	70 (7–57)	Increased accuracy associated with increased fronto-striatal network activation
Bunge et al.	2002	fMRI	Flanker task	16 (8–12), 16 (19–33)	Decreased left VLPFC associated with interference and decreased parietal and temporal cortices associated with inhibition; increased right VLPFC overall
Rubia et al.	2006	fMRI	Go-no-go, switch, Simon tasks	23 (10–17), 29 (20–43)	Increased prefrontal, temporo-parietal, striatal regions
Error-processing					
Davies et al.	2004	EEG	Flanker task	124 (7–18), 27 (19–25)	Decreased ERN-amplitude

Notes:

ACC, anterior cingulate cortex; BG, basal ganglia; CN, caudate nucleus; conjunction search, a search task in which the distractors share a common feature with the target; DTI, diffusion tensor imaging; DLPFC, dorsolateral prefrontal cortex; EEG, electroencephalography; FEF, frontal eye field; fMRI, functional magnetic resonance imaging; IFG, inferior frontal gyrus; IPS, intraparietal sulcus; LPFC, lateral prefrontal cortex; MCC, middle cingulate cortex; MFG, middle frontal gyrus; ODR, oculomotor delayed response; PMC, primary motor cortex; PPC, posterior parietal cortex; SC, superior colliculus; SFG, superior frontal gyrus; SMA, supplementary motor area; SMG, superior middle gyrus; SPL, superior parietal lobe; STG, superior temporal gyrus; VSWM, visuospatial working memory; VLPFC, ventrolateral prefrontal cortex.

Children around the age of 13 years exhibited greater SFS activation than adults during this period of distraction (likely associated with spatial processing of the distractors), despite having shown reduced activation in this region in VSWM studies that did not involve distraction. This finding suggests that children are less effective than adults at ignoring the irrelevant stimuli.

In contrast, adults more strongly engaged right DLPFC and bilateral intraparietal cortex than children while they maintained relevant information online. This finding is potentially significant in light of a prior study in adults (Sakai et al., 2002). Using a similar task with adults, Sakai and colleagues showed that engagement of a slightly anterior region of right DLPFC was associated with better performance on the working memory task. Thus, in the study by Olesen and colleagues, children showed weaker activation during VSWM in a region of the right DLPFC, which, in adults, may serve to create a distractor-resistant memory trace (Olesen et al., 2007). It would be of great interest to examine developmental changes in the functional interactions between DLPFC, SFS, and IPS, and how these changes affect the ability to suppress interference (see Sakai et al., 2002).

Manipulation

As noted above, improvements in the ability to maintain information online are observed during childhood, and – with highly sensitive measures – throughout adolescence (Luna and Sweeney, 2004). However, developmental changes are more dramatic when one must manipulate, or work with, information held in working memory (Hitch, 2002). Our working memory study, described above (Crone et al., 2006c), provided evidence for protracted neurodevelopmental changes in regions involved in manipulation relative to pure maintenance. Prior imaging research in adults had implicated DLPFC and superior parietal cortex in manipulation (Wager and Smith, 2003). In our study, 13- to 17-year-olds and 18- to 25-year-olds – but not 8- to 12-year-olds – engaged right DLPFC and bilateral

superior parietal cortex when it was necessary to reverse the order of items held in working memory (Figure 2.1). Unlike the older age groups, 8- to 12-year-olds did not recruit additional regions for manipulation relative to pure maintenance; this reliance on maintenance circuitry was associated with suboptimal manipulation ability. The finding of lower DLPFC engagement in children compared to adults has also been observed in other studies of working memory and cognitive control (Konrad et al., 2005; Olesen et al., 2007).

These data do not clarify whether children fail to recruit brain regions involved in manipulation as a result of: (1) *maturational constraints* associated with immature neural circuitry and/or (2) *limited practice* with this type of task. Interestingly, in our study, the children did recruit these DLPFC and superior parietal regions during encoding and response selection – just not during the delay period, when manipulation was required. A pattern of mature DLPFC activation has also been observed in 8- to 12-year-olds performing a simple gambling task, even though age-related differences relating to the processing of uncertainty and negative feedback were observed in other parts of PFC (van Leijenhorst et al., 2006).

These observations highlight a general point about developmental changes in brain function: a region may exhibit adult-like patterns of activation in one task but not in another. As another example, VLPFC showed a mature pattern of activation in our non-spatial working memory task (Crone et al., 2006c), but not in tasks involving response inhibition (Bunge et al., 2002; Durston et al., 2002; Rubia and Smith, 2007) or rule representation (Crone et al., 2006a). Thus, a region may contribute effectively to a neural circuit underlying one task or cognitive function, but not to another.

Response inhibition

Improvements in the ability to control one's actions are observed during childhood (Davidson et al., 2006). Control is needed when one must inhibit a tendency to respond to a stimulus (termed response

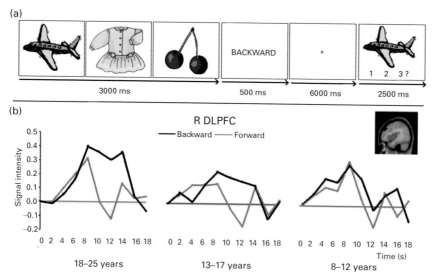

Figure 2.1a, b. Manipulating items in working memory (Crone, Wendelken et al., 2006). (a) Example of trial sequence with backward instruction. Each trial started with the consecutive presentation of three items, followed by a "forward" or a "backward" instruction. In the case of a forward instruction, participants had to maintain the information in working memory during the delay period. In the case of a backward instruction, participants had to reorder the items in working memory. The probe required a response according to whether the picture was presented first, second or third in the forward or in the backward condition. (b) Timecourses for participants aged 18–25 years, 13–17 years, and 8–12 years for right dorsolateral prefrontal cortex (DLPFC). Unlike adults and adolescents, 8- to 12-year-old children failed to recruit this region during the delay period of the backward trials, as indicated by the lack of difference between forward and backward trials.

inhibition). This ability is typically measured with go/no-go and stop-signal paradigms (Booth et al., 2003; Bunge et al., 2002; Durston et al., 2002; Lamm et al., 2006; Rubia et al., 2001; Rubia and Smith, 2007; Tamm et al., 2002).

In the go/no-go paradigm, participants must respond to visual stimuli presented in rapid sequence, thereby building up a prepotent response tendency. Participants are also instructed to withhold their response to a particular stimulus (the "no-go" stimulus), which is presented without warning from time to time. Functional MRI studies involving the go/no-go paradigm have revealed a network of largely right-hemispheric brain regions associated with successful response inhibition, including right VLPFC (Garavan et al., 1999).

In the stop-signal task, by contrast, participants press a button in response to visual stimuli, but must

on some occasions (e.g., if they hear a tone) inhibit their response at the last moment. This task is thought to tax response inhibition more heavily than the go/no-go paradigm because participants must override a motor response that they have already initiated. The stop-signal paradigm has an advantage over other response inhibition paradigms for use in clinical and developmental brain imaging studies: the task difficulty varies dynamically on a trial-by-trial basis to ensure a fixed level of performance across participants. As such, differences in brain activation between groups cannot simply be attributed to differences in performance.

Response selection

Response control is needed not only when one must inhibit an inappropriate response, but also when one must select between competing response alternatives. Neurodevelopmental changes in response selection have been studied with a variety of paradigms, including the Simon task (Rubia et al., 2006), antisaccade task (Luna et al., 2001), Eriksen flanker task (Booth et al., 2003; Bunge et al., 2002; Rubia et al., 2006), and the Stroop task (Adleman et al., 2002; Marsh et al., 2006; Schroeter et al., 2004). Although these tasks differ in many ways and therefore likely rely on distinct cognitive processes, many of them may tax common underlying neural substrates supporting controlled responding.

As such, to better understand the development of brain networks underlying response control, it is critical to determine which age-related differences are task-specific and which generalize across several paradigms. An earlier fMRI study combining elements of the go/no-go and flanker tasks revealed that children aged 8–12 years failed to engage a region in the right VLPFC that young adults recruited for both response selection and inhibition (Bunge et al., 2002). It has since been shown that adults with damage to the right VLPFC have difficulty with several tasks involving response control (Aron et al., 2004). A new developmental study by Rubia and colleagues further shows a positive correlation between right VLPFC activation and age (between 10 and 42 years) during successful versus unsuccessful inhibitions on the stop-signal paradigm (Rubia and Smith, 2006; for a similar correlation in a developmental Stroop study, see Marsh et al., 2006). These findings indicate that suboptimal response control in children and adolescents is associated with insufficient recruitment of the right VLPFC and the functionally connected regions, including the thalamus, caudate, and cerebellum (Rubia and Smith, 2007).

Another recent study by Rubia and colleagues combined three tasks involving response control: go/no-go, Simon, and attentional set-shifting tasks (Rubia et al., 2006). The investigators compared activation on all three tasks for youths aged 10–17 and adults aged 20–43 years. In all three tasks, adults recruited portions of PFC, anterior cingulate cortex, and striatum more strongly than the youths,

and there was a positive linear correlation with age in task-relevant frontal and striatal regions. Additionally, adults engaged inferior parietal cortex more strongly than youths on the Simon and set-shifting tasks (Rubia et al., 2006), and a similar finding has previously been reported for the go/no-go task (Bunge et al., 2002). However, the lateralization and precise location of these age-related changes depended on the task at hand, further highlighting the need to seek converging evidence from multiple tasks.

Error- and feedback-based performance adjustments

Besides the execution of controlled behavior, the ability to monitor ongoing performance in a dynamically changing environment is also an important aspect of goal-directed behavior (Miller and Cohen, 2001), especially when behavior needs to be adjusted after the detection of errors or changes in the environment. Neuropsychological and neuroimaging studies have demonstrated that the DLPFC is important for performance monitoring and updating, and that the DLPFC works closely together with the anterior cingulate cortex (ACC), which signals the requirement for increased control after an erroneous response (Gehring and Knight, 2000; Kerns et al., 2004).

The detection of errors, or error-based learning, has been studied extensively in the adult literature using electrophysiological techniques. These studies have incorporated a focus on the error-related negativity (ERN), a scalp potential that is observed when participants make errors in choice reaction time tasks (Falkenstein et al., 1991; Scheffers et al., 1996). Source localization studies have placed this potential at the ACC (Holroyd et al., 2004; Van Veen and Carter, 2002). Studies examining error processing in children using simple error-detection tasks have demonstrated that the ERN is minimal at around 7–8 years of age and increases in magnitude during adolescence (Davies et al., 2004). Despite the slowly developing ERN, young children already show evidence of error detection at a behavioral level, because they slow down following an erroneous response, as do adults. Furthermore, children as young as 7 years of age show evidence of a positive brain potential following the ERN, the so-called error-positivity. This potential has been linked to the awareness of an error (Nieuwenhuis et al., 2001). This different developmental pattern of immature monitoring versus mature performance adjustment warrants further investigation.

Performance monitoring is also required following the presentation of performance feedback. We have studied developmental changes in feedback-based learning in a series of experiments, which were based on the experimental variations of the classic Wisconsin Card Sorting Task (WCST). In these studies, we have made use of several psychophysiological indices, including heart rate slowing. These studies were inspired by an initial finding reported by Somsen et al., showing that the shift toward heart rate acceleration that is usually observed following response onset was delayed when the response was followed by negative feedback (Somsen et al., 2000).

In subsequent experiments, we demonstrated that heart rate was a sensitive index of feedback-based performance monitoring (Crone et al., 2003) and that the development of the feedback-based monitoring system was dependent on the type of information that was signaled by the feedback. When the feedback was unexpected and unrelated to performance (i.e., presented randomly), even 8-year-old children demonstrated cardiac slowing (Crone et al., 2006b). However, when the feedback was informative for future performance adjustments, 8-year-old children showed reduced heart rate slowing relative to 12-year-old children and adults, and the amount of heart rate slowing was correlated with subsequent performance adjustments (Crone et al., 2006b). Together, this series of studies led us to hypothesize that the system underlying performance-based feedback learning (i.e., feedback that can be used to adjust your performance) matures more slowly than the system underlying the signaling of unexpected feedback (i.e., feedback that is not associated with performance and therefore unexpected).

Recently, we have started to examine the neural basis of feedback-based learning using fMRI (Crone et al., 2008a). Children aged 9–11 years, adolescents aged 14–15 years, and young adults were asked to perform a rule-switching task in which they had to switch between three possible location rules based on positive and negative feedback. The location rules were taught beforehand, and participants were informed that rules could change unexpectedly while they performed the task. The location rules were chosen such that there were no ambiguous trials; therefore, location sorts could only be correct according to one of the three rules. Based on a scoring method devised by Barcelo and colleagues (Barcelo and Knight, 2002), we distinguished between three types of negative feedback. The first-warning error was the first indication that prior performance was no longer correct (similar to the rule changes in the WCST). The efficient negative feedback was a cue that could be used to find the correct sorting rule. Following a first warning error, participants had a 50% chance of identifying the new rule on the subsequent trial. In case they first chose the incorrect rule, followed by the correct rule, then this negative feedback was coined "efficient," because it guided the participants towards finding the correct rule. Finally, the third type of negative feedback was related to performance errors.

In this study, negative performance feedback was associated with increased activation in both DLPFC and ACC relative to positive feedback. However, DLPFC was more active following efficient negative feedback, whereas the ACC was more active following first-warning negative feedback. Therefore, DLPFC and ACC may have different roles in feedback-based learning. Developmental comparisons indicated that children aged 9–11 showed reduced activation in DLPFC and ACC following negative feedback, whereas adolescents aged 14–15 years did not differ from adults in ACC activation, but failed to recruit DLPFC to the same extent as adults following negative feedback (Zanolie et al., 2006). These results suggest that the monitoring of performance feedback continues to develop until

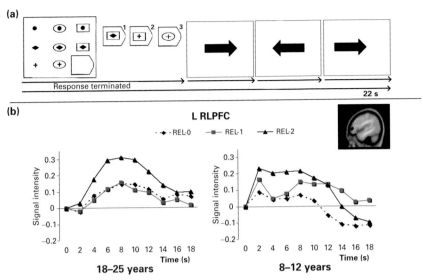

Figure 2.2a, b. Integrating visuospatial relations (Crone et al., 2008b). (a) Example of a two-relational matrix problem. Participants were instructed to select one of the three items that had the best fit in the matrix within a 22-s trial period. The presentation of the problem was terminated by the subject's response and was followed by left and right pointing arrows that required left or right button presses, respectively. (b) Time courses for 18- to 25-year-old adults and 8- to 12-year-old children for right rostrolateral prefrontal cortex (RLPFC). Adults selectively activate for two-relational problems, whereas children show a more variable response pattern.

middle adolescence and that the late developmental progression is mostly associated with functional development of DLPFC, a region important for implementing cognitive control (e.g., Braver et al., 2002). These findings are also consistent with our prior fMRI work demonstrating slow DLPFC development in other domains of cognitive control, for example in the ability to manipulate information in working memory (Crone et al., 2006b).

Relational reasoning

Reasoning, an important component of fluid intelligence, represents the capacity to think logically and solve problems in novel situations (Cattell, 1987). This capacity is thought to be instrumental in the learning of tasks requiring complex spatial, numerical, or conceptual operations. Reasoning ability emerges in the first 2 or 3 years of life, after the development of general perceptual, attentional,

and motoric capabilities. Children begin to consider multiple relationships between mental representations (i.e., relations) at about 5 years of age (Halford et al., 1998), but improvements in relational reasoning are observed throughout childhood (Sternberg and Rifkin 1979; Vodegel Matzen, 1994).

Reasoning has long been studied in cognitive and developmental psychology, but has only recently been examined from a cognitive neuroscience perspective. Brain imaging studies involving Raven's Progressive Matrices (RPM) (Figure 2.2a) reveal that relational integration in the visuospatial domain places high demands on DLPFC and the most anterior part of lateral PFC, referred to as rostrolateral PFC (RLPFC) or the lateral part of Brodmann area 10 (Christoff et al., 2001). Rostrolateral PFC is particularly sensitive to the number of relations under consideration, whereas DLPFC activation is additionally sensitive to non-relational factors that influence task difficulty, such as the number of

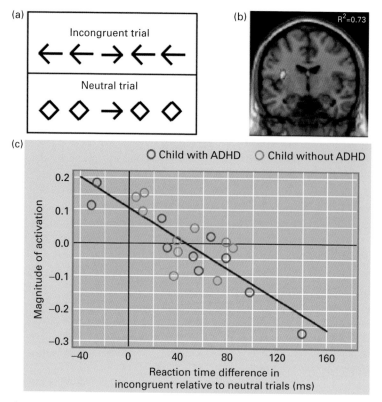

Figure 2.3a–c. Interference effects on flanker task. (a) Sample incongruent and neutral stimulus arrays from the flanker task used by Vaidya et al. (2005) to compare fronto-striatal function in children with and without ADHD. On incongruent trials, the flanking stimuli were arrows in the opposite direction from the central arrow, and were therefore associated with the opposite response relative to the target stimulus, leading to response competition. On neutral trials, the flanking stimuli were diamonds, which were not associated with any response. (b) Region in left ventrolateral prefrontal cortex (VLPFC) exhibiting greater activation in participants whose performance was less negatively affected by interference, as identified by a whole-brain regression analysis. (c) Plotted for this region in left VLPFC is the average correlation between the level of activation for incongruent > neutral trials, across participants, and the amount of slowing of response times for incongruent relative to neutral trials. Figure adapted from Vaidya et al. (2005) and reprinted with permission.

distractors in the set of possible responses to an RPM problem (Kroger et al., 2002).

In a recent study, we addressed the question of why children have difficulty considering multiple relations (Crone et al., 2008b). With fMRI, we examined whether 8- to 12-year-old children recruit RLPFC and DLPFC differently from adults during performance of a modified RPM task. Participants were asked to solve three types of problems (Figure 2.3b). Zero-relational (REL-0) matrix problems required

perceptual matching, and served to control for the visual and motoric demands of the relational problems. One-relational (REL-1) matrix problems required reasoning about a single dimension of change (vertical or horizontal) to identify the correct answer, and two-relational (REL-2) matrix problems required participants to integrate across two dimensions of change.

Consistent with prior behavioral studies, 8- to 12-year-olds were disproportionately less accurate on REL-2 problems than adults, suggesting that developmental changes were most evident for matrix problems that required relational integration. It is likely that children gave up more easily and resorted to guessing on the REL-2 problems compared to adults, resulting in lower accuracy (Sternberg and Rifkin, 1979; Vodegel Matzen, 1994). Functional MRI analyses showed that, like adults, children engaged both RLPFC and DLPFC on REL-2 trials relative to the other conditions. Unlike adults, however, this increased activation on REL-2 trials was not sustained over the entire duration of the trial, consistent with the idea that children tend to select a response without having considered both dimensions of change. To our knowledge, this experiment constitutes the first fMRI study of reasoning in children and also the first to compare RLPFC activation in children and adults. The results indicate that classic developmental questions can benefit from the use of neuroimaging methods.

Summary and future directions

In summary, a growing literature indicates that the increased recruitment of task-related regions in frontal, parietal, and striatal regions underlie improvements in working memory and cognitive control over middle childhood and adolescence. The pattern of developmental changes in brain activation has been generally characterized as a shift from diffuse to focal activation (Durston et al., 2006) and from posterior to anterior activation (Brown et al., 2005; Rubia et al., 2006). Differences can be quantitative, with one age group engaging a region more strongly or extensively than another, and/or qualitative, with a shift in reliance on one set of brain regions to another (Brown et al., 2005, 2006; Rubia et al., 2006; Scherf et al., 2006). Importantly, the precise pattern of change observed depends on the task, the ages being examined, and the brain region in question. By further characterizing neurodevelopmental changes in cognitive control processes within subjects and across a range of tasks, we hope to better understand the development of the human mind.

By middle childhood (around age 12), the ability to hold goal-relevant information in mind and to use it to select appropriate actions is already adequate. It is of great interest to track brain function associated with working memory and cognitive control earlier in childhood, when these abilities are first acquired. Optical imaging studies can be conducted from infancy onward, although the spatiotemporal resolution of this method is suboptimal. It is now possible to acquire fMRI data in children as young as 4 years of age (Cantlon et al., 2006), although not without challenges such as head motion, low accuracy, and limited attention span.

An important future direction is to determine the extent to which observed age differences in brain activation reflect hard developmental constraints (e.g., the required anatomical network is simply not in place yet at a given age) as opposed to lack of experience with a given type of task or cognitive strategy. Training studies involving several age groups would allow us to investigate effects of age and effects of practice independently and to test whether inherent age differences in performance and brain activation are still present after substantial practice (Luna and Sweeney 2004; Qin et al., 2004).

Thus far, all but one (Durston et al., 2006) of the published developmental fMRI studies on working memory or cognitive control have compared groups of individuals at different ages. While these cross-sectional studies are valuable, they provide only a coarse indicator of developmental change. It is also important to conduct longitudinal studies to characterize intraindividual changes in brain function with age.

To understand how cognitive control is achieved, it will be necessary to know how PFC interacts with other brain regions. Some information about these interactions can be gleaned from functional connectivity analyses of fMRI data, although fMRI provides little information about the timing of these interactions. Another approach is to acquire fMRI and EEG data in the same group of participants, either in separate sessions or simultaneously (Debener et al., 2005). An important current and future direction for developmental neuroimaging studies is to examine changes in interactions among brain regions over childhood.

Implications for the study of neurodevelopmental disorders

A better understanding of how cognitive control develops in healthy children will lead to insights into the causes of impoverished control in a number of neurodevelopmental disorders. For example, prior studies have indicated that impulsivity disorders, such as attention deficit/hyperactivity disorder (ADHD), are characterized by deficits in inhibitory control (Barkley, 1997). This assumption was based on the deficits that were observed in go/no-go tasks and interference suppression tasks, but the neural circuitry underlying these deficits remained unclear. Several studies have implicated the fronto-striatal circuitry in this pathology (Durston et al., 2006; Vaidya et al., 2005). One study has shown that during performance on a response selection task known as the flanker task (Figure 2.3a), a region in left VLPFC previously associated with efficient response selection in healthy children (Bunge et al., 2002) was insufficiently engaged by children with ADHD (Vaidya et al., 2005); (Figure 2.3b, c). Importantly, the children diagnosed as having ADHD were not qualitatively different from the healthy children; they were merely at the lower end of a continuum in terms of both performance and engagement of left VLPFC.

The examination of the normal developmental pathways of distinct control functions will be important for understanding sensitive periods in brain development. Indeed, damage to PFC in childhood has a much greater impact than does damage in adulthood, likely because this region is important for acquiring skills and knowledge during childhood (Eslinger et al., 2004). Further research is needed to help mitigate the impact of early prefrontal damage. Likewise, investigations of developmental disorders have the potential to inform us on the role of specific brain regions within cognition. As mentioned above, the role of

experience in shaping neural responses to cognitive challenges and the elucidation of functional/effective connectivity are two research priorities that emerge at the forefront of developmental cognitive neuroscience.

ACKNOWLEDGMENTS

Funding was provided by the National Science Foundation (NSF 00448844) and the Dutch Science Foundation (VENI-NWO 055-142). We thank Samantha Wright for preparing Table 2.1, and Carol Baym and Zdema Op de Macks for helpful comments on a prior version of the manuscript.

REFERENCES

Adleman, N. E., Menon, V., Blasey, C. M. et al. (2002) A developmental fMRI study of the Stroop color-word task. *NeuroImage*, **16**(1), 61–75.

Aron, A. R., Robbins, T. W., and Poldrack, R. A. (2004) Inhibition and the right inferior frontal cortex. *Trends Cogn Sci*, **8**(4), 170–7.

Barcelo, F. and Knight, R. T. (2002) Both random and perseverative errors underlie WCST deficits in prefrontal patients. *Neuropsychologia*, **40**(3), 349–56.

Barkley, R. A. (1997) Behavioral inhibition, sustained attention, and executive functions: constructing a unifying theory of ADHD. *Psychol Bull*, **121**(1), 65–94.

Booth, J. R., Burman, D. D., Meyer, J. R., et al. (2003) Neural development of selective attention and response inhibition. *NeuroImage*, **20**(2), 737–51.

Braver, T. S., Cohen, J. D., and Barch, D. M. (2002) The role of prefrontal cortex in normal and disordered cognitive control: a cognitive neuroscience perspective. In: Stuss, D. T., Knight, R. T. (eds.) *Principles of Frontal Lobe Function.* New York: Oxford University Press, pp. 428–47.

Brocki, K. C. and Bohlin, G. (2004) Executive functions in children aged 6 to 13: a dimensional and developmental study. *Dev Neuropsychol*, **26**(2), 571–93.

Brown, T. T., Lugar, H. M., Coalson, R. S. et al. (2005) Developmental changes in human cerebral functional organization for word generation. *Cereb Cortex*, **15**(3), 275–90.

Brown, T. T., Petersen, S. E., and Schlaggar, B. L. (2006) Does human functional brain organization shift from diffuse to focal with development? *Dev Sci*, **9**(1), 9–11.

Bunge, S. A. and Zelazo, P. D. (2006) A brain-based account of the development of rule use in childhood. *Curr Dir Psychol Sci*, **15**(3), 118–21.

Bunge, S. A., Dudukovic, N. M., Thomason, M. E. et al. (2002) Immature frontal lobe contributions to cognitive control in children: evidence from fMRI. *Neuron*, **33**(2), 301–11.

Cantlon, J. F., Brannon, E. M., Carter, E. J., Pelphrey, K. A. (2006) Functional imaging of numerical processing in adults and 4-y-old children. *PLoS Biol*, **4**(5), e125.

Case, R. (1995) Capacity based explanations of working memory growth: a brief history and a reevaluation. In: Weinert, F. M., Schneider, W. (eds.) *Memory Performance and Competencies: Issues in Growth and Development*. Hillsdale, NJ: Lawrence Erlbaum.

Casey, B. J., Cohen, J. D., Jezzard, P. et al. (1995) Activation of prefrontal cortex in children during a nonspatial working memory task with functional MRI. *NeuroImage*, **2**, 221–9.

Casey, B. J., Galvan, A., and Hare, T. A. (2005a) Changes in cerebral functional organization during cognitive development. *Curr Opin Neurobiol*, **15**(2), 239–44.

Casey, B. J., Tottenham, N., Liston, C., Durston, S. (2005b) Imaging the developing brain: what have we learned about cognitive development? *Trends Cogn Sci*, **9**(3), 104–10.

Casey, B. J., Jones, R. M., and Hare, T. A. (2008) The adolescent brain. *Ann N Y Acad Sci*, **1124**, 111–26.

Cattell, R. B. (1987) *Intelligence: Its Structure, Growth and Action*. Amsterdam: North-Holland.

Christoff, K., Prabhakaran, V., Dorfman, J. et al. (2001) Rostrolateral prefrontal cortex involvement in relational integration during reasoning. *NeuroImage*, **14**(5), 1136–49.

Crone, E. A., Donohue, S., Honomichl, R., Wendelken, C., Bunge, S. A. (2006a) Brain regions mediating flexible rule use during development. *J Neurosci*, **26**, 11239–47.

Crone, E. A., Jennings, J. R., and Van der Molen, M. W. (2004a) Developmental change in feedback processing as reflected by phasic heart rate changes. *Dev Psychol*, **40**(6), 1228–38.

Crone, E. A., Ridderinkhof, K. R., Worm, M., Somsen, R. J., Van der Molen, M. W. (2004b) Switching between spatial stimulus-response mappings: a developmental study of cognitive flexibility. *Dev Sci*, **7**(4), 443–55.

Crone, E. A., Somsen, R. J., Zanolie, K., Van der Molen, M. W. (2006b) A heart rate analysis of developmental change in feedback processing and rule shifting from childhood to early adulthood. *J Exp Child Psychol*, **95**(2), 99–116.

Crone, E. A., van der Veen, F. M., Van der Molen, M. W., Somsen, R. J., van Beek, B., Jennings, J. R. (2003) Cardiac concomitants of feedback processing. *Biol Psychol*, **64**(1–2), 143–56.

Crone, E. A., Wendelken, C., Donohue, S., van Leijenhorst, L., Bunge, S. A. (2006c) Neurocognitive development of the ability to manipulate information in working memory. *Proc Natl Acad Sci U S A*, **103**(24), 9315–20.

Crone, E. A., Zanolie, K., Van Leijenhorst, L., Westenberg, P. M. and Rombouts, S. A. (2008a) Neural mechanisms supporting flexible performance adjustment during development. *Cogn Affect Behav Neurosci*, **8**(2), 165–77.

Crone, E. A., Wendelken, C., van Leijenhorst, L., Honomichl, R. D., Christoff, K., and Bunge, S. A. (2008b) Neurocognitive development of relational reasoning. *Dev Sci*, **11**(6).

Davidson, M. C., Amso, D., Anderson, L. C., Diamond, A. (2006) Development of cognitive control and executive functions from 4 to 13 years: evidence from manipulations of memory, inhibition, and task switching. *Neuropsychologia*, **44**(11), 2037–78.

Davies, P. L., Segalowitz, S. J., and Gavin, W. J. (2004) Development of response-monitoring ERPs in 7- to 25-year-olds. *Dev Neuropsychol*, **25**(3), 355–76.

Debener, S., Ullsperger, M., Siegel M., Fiehler, K., von Cramon, D. Y., Engel, A. K. (2005) Trial-by-trial coupling of concurrent electroencephalogram and functional magnetic resonance imaging identifies the dynamics of performance monitoring. *J Neurosci*, **25**(50), 11730–7.

Dempster, F. N. (1993) Resistance to interference: developmental changes in a basic processing mechanism. In: Howe, M. L., Pasnak, R. (eds.) *Emerging Themes in Cognitive Development*. Vol. 1: *Foundations*. New York: Springer-Verlag, pp. 3–27.

Diamond, A. (2002) Normal development of prefrontal cortex from birth to young adulthood: cognitive functions, anatomy and biochemistry. In: Stuss, D. T., Knight, R. T. (eds.) *Principles of Frontal Lobe Function*. London: Oxford University Press, pp. 466–503.

Durston, S., Davidson, M. C., Tottenham, N. et al. (2006) A shift from diffuse to focal cortical activity with development. *Dev Sci*, **9**(1), 1–8.

Durston, S., Thomas, K. M., et al. (2002) A neural basis for the development of inhibitory control. *Dev Sci*, **5**(4), F9–F16.

Edin, F., Macoveanu, J., Olesen, P., Tegnér, J., Klingberg, T. (2007) Stronger synaptic connectivity as a mechanism behind development of working memory-related brain activity during childhood. *J Cogn Neurosci*, **19**(5), 750–60.

Eslinger, P. J., Flaherty-Craig, C. V., and Benton, A. L. (2004) Developmental outcomes after early prefrontal cortex damage. *Brain Cogn*, **55**(1), 84–103.

Falkenstein, M., Hohnsbein, J., Hoormann, J., Blanke, L. (1991) Effects of crossmodal divided attention on late ERP components. II. Error processing in choice reaction tasks. *Electroencephalogr Clin Neurophysiol*, **78**(6), 447–55.

Fuster, J. M. (1997) *The Prefrontal Cortex: Anatomy, Physiology, and Neuropsychology of the Frontal Lobe*. Philadelphia: Lippincott-William & Wilkins.

Garavan, H., Ross, T. J., and Stein, E. A. (1999) Right hemispheric dominance of inhibitory control: an event-related functional MRI study. *Proc Natl Acad Sci U S A*, **96**(14), 8301–6.

Gehring, W. J. and Knight, R. T. (2000) Prefrontal-cingulate interactions in action monitoring. *Nature Neurosci*, **3**(5), 516–520.

Gogtay, N., Giedd, J. N., Lusk, L., et al. (2004) Dynamic mapping of human cortical development during childhood through early adulthood. *Proc Natl Acad Sci U S A*, **101**(21), 8174–9.

Halford, G. S., Wilson, W. H., and Phillips, S. (1998) Processing capacity defined by relational complexity: implications for comparative, developmental, and cognitive psychology. *Behav Brain Sci*, **21**(6), 803–31; discussion 831–64.

Hitch, G. J. (2002) Developmental changes in working memory: a multicomponent view. In: Graf, P., Ohta, N. (eds.) *Lifespan Development of Human Processing*. Cambridge, MA: MIT Press.

Holroyd, C. B., Nieuwenhuis, S., Yeung, N., et al. (2004) Dorsal anterior cingulate cortex shows fMRI response to internal and external error signals. *Nature Neurosci*, **7**, 497–8.

Huizinga, M., Dolan, C. V., and Van der Molen, M. W. (2006) Age-related change in executive function: developmental trends and a latent variable analysis. *Neuropsychologia*, **44**(11), 2017–36.

Kerns, J. G., Cohen, J. D., MacDonald, A. W. 3rd, et al. (2004) Anterior cingulate conflict monitoring and adjustments in control. *Science*, **303**(5660), 1023–6.

Klingberg, T., Forssberg, H., and Westerberg, H. (2002) Increased brain activity in frontal and parietal cortex underlies the development of visuospatial working memory capacity during childhood. *J Cogn Neurosci*, **14**(1), 1–10.

Konrad, K., Neufang, S., Thiel, C. M., et al. (2005) Development of attentional networks: an fMRI study with children and adults. *NeuroImage*, **28**(2), 429–39.

Kroger, J. K., Sabb, F. W., Fales, C. L., et al. (2002) Recruitment of anterior dorsolateral prefrontal cortex in human reasoning: a parametric study of relational complexity. *Cereb Cortex*, **12**(5), 477–85.

Kwon, H., Reiss, A. L., and Menon, V. (2002) Neural basis of protracted developmental changes in visuo-spatial working memory. *Proc Natl Acad Sci U S A*, **99**(20), 13336–41.

Lamm, C., Zelazo, P. D., and Lewis, M. D. (2006) Neural correlates of cognitive control in childhood and adolescence: disentangling the contributions of age and executive function. *Neuropsychologia*, **44**(11), 2139–48.

Luna, B. and Sweeney, J. A. (2004) The emergence of collaborative brain function: FMRI studies of the development of response inhibition. *Ann N Y Acad Sci*, **1021**, 296–309.

Luna, B., Thulborn, K. R., Munoz, D. P., et al. (2001) Maturation of widely distributed brain function subserves cognitive development. *NeuroImage*, **13**(5), 786–93.

Marsh, R., Zhu, H., Schultz, R. T., et al. (2006) A developmental fMRI study of self-regulatory control. *Hum Brain Mapp*, **27**, 848–63.

Miller, E. K. and Cohen, J. D. (2001) An integrative theory of prefrontal cortex function. *Annu Rev Neurosci*, **24**, 167–202.

Nieuwenhuis, S., Ridderinkhof, K. R., Blom, J., Band, G. P., Kok, A. (2001) Error-related brain potentials are differentially related to awareness of response errors: evidence from an antisaccade task. *Psychophysiology*, **38**(5), 752–60.

Olesen, P. J., Macoveanu, J., Tegnér, J., and Klingberg, T. (2007) Brain activity related to working memory and distraction in children and adults. *Cereb Cortex*, **17**(5), 1047–54.

Olesen, P. J., Nagy, Z., Westerberg, H., Klingberg, T. (2003) Combined analysis of DTI and fMRI data reveals a joint maturation of white and grey matter in a fronto-parietal network. *Brain Res Cogn Brain Res*, **18**(1), 48–57.

Qin, Y., Carter, C. S., Silk, E. M., et al. (2004) The change of the brain activation patterns as children learn algebra equation solving. *Proc Natl Acad Sci U S A*, **101**(15), 5686–91.

Rubia, K., Smith, A. B., Taylor, E., and Brammer, M. (2007) Linear age-correlated functional development of right inferior fronto-striato-cerebellar networks during response inhibition and anterior cingulate during error-related processes. *Hum Brain Mapp*, **28**(11), 1163–77.

Rubia, K., Russell, T., Overmeyer, S., et al. (2001) Mapping motor inhibition: conjunctive brain activations across different versions of Go/No-Go and stop tasks. *NeuroImage*, **13**(2), 250–61.

Rubia, K., Smith, A. B., Woolley, J., et al. (2006) Progressive increase of frontostriatal brain activation from childhood to adulthood during event-related tasks of cognitive control. *Hum Brain Mapp*, **27**(12), 973–93.

Sakai, K., Rowe, J. B., and Passingham, R. E. (2002) Active maintenance in prefrontal area 46 creates distractor-resistant memory. *Nat Neurosci*, **5**(5), 479–84.

Scheffers, M. K., Coles, M. G., Bernstein, P., Gehring, W. J., Donchin, E. (1996) Event-related brain potentials and error-related processing: an analysis of incorrect responses to go and no-go stimuli. *Psychophysiology*, **33**(1), 42–53.

Scherf, K. S., Sweeney, J. A., and Luna, B. (2006) Brain basis of developmental change in visuospatial working memory. *J Cogn Neurosci*, **18**(7), 1045–58.

Schroeter, M. L., Zysset, S., Wahl, M., von Cramon, D. Y. (2004) Prefrontal activation due to Stroop interference increases during development – an event-related fNIRS study. *NeuroImage*, **23**(4), 1317–25.

Shaw, P., Greenstein, D., Lerch, J., et al. (2006) Intellectual ability and cortical development in children and adolescents. *Nature*, **440**(7084), 676–9.

Somsen, R. J., Van der Molen, M. W., Jennings, J. R., van Beek, B. (2000) Wisconsin Card Sorting in adolescents: analysis of performance, response times and heart rate. *Acta Psychol (Amst)*, **104**(2), 227–57.

Sowell, E. R., Peterson, B. S., Thompson, P. M., Welcome, S. E., Henkenius, A. L., Toga, A. W. (2003) Mapping cortical change across the human life span. *Nat Neurosci*, **6**(3), 309–15.

Sternberg, R. J., Rifkin, B. (1979) The development of analogical reasoning processes. *J Exp Child Psychol*, **27**(2), 195–232.

Tamm, L., Menon, V., and Reiss, A. L. (2002) Maturation of brain function associated with response inhibition. *J Am Acad Child Adolesc Psychiatry*, **41**(10), 1231–8.

Tsujimoto, S., Yamamoto, T., Kawaguchi, H., Koizumi, H., and Sawaguchi, T. (2004) Prefrontal cortical activation associated with working memory in adults and preschool children: an event-related optical topography study. *Cereb Cortex*, **14**(7), 703–12.

Vaidya, C. J., Bunge, S. A., Dudukoric, N. M., Zalecki, C. A., Elliott, G. R., Gabrieli, J. D. (2005) Altered neural substrates of cognitive control in childhood ADHD: evidence from functional magnetic resonance imaging. *Am J Psychiatry*, **162**(9), 1605–13.

Van den Wildenberg, W. P. M., and Van der Molen, M. W. (2004) Developmental trends in simple and selective inhibition of compatible and incompatible responses. *J Exp Child Psychol*, **87**, 201–20.

van Leijenhorst, L., Crone, E. A., and Bunge, S. A. (2006) Neural correlates of developmental differences in risk estimation and feedback processing. *Neuropsychologia*, **44**(11), 2158–70.

Van Veen, V. and Carter, C. S. (2002) The timing of action-monitoring processes in the anterior cingulate cortex. *J Cogn Neurosci*, **14**(4), 593–602.

Vodegel Matzen, L. (1994) Performance on Raven's Progressive Matrices: What makes a difference? Unpublished master's thesis, Faculty of Psychology, University of Amsterdam.

Wager, T. D. and Smith, E. E. (2003) Neuroimaging studies of working memory: a meta-analysis. *Cogn Affect Behav Neurosci*, **3**(4), 255–74.

Welsh, M. C., Pennington, B. F., and Groisser, D. B. (1991) A normative-developmental study of executive function in children. *Dev Neuropsychol*, **7**, 131–49.

Zanolie, K., Van Meel, C. S., et al. (2006) Development of feedback processing. Presented at the Cognitive Neuroscience Meeting, San Francisco, April 2006.

Zelazo, P. D. (2004) The development of conscious control in childhood. *Trends Cogn Sci*, **8**(1), 12–17.

3

Neurobiology of emotion regulation in children and adults

Ian H. Gotlib and Jutta Joormann

Introduction

The construct of emotion regulation, which evolved from the broader concept of coping with stress, involves the utilization of behavioral and cognitive strategies in an effort to modulate affect intensity and duration (Thompson, 1994). The assessment of emotion regulation is not an easy endeavor because it includes both voluntary and automatic processes. Nevertheless, given the importance of emotion regulation for well-being, this construct has been the focus of increasing attention in the developmental literature (Cole et al., 2004). Indeed, children's development of emotion regulation is considered to be one of the cornerstones of socialization and moral development (Eisenberg, 2000). Investigators examining emotion regulation in response to stress in young children have identified a developmental sequence, progressing from a basic limited self-regulatory capacity for managing emotion (as demonstrated in studies on infant temperament and behavioral inhibition), through toddlers engaging in mutually regulatory interactions with their mothers, culminating in the development of a wider array of self-regulatory strategies in adolescents. Importantly, the developmental literature continues to document the long-term negative consequences related to difficulties in these stages of the development of emotion regulation and stress responsivity.

Cole et al. (2004) discussed a number of methodological issues in the study of emotion regulation from a developmental perspective. They described the difficulty of distinguishing between emotional response and emotion regulation and suggested that developmental research on emotion regulation requires a multi-method approach. Most of the empirical research on emotion regulation has focused on infancy and toddlerhood and has relied primarily on observer ratings. These measures are susceptible to demand characteristics and, perhaps more importantly, do not permit the assessment of automatic regulation processes. Goldsmith and Davidson (2004) argued that, in the future, research examining emotion regulation from a developmental perspective must include behavioral measures, measures of hypothalamic-pituitary-adrenocortical (HPA) axis function, and regional brain function (specifically the amygdala, prefrontal cortex, hippocampus, and anterior cingulate cortex). Only recently, however, have researchers begun to examine the neural correlates of negative mood states and emotions. Importantly, few studies have examined patterns of brain activation that are related to mechanisms underlying recovery from negative affect.

In this chapter we describe current research on emotion regulation in response to stress that focuses on these specific domains of functioning. We present findings of studies that have investigated emotion

Neuroimaging in Developmental Clinical Neuroscience, eds. Judith M. Rumsey and Monique Ernst. Published by Cambridge University Press. © Cambridge University Press 2009.

regulation from a neural and neuroendocrine perspective. We then discuss the implications of these findings for research on emotional disorders, and, in particular, depression in children.

Emotion regulation and stress reactivity

Cole et al. (2004) raised concerns about the broad range of studies and measures that are subsumed under the label of emotion regulation. They argue that most studies use the term "emotion regulation" without providing an explicit definition of the construct. Thus, it is not clear whether emotion regulation pertains primarily to the regulation of negative emotions or more generally to the regulation of all affective states. And even if the latter definition is used, it is important to specify whether only recovery from a mood state is included or whether strategies that maintain or intensify a desired affective state are also implicated. In addition, as noted earlier, emotional response and emotion regulation are so intricately linked that it is difficult to differentiate between the two constructs. Indeed, there is a large body of research in which investigators have presented positive and negative material to participants and interpreted any response to this material as an instance of emotion regulation without providing evidence that an affective state was induced and that regulation actually occurred.

Given these important and pervasive methodological and definitional problems in research on emotion regulation, it is critical that we be clear from the outset in this chapter. We are including studies that are relevant to a relatively circumscribed conceptualization of emotion regulation. We are interested primarily in recovery from negative affect. Because research on emotion regulation evolved, in part, from the broader concept of coping, we follow its focus on the constructs of stress exposure, stress reactivity and recovery, and other self-regulatory processes that are engaged in response to stress. We are also restricting our discussion of empirical findings to studies that help us understand biological mechanisms that might contribute to individual differences in emotion regulation. For example, baseline HPA-axis functioning and HPA reactivity and recovery, which likely reflect reactions to stress and the ability to regulate negative emotions, have recently been found to be aberrant in samples of depressed adults (see Burke et al., 2005 for a review) and children (Goodyer et al., 2001). Similarly, investigators examining patterns of brain activation have begun to move from studies of neural responses to emotional faces or to rewarding stimuli without an assessment of mood changes (which are not necessarily, we argue, examinations of regulatory processes) to studies in which they attempt to delineate neural correlates of emotion regulation (Cooney et al., 2007; Ochsner et al., 2002). In the following sections we briefly describe the assessment of emotion regulation and stress reactivity and recovery from the perspectives of patterns of neural activation and neuroendocrine functioning, and we present an overview of relevant studies.

Neurobiological correlates of emotion regulation

Healthy adults

Neurobiological approaches to the study of emotion regulation are still in their early stages (Goldsmith and Davidson, 2004). Investigators examining the neuroimaging correlates of emotion induction, specifically induction of sad mood, have reported activations in the medial prefrontal cortex (MPFC), medial temporal lobes, anterior cingulate cortex (ACC), and, in particular, in the subgenual ACC (Brodmann area (BA) 25), which has been postulated to play a critical role in the experience and modulation of sad mood (Lévesque et al., 2003a). Phan et al. (2002) presented a meta-analysis of neuroimaging studies of emotion in which they reported that, whereas MPFC played a general role in emotion processing, fear was specifically related to activation in the amygdala and sadness to activation in the subgenual ACC. More recent studies, however, have found sadness to be related to amygdala activation (Lévesque et al., 2003a), so the specificity of processing of different emotions to particular neural structures is still an unresolved question.

The few studies that have been conducted investigating the neural substrates of the regulation of negative affect have focused primarily on the neural correlates of reappraisal (Ochsner et al., 2002) and suppression (Ochsner et al., 2004; Ohira et al., 2006; Phan et al., 2005; see Table 3.1). In general, these studies have implicated the recruitment of dorsal regions of both the prefrontal cortex (PFC) and the ACC in the successful regulation of affect. In one of the first studies to examine neural aspects of the regulation of affective states, Beauregard et al. (2001) found that the suppression of sexual arousal was associated with activation in right lateral PFC (LPFC) and right ACC. Ochsner et al. (2002, 2004) asked participants to increase, decrease or maintain their affective response to positive, negative and neutral pictures and found that the dorsal ACC and PFC systems appeared to regulate limbic regions such as the amygdala and insula. Both increasing and decreasing emotional response were associated with activation in the left LPFC, which has been implicated in cognitive control and working memory, and with activation in the dorsal ACC and dorsal MPFC. Interestingly, both instructions also modulated activation of the left amygdala, which decreased or increased in activation with the regulation instructions (see also Schaefer et al., 2002). Finally, Hariri et al. (2003) reported that, whereas processing of threatening and fearful pictures was associated with bilateral amygdala response, reappraisal was associated with a reduction in amygdala activation and an increase in activation in the ACC and the right PFC.

Drawing on this pattern of results, contemporary accounts of neural aspects of emotion regulation posit a top-down modulation of ventral limbic structures by more dorsal structures, such as the dorsolateral prefrontal cortex (DLPFC; e.g., Ochsner and Gross, 2005). Interestingly, these regions have also been found to be activated in tasks that involve the cognitive inhibition of prepotent behavioral responses (e.g., Sylvester et al., 2003), an observation that led researchers to propose that an overlapping set of prefrontal regions may also play an important role in the cognitive control of emotion; for example, through such effortful cognitive strategies as reappraisal or the cognitive suppression of emotional responses. It is important to note, however, that recent studies have identified several additional brain areas that seem to be associated with the generation and regulation of emotional responses; in particular, ventromedial and ventrolateral areas, including the orbitofrontal cortex (OFC; BA 11), subgenual ACC (BA 25, 32), and BA 10 (e.g., Harenski and Hamann, 2006). Given these findings, researchers have suggested that, despite some similarities, different emotion regulatory strategies recruit different brain regions. Indeed, Ochsner and Gross (2005) have proposed that, in addition to the classic network of top-down regulation of ventral and limbic structures by dorsal structures such as DLPFC, another more ventral system is also implicated in mood and emotion regulation. It is possible, therefore, that the inconsistencies in the literature reflect differences in the level of involvement of these two systems in the various regulatory strategies examined in these studies.

In this context, it is important to note that there are at least two broad but distinct classes of cognitive affect-regulation strategies. One class of strategies reduces negative affect by generating thoughts and appraisals that directly alter the perceptions of the situation that elicited the emotion or the experience of the negative affect (e.g., reappraisal). In contrast, a different class of strategies alleviates negative affect less directly by strengthening mood-incongruent (i.e., positive) thoughts and associations and by evoking competing affective states. Cognitive models of mood and emotion propose that affective states are initiated and maintained by the activation of mood-congruent cognition (Bower, 1987). Activating positive thoughts while in a negative mood state, therefore, could prove to be effective in regulating negative affect. Importantly, these latter strategies can be useful in modulating affective states that have no identifiable

Table 3.1 Summary of brain imaging studies on emotion regulation in adults and children

Study	Paradigm	Sample size	Age group	Results	Additional results
1. Adults					
Beauregard et al. (2001)	Suppress emotional arousal to erotic film clips	10 males	Adults	Right DLPFC ↑; right ACC ↑	
Ochsner et al. (2002)	Reappraise vs. attend	15 females	Adults	LPFC ↑, MPFC ↑, amygdala and MOFC ↓	
Schaefer et al. (2002)	Maintain vs. passive viewing	7 females	Adults	Amygdala ↑	
Lévesque et al. (2003b)	Suppress emotional response to sad film clip	20 females	Adults	Right DLPFC and right OFC ↑	BOLD signal changes in both areas correlated with self-reported sadness
Ochsner et al. (2004)	Up- and down-regulate negative affect	24 females	Adults	Left LPFC ↑; dACC ↑; dorsal medial PFC ↑; up-regulate: amygdala ↑; or down-regulate: amygdala ↓	
Phan et al. (2005)	Suppress vs. maintain	14 participants	Adults	dACC ↑, dorsal medial prefrontal and lateral prefrontal cortices ↑, NAcc and amygdala ↓	dACC activity inversely correlated with negative affect, amygdala activation positively correlated with negative affect
Ohira et al. (2006)	Suppress vs. attend	10 females	Adults	Left LPFC, MPFC, and medial OFC ↑	Skin conductance in the suppress condition was correlated with OFC activation
Cooney et al. (2007)	Recall of mood-incongruent memories	14 females	Adults	VLPFC, VMPFC, OFC, subgenual ACC ↑	
2. Children					
Lévesque et al. (2004)	Suppression of emotional response to film clip vs. attend	14 girls	8–10 years	LPFC, OFC, medial PFC, right ACC, right VLPFC ↑	

Notes:
↑ increased activation; ↓ decreased activation; ACC, anterior cingulate cortex; dACC, dorsal anterior cingulate cortex; DLPFC, dorsolateral prefrontal cortex; LPFC, lateral prefrontal cortex; MOFC, medial orbitofrontal cortex; MPFC, medial prefrontal cortex; NAcc, nucleus accumbens; OFC, orbitofrontal cortex; PFC, prefrontal cortex; VLPFC, ventrolateral prefrontal cortex; VMPFC, ventromedial prefrontal cortex.

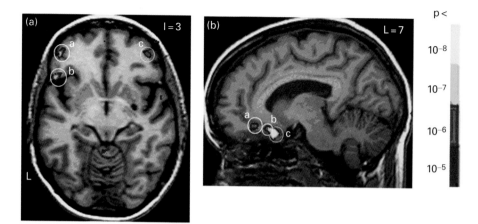

Figure 3.1. Neural correlates of emotion regulation. Participants see a fixation cross, followed by a cue to recall a positive memory, followed by a negative mood induction, followed by another cue to recall a positive memory. This figure depicts changes in activation in response to the first recall of positive autobiographical memories compared to a baseline scan (fixation cross) and compared to a second recall of positive autobiographical memories following a negative mood induction in adults. (a) Second positive recall versus baseline. Neural activation in response to recalling a positive autobiographical memory. Peak ventrolateral prefrontal cortex (VLPFC) activations centered at left middle frontal gyrus/BA 10 (a; -45, 42, -3; $Z = 5.03$); left inferior frontal gyrus/BA 47 (b; -50, 20, -1; $Z = 5.5$); and right inferior frontal gyrus/BA 10 (c; 42, 42, -1; $Z = 5.2$). (b) Second positive recall versus first positive recall. Emotion regulation: neural activation in response to recalling a positive memory after a negative mood induction. Peak activations centered at left anterior cingulate/BA 32 (a; -7, 34, -10; $Z = 4.4$); left subcallosal gyrus/BA 25 (b; -7, 20, -13; $Z = 5.3$), and left medial frontal gyrus/BA 11 (c; -7, 14, -18; $Z = 6.1$). Reprinted from Cooney et al. (2007) with permission from Lippincott Williams & Wilkins.

triggers or cause; consequently, they may be particularly effective in regulating the normal fluctuations in mood states that people experience throughout the day.

Recently, investigators have suggested that memory might play an important role in these indirect attempts to regulate mood and emotion. Specifically, researchers have found that people often recall positive memories when they are experiencing a negative mood (e.g., Josephson, 1996). In this context, Joormann and her colleagues (e.g., Joormann and Siemer, 2004) examined the effectiveness of having individuals recall positive autobiographical memories to regulate induced or naturally occurring negative mood states. These investigators found that recalling positive memories reduced participants' levels of negative affect. While these studies suggest that recalling mood-incongruent memories is an effective affect-regulation strategy, the mechanisms

underlying the mood-enhancing quality of this strategy are not well understood. In contrast to affect-regulation strategies such as reappraisal or suppression, positive recall is likely to be associated less strongly with direct suppression of mood-congruent thoughts, with direct attempts to change the negative mood state, or with cognitive reinterpretation of the affect-eliciting situation.

To examine these possibilities, Cooney et al. (2007) investigated the neural correlates of mood-incongruent recall. These investigators induced sad mood in participants and had them use positive autobiographical memories to regulate that negative affect. Cooney et al. found less involvement of dorsal brain regions, such as DLPFC and dorsal ACC, in recovery from the sad mood; instead, activation was found in OFC and ventromedial prefrontal cortex (VMPFC) (see Figure 3.1). These findings are consistent with the formulation that different emotion-regulation strategies are associated with activation in different brain areas. Whereas voluntary suppression of affect and strategic reappraisal of the emotion-eliciting situation might be more strongly associated with dorsal areas of the PFC that have traditionally been linked to cognitive control and working memory, more automatic regulation strategies such as changes in the accessibility of mood-congruent thoughts may be associated with activation in more ventral areas of the PFC that have traditionally been linked to emotional processing and associative learning. Clearly, future research is needed to investigate this proposition more explicitly.

Healthy children

Neuroimaging studies suggest that children do not differ from adults when considering which brain regions are activated when a sad mood is induced: for both adults and children, sadness is associated with the anterior temporal pole (BA 21), the MPFC (BA 10) and the right ventrolateral prefrontal cortex (VLPFC; BA 47). Lévesque et al. (2003a), for example, presented sad movie clips to 8- to 10-year-old girls and found activation in MPFC, anterior temporal lobe, and right VLPFC, similar to the sadness-related activations that have been found in adults. The finding that the neural substrates of the primary emotions are similar in adults and children is surprising given that there are clear behavioral differences between adults' and children's emotional responses to stressful situations. In this section we examine whether the regulation of emotion is also associated with activation in similar brain regions in children as in adults or whether differences between adults and children in the neural correlates of emotion regulation might help explain the developmental differences in behavioral responsivity to stress.

Our review of the adult literature suggests that the development of emotion regulation depends critically on the development of the PFC. This hypothesis, first proposed by Stuss (1992), has received support in the neuroimaging literature (see Casey et al., 2000,

for a review). Recent longitudinal MRI studies show a clear sequence of structural changes from childhood to adolescence through adulthood, and PFC and lateral temporal cortices are among the last areas to mature (see Chapter 1). Studies examining the development of brain structure also suggest that maturation is associated with gray matter loss, pruning, and the elimination of connections, reflecting a sculpting process that leads to the mature brain. Changes in subcortical regions have also been noted, including parts of the basal ganglia to which the PFC projects directly. Protracted development of the prefrontal regions with a transition from diffuse to focal recruitment is consistent with MRI-based neuroanatomical (Sowell et al., 2003) and fMRI studies (Casey et al., 2002; Durston et al., 2006) of prefrontal development.

The maturation of the PFC is proposed to play an important role in the development of higher-order cognitive abilities (e.g., Bunge et al., 2002; Casey et al., 2000). Maturation is characterized by an increased ability to filter and suppress irrelevant information and actions and to respond more readily to relevant material. Increased cognitive control has been found to be associated with the recruitment of both ventral and dorsal PFC. Studies on interference, inhibition and the development of cognitive control in children and adults have shown that children are more prone to interference effects. Interestingly, Bunge et al. (2002) found that effective interference inhibition in children was associated with PFC activation, but in the opposite hemisphere as has been found with adults; the authors concluded that these differences are due to the immaturity of the PFC. In fact, studies using Stroop and go/no-go paradigms have found that children recruit distinct but often different or more prefrontal regions than do adults (Durston et al., 2004). With increasing age, regions in which brain activity is related to task performance become more fine-tuned, while activity decreases in regions that are not correlated with task performance. Consistent with these findings, Fox and Calkins (2003) documented similarities between children and adults with damage to the PFC in emotional reactivity and difficulties in regulation. In addition, Eslinger et al. (1997) found that early damage to the OFC is associated with the disruption of social and emotional behavior. Finally, investigators have underscored the importance of inhibitory control of limbic structures by the PFC.

Although these studies on the development of brain structure and function permit the formulation of specific predictions about the neural correlates of emotion regulation in children, there is surprisingly little research that tests these predictions. In one of the few studies to investigate neural correlates of emotion regulation in children, Lévesque et al. (2004) reported that voluntary suppression of induced sadness was related to activation in LPFC in 8- to 10-year-old girls. Lévesque et al. also found activation in the OFC, rostral ACC, MPFC, and VLPFC, regions that have been identified in the adult literature on emotion regulation. Specifically, the activation in the LPFC fits nicely with evidence we reviewed earlier suggesting that this prefrontal region plays an important role in the ability to monitor and control cognition and to inhibit irrelevant material and override prepotent responses. Furthermore, Lévesque et al.'s finding of activation of the OFC is consistent with research demonstrating its important role in protecting goal-directed behavior from interference. The OFC is at the junction of the prefrontal associative cortex and the limbic system and has strong links with the LPFC and the insula. It also sends extensive projections to the amygdala. Damage to the OFC is associated with impulsivity, aggressiveness and emotional outbursts, findings that support the formulation that the OFC plays an important role in emotion regulation. Lévesque et al. also found activation in the rostral ACC, an area that is closely connected to brain regions implicated in the modulation of autonomic and neuroendocrine responses, and postulated that this activation is related to the suppression of the autonomic and neuroendocrine aspects of sad mood.

Additional empirical support for the important role of the development of the PFC comes from studies that have linked children's emotion regulation

to the approach-withdrawal model of frontal electroencephalogram (EEG) asymmetry (Davidson et al., 2000). Whereas greater relative left frontal activation is proposed to be associated with approach behavior, including both anger and positive affect, greater relative right frontal activity is postulated to be associated with withdrawal behavior and sadness. Forbes et al. (2006) recently tested 3- to 5-year-old children's responses to receiving a disappointing toy and reported that those children with relatively greater right frontal activity at rest exhibited more withdrawal behaviors when receiving the unwanted toy. Another model akin to this approach-withdrawal model is the triadic model proposed by Ernst et al. (2005). This model suggests that the peak onset of mood disorders in adolescence may reflect a unique functional balance among the systems underlying approach behavior (e.g., ventral striatum), avoidance behavior (e.g., amygdala) and supervisory control (medial PFC).

Depression

Major depressive disorder (MDD) is arguably the prototypical disorder of emotion regulation. Indeed, the hallmark characteristic of MDD is sustained negative affect. In attempting to understand the neural foundations of emotion dysregulation in depression, investigators have examined both the structure and function of specific areas of the brain. For example, a number of researchers using MRI with individuals diagnosed with MDD have reported atrophy in the hippocampus, subgenual ACC, and amygdala [see Gotlib and Hamilton (2008), for a review of this literature].

In addition to these structural anomalies, researchers have also documented depression-associated abnormalities in the function of these brain areas. Using positron emission tomography (PET) and functional MRI (fMRI), investigators have identified abnormalities in limbic and prefrontal cortical areas in depression (e.g., Mayberg et al., 2005). Although relevant research is still in its infancy, investigators have suggested that limbic-cortical dysregulation might result in sustained negative affect, rumination, and impaired reward processing, formulations that parallel behavioral findings in depression of biased processing of negative information, rumination, and a lack of responsivity to positive stimuli.

Theorists have posited that emotional behavior is linked to the functioning of two neural systems, a ventral system and a dorsal system (Ochsner and Gross, 2005; Phan et al., 2005). Whereas the ventral system involves brain regions that are important for identifying the emotional significance of a stimulus and producing an affect state, the dorsal system is important for executive function, including selective attention, planning, and effortful regulation of affective states. Adaptive emotional behavior is posited to depend on the integrity and balanced interaction of these systems. Importantly, both animal and human research suggests that these systems are dysregulated in depression (see Mayberg, 1997).

Essentially the dorsal system consists of dorsal aspects of the frontal lobe, cingulate cortex, and parietal regions, which include DLPFC, dorsal ACC, and inferior parietal cortex. The ventral system consists primarily of subcortical nuclei, ventral and orbital prefrontal cortices, and the insula. It is important to note that the rostral ACC lies at the intersection of the dorsal and ventral subsystems. Its special role in depression stems from its connections with the ventral and dorsal ACC; indeed, recent findings indicate that activation in this area predicts response to antidepressant treatment in depression (e.g., Mayberg et al., 2005). Thus, the rostral ACC may serve an important regulatory role by facilitating interactions between the dorsal and ventral systems. Abnormal elevations in cerebral blood flow, blood oxygenation, and glucose metabolism have been found in the amygdala and the subgenual ACC in depressed individuals (Drevets et al., 1999; Siegle et al., 2002). Finally, further implicating limbic dysfunction in emotion-regulation difficulties in depression, a number of investigators have reported depression-associated amygdala responsivity in response to fearful or negative faces, a pattern that also appears to diminish with pharmacotherapy

[see Gotlib and Hamilton (2008) for a discussion of this literature].

Siegle et al. (2002) provided an early demonstration that the amygdala may be involved in sustained processing of negative emotional material in depression that interferes with responsivity to neutral material. Interestingly, Siegle et al. also found self-reported levels of rumination to be significantly correlated with left amygdala activity. Elliott et al. (2002) identified medial and orbital prefrontal regions as critical to mood-congruent processing biases. During an emotional go/no-go task, Elliott et al. found attenuated engagement of the posterior OFC in depressed participants for emotionally valenced trials. The likelihood that orbitofrontal areas are distinctly involved in mediation of emotional processing is buttressed by neuroanatomical data. The OFC is highly interconnected with both the hippocampus and amygdala and is thought to inhibit posterior structures, particularly structures that are postulated to be involved in inappropriate emotional responses (e.g., Shimamura, 1995). The results of Siegle et al.'s study, in which the processing of negative words was associated with increased amygdala activation and decreased DLPFC activity, is consistent with a model of inverse functionality between limbic and prefrontal activity. Indeed, because the DLPFC may modulate activity in the amygdala, disinhibition or decreased activity of the DLPFC may be related to patterns of hyper-responsivity in limbic regions such as the amygdala. From a similar perspective, Pochon et al. (2001) have suggested that limbic inhibition of the DLPFC may serve to override top-down emotional control processes by selecting which representations to maintain.

In contrast to this growing literature in adults, neuroimaging studies of depressed children are rare, and imaging studies of children at risk for depression are virtually non-existent. MacMillan et al. (2003) reported marginally significant elevations in amygdala volume among depressed adolescents. Using a clinical sample, Thomas et al. (2001) examined patterns of neural activation in depressed and anxious children, relative to healthy controls, as they viewed fearful and neutral faces. Thomas et al. found that whereas the anxious children exhibited an exaggerated amygdala response to the fearful faces, the five depressed girls in this study demonstrated a blunted amygdala response to the fearful faces. This blunted neural response is intriguing, because it mirrors findings of blunted psychophysiological reactivity to emotional stimuli among depressed adults (e.g., Rottenberg et al., 2002a, 2002b). In contrast, Roberson-Nay et al. (2006) reported greater amygdala activation in depressed children during successful encoding of emotional faces.

In sum, although there are few empirical investigations of neural correlates of the development of emotion-regulation abilities in children and adolescents, the existing studies provide converging evidence that changes in PFC functioning are important in this process. Interestingly, children and adults do not appear to differ significantly in the brain areas that are activated when affect is induced. They do differ, however, in their activation of PFC when trying to regulate the ensuing affective states. Importantly, the results of the studies we reviewed above suggest that immaturity of the PFC is associated with greater but diffuse activation of the PFC in these tasks, and that the development of emotion-regulation abilities occurs in concert with more fine-tuned functioning of this important brain region. The studies also suggest that emotion regulation and maturation of PFC functioning are intimately related to the development of cognitive abilities over the life span, and specifically to the development of cognitive control.

As we noted above, a second important biological system that is involved in stress response and emotion regulation is the HPA axis. Indeed, HPA axis functioning is a central component of the body's neuroendocrine response to stressful events. A discussion of the development of biological systems of stress reactivity and recovery and emotion regulation in children would be incomplete without considering the role of the HPA axis.

Neuroendocrine aspects of emotion regulation

Healthy adults and children

The HPA system and the resultant production of cortisol are conceptualized as the psychophysiological substrates of the regulation and coping systems. In fact, investigators have argued that levels of cortisol produced under stress reflect the individual's ability to regulate and cope (Gunnar et al., 1992). Even tonic HPA functioning is of interest to researchers examining emotion regulation because basal cortisol varies across the day, promotes various adaptive functions, and may reflect an individual's baseline physiological and emotional arousal. From another perspective, the investigation of neuroendocrine responses is important because they may provide an important link to the neurobiological research we described earlier in this chapter. For example, the hippocampus and the PFC are the brain regions with the highest density of glucocorticoid receptors, and chronic exposure to stressful events that leads to increased cortisol secretion might lead to neurotoxicity in these areas. The glucocorticoid data are intriguing because the PFC appears to play an important role in the modulation of limbic regions, specifically the amygdala, and, therefore, has been implicated in emotion regulation. Indeed, investigators have reported elevated activation in the amygdala and the hippocampus during the encoding of emotional faces in adolescents with Cushing syndrome and with congenital adrenal hyperplasia, both of which are characterized by altered HPA-axis functioning (Ernst et al., 2007; Maheu et al., 2008). Given that the functioning of the HPA system is so integrally related to the human stress response and to emotion regulation, it is not surprising that atypical patterns of both basal HPA functioning and HPA reactivity have been documented in various psychiatric disorders, including depression (cf. Gunnar and Cheatham, 2003).

While cortisol levels in newborns are higher than those in adults, this difference slowly decreases over the first year of life (Gunnar et al., 1989). Cortisol response to stressful events has been found in newborns, but changes considerably over the lifespan. In their longitudinal study, for example, Lewis and Ramsey (1999) found no increase in cortisol levels in response to an injection across the ages of 2, 4, and 6 months, but found a moderate increase in cortisol levels between 6 and 18 months. Similarly, Goldberg et al. (2003) examined the 1-week stability of baseline cortisol and cortisol response to a stressor in 12- to 18-month-old infants and reported considerable stability across time and location (home versus laboratory), as well as stable cortisol changes in response to stressors.

When children start attending school, they respond to tests and social interactions with increased cortisol secretion. Importantly, cortisol response in these situations is related to temperament (Tennes and Kreye, 1987) and, perhaps, to pubertal status (Ronsaville et al., 2006) and gender (Angold et al., 1998). Indeed, Walker et al. (2001) proposed that an increase in basal cortisol levels with puberty is related to the increased incidence of emotional disorders after this developmental period.

Adverse experiences in early life have been associated with alterations in HPA functioning. Both in animal studies and in studies with humans, early exposure to stressful life circumstances, separation from caregivers, and differences in caregiving behavior have been found to be associated with variations in cortisol levels. Rhesus monkeys that were separated from their mothers and reared with peers, for example, exhibited increases in cortisol responses to a variety of stressful situations through adolescence (Fahlke et al., 2000). Gunnar et al. (2001) found that Romanian orphans raised under suboptimal conditions for several months early in life had elevated levels of cortisol, even after adoption. In a similar vein, caregiving behavior has been shown to buffer neuroendocrine responses to stress in infants separated from their mothers for short periods of time; similarly, secure attachment has been found to buffer cortisol response to injections (Gunnar et al., 1992). Finally, a number of investigators have examined associations between cortisol

levels and emotional problems in children and have found children's internalizing problems, including symptoms of mood disorders, to be associated with low basal cortisol (e.g., Granger et al., 1998). In turn, low basal cortisol has been found to be associated with increasing externalizing problems, such as aggression and impulsivity. In addition, at a 5-year follow-up assessment when children reached adolescence (Shoal et al., 2003). Finally, there have been reports of decreased cortisol reactivity in children with externalizing disorders (e.g., Snoek et al., 2004).

Depression

Major cognitive and biological theories of depression implicate stress, often in the form of stressful life events, in the onset of this disorder. Given this emphasis on the association between stress and depression, it is not surprising that the potential role of cortisol irregularity in adult depression has long been the focus of research. Dinan (1996) hypothesized that stressful life events lead to the production of elevated levels of cortisol that, in turn, reduce brain serotonin function, resulting in the onset of depression. Consistent with this hypothesis and with theoretical postulates emphasizing the elevated levels of stress in depression, the results of empirical studies generally indicate that MDD is associated with elevated HPA-axis activity, reflected by both high tonic levels of cortisol and cortisol hypersecretion following an acute stressor (Parker et al., 2003).

There is considerable evidence, therefore, that MDD is associated with HPA-axis dysfunction, presumably reflecting difficulties in emotion regulation and coping with the effects of stress. Moreover, results of studies examining the suppression of dexamethasone, a synthetic glucocorticoid that suppresses cortisol, indicate that HPA-axis dysregulation may play a role in the recurrence of depression (Targum, 1984). Indeed, consistent with a diathesis-stress perspective, HPA-axis dysfunction may represent a trait-like vulnerability factor for depression, interacting with environmental stress to precipitate recurrences of depressive episodes.

In part to examine the veracity of this formulation, investigators have begun to assess HPA-axis functioning in samples of depressed children and, more recently, in samples of children who are at elevated risk for depression.

In contrast to the relatively consistent findings of hypercortisolemia in depressed adults, results of studies examining HPA-axis functioning in depressed children and adolescents are equivocal. For example, a number of researchers have reported that, unlike depressed adults, depressed youth are not characterized by general cortisol hypersecretion over the course of the diurnal cycle (e.g., Birmaher et al., 1992). Instead, they exhibit high cortisol levels during the evening, when the HPA axis is typically quiescent (Goodyer et al., 1991, 2001). Interestingly, Goodyer et al. (1991) found that elevated levels of evening cortisol secretion among depressed adolescents normalize following recovery from a depressive episode, suggesting that cortisol hyper-secretion is a state characteristic of depression. Countering this position, however, are data indicating that levels of cortisol secretion among depressed youth during a depressive episode have predictive utility. Rao et al. (1996), for example, found that elevations in evening cortisol levels among depressed children predicted a recurrent course of the disorder at a 7-year follow-up assessment. Similarly, Mathew et al. (2003) found that elevations of daytime cortisol in depressed adolescents predicted subsequent suicide attempts over a 10-year follow-up period. Thus, although depression-associated HPA-axis dysfunction has not been found consistently in children and adolescents in cross-sectional investigations, there is evidence that basal levels of diurnal cortisol secretion predict recurrence of depression and suicide attempts.

These longitudinal findings may reflect the relation between HPA-axis functioning and coping under stress, indicating that depressed adolescents are less capable than are their non-depressed counterparts of coping effectively with environmental stressors or regulating their emotion. Consistent with this formulation, several investigators have reported heightened HPA-axis reactivity to stress in children and adolescents with clinically significant internalizing symptoms. For example, Ashman et al. (2002) found that although children with clinically significant internalizing symptoms and non-psychiatric control children did not differ with respect to levels of basal cortisol, internalizing children exhibited a greater cortisol response to a laboratory stressor. From a different perspective, Granger et al. (1996) found that increased cortisol reactivity to a laboratory stressor predicted internalizing symptoms and the presence of anxiety disorders among clinic-referred children at a 6-month follow-up assessment. Finally, Luby et al. (2003) found that depressed preschoolers secrete high levels of cortisol in response to a discrete stressor, and Coplan et al. (2002) found elevated levels of cortisol in anticipation of a laboratory stressor (carbon-dioxide inhalation) among children and adolescents who were diagnosed with depression and/or anxiety and who were sensitive to the anxiogenic effects of the challenge. These findings indicate that depressed youth are characterized by relatively high levels of cortisol reactivity in response to relevant stressors and that elevated cortisol reactivity may be a significant predictor of increases in depressive symptoms.

Little research has examined HPA dysregulation as a risk factor for the onset of a depressive episode in youth. Ashman et al. (2002) found abnormal HPA reactivity in response to a laboratory stressor in non-depressed children of depressed mothers. More recently, Halligan et al. (2004) reported that having a depressed mother was associated with higher and more variable morning cortisol levels in their children. In a prospective study, Essex et al. (2002) found that preschool-age children of depressed mothers exhibited elevated cortisol levels that were associated with subsequent higher levels of psychological symptoms in first grade. Finally, in a 3-year longitudinal study, Goodyer and his colleagues (2000) demonstrated that peak levels of morning cortisol and late-afternoon levels of dehydroepiandrosterone (DHEA; another hormone that indexes HPA activity) among adolescents predicted the subsequent onset of MDD. In fact, DHEA hypersecretion was found in this sample to precede the onset of the depressive episode.

In sum, although few studies have been conducted, children at high risk for depression have been found to be characterized by elevated and/or more variable levels of cortisol secretion than have their low-risk counterparts. It appears, therefore, that dysregulated HPA-axis functioning, both tonic and in response to a stressor, may serve as a risk factor for the development of an episode of major depression in children. Indeed, these data are consistent with Goodman and Gotlib's (1999) formulation that a critical consequence for children of depressed mothers is chronic activation of the HPA-axis as a result of the stress of living with a depressed parent, reflecting an impaired ability to cope effectively with stress and regulate negative affect. It is important to note here that chronic activation of the HPA-axis, resulting in high levels of cortisol, can disrupt functioning in regions of the brain that are responsible for the regulation of emotion (e.g., the PFC, the ACC, and the amygdala). This may represent a mechanism through which HPA-axis activation is related to a child's ability to cope with stress. No studies, however, have examined cortisol secretion in response to stress in school-age or adolescent offspring of depressed parents.

Implications and directions for future research

There are now substantial publications delineating the psychological and biological aspects of emotion regulation in adults and children, factors involved in the development of stress reactivity and recovery, and the regulation of negative affect. In this chapter we have briefly reviewed these publications and have presented preliminary results from a study designed to examine psychobiological aspects of risk for depression. We summarized exciting findings underscoring the importance of investigating both the role of activation in specific brain areas and variation in functional connectivity, as well as research examining neuroendocrine responses to stressful events and factors that moderate these responses. Recently, researchers have suggested that neural activation and functional connectivity might be influenced by variations in genotype. We did not review this promising new area of research in this chapter because, to date, the designs used in these studies do not allow an unequivocal interpretation of the findings in terms of emotion regulation. More specifically, investigators have assessed responses to emotional stimuli without any instructions to regulate the emotional response and, therefore, do not meet the criteria that we described at the beginning of this chapter. Nevertheless, it is clear that future research in emotion regulation must consider genetic factors.

Understanding the associations among these various domains of functioning as they affect the development of recovery from stress and emotion regulation continues to be a critical goal for future research. Indeed, it is only through such an integration that we can generate cogent new experimental questions and methodologies in this area. Findings from such future work promise to contribute to the development of an integrative psychobiological theory of the development of emotion regulation and help us to gain a better understanding of factors involved in typical and atypical development.

ACKNOWLEDGMENTS

Preparation of this chapter was facilitated by grants MH59259 and MH074849 from the National Institute of Mental Health.

REFERENCES

Angold, A., Costello, E. J., and Worthman, C. M. (1998) Puberty and depression: the roles of age, pubertal status and pubertal timing. *Psychol Med*, **28**(1), 51–61.
Ashman, S. B., Dawson, G., Panagiotides, H., et al. (2002) Stress hormone levels of children of depressed mothers. *Dev Psychopathol*, **14**(2), 333–49.

Beauregard, M., Lévesque, J., and Bourgouin, P. (2001) Neural correlates of conscious self-regulation of emotion. *J Neurosci*, **21**, 165, 1–6.

Birmaher, B., Ryan, N. D., Dahl, R., et al. (1992) Dexamethasone suppression test in children with major depressive disorder. *J Am Acad Child Adolesc Psychiatry*, **31**(2), 291–7.

Bower, G. H. (1987) Commentary on mood and memory. *Behav Res Ther*, **25**(6), 443–55.

Bunge, S. A., Dudukovic, N. M., Thomasson, M. E., et al. (2002) Immature frontal lobe contributions to cognitive control in children: evidence from fMRI. *Neuron*, **33**, 301–11.

Burke, H. M., Davis, M. C., Otte, C., and Mohr, D. C. (2005) Depression and cortisol responses to psychological stress: a meta-analysis. *Psychoneuroendocrinology*, **30**(9), 846–56.

Casey, B. J., Giedd, J., Thomas, K. M. (2000) Structural and functional brain development and its relation to cognitive development. *Biol Psychol*, **54**, 241–57.

Casey, B. J., Thomas, K. M., Davidson, M. C., et al. (2002) Dissociating striatal and hippocampal function developmentally with a stimulus-response compatibility task. *J Neurosci*, **22**(19), 8647–52.

Cole, P. M., Martin, S. E., and Dennis, T. A. (2004) Emotion regulation as a scientific construct: methodological challenges and directions for child development research. *Child Dev*, **75**(2), 317–33.

Cooney, R. E., Joormann, J., Atlas, L. Y., et al. (2007) Remembering the good times: neural correlates of affect regulation. *NeuroReport*, **18**(17), 1771–4.

Coplan, J. D., Moreau, D., Chaput, F., et al. (2002) Salivary cortisol concentrations before and after carbon-dioxide inhalations in children. *Biol Psychiatry*, **51**, 326–33.

Davidson, R. J., Jackson, D. C., and Kalin, N. H. (2000) Emotion, plasticity, context, and regulation: perspectives from affective neuroscience. *Psychol Bull*, **126**(6), 890–9.

Dinan, T. (1996) Serotonin and the regulation of hypothalamic-pituitary-adrenal axis function. *Life Sci*, **58**(20), 1683–94.

Drevets, W. C., Frank, E., Price, J. C., et al. (1999) PET imaging of serotonin 1A receptor binding in depression. *Biol Psychiatry*, **46**, 1375–87.

Durston, S., Davidson, M. C., Tottenham, N., et al. (2006) A shift from diffuse to focal cortical activity with development. *Dev Sci*, **9**(1), 1–8.

Durston, S., Hulshoff, P., Hilleke, E., et al. (2004) Magnetic resonance imaging of boys with attention-deficit/hyperactivity disorder and their unaffected siblings. *J Am Acad Child Adolesc Psychiatry*, **43**(3), 332–40.

Eisenberg, N. (2000) Emotion, regulation, and moral development. *Annu Rev Psychol*, **51**, 665–97.

Elliott, R., Rubinsztein, J. S., Sahakian, B. J., et al. (2002) The neural basis of mood-congruent processing biases in depression. *Arch Gen Psychiatry*, **59**(7), 597–604.

Ernst, M., Jazbec, S., McClure, E. B., et al. (2005) Amygdala and nucleus accumbens activation in response to receipt and omission of gains in adults and adolescents. *NeuroImage*, **25**, 1279–91.

Ernst, M., Maheu, F. S., Schroth, E. A., et al. (2007) Amygdala function in adolescents with congenital adrenal hyperplasia: a model for the study of early steroid abnormalities. *Neuropsychologia*, **45**, 2104–13.

Eslinger, P. J., Biddle, K. R., and Grattan, L. M. (1997) Cognitive and social development in children with prefrontal cortex legions. In: *Development of the Prefrontal Cortex: Evolution, Neurobiology, and Behavior*. Baltimore, MD: Paul H. Brookes Publishing.

Essex, M. J., Klein, M. H., Cho, E., et al. (2002) Maternal stress beginning in infancy may sensitize children to later stress exposure: effects on cortisol and behavior. *Biol Psychiatry*, **52**(8), 776–84.

Fahlke, C., Lorenz, J. G., Long, J., et al. (2000) Rearing experiences and stress-induced plasma cortisol as early risk factors for excessive alcohol consumption in nonhuman primates. *Alcohol Clin Exp Res*, **24**, 644–50.

Forbes, E. E., Fox, N. A., Cohn, J. F., Galles, S. F., and Kovacs, M. (2006) Children's affect regulation during a disappointment: Psychophysiological responses and relation to parent history of depression. *Biol Psychol*, **71**(3), 264–77.

Fox, N. A. and Calkins, S. D. (2003) The development of self-control of emotion: intrinsic and extrinsic influences. In: *Motivation and Emotion*. Special issue. *Developmental Aspects of Emotion Regulation Across the Lifespan: Integrating Diverse Developmental Perspectives*. Part I. *Motiv Emot*, **27**(1), 7–26.

Goldberg, S., Levitan, R., Leung, E., et al. (2003) Cortisol concentrations in 12- to 18-month-old infants: Stability over time, location and stressor. *Biol Psychiatry*, **54**(7), 719–26.

Goldsmith, H. H. and Davidson, R. J. (2004) Disambiguating the components of emotion regulation. *Child Dev*, **75**(2), 361–5.

Goodman, S. H. and Gotlib, I. H. (1999) Risk for psychopathology in the children of depressed mothers: a developmental model for understanding mechanisms of transmission. *Psychol Rev*, **106**, 458–90.

Goodyer, I., Herbert, J., Moor, S., et al. (1991) Cortisol hypersecretion in depressed school-aged children and adolescents. *Psychiatry Res*, **37**(3), 237–44.

Goodyer, I. M., Herbert, J., Tamplin, A., et al. (2000) Recent life events, cortisol, dehydroepiandrosterone and the onset of major depression in high-risk adolescents. *Br J Psychiatry*, **177**, 499–504.

Goodyer, I. M., Park, R. J., and Herbert, J. (2001) Psychosocial and endocrine features of chronic first-episode major depression in 8–16 year olds. *Biol Psychiatry*, **50**(5), 351–7.

Gotlib, I. H. and Hamilton, J. P. (2008) Neuroimaging and depression: current status and unresolved issues. *Curr Direct Psychol Sci*, **17**(2), 159–63.

Granger, D. A., Serbin, L. A., Schwartzman, A., et al. (1998) Children's salivary cortisol, internalising behaviour problems, and family environment: results from the Concordia Longitudinal Risk Project. *Int J Behav Dev*, **22**(4), 707–28.

Granger, D. A., Weisz, J. R., McCracken, J. T., et al. (1996) Reciprocal influences among adrenocortical activation, psychosocial processes, and the behavioral adjustment of clinic-referred children. *Child Dev*, **67**(6), 3250–62.

Gunnar, M. R. and Cheatham, C. L. (2003) Brain and behavior interfaces: stress and the developing brain. *Infant Mental Health J*, **24**(3), 195–211.

Gunnar, M. R., Larson, M. C., Hertsgaard, L., and Harris, M. L. (1992) The stressfulness of separation among nine-month-old infants: effect of social context variables and infant temperament. *Child Dev*, **63**(2), 290–303.

Gunnar, M., Marvinney, D., Isensee, J., et al. (1989) Coping with uncertainty: new models of the relations between hormonal, behavioral, and cognitive processes. In: Palermo, D. S. (ed.) *Coping with Uncertainty: Behavioral and Developmental Perspectives*. Hillsdale, NJ: Lawrence Erlbaum Associates, Inc., pp. 101–29.

Gunnar, M. R., Morrison, S. J., Chisholm, K., and Schuder, M. (2001) Salivary cortisol levels in children adopted from Romanian Orphanages. *Dev Psychopathol*, **13**(3), 611–28.

Halligan, S. L., Herbert, J., Goodyer, I. M., et al. (2004) Exposure to postnatal depression predicts elevated cortisol in adolescent offspring. *Biol Psychiatry*, **55**(4), 376–81.

Harenski, C. L. and Hamann, S. (2006) Neural correlates of regulating negative emotions related to moral violations. *NeuroImage*, **30**, 313–24.

Hariri, A. R., Mattay, V. S., Tessitore, A., et al. (2003) Neocortical modulation of the amygdala response to fearful stimuli. *Biol Psychiatry*, **53**, 494–501.

Joormann, J. and Siemer, M. (2004) Memory accessibility, mood regulation, and dysphoria: difficulties in repairing sad mood with happy memories? *J Abnormal Psychol*, **113**(2), 179–88.

Josephson, B. R. (1996) Mood regulation and memory: repairing sad moods with happy memories. *Cognit Emot*, **10**, 437–44.

Lévesque, J., Joanette, Y., Mensour, B., et al. (2003a) Neural correlates of sad feelings in healthy girls. *Neuroscience*, **121**(3), 545–51.

Lévesque, J., Eugène, F., Joanette, Y., et al. (2003b) Neural circuitry underlying voluntary suppression of sadness. *Biol Psychiatry*, **53**, 502–10.

Lévesque, J., Joanette, Y., Mensour, B., et al. (2004) Neural correlates of sad feelings in healthy girls. *Neuroscience*, **129**, 361–9.

Lewis, M. and Ramsay, D. S. (1999) Effect of Maternal Soothing on infant stress response. *Child Dev*, **70**(1), 11–20.

Luby, J. L., Heffelfinger, A., Mrakotsky, C., et al. (2003) Alteration in stress cortisol reactivity in depressed preschoolers relative to psychiatric and no-disorder comparison groups. *Arch Gen Psychiatry*, **60**(12), 1248–55.

MacMillan, S., Szeszko, P. R., Moore, G. J., et al. (2003) Increased amygdala: hippocampal volume ratios associated with severity of anxiety in pediatric major depression. *J Child Adolesc Psychopharmacol*, **13**(1), 65–73.

Maheu, F. S., Mazzone, L., Merke, D. P., et al. (2008) Altered amygdala and hippocampus function in adolescents with hypercortisolemia: an fMRI study of Cushing Syndrome. *Dev Psychopathol*, **20**(4), 1177–89.

Mathew, S. J., Coplan, J. D., Goetz, R. R., et al. (2003) Differentiating depressed adolescent 24h cortisol secretion in light of their adult clinical outcome. *Neuropsychopharmacology*, **28**(7), 1336–43.

Mayberg, H. S. (1997) Limbic-cortical dysregulation: a proposed model of depression. *J Prof Neuropsychiatry Clin Neurosci*, **9**(3), 471–81.

Mayberg, H. S., Lozano, A. M., Voon, V., et al. (2005) Deep brain stimulation for treatment-resistant depression. *Neuron*, **45**(5), 651–60.

Nyklicek, I., Vingerhoets, A., and Denollet, J. (2002) Emotional (non-)expression and health: data, questions, and challenges. *Psychol Health*, **17**, 517–28.

Ochsner, K. N. and Gross, J. J. (2005) The cognitive control of emotion. *Trends Cogn Sci*, **9**(5), 242–9.

Ochsner, K. N., Bunge, S. A., Gross, J. J., et al. (2002) Rethinking feelings: an fMRI study of the cognitive regulation of emotion. *J Cogn Neurosci*, **14**(8), 1215–29.

Ochsner, K. N., Ray, R. D., Cooper, J. C., et al. (2004) For better or for worse: neural systems supporting the cognitive down- and up-regulation of negative emotion. *NeuroImage*, **23**, 483–99.

Ohira, H., Nomura, M., Ichikawa, N., et al. (2006) Association of neural and physiological responses during voluntary emotion suppression. *NeuroImage*, **29**, 721–33.

Parker, K. J., Schatzberg, A. F., and Lyons, D. M. (2003) Neuroendocrine aspects of hypercortisolism in major depression. *Hormones Behav*, **43**(1), 60–6.

Phan, K. L., Fitzgerald, D. A., Nathan, P. J., et al. (2005) Neural substrates for voluntary suppression of negative affect: a functional magnetic resonance imaging study. *Biol Psychiatry*, **57**, 210–19.

Phan, K. L., Wager, T., Taylor, S. F., et al. (2002) Functional neuroanatomy of emotion: a meta-analysis of emotion activation studies in PET and fMRI. *NeuroImage*, **16**, 331–48.

Pochon, J., Levy, R., Poline, J., et al. (2001) The role of dorsolateral prefrontal cortex in the preparation of forthcoming actions: an fMRI study. *Cereb Cortex*, **11**(3), 260–6.

Rao, U., Dahl, R. E., Ryan, N. D., et al. (1996) The relationship between longitudinal clinical course and sleep and cortisol changes in adolescent depression. *Biol Psychiatry*, **40**(6), 474–84.

Reus, V., Wolkowitz, O., and Frederick, S. (1997) Antiglucocorticoid treatments in psychiatry. *Psychoneuroendocrinology [Special Issue: Cortisol and Anti-cortisol]* **22**(Suppl 1), S121–S124.

Roberson-Nay R., McClure E. B., Monk C. S., et al. (2006) Increased amygdala activity during successful memory encoding in adolescent major depressive disorder: an FMRI study, *Biol Psychiatry*, **60**, 966–73.

Ronsaville, D. S., Municchi, G., Laney C., et al. (2006) Maternal and environmental factors influence the hypothalamic-pituitary-adrenal axis response to corticotropin-releasing hormone infusion in offspring of mothers with and without mood disorders. *Dev Psychopathol*, **18**, 173–94.

Rottenberg, J., Gross, J. J., Wilhelm, F. H., et al. (2002a) Crying threshold and intensity in major depressive disorder. *J Abnormal Psychol*, **111**, 302–12.

Rottenberg, J., Kasch, K. L., Gross, J. J., et al. (2002b) Sadness and amusement reactivity differentially predict concurrent and prospective functioning in major depressive disorder. *Emotion*, **2**, 135–46.

Schaefer, S. M., Jackson, D. C., Davidson, R. J., et al. (2002) Modulation of amygdalar activity by the conscious regulation of negative emotion. *J Cogn Neurosci*, **14**, 913–21.

Shimamura, A. P. (1995) Memory and the prefrontal cortex. *Ann N Y Acad Sci*, **769**, 151–9.

Shoal, G. D., Giancola, P. R., and Kirillova, G. P. (2003) Salivary cortisol, personality, and aggressive behavior in adolescent boys: a 5-year longitudinal study. *J Am Acad Child Adolesc Psychiatry*, **42**(9), 1101–7.

Siegle, G. J., Steinhauer, S. R., Thase, M. E., et al. (2002) Can't shake that feeling: event-related fMRI assessment of sustained amygdala activity in response to emotional information in depressed individuals. *Biol Psychiatry*, **51**(9), 693–707.

Snoek, H., Van Goozen, S. H. M., and Matthys, W. (2004) Stress responsivity in children with externalizing behavior disorders. *Dev Psychopathol*, **16**, 389–406.

Sowell, E. R., Peterson, B. S., Thompson, P. M., et al. (2003) Mapping cortical change across the human life span. *Nature Neurosci*, **6**(3), 309–15.

Stuss, D. T. (1992) Biological and psychological development of executive functions. *Brain Cogn*, **20**(1), 8–23.

Sylvester, C. C., Wager, T. D., Lacey, S. C., et al. (2003) Switching attention and resolving interference: fMRI measures of executive functions. *Neuropsychologia*, **41**, 357–70.

Targum, S. D. (1984) Persistent neuroendocrine dysregulation in major depressive disorder: a marker for early relapse. *Biol Psychiatry*, **19**(3), 305–18.

Tennes, K. and Kreye, M. (1987) Children's adrenocortical responses to classroom activities and tests in elementary school. *Psychosom Med*, **47**(5), 451–60.

Thomas, K. M., Drevets, W. C., Dahl, R. E., et al. (2001) Amygdala response to fearful faces in anxious and depressed children. *Arch Gen Psychiatry*, **58**, 1057–63.

Thompson, R. A. (1994) Emotion regulation: a theme in search of definition. *Monogr Soc Res Child Dev*, **59**(2–3), 25–52, 250–283.

Walker, E. F., Walder, D. J., and Reynolds, F. (2001) Developmental changes in cortisol secretion in normal and at-risk youth. *Dev Psychopathol*, **13**, 721–32.

4

Goal-directed behavior: evolution and ontogeny

Monique Ernst and Michael G. Hardin

Introduction

The study of goal-directed behavior spans an extraordinarily broad range of disciplines, including philosophy, anthropology, economics, neurobiology, and sociology, to name but a few. Within biology, an evolutionary perspective views patterns of goal-directed behaviors as having been *selected* over time to maximize *survival* of the species by prioritizing basic goals such as feeding and reproduction. Whereas this scheme underlies most of the behavioral repertoire of lower level species, it oversimplifies primate behavior. It is particularly restrictive for such complex processes as metacognition. Nonetheless, evolutionary theories provide a starting point for study of the neural underpinnings of complex behavior. For example, microeconomics has used models of animal foraging to explain patterns of economic behavior and formulate mathematical models of economic decisions, such as those made in the stock market (e.g., Schultz et al., 1997). As discussed in more detail below, an evolutionary framework has also been used to explain typical ontogenic changes in patterns of goal-directed behavior, particularly during the adolescent period.

Adolescence through the lens of evolutionary theory

Adolescence is the developmental period during which individuals transition from a dependent, family-oriented state to an independent, peer oriented state. This fundamental shift is accompanied by refinements in cognitive, emotional and social skills, as well as the acquisition of sexual maturity. Despite wide interindividual variability at any one point in time, the developmental changes associated with adolescence follow a fairly stereotypical sequence. The consistency of this sequence suggests "hard-wired" mechanisms that are most likely selected through evolution to serve species reproduction and survival.

Successful species reproduction requires the minimization of genetic inbreeding which would narrow genetic diversity in the offspring of biologically related parents. To this end, individuals must move away from the familial cell, a situation that favors the evolutionary selection of genotypes biased toward risk taking, novelty seeking, and peer primacy in social interest during adolescence (e.g., Ernst et al., 2006; Spear, 2000).

Here, we narrow the scope of this review to the neural mechanisms underlying goal-directed behaviors and the ways in which their maturation contributes to adolescent behavior. We first define the concept of goal-directed behavior. We then focus on the more restricted theme of one aspect of goal-directed behavior – decision-making – and its neural substrates. Next we describe the fractal triadic model, a heuristic model of the neurobiological systems supporting decision-making. Finally, we summarize the neuroimaging studies that have

Neuroimaging in Developmental Clinical Neuroscience, eds. Judith M. Rumsey and Monique Ernst. Published by Cambridge University Press. © Cambridge University Press 2009.

begun to examine the neurodevelopmental changes in these systems in healthy, typically developing human subjects.

Goal-directed behavior

Definition

Goal-directed behavior involves a self-determined response to a stimulus that serves the purpose of optimizing a state (physical, psychological, or social) of well-being. In the extant literature, terms such as *motivated behavior, directed action, conscious behavior*, or *decision-making* are synonymous with goal-directed behavior. Schematically, there are three possible behavioral responses to a stimulus: (1) not responding (inhibition), (2) moving toward a stimulus/event/situation (approach), or (3) moving away from a stimulus/event/situation (avoidance).

The determinants of a goal-directed behavior include the individual (agent), the stimulus, and the environment. Critical characteristics of these determinants include: (1) the physical, physiological, cognitive, and emotional *state of the agent* at the time of the behavior; (2) the cognitive and emotional *traits of the agent* (e.g., temperament); (3) the physical, probabilistic, and temporal *features of the stimuli*, as well as previous experience with the stimulus (all of which confer valence and salience to the stimulus); and (4) the *environmental context* (e.g., social versus non-social).

These factors interact at multiple levels and have given rise to a wealth of research (e.g., Ernst and Paulus, 2005). Within this research, the biological significance and neural underpinnings of innate versus learned behavior is particularly relevant to ontogeny. Similarly, dual-system models of behavior (i.e., approach versus avoidance) have been addressed extensively in psychology and neuroscience research and used to classify behavioral patterns into discrete, homogeneous categories (i.e., temperaments) well tailored to scientific scrutiny (Corr, 2004; Davidson, 2003).

Innate versus learned behaviors

The difference between innate and learned responses can inform the neural substrates of respective behaviors. For example, an avoidance response to a negative stimulus that has an innate value, such as a snake, may more closely hinge on the automatic response of subcortical structures, such as the amygdala, rather than on a more complex network involving higher-level (i.e., cortical) regions. A learned response calls upon acquired internal representations of stimuli. Accordingly, the orbitofrontal cortex (OFC) has been shown to be the seat of representations of learned affective values of stimuli (for review, Ernst and Fudge, in press).

Behavioral patterns may also reflect biological mechanisms rooted in genetics. For example, temperaments reflect enduring behavioral traits that are hypothesized to be innate and dependent on genotype (Schmidt and Fox, 2002). Differentiating innate from acquired responses may be more relevant early in development and may underlie some age-related differences in the neural correlates of behaviors. It will be important to determine whether and how innate and learned responses of approach, avoidance or inhibition are differently processed by the brain.

Approach and avoidance behavior

Concepts of approach and avoidance have been used to characterize temperament, abnormal behavior, and animal learning. The literature on temperament has focused heavily on *behavioral inhibition*, characterized by a persistent propensity to avoid and withdraw from social and novel stimuli (Fox et al., 2005a). This behavioral pattern has been associated with enhanced amygdala and striatal responses to affectively laden stimuli (Guyer et al., 2006; Perez-Edgar et al., 2007).

Neuropsychiatric disorders have been conceptualized in terms of biases toward approach and avoidance. Substance abuse is characterized by uncontrolled excessive approaches to drugs and drug-related stimuli that are said to "hijack" the reward system (Hyman et al., 2006; Koob, 2006). In contrast to a behaviorally inhibited temperament, which is thought to be highly heritable (Fox et al., 2005b; Schmidt and Fox, 2002), addictive behavior is mostly acquired, although levels of heritability (ranging from 0.27 to 0.65) and genetic associations indicate potent biological risk factors (Goldman et al., 2005). Neurobiological models of substance abuse have implicated striatum, amygdala and pre-frontal cortex (PFC) (see, Volkow et al., 2005).

Williams syndrome, a rare genetic disorder, pre-sents with an inappropriate approach to social stimuli (e.g., Bellugi et al., 1999; Jones et al., 2000), combined with an excessive avoidance of non-social stimuli (Dykens, 2003; Leyfer et al., 2006). This dis-sociation supports the notion of separate neural systems that code for the affective value of social and non-social stimuli. The amygdala and PFC have been identified as key contributors to these behav-ioral manifestations (Meyer-Lindenberg et al., 2006).

Animal studies of approach and avoidance have provided behavioral assays with which to probe the neural systems implicated in behavioral control. Several fields of investigations have studied behav-ioral responses according to distinct frameworks such as learning and conditioning (Cahill et al., 2001; Kim and Jung, 2006), motivation and reward (Cardinal et al., 2002; Kelley, 2004), and cognitive and affective regulation (Miller and Cohen, 2001). Across these domains, findings again implicate the amygdala, striatum, and PFC as substrates.

The multiplicity of factors influencing goal-directed behaviors, as described thus far, coalesce into a complex and unwieldy body of research. By way of simplification, we further constrain our framework to decision-making.

Decision-making

The decision-making template focuses on the act of selecting an option from a number of possible options. The selected option is paired with an out-come that will modify the physical, physiological, and psychological state of the decision-maker.

Extensive lines of research devoted to decision-making, including microeconomics and cognitive neuroscience, have provided quantitative models of behavior and organizational schemes that decom-pose decision-making into elemental cognitive processes (see Ernst and Paulus, 2005). Although ontogeny is virtually absent in this research, recent forays are emerging (see Ernst et al., 2006). For example, questions regarding developmental pat-terns of loss aversion (Crone and van der Molen, 2007; Crone et al., 2005), intolerance to uncertainty (Krain et al., 2007), and delay aversion (Overman, 2004) are beginning to be investigated within the framework of decision-making. In other words, a blueprint for the neurobiology of decision-making now provides a framework within which to study neurodevelopment.

Decision-making evolves in three main stages: (1) the *evaluation* of options (i.e., cues or stimuli) and formulation of a preference (selected option/ action), (2) the expression and *execution* of this preference, and (3) the *experience of an outcome* and its integration into the value of the option (consumption and learning) (Figure 4.1). Anticipa-tion is an affective state that runs across the first and second stages of decision-making. In neuro-imaging studies, anticipation is most clearly appre-hended when a delay is inserted between the execution and outcome stages.

Cognitive, affective, and regulatory processes are engaged during each stage of decision-making; however, the balance and nature of these processes can be stage specific (Ernst and Paulus, 2005). Cognitive processes engaged in these three stages include: (1) attention to direct cognitive/affective resources; (2) coding of the somatosensory proper-ties of options; (3) activation and integration of memory representations; (4) tagging of subjective values onto available options or actions paired with these options; (5) engagement of inhibitory/ excitatory systems to reject/select options; (6) initi-ation and execution of motor actions; (7) coding of somatosensory features of outcome stimuli; (8) formation of memory representations; and (9) elaboration of behavioral rules upon repetition

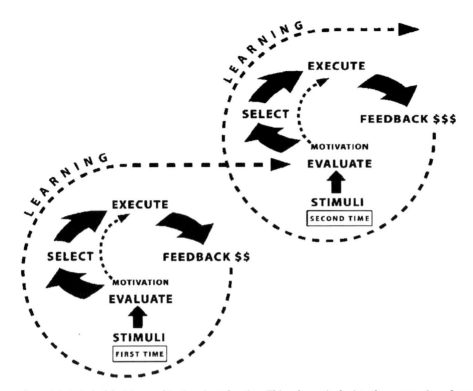

Figure 4.1. Spiral of decision-making/motivated action. This schematic depicts the progression of processes that take place in a simple and completed motivated action. Individuals are first exposed to stimuli that represent options from which one needs to be selected. Upon exposure, individuals evaluate the stimuli options, form a preference, and select a course of action (stage 1). The action is then executed (stage 2). If the result of their action occurs with a delay, anticipation of the action outcome occurs. Feedback is then received and the outcome is experienced (stage 3). The experience of feedback/outcome will inform the value of the option selected in the first stage of the motivated action. This occurs through learning and will influence the evaluation and selection of similar stimuli during subsequent experiences. Reprinted from Ernst et al. (2006) with permission. (Graphic designed by Cynthia Friedman.)

of the decision-making template. As illustrated in Figure 4.2, key neural systems work dynamically in concert to complete the decision-making process. Here, we focus on the amygdala, striatum, and PFC as the neural substrates that are most consistently recruited across the decision-making process and that are most prominent in behavioral control. These structures form the main nodes of the triadic model.

Fractal triadic model

The *fractal triadic model* proposes a tri-dimensional organization of the brain systems devoted to

behavioral control (Ernst and Spear, in press; Ernst et al., 2006). This model is a re-formulation of the previous "triadic model" (Ernst et al., 2006), which, as a first step, did not specify the heterogeneity of the neural systems involved.

These systems are supported by neural networks whose representative core structures are: (1) the striatum, dominant for coding approach behavior; (2) the amygdala, dominant for coding avoidance behavior; and (3) PFC (particularly medial PFC), acting to regulate/modulate the approach and avoidance systems. Each structure is composed of a number of functional subunits. These subunits

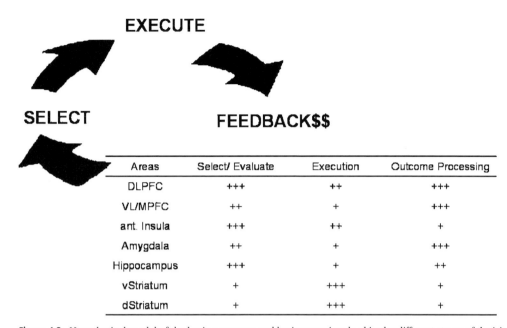

EXECUTE

SELECT   **FEEDBACK$$**

Areas	Select/ Evaluate	Execution	Outcome Processing
DLPFC	+++	++	+++
VL/MPFC	++	+	+++
ant. Insula	+++	++	+
Amygdala	++	+	+++
Hippocampus	+++	+	++
vStriatum	+	+++	+
dStriatum	+	+++	+

Figure 4.2. Hypothetical model of the basic processes and brain areas involved in the different stages of decision-making. Decision-making is divided into three stages (top): (1) the *evaluation*/formation of options and formulation of preferences, including the *selection* of a specific option; (2) the *execution* of a preference; and (3) receipt of *feedback*, experience and evaluation of outcome. Table of neural circuitry (bottom): a distributed network of brain areas is differentially involved in the processes of each stage. The degree of involvement for each region at each stage is reflected by the number of plus signs. In addition to each region being differentially recruited during each decision-making stage, the balance of recruitment of each region at each stage may shift depending on the approach, avoidance, or regulation goal of the behavior. DLPFC, dorsolateral prefrontal cortex; VL/MPFC, ventrolateral/medial prefrontal cortex; ant. Insula, anterior insula; vStriatum, ventral striatum; dStriatum, dorsal striatum.

may themselves be polarized toward the support of approach, avoidance, or regulation of behavior, thereby forming a fractal representation of the triadic model (Figure 4.3).

From an ontogenic perspective, each portion of these networks undergoes developmental change according to a predetermined time table that can be altered either by disease states or responses to extreme environments. These ontogenic changes manifest themselves along unique trajectories and exert quantitative and qualitative influences on behavioral patterns. Ultimately, these influences mold behavior to satisfy the demands of evolutionary fitness. The following describes the structural basis of the fractal triadic model.

Striatum

The striatum, composed of dorsal (caudate nucleus, putamen and globus pallidus) and ventral (nucleus accumbens, NAcc) structures, lies inferior to the corpus callosum along its entire length. The term "ventral striatum" is typically used to refer to the most ventral parts of the striatum, including the NAcc. The output of the *striatum*, through the brainstem nuclei via the globus pallidus, promotes predominantly *approach* responses to *appetitive* stimuli (Cardinal et al., 2002; Haber et al., 2006). However, this region has also been shown to a lesser extent to modulate withdrawal responses to aversive stimuli (Seymour et al., 2007).

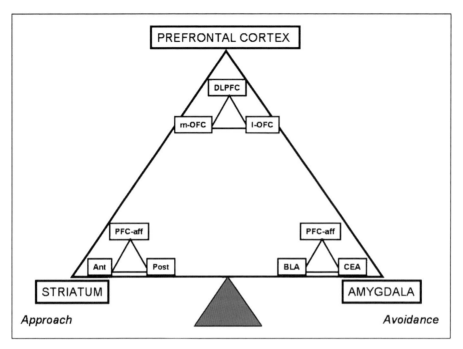

Figure 4.3. Schematic representation of the fractal triadic model. In this model, each of the three representative nodes of approach, avoidance and regulatory control are composed of triadic formations that mirror the overall organization. Examples of regions previously identified as associated with either appetitive or aversive responses are provided. DLPFC, dorsolateral prefrontal cortex; m-OFC, medial orbital frontal cortex; l-OFC, lateral orbital frontal cortex; PFC-aff, prefrontal cortical afferents; Ant, anterior striatum; Post, posterior striatum; BLA, basolateral amygdala; CEA, central amygdala.

Different functional constructs of the striatum have been proposed. A limbic-cognitive-motor functional gradient has been drawn along a ventral-dorsal vector (Alexander et al., 1990; Haber et al., 2000); and an appetitive-aversive gradient, along a rostral-caudal vector (Reynolds and Berridge, 2001, 2002). The NAcc comprises a core and a shell. The core has been shown to be involved with the acquisition of Pavlovian approach behavior (Martin-Soelch et al., 2007; Parkinson et al., 1999), whereas the integrity of the shell does not seem to be essential to this process. The ways in which appetitive and aversive stimuli, or approach and withdrawal behavioral responses, may be differentially processed in the striatum are not known. However, a balance between these two opposite processes may follow some homeostatic equilibrium that is determined or maintained by PFC inputs.

Amygdala

The amygdala, located in the anterior medial temporal lobe, is recognized as a key center for emotion and social-related processes (see LeDoux, 2000). The output of the *amygdala*, through the hypothalamus and brainstem nuclei, dominantly promotes *withdrawal* responses to *aversive* or threat stimuli. Viewed as the heart of the "limbic" or emotional coding system, the amygdala has also been conferred the role of a "fear center." However, conditioning (Martin-Soelch et al., 2007), electrophysiological (Grace et al., 2007), lesion (Baxter and Murray, 2002), and neuroimaging (see Ernst and Mueller, 2008) studies have also identified amygdala contributions to coding appetitive stimuli.

The amygdala comprises several nuclei. These include the basolateral (BLA) and central (CeN) nuclei (see Ernst and Fudge, in press), which subserve qualitatively different functions in conditioning (see Martin-Soelch et al., 2007). The BLA has been uniquely associated with the encoding of stimulus-affective value associations (evaluative conditioning). Conversely, the CeN appears to be prominently involved in encoding stimulus-response associations (operant conditioning).

The CeN is also thought to indirectly regulate, via the midbrain ventral tegmental area, dopamine release in the NAcc in response to food or amphetamine administration (e.g., Robledo et al., 1996). It is through this indirect regulation that the CeN may facilitate appetitive responses. In contrast, the BLA appears to inhibit dopamine responses to acute stimuli within the NAcc (see Ernst and Spear, in press). A proposed mechanism invokes an interaction between the tonic and phasic dopamine activity in the striatum. The BLA sends direct excitatory projections to the NAcc that enhance tonic dopamine activity, in turn inhibiting phasic dopamine activity (Grace, 1991). Thus, CeN can facilitate appetitive responses, whereas BLA can reduce striatal responses to appetitive stimuli.

Finally, reciprocal connections with the PFC (particularly orbital and medial) modulate the output of the amygdala (see Ernst and Fudge, in press). This organization suggests that PFC may control the balance between appetitive and aversive coding within the amygdala, maintaining a homeostatic state between activity linked to approach and/or avoidance. Such modulatory roles of PFC on amygdala function have been supported by functional neuroimaging studies (e.g., Beauregard et al., 2001; Ochsner et al., 2004), most convincingly those of emotion appraisal/reappraisal (Ochsner and Gross, 2005).

Prefrontal cortex

Orbital and medial PFC (sometimes coined "limbic" or emotional PFC) have reciprocal projections with the amygdala and send direct afferents to the ventral striatum (Ferry et al., 2000; Fudge et al., 2005). The ventral striatum indirectly influences these cortical regions through striato-pallidal-cortical loops (for overview Ernst and Fudge, in press).

These orbital and medial prefrontal regions have been implicated in the representations of affective values (e.g., O'Doherty, 2007; Rolls, 2004), inhibition (Elliott and Deakin, 2005), response reversal (Schoenbaum et al., 2007), and conflict resolution (Yeung et al., 2004). The anterior medial PFC has also been associated more specifically with metacognition (Fletcher et al., 1995), self-evaluation (see Amodio and Frith, 2006), and rule formation (Bunge et al., 2005). To be efficient, these high-level cognitive functions require integrated information about the endogenous (physical and emotional state) and exogenous environment. Some of this information is provided by the amygdala and striatum.

The specific content and role of the direct amygdala and indirect striatal inputs to the cortex are yet not well worked out. The fractal triadic model contends that these inputs from the amygdala favor information important to avoid aversive or dangerous stimuli, whereas those from the striatum favor information relevant to approach, appetitive, or habitual stimuli. Both types of information need to be integrated to generate valid self and environmental assessment, adaptive behavioral rules, and preference formation in decision-making processes. These operations are further facilitated or biased toward one or the other pole (approach versus withdrawal) by regulatory inputs providing a biologically defined homeostatic equilibrium.

The loci of these operations may be mapped onto distinct or overlapping neural circuitries. Alternatively, they may occur within the same neural space, supported by different neurochemical mechanisms. Several instances of cortical somatotopic schemes have been proposed in relation to the coding of, or response to, appetitive versus aversive stimuli. For example, the role of the right hemisphere for coding negatively valenced emotions and of the left hemisphere for coding positively valenced emotions has been the object of much attention, particularly in work concerning temperament and psychopathology (Davidson et al., 2000;

Fox et al., 1995). Another scheme attributes a role of the lateral orbitofrontal regions to the coding of punishments and the medial orbitofrontal regions to the coding of rewards (Liu et al., 2007).

In summary, the fractal re-formulation of the triadic model (Ernst et al., 2006) adds additional processing layers to the coding of goal-directed behavior. These layers provide a larger capacity for adaptive modulations, including compensatory mechanisms, but may also present more opportunities for perturbations. Functional neuroimaging in humans combined with animal work will be able to test the various facets of the fractal triadic model.

Functional neuroimaging

Functional magnetic resonance imaging paradigms of reward-related decision-making

The development of valid paradigms is perhaps the most challenging aspect of designing developmental functional neuroimaging studies of decision-making. A major issue concerns the limits imposed by an acceptable task duration that can be tolerated by children. As previously discussed, decision-making involves multiple processes that occur both serially across the three decision-making stages and concurrently. Because these stages are interdependent, the functional magnetic resonance imaging (fMRI) signals in response to one stage can be difficult to disentangle from the signals associated with the preceding or following stages. One solution for this problem is to insert time intervals between the stages. This, however, lengthens the paradigm leading to potential alterations in the psychological significance of the task (e.g., boredom).

Additionally, decision-making is influenced by the multiple characteristics of the possible option choices, including the *probability, magnitude,* and *timing* of the outcomes. Manipulations of these variables require multiple trials for meaningful statistical analyses, thus resulting in longer paradigms. Probability is perhaps the most complex but important of these variables for the study of decision-making

in adolescence. The probability of outcomes, along with their magnitude, define risk-taking and are central to adolescent motivated behavior. The primary concern (for adolescents or adults) is that high-risk options (low probability of high magnitude) are selected much less often than low-risk options. To have sufficient statistical power to analyze high-risk selections, there must be a large enough number of repeats of this condition. There is also wide interindividual variability in patterns of selections. Different selection patterns can lead to different "contexts" of decision-making (e.g., prominent favorable outcomes with the consistent selection of low-risk options, and prominent unfavorable outcomes with high-risk options). Adjusting the expected values (EV = probability × magnitude) of outcomes can be one way to address this issue.

A third challenge concerns the type of outcome (e.g., primary versus secondary rewards). Monetary feedback has been favored in tasks of decision-making because it is quantitative by nature and an unequivocal reward for all individuals. However, it is a learned reward and may lead to findings that differ from those relating to innate rewards (such as food). Money may also hold different values for different subjects. To address this issue, subjective responses to feedback can be used as covariates to control for such potential bias.

Effects of learning during task performance are also critical to consider. Outcomes serve to update the values of the stimuli (options) with which they are paired (see Figure 4.1). To complicate matters, learning processes evolve with development, and age differences in regional activations may be related to differences in learning processes rather than in the cognitive or emotional aspects under investigation. Thus, to obviate learning effects, participants need to be fully aware of the value of the options and familiar with the task before performing it in the scanner.

Some of these complexities can be avoided altogether by narrowing the scope of the question under scrutiny. For example, it could be possible to study only one stage of the decision-making process. One could study only the outcome stage, by skipping the preference formation and selection/

execution stages. Constructing a paradigm where outcomes are given without a selection phase would miss the "contingency" effect that is central to the decision-making process. However, using a paradigm focusing on preference/formation and not presenting outcome would bypass outcome reinforcement, which is critical to the selection stage. Issues such as these will be illustrated in the developmental studies presented below (see Table 4.1).

fMRI studies of decision-making in adolescence

At present, only seven studies focusing on decision-making and reward processes in adolescents have been published (Bjork et al., 2004, 2007; Ernst et al., 2005; Eshel et al., 2007; Galvan et al., 2006; May et al., 2004; van Leijenhorst et al., 2006). With the exception of May et al. (2004), which included only adolescents, these studies have directly compared different age groups (adolescents, children, and adults). In addition to the summaries provided in Table 4.1, these studies are addressed below. These studies are organized as a function of the focus of their analyses, i.e., across the decision-making stages, during reward anticipation, or during selection.

Global analysis of the decision-making processes

Two studies have examined the hemodynamic signal as it evolved over a whole performance trial, encompassing all stages of the decision-making process. May et al. (2004) used a guessing task (50% probability of being correct), associated with potential monetary gains, losses, and null outcome, to study decision-making in a sample of adolescents. Each trial consisted of a cue (preference formation), followed by a delay period (anticipation), then an outcome. The temporal pattern of activation of a given region was interpreted as informing the stage at which this region was most contributory. However, as previously mentioned, such interpretation is problematic because of the difficulty separating (deconvolving) the signal of one event from

adjacent events. In this study, the analysis revealed a regional activation pattern that was very similar to that found in a previous study of adults using the same task. This pattern included activation of lateral OFC that peaked during reward anticipation, ventral striatum that peaked during reward outcome, and medial OFC that deactivated during the neutral condition. Other regional activations included medial PFC and dorsal anterior cingulate cortex (ACC). Overall, this study confirms that adolescents and adults activate a similar neural network during decision-making.

Galvan et al. (2006) compared adults, children, and adolescents on a paradigm that used rewards of variable magnitudes in a delayed response two-choice task. The task did not include preference formation or choice selection stages, but probed reward anticipation and response to outcomes. Unlike other tasks employed in adolescent decision-making studies, the outcome in this task was represented only in relative terms, as pictures representing a small, medium or large treasure chest, without an attached monetary value. Similar to May et al. (2004), the blood-oxygen-level-dependent (BOLD) signal change was analyzed over the entire trial. The level of peak activation in ventral striatum was greater for adolescents compared to both adults and children. The OFC also showed an age-related difference with the greatest activity occurring in children compared to adults and adolescents. The OFC seemed to show a similar extent of activation in adolescents and adults, whereas the NAcc exhibited a larger extent of activation in adolescents than in adults. The authors concluded that the OFC and NAcc differed in maturational profiles.

Analysis of reward anticipation

Two studies have examined the anticipation of monetary rewards. Bjork et al. (2004) used a reaction time task, in which subjects pressed a button during the brief presentation of a target. The task included an anticipation phase, during which a cue announced the type of reward trial, followed by a variable delay interval before the appearance of

Table 4.1 Developmental studies of decision-making and reward processing

A. Samples and paradigms

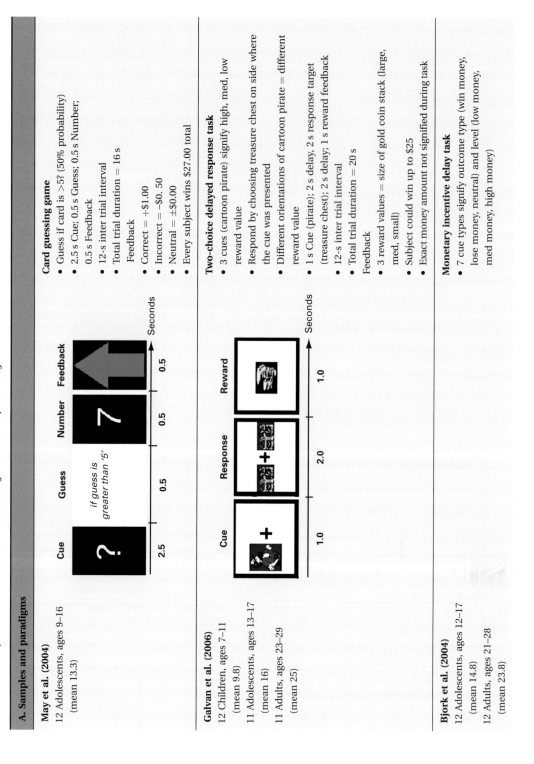

May et al. (2004)

12 Adolescents, ages 9–16
(mean 13.3)

Cue | Guess | Number | Feedback
2.5 | 0.5 | 0.5 | 0.5 → Seconds

Card guessing game
- Guess if card is >5? (50% probability)
- 2.5 s Cue; 0.5 s Guess; 0.5 s Number; 0.5 s Feedback
- 12-s inter trial interval
- Total trial duration = 16 s Feedback
- Correct = +$1.00
- Incorrect = –$0. 50
- Neutral = ±$0.00
- Every subject wins $27.00 total

Galvan et al. (2006)

12 Children, ages 7–11
(mean 9.8)

11 Adolescents, ages 13–17
(mean 16)

11 Adults, ages 23–29
(mean 25)

Cue | Response | Reward
1.0 | 2.0 | 1.0 → Seconds

Two-choice delayed response task
- 3 cues (cartoon pirate) signify high, med, low reward value
- Respond by choosing treasure chest on side where the cue was presented
- Different orientations of cartoon pirate = different reward value
- 1 s Cue (pirate); 2 s delay; 2 s response target (treasure chest); 2 s delay; 1 s reward feedback
- 12-s inter trial interval
- Total trial duration = 20 s Feedback
- 3 reward values = size of gold coin stack (large, med, small)
- Subject could win up to $25
- Exact money amount not signified during task

Bjork et al. (2004)

12 Adolescents, ages 12–17
(mean 14.8)

12 Adults, ages 21–28
(mean 23.8)

Monetary incentive delay task
- 7 cue types signify outcome type (win money, lose money, neutral) and level (low money, med money, high money)

- Respond by making button press while target is on screen
- 5 s Cue; 4–4.5 s anticipation; 160–260 ms target; 0.5 s feedback
- No inter trial interval
- 6 s total trial duration

Feedback
- Win money = +$0.20; +$1.00; and +$5.00
- Lose money = −$0.20; −$1.00; and −$5.00
- Neutral = ±$0.00

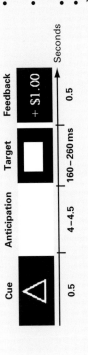

Cue	Anticipation	Target	Feedback
△			+ $1.00
0.5	4–4.5	160–260 ms	0.5 Seconds

Bjork et al. (2007)

20 Adolescents, ages 12–17 (mean 14.3)

20 Adults, ages 23–33 (mean 28.5)

Game of chicken

- Button press begins the accumulation of money
- If "bust" during accumulation, money is either not won or lost
- 4 conditions:
  (1) Motor control – just make button press, no reward and no penalties
  (2) No-penalty – money always accumulates and never a penalty
  (3) Low penalty – money accumulates until either button press or "bust" – at "bust" no money is won
  (4) High penalty – money accumulates until either button press or "bust" – at "bust" money is lost from total
- 2 s Start button press period; 4–10 s accumulation period; feedback starts at stop button press or "bust" and last until total trial = 14 s
- No inter trial interval
- 14 s total trial duration

Feedback
- Display includes trial outcome and cumulative money amount

Start	Accumulation		Feedback
Button press at $	Maker button press to stop		
Total earnings $4.53	Total earnings $4.53	Total earnings $5.56	Total earnings $5.56
$	→		You just won! $1.53
2.0	4.0 – 10.0		Until 14.0
	Seconds		

Cake task

- 2 cue types signify low or high risk choice
- Low risk = 2 of 9 cake pieces; high risk = 4 or 5 of 9 cake pieces
- Respond by picking piece type that is most likely to be selected by the computer
- 0.5 s Fix; 2.0 s cue; 1.5 s response; 2.0 s feedback

van Leijenhorst et al. (2006)

12 Adolescents, ages 9–12 (mean 11.3)

14 Adults, ages 18–26 (mean 21.5)

Table 4.1 (*cont.*)

A. Samples and paradigms

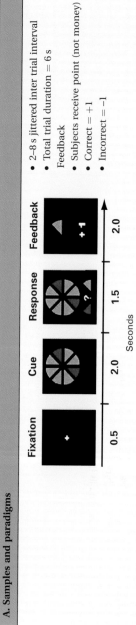

Fixation	Cue	Response	Feedback
+		?	▲ +1

0.5 2.0 1.5 2.0

Seconds

- 2–8 s jittered inter trial interval
  - Total trial duration = 6 s
  Feedback
  - Subjects receive point (not money)
  - Correct = +1
  - Incorrect = −1

Eshel et al. (2007)
Ernst et al. (2005)
14 Adolescents,
ages 9–17 (mean 13.3)
16 Adults, ages 20–40
(mean 26.7)

Selection	Anticipation	Outcome
$7 $1	Unsure How sure are you? Sure 1 2 3 4 5	You Win $7 Total $8 How do you feel?

3.0 4.0 4.0

Seconds

Wheel of fortune task

- Respond by selecting piece of wheel you think pointer will land on
- 4 conditions vary probability of wheel pieces and money amount
  (1) 10/90 wheel: 10% chance of winning $7 vs. 90% chance of winning $1
  (2) 30/70 wheel: 30% chance of winning $2 vs. 70% chance of winning $1
  (3) 50/50 high money: 50% chance of winning $4
  (4) 50/50 low money: 50% chance of winning $0.50
- If pointer lands on selected piece = win corresponding money amount
- If pointer does not land on selected piece = win no money
- During anticipation, denote how sure you are of choice
- During feedback, denote how happy you are about the outcome
- 3 s selection; 4 s anticipation; 4 s outcome
- No inter trial interval
- 11 s total trial duration
  Feedback
- Display includes trial outcome and cumulative money amount

B. Findings

Study	Performance	Imaging
		Across whole trial
May et al. (2004)	• No correlation with age (9–16 years) • From trial to trial, subjects tended to exhibit win-stay or lose-shift strategies	**Condition (reward, loss, neutral) × scan (cue, execution, anticipation, feedback)** • VS (–10, 7, –5) (reward: more sustained activation) • lat-OFC (–43, 45, –7) and m-OFC (–3, 51, –8) (reward: more sustained activation) • ACC: reduced with feedback from neutral trials

Galvan et al. (2006)	• No group accuracy differences (all > 95%)
	• Reaction time fastest for highest reward:
	  • Adult: large > med > low
	  • Adol: large > med = low
	  • Child: large = med = low
		Peak activation
		• NAcc (6, 5, –2) and (–8, 6, –2): Adol > Adult & Adol > Child
		• OFC (46, 31, 1) and (–8, 6, –2): Adol = Adult & Adol < Child
		Extent of activation
		• NAcc: Child > Adol = Adult
		• OFC: Child > Adol > Adult
		Learning (early vs. middle vs. late trials)
		• More effects of reward in Adol in the late trials

	Preference formation/ execution stage	Anticipation stage	Feedback stage

van Leijenhorst et al. (2006)	Accuracy
	• Adol = 91%
	• Adult = 98%
	Errors on high risk
	• Adol > adult (p = 0.07)
		High risk vs. low risk
		• OFC (22, 50, –14)
		• DLPFC (42, 30, 18)
		  • Both regions significant in both groups but no Adol–Adult difference
		• m-PFCC/ACC (0, 6, 20)
		  • Significant only in Adol
		• midbrain (0, –15, –9)
		  • Not significant
				Negative (high-risk) > positive (high-risk)
				• m-PFC (–22, 18, 50)
				• R-VLPFC (IFG/insula, –20, 12, –20)
				  • Both regions more active in neg > pos
				  • No Adol–Adult difference
				• R-lat-OFC (40, 46, –12)
				  • More activation for negative in Adol than Adult

Table 4.1 (*cont.*)

	Preference formation/execution stage	Anticipation stage	Feedback stage

Bjork et al. (2004)	• No differences between Adol and Adults for % correct or reaction time
	• Overall accuracy rate = 69.9%
		Whole-brain analysis
		Gain-cue vs. neutral-cue
		• VS (9, 17, −2) extends to R-NAcc, R-a-insula, R-amygdala
		  • Adult > Adol
		• d-ACC
		  • Adult only
		Loss-cue vs. neutral-cue
		• R-insula
		  • Adol only
		• m-PFC (−5, 53, 31)
		  • Deactivation
		  • Adult only
		• R-m-caudate nucleus
		  • Adol and Adults
		ROI analysis
		Group × incentive in gain trials
		• R-NAcc
		  • Lower signal increase while anticipating response to win large gains ($5) in Adol
			Whole-brain analysis
			Gain vs. no gain
			• NAcc (12, 45, −11)
			• m-PFC (−1, 41, −11)
			• Amygdala and hippocampus
			  • All regions significant for Adol and Adults
			Loss vs. no loss
			• Putamen
			  • No loss > loss for Adol and Adults
			• m-PFC
			  • Deactivation
			  • Adult only
			ROI analysis
			• No Adol–Adult differences
			• m-PFC
			  • Activated only in gain vs. no gain, and not in loss vs. no loss

Bjork et al. (2007)	• Adol slower than adult in all conditions
	• No group × condition interactions
		Motor response vs. reward-only
		• p-m-PFC
		  • No Adol–Adult differences
		Penalties vs. no-penalties
		• p-m-PFC (0, 15, 34) and (8, 11, 44)
		  • Adult > Adol,
		  • Low penalties: Adult > Adol
		  • High penalties: Adult > Adol	

Ernst et al. (2005)	• Both groups happier to win $4 than $0.50	**Win vs. no win**
	• More intense ratings for win than no win	• Amygdala (−26, −4, −14)
	• Adol happier than adult during win	• Adult > Adol
		• L-NAcc (−16, 20, −4)
		• Adol > Adult
Eshel et al. (2007)	• 30% probability selection: Adol > Adult (suggests Adol more risk-taking)	**High risk (10/30) vs. low risk (70/90) risk**
		• OFC/L-v-PFC (−44, 14, −4)
		• Brodmann's area 47
		• ACC (2, 26, 30)
		• Ventral ACC (−2, 38, 2)
		• All regions significant for Adult > Adol
		• NAcc
		• Amygdala
		• DLPFC
		• No Adol–Adult differences

Notes:

Illustrations of stimuli are reprinted above with permission from the references as listed under "Study." Numbers in parenthesis = (x, y, z) coordinates in Talairach space. Abbreviations: a, anterior; ACC, anterior cingulate cortex; Adol, adolescent; d, dorsal; DLPFC, dorsolateral prefrontal cortex; L, left; lat, lateral; m, medial; NAcc, nucleus accumbens; OFC, orbital frontal cortex; p, posterior; PFC, prefrontal cortex; R, right; RT, reaction time; v, ventral; VLPFC, ventrolateral prefrontal cortex; VS, ventral striatum.

the target. Outcome was presented after the target disappeared from the screen. The cues announced that good performance would be rewarded with a monetary gain, avoidance of monetary loss, or no contingency (neutral). The paradigm did not include a preference formation phase or a selection phase. The analysis of the variable delay period was understood as reflecting anticipation. Adults showed more ventral striatal activation than adolescents during anticipation of gain. The authors proposed that adolescents have a hypofunctional reward system that requires extra-stimulation to be maintained at a homeostatic level, thereby contributing to high levels of risk-seeking behavior.

In a more recent study, Bjork et al. (2007) used a decision-making paradigm that manipulated penalties. The paradigm included a cue announcing the trial type, a button press to begin accumulation of monetary reward over a fixed duration, a second button press to stop the accumulation of reward in order to avoid a potential penalty, and finally the presentation of the monetary outcome. The cue announced whether the trial would have monetary gain without the possibility of a penalty, with the possibility of a small penalty (no monetary gain), or with the possibility of a large penalty (monetary loss). A fourth condition included no reward or penalty. Reward was invariant across conditions and depended upon the subjects' willingness to take risks. In the penalty conditions, the longer subjects waited to press a button as reward accumulated, the more likely they were to have the trial interrupted by a penalty. Only the cue phase was analyzed. Findings revealed greater deactivation in medial PFC in adults relative to adolescents during the penalty conditions, suggesting that adolescents were less able to engage prefrontal circuits to control behavior during punishment. This may contribute to the notion of lower sensitivity to punishment in adolescents (Spear, 2000).

Analysis of preference formation/selection

Two adolescent studies have reported on the preference formation/selection phase. Van Leijenhorst

et al. (2006) manipulated reward probabilities while keeping reward magnitude constant. In this study, points (+1, −1) served as rewards. Adolescents tended to show activation of the ACC during the selection phase of more uncertain outcomes compared to safer outcomes; however, no significant differences between activation patterns for adolescents versus adults were reported.

The second study (Eshel et al., 2007) used a paradigm that manipulated both reward probability and magnitude to generate risky versus safe choices. This paradigm comprised a cue phase during which subjects formed a preference and executed a selection, an anticipation phase that involved a delay prior to the outcome, and an outcome phase. Outcomes were either monetary gains or a failure to obtain a monetary gain. Adults showed greater activation of the ventrolateral PFC and dorsal ACC than adolescents when risky choices were compared to safe choices (Figure 4.4). In line with predictions of developmental trajectories of the PFC (see Chapters 1 and 2), this finding suggested that behavioral control from prefrontal regions is lagging behind in adolescents, who do not regulate risk-taking as efficiently as adults.

The contrasting ACC findings of these studies raise interesting questions. In the case of the manipulation of uncertainty with no monetary incentives, adolescents seem to recruit parts of the medial PFC (including the implicated regions of ACC) more strongly than adults, whereas they recruit these regions less than adults when reward magnitude and monetary outcomes are involved. Thus, this region shows developmental changes that manifest differently as a function of the task demands.

Analysis of outcome

Three studies have reported on adolescent responses to outcome (Bjork et al., 2004; Ernst et al., 2005; van Leijenhorst et al., 2006). In their study described above, van Leijenhorst et al. (2006) reported greater lateral OFC activation in adolescents compared to adults for losses when compared to gains. When a large monetary gain was compared to the failure to

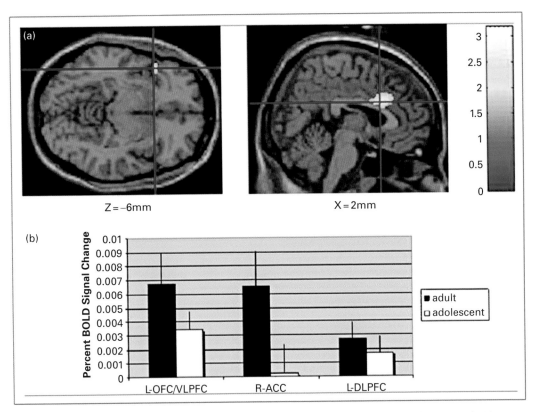

Figure 4.4a, b. Age-related differences in activation during decision-making. Illustrated in the top panels (a) are greater regional activations in the lateral orbitofrontal/ventrolateral prefrontal cortex (L-OFC/VLPFC) (MNI x, y, z: −44, 14, −6 mm) and right anterior cingulate cortex (R-ACC) (MNI x, y, z: 2, 26, 30 mm) in adults compared to adolescents when making risky compared to safe decisions. (b) The bottom panel represents the group mean fraction of blood-oxygen-level-dependent (BOLD) signal changes in adults and adolescents at the peak voxels identified on panel a. These activation patterns likely reflect underdeveloped regulation and conflict resolution in adolescents, both of which may contribute to adolescents' increased propensity for risk-taking and approach behavior. Reprinted from Eshel et al. (2007) with permission.

make a large monetary gain, Ernst et al. (2005) reported greater activation of dorsolateral PFC and amygdala in adults than in adolescents, but greater activation of the NAcc in adolescents compared to adults. Contrary to both of these studies, Bjork et al. (2004) failed to detect any activation differences between adolescents and adults in any of the feedback contrasts. These findings, although still preliminary, suggest that the neural network engaged during the receipt of rewards or punishments may differ between adolescents and adults. The nature of these differences obviously warrants a body of

work to assess all possible scenarios. The use of a neurobiological model can help guide a systematic approach to this question.

Conclusions

The neural correlates of goal-directed behaviors are multifaceted. The goal of delineating the trajectory of ontogenic development increases the complexity of this research and requires theoretical models to constrain hypotheses and guide the development of experimental paradigms for a step-wise systematic approach. Apparent in the review of functional neuroimaging studies is the lack of consistent analyses of all key regions critically involved in goal-directed behavior. For example, the response of the amygdala is not systematically reported. Such omission makes it difficult to test a comprehensive model of the neurobiology of goal-directed behavior from a developmental perspective.

As illustrated by the contributions of temperament to patterns of goal-directed behavior, an examination of genetic determinants will need to be integrated into this research. Large samples are needed for traditional association and linkage studies. However, sample sizes can be significantly reduced through the use of intermediate phenotypes identifiable using neuroimaging (see Goldman et al., Chapter 22).

Finally, it is important to recognize the potential implications of this work for legal, educational, and clinical disciplines. Questions regarding societal rules encompass issues such as age limits for alcohol consumption or drivers' licenses. With regard to education, informing the way(s) to motivate children and adolescents and the critical developmental periods at which different teaching strategies are best implemented may bring significant improvements to our educational system. Within the clinical domain, the recognition that most psychiatric problems that involve perturbations in goal-directed behavior can energize research aimed at understanding the potential neural mediators of such perturbations and at developing preventive and therapeutic intervention strategies.

REFERENCES

Alexander, G. E., Crutcher, M. D., and DeLong, M. R. (1990) Basal ganglia-thalamocortical circuits: parallel substrates for motor, oculomotor, "prefrontal" and "limbic" functions. *Prog Brain Res*, **85**, 119–46.

Amodio, D. M. and Frith, C. D. (2006) Meeting of minds: the medial frontal cortex and social cognition. *Nat Rev Neurosci*, **7**(4), 268–77.

Baxter, M. G. and Murray, E. A. (2002) The amygdala and reward. *Nat Rev Neurosci*, **3**(7), 563–73.

Beauregard, M., Levesque, J., and Bourgouin, P. (2001) Neural correlates of conscious self-regulation of emotion. *J Neurosci*, **21**(18), RC165.

Bellugi, U., Adolphs, R., Cassady, C., and Chiles, M. (1999) Towards the neural basis for hypersociability in a genetic syndrome. *NeuroReport*, **10**(8), 1653–7.

Bjork, J. M., Knutson, B., Fong, G. W., et al. (2004) Incentive-elicited brain activation in adolescents: similarities and differences from young adults. *J Neurosci*, **24**(8), 1793–802.

Bjork, J. M., Smith, A. R., Danube, C. L., and Hommer, D. W. (2007) Developmental differences in posterior mesofrontal cortex recruitment by risky rewards. *J Neurosci*, **27**(18), 4839–49.

Bunge, S. A., Wallis, J. D., Parker, A., et al. (2005) Neural circuitry underlying rule use in humans and nonhuman primates. *J Neurosci*, **25**(45), 10347–50.

Cahill, L., McGaugh, J. L., and Weinberger, N. M. (2001) The neurobiology of learning and memory: some reminders to remember. *Trends Neurosci*, **24**(10), 578–81.

Cardinal, R. N., Parkinson, J. A., Hall, J., and Everitt, B. J. (2002) Emotion and motivation: the role of the amygdala, ventral striatum, and prefrontal cortex. *Neurosci Biobehav Rev*, **26**(3), 321–52.

Corr, P. J. (2004) Reinforcement sensitivity theory and personality. *Neurosci Biobehav Rev*, **28**(3), 317–32.

Crone, E. A. and van der Molen, M. W. (2007) Development of decision making in school-aged children and adolescents: evidence from heart rate and skin conductance analysis. *Child Dev*, **78**(4), 1288–301.

Crone, E. A., Bunge, S. A., Latenstein, H., and van der Molen, M. W. (2005) Characterization of children's decision making: sensitivity to punishment frequency, not task complexity. *Child Neuropsychol*, **11**(3), 245–63.

Davidson, R. J. (2003) Affective neuroscience and psychophysiology: toward a synthesis. *Psychophysiology*, **40**(5), 655–65.

Davidson, R. J., Jackson, D. C., and Kalin, N. H. (2000) Emotion, plasticity, context, and regulation: perspectives from affective neuroscience. *Psychol Bull*, **126**(6), 890–909.

Dykens, E. M. (2003) Anxiety, fears, and phobias in persons with Williams syndrome. *Dev Neuropsychol*, **23**(1–2), 291–316.

Elliott, R. and Deakin, B. (2005) Role of the orbitofrontal cortex in reinforcement processing and inhibitory control: evidence from functional magnetic resonance imaging studies in healthy human subjects. *Int Rev Neurobiol*, **65**, 89–116.

Ernst, M. and Fudge, J. L. (In Press) Adolescence: on the neural path to adulthood. In M. Potenza & J. Grant (Eds.), *Young Adult Mental Health*: Oxford University Press.

Ernst, M. and Mueller, S. (2008) The adolescent brain: insights from functional neuroimaging research. *Dev Neurobiol*, **68**(6); 729–43.

Ernst, M. and Paulus, M. P. (2005) Neurobiology of decision making: a selective review from a neurocognitive and clinical perspective. *Biol Psychiatry*, **58**(8), 597–604.

Ernst, M. and Spear, L. P. (In Press) Development of reward systems in adolescence. In de Haan, M., Gunnar, M. R. (eds.) *Handbook of Developmental Social Neuroscience*: New York: Guilford Press.

Ernst, M., Nelson, E. E., Jazbec, S., et al. (2005) Amygdala and nucleus accumbens in responses to receipt and omission of gains in adults and adolescents. *NeuroImage*, **25**(4), 1279–91.

Ernst, M., Pine, D. S., Hardin, M. (2006) Triadic model of the neurobiology of motivated behavior in adolescence. *Psychol Med*, **36**(3), 299–312.

Eshel, N., Nelson, E. E., Blair, R. J., et al. (2007) Neural substrates of choice selection in adults and adolescents: development of the ventrolateral prefrontal and anterior cingulate cortices. *Neuropsychologia*, **45**(6), 1270–9.

Ferry, A. T., Ongur, D., An, X., and Price, J. L. (2000) Prefrontal cortical projections to the striatum in macaque monkeys: evidence for an organization related to prefrontal networks. *J Comp Neurol*, **425**(3), 447–70.

Fletcher, P. C., Happe, F., Frith, U., et al. (1995) Other minds in the brain: a functional imaging study of "theory of mind" in story comprehension. *Cognition*, **57**(2), 109–28.

Fox, N. A., Henderson, H. A., Marshall, P. J., et al. (2005a) Behavioral inhibition: linking biology and behavior within a developmental framework. *Annu Rev Psychol*, **56**, 235–62.

Fox, N. A., Nichols, K. E., Henderson, H. A., et al. (2005b) Evidence for a gene-environment interaction in predicting behavioral inhibition in middle childhood. *Psychol Sci*, **16**(12), 921–6.

Fox, N. A., Rubin, K. H., Calkins, S. D., et al. (1995) Frontal activation asymmetry and social competence at four years of age. *Child Dev*, **66**(6), 1770–84.

Fudge, J. L., Breitbart, M. A., Danish, M., and Pannoni, V. (2005) Insular and gustatory inputs to the caudal ventral striatum in primates. *J Comp Neurol*, **490**(2), 101–18.

Galvan, A., Hare, T. A., Parra, C. E., et al. (2006) Earlier development of the accumbens relative to orbitofrontal cortex might underlie risk-taking behavior in adolescents. *J Neurosci*, **26**(25), 6885–92.

Goldman, D., Oroszi, G., and Ducci, F. (2005) The genetics of addictions: uncovering the genes. *Nat Rev Genet*, **6**(7), 521–32.

Grace, A. A. (1991) Phasic versus tonic dopamine release and the modulation of dopamine system responsivity: a hypothesis for the etiology of schizophrenia. *Neuroscience*, **41**, 1–24.

Grace, A. A., Floresco, S. B., Goto, Y., and Lodge, D. J. (2007) Regulation of firing of dopaminergic neurons and control of goal-directed behaviors. *Trends Neurosci*, **30**(5), 220–7.

Guyer, A. E., Nelson, E. E., Perez-Edgar, K., et al. (2006) Striatal functional alteration in adolescents characterized by early childhood behavioral inhibition. *J Neurosci*, **26**(24), 6399–405.

Haber, S. N., Fudge, J. L., and McFarland, N. R. (2000) Striatonigrostriatal pathways in primates form an ascending spiral from the shell to the dorsolateral striatum. *J Neurosci*, **20**(6), 2369–82.

Haber, S. N., Kim, K. S., Mailly, P., and Calzavara, R. (2006) Reward-related cortical inputs define a large striatal region in primates that interface with associative cortical connections, providing a substrate for incentive-based learning. *J Neurosci*, **26**(32), 8368–76.

Hyman, S. E., Malenka, R. C., and Nestler, E. J. (2006) Neural mechanisms of addiction: the role of reward-related learning and memory. *Annu Rev Neurosci*, **29**, 565–98.

Jones, W., Bellugi, U., Lai, Z., et al. (2000) II. Hypersociability in Williams Syndrome. *J Cogn Neurosci*, **12** [Suppl 1], 30–46.

Kelley, A. E. (2004) Ventral striatal control of appetitive motivation: role in ingestive behavior and reward-related learning. *Neurosci Biobehav Rev*, **27**(8), 765–76.

Kim, J. J. and Jung, M. W. (2006) Neural circuits and mechanisms involved in Pavlovian fear conditioning: a critical review. *Neurosci Biobehav Rev*, **30**(2), 188–202.

Koob, G. F. (2006) The neurobiology of addiction: a neuroadaptational view relevant for diagnosis. *Addiction*, **101** [Suppl 1], 23–30.

Krain, A. L., Gotimer, K., Hefton, S., et al. (2007) A functional magnetic resonance imaging investigation of uncertainty in adolescents with anxiety disorders. *Biol Psychiatry*, **63**(6), 563–8.

LeDoux, J. E. (2000) Emotion circuits in the brain. *Annu Rev Neurosci*, **23**, 155–84.

Leyfer, O. T., Woodruff-Borden, J., Klein-Tasman, B. P., et al. (2006) Prevalence of psychiatric disorders in 4 to 16-year-olds with Williams syndrome. *Am J Med Genet B Neuropsychiatr Genet*, **141**(6), 615–22.

Liu, X., Powell, D. K., Wang, H., et al. (2007) Functional dissociation in frontal and striatal areas for processing of positive and negative reward information. *J Neurosci*, **27**(17), 4587–97.

Martin-Soelch, C., Linthicum, J., and Ernst, M. (2007) Appetitive conditioning: neural bases and implications for psychopathology. *Neurosci Biobehav Rev*, **31**(3), 426–40.

May, J. C., Delgado, M. R., Dahl, R. E., et al. (2004) Event-related functional magnetic resonance imaging of reward-related brain circuitry in children and adolescents. *Biol Psychiatry*, **55**(4), 359–66.

Meyer-Lindenberg, A., Mervis, C. B., and Berman, K. F. (2006) Neural mechanisms in Williams syndrome: a unique window to genetic influences on cognition and behaviour. *Nat Rev Neurosci*, **7**(5), 380–93.

Miller, E. K. and Cohen, J. D. (2001) An integrative theory of prefrontal cortex function. *Annu Rev Neurosci*, **24**, 167–202.

O'Doherty, J. P. (2007) Lights, camembert, action! The role of human orbitofrontal cortex in encoding stimuli, rewards and choices. *Ann N Y Acad Sci*, **1121**, 254–72.

Ochsner, K. N. and Gross, J. J. (2005) The cognitive control of emotion. *Trends Cogn Sci*, **9**(5), 242–9.

Ochsner, K. N., Ray, R. D., Cooper, J. C., et al. (2004) For better or for worse: neural systems supporting the cognitive down- and up-regulation of negative emotion. *NeuroImage*, **23**(2), 483–99.

Overman, W. H. (2004) Sex differences in early childhood, adolescence, and adulthood on cognitive tasks that rely on orbital prefrontal cortex. *Brain Cogn*, **55**(1), 134–47.

Parkinson, J. A., Olmstead, M. C., Burns, L. H., et al. (1999) Dissociation in effects of lesions of the nucleus accumbens core and shell on appetitive Pavlovian approach behavior and the potentiation of conditioned reinforcement and locomotor activity by D-amphetamine. *J Neurosci*, **19**(6), 2401–11.

Perez-Edgar, K., Roberson-Nay, R., Hardin, M. G., et al. (2007) Attention alters neural responses to evocative faces in behaviorally inhibited adolescents. *NeuroImage*, **35**(4), 1538–46.

Reynolds, S. M. and Berridge, K. C. (2001) Fear and feeding in the nucleus accumbens shell: rostrocaudal segregation of GABA-elicited defensive behavior versus eating behavior. *J Neurosci*, **21**(9), 3261–70.

Reynolds, S. M. and Berridge, K. C. (2002) Positive and negative motivation in nucleus accumbens shell: bivalent rostrocaudal gradients for GABA-elicited eating, taste "liking"/"disliking" reactions, place preference/avoidance, and fear. *J Neurosci*, **22**(16), 7308–20.

Robledo, P., Robbins, T. W., and Everitt, B. J. (1996) Effects of excitotoxic lesions of the central amygdaloid nucleus on the potentiation of reward-related stimuli by intra-accumbens amphetamine. *Behav Neurosci*, **110**(5), 981–90.

Rolls, E. T. (2004) The functions of the orbitofrontal cortex. *Brain Cogn*, **55**(1), 11–29.

Schmidt, L. A. and Fox, N. A. (2002) Molecular genetics of temperamental differences in children. In: Benjamin, J., Ebstein R. P., and Belmaker, R. H. (eds.) *Molecular Genetics and the Human Personality*. Washington, DC: American Psychiatric Publishing, Inc., pp. 245–256.

Schoenbaum, G., Saddoris, M. P., and Stalnaker, T. A. (2007) Reconciling the roles of orbitofrontal cortex in reversal learning and the encoding of outcome expectancies. *Ann N Y Acad Sci*, **1121**, 320–35.

Schultz, W., Dayan, P., and Montague, P. R. (1997) A neural substrate of prediction and reward. *Science*, **275**(5306), 1593–9.

Seymour, B., Daw, N., Dayan, P., et al. (2007) Differential encoding of losses and gains in the human striatum. *J Neurosci*, **27**(18), 4826–31.

Spear, L. P. (2000) The adolescent brain and age-related behavioral manifestations. *Neurosci Biobehav Rev*, **24**(4), 417–63.

van Leijenhorst, L., Crone, E. A., and Bunge, S. A. (2006) Neural correlates of developmental differences in risk estimation and feedback processing. *Neuropsychologia*, **44**(11), 2158–70.

Volkow, N. D., Wang, G. J., Ma, Y., et al. (2005) Activation of orbital and medial prefrontal cortex by methylphenidate in cocaine-addicted subjects but not in controls: relevance to addiction. *J Neurosci*, **25**(15), 3932–9.

Yeung, N., Cohen, J. D., and Botvinick, M. M. (2004) The neural basis of error detection: conflict monitoring and the error-related negativity. *Psychol Rev*, **111**(4), 931–59.

5

Charting brain mechanisms for the development of social cognition

Kevin A. Pelphrey and Susan B. Perlman

Introduction

In this chapter, we use the term "social cognition" to refer to the fundamental abilities to perceive, categorize, remember, analyze, reason with and behave toward other conspecifics (Adolphs, 2001; Nelson et al., 2005; Pelphrey et al., 2004). This definition is wide-ranging, so as to emphasize the multidisciplinary quality of work in this area. However, scientific disciplines vary in their emphasis on different aspects of this multifaceted construct. In social psychology, social cognition describes a range of phenomena including moral reasoning, attitude formation, stereotyping, and related topics (Kunda, 1999). In neuroscience, social cognition is defined more narrowly as the ability to perceive the intentions and dispositions of others (Brothers, 1990). In developmental psychology, the characterization of social cognition has focused most frequently upon the study of "theory of mind" (ToM), the awareness that other people have beliefs and desires different from our own and that behavior can be explained by reference to these beliefs and desires (Frith and Frith, 1999; Premack and Woodruff, 1978). Across disciplines, definitions of social cognition commonly link this construct to social behavior and include social perception (the initial stages of evaluating intentions and dispositions of others by analysis of gaze direction, body movement, and other types of biological motion[1]), and attributional style (the way one tends to explain other people's behavior).

Much of the neural circuitry thought to support social cognition consists of mechanisms that are relatively old in evolutionary terms. The world in which the human brain carries out its functions, however, has changed dramatically over a relatively short period. For example, a child born 50 000 years ago would have very similar mental abilities as a child today, but a decidedly different set of demands, aspirations, and opportunities, not to mention rights, responsibilities, chances of survival, and definitions of success. Nonetheless, the extent to which social cognitive processes functioned successfully helped to determine the fate of individual humans in the past and continues to do so today. For much of the twentieth century, psychologists produced theories and a wealth of empirical evidence concerning the basic building blocks of social

---

[1] Biological motion refers to the visual perception of an entity engaged in a recognizable activity. This definition includes the observation of humans walking and making eye and mouth movements, but the term can also refer to the visual system's ability to recover information about another's motion from sparse input. The latter is well illustrated by the discovery that point-light displays (moving images created by placing lights on the major joints of a walking person and filming them in the dark), while being relatively impoverished stimuli, contain the information necessary to identify the agent of motion and the kind of motion produced by the agent (Johansson, 1973).

Neuroimaging in Developmental Clinical Neuroscience, eds. Judith M. Rumsey and Monique Ernst. Published by Cambridge University Press. © Cambridge University Press 2009.

cognition. This understanding has now begun to benefit from the use of non-invasive brain imaging techniques, such as functional magnetic resonance imaging (fMRI), to identify a network of brain regions supporting various facets of social cognition.

In the past decade, fMRI studies of how we think about our and others' minds, how we mimic others, change our attitudes, and how we perceive emotions have allowed us to learn about the paths toward better and worse mental health outcomes. However, much of this work has focused on the mature minds of adult humans. Few studies have examined brain mechanisms for social cognition in children. At the same time, within developmental psychology, behavioral research has examined the age-related development of social cognition, but often neglected the goal of identifying mechanisms of change (Siegler, 2000). Regrettably, few studies have addressed the crucial question of the processes and mechanisms by which social cognition develops, leaving this area wide open to investigation.

Advances in techniques for imaging the developing brain are now providing exciting opportunities for studying the brain mechanisms involved in the development of social cognition abilities, opening up a new field of research called *developmental social cognitive neuroscience*. Work in this emerging field stems from direct collaboration among cognitive neuroscientists, developmental psychologists, and social psychologists. In this chapter, we review studies that have examined the development of neural mechanisms for selected social cognitive processes. In particular, we focus on those studies that have employed fMRI to study the brain mechanisms involved in the development of social perception, emotion understanding, and ToM. To provide a context for these neuroscientific studies, we begin by outlining selected studies from a large body of rich and elegant behavioral research describing the developmental trajectories of these key facets of social cognition. A second section reviews the brain mechanisms studied in adult social cognition. Finally, we review studies of the developing social brain and present (Table 5.1) neuroimaging studies that have investigated social

perception, face processing, emotion understanding, and ToM in typically developing children.

Development of social cognition

Understanding others' goals and intentions

Social perception is one of the earliest developing aspects of social cognition. Very young infants are sensitive to the goal-directed structure of other people's actions, as well as their direction of gaze. Within the first 6 months of life, infants develop a basic understanding of the intentions and goal-directed structure underlying biological motion (Sommerville et al., 2005; Woodward, 1998). For example, Woodward (1998) habituated 5- to 8-month-old infants to a reaching and grasping human hand. The infants looked longer when the object of the motion changed (intentional aspect) than when the direction of the reaching changed (physical aspect), suggesting an awareness and attribution of goals. Other research indicates that the infant's attribution of goals is not restricted to human actors, but includes non-human, machine-like agents capable of mimicking biological (animate) movement (Gergely and Csibra, 1997; Johnson et al., 2001).

By 18 months, toddlers distinguish between the intended goal of an action and its accidental consequences and become more discerning with regard to their attributions of intention (Carpenter et al., 1998; Meltzoff, 1995). Meltzoff (1995) recorded the reactions of 18-month-olds to an experimenter's actions upon a novel toy. The experimenter attempted to pull off the end of a dumbbell-shaped toy, but failed when his hand slipped. When encouraged to imitate, the infants typically imitated the intended, rather than the observed, action. In contrast, when a non-human apparatus, consisting of two arm-like parts, attempted the same action and goal with the toy, yet failed in the same fashion as the experimenter, the children imitated the observed action rather than the desired goal. The author argued that the children were able to attribute the behavior of the human to deeper goals and intentions.

Table 5.1 Neuroimaging studies of social cognition in children

Author	Year	Age range	ROI	Task	Major findings
Development of facial perception					
Tzourio-Mazoyer et al.	2002	2-month-old infants $n = 6$	FFG (PET study)	Viewing faces and shapes	Specialization for faces may be apparent in early development
Taylor et al.	1999	4–14 years $n = 53$	FFG (ERP study)	Viewing faces, cars, scrambled faces, scrambled cars, and butterflies	Neural basis for face processing matures gradually
Taylor et al.	2001	4–15 years $n = 128$	FFG (ERP study)	Viewing faces and eyes alone	Development of processing of the eyes occurs more rapidly than that of the entire face
Scherf et al.	2007	6–14 years $n = 10$	FFA, OFA, STS	Viewing of movie clips involving faces, places, and objects	Activation to faces is not localized in children as it is in adults. However, it is localized for objects and places
Golijeh et al.	2007	7–16 years $n = 30$	FFA, PPA	One-back task while viewing faces, scenery, abstract sculpture, and textures presented with a recognition task outside of scanner	Volume of activation in the FFA and PPA to faces and places increases with age and correlates with improved recognition memory
Alyward et al.	2005	8–14 years $n = 18$	FFG	Viewing neutral upright and inverted faces and houses	Older children show more activation to faces in the FFG than younger children
Development of social perception and theory of mind					
Carter and Pelphrey	2007	7–10 years $n = 9$	STS	Viewing animated characters portraying biological and non-biological motion	Domain specificity to biological motion in the STS increases with age
Pfeifer et al.	2007	9–10 years $n = 12$	MPFC	Rating of personality traits for self and a fictional character	There is greater MPFC activity during self-knowledge retrieval in children as compared to adults
Moriguchi et al.	2007	9–16 years $n = 16$	MPFC	Perception of animated geometric figures	There are significant correlations between age and MPFC activation in a theory of mind task
Ohnishi et al.	2004	10–12 years $n = 11$	Mirror neuron system	Perception of object-directed hand movements and moving geometric figures	Children have the same neural networks for the mirror system as adults

Table 5.1 (*cont.*)

Author	Year	Age range	ROI	Task	Major findings
Development of emotion					
Mosconi et al.	2005	7–10 years $n = 8$	STS	Viewing animated character shifting gaze toward or away from moving object	The posterior STS region is implicated in processing of eye gaze shifts in children as young as 7 years
Killgore and Yurgelun-Todd	2004	8–16 years $n = 19$	AMY, PFC	Viewing facial expression of fear versus neutral expressions	Male and female children differ in asymmetry of activation of the AMY and PFC across age
Monk et al.	2003	9–17 years $n = 17$	AMY, ACC, OFC	Rating of emotional facial expressions for subjective fear or width of nose	Adolescents show more activation in fear circuitry in response to fearful faces than do adults
Nelson et al.	2003	9–17 years $n = 17$	ACC, temporal pole, hippocampus	Viewing emotional facial expressions followed by a surprise memory test	Adolescents display more activity in the ACC when viewing previously seen angry faces and more activity in the right temporal pole when viewing previously seen fearful faces relative to adults. Adults display more activity in the ACC for previously viewed happy faces and the posterior hippocampus for neutral faces
McClure et al.	2004	9–17 years $n = 17$	AMY, OFC, ACC	Rating emotional faces for subjective fear and threat, and nose width	There was no difference in activation to a threatening facial expression between adolescent males and females. However, sex differences were present in adults
Lobaugh et al.	2006	10 years $n = 10$	AMY, insula, ACC, FFG, STG	Viewing facial expressions of fear, disgust, and sadness as well as neutral faces	Facial expressions of fear, disgust, and sadness recruit distinct neural systems in 10-year-old children as in adults

Thomas et al.	2001	11 years $n = 11$	AMY	Viewing fearful and neutral faces	Children show greater activation to neutral than to fearful faces
Killgore et al.	2001	11–18 years $n = 19$	AMY	Viewing fearful and neutral faces	Left AMY activity decreases over adolescence in females, but not in males
Baird et al.	1999	12–17 years $n = 12$	AMY	Recognition test of emotional facial expressions	The AMY activates to fearful faces in adolescents

Notes:

ACC, anterior cingulate; AMY, amygdala; ERP, event-related potentials; FFA, fusiform face area; FFG, fusiform gyrus; MPFC, medial prefrontal cortex; OFA, occipital facial area; OFC, orbital frontal cortex; PET, positron emission tomography; PFC, prefrontal cortex; PPA, parahippocampal place area, ROI, region of interest; STG, superior temporal gyrus; STS, superior temporal sulcus region.

Perceiving eye gaze: linking actions and intentions

In addition to appearing to take an "intentional stance" (Dennet, 1987) toward the analysis of hand and arm movements, over the first year of life infants begin to link goals and intentions to eye gaze. A referential concept of gaze (i.e., the understanding that perception depends on a relationship between a person and an object) emerges toward the end of the first year (Brooks and Meltzoff, 2002; Caron et al., 2002, but see Doherty and Anderson, 1999). Brooks and colleagues (described in Caron et al., 2002) adapted Woodward's (1998) paradigm to measure referential understanding of perception. When 14-month-old infants were habituated to an adult turning toward an object, they looked longer when the adult later looked to the same location, rather than to the new location when the object was moved, indicating possible surprise that the adult was no longer pursuing the goal object. This effect, however, only occurred if the actor's eyes were open during habituation, suggesting that the infants understood that looking, but not necessarily direction of movement, is goal-directed.

Within the vast repertoire of human behavior, gaze is among the most potent of social cues, with mutual gaze often signaling threat or approach and averted gaze conveying submission or avoidance (e.g., Argyle and Cook, 1976; Darwin, 1872; Kleinke, 1986). Processing of eye gaze is fundamental to social interactions, as illustrated by its early emergence in ontogeny and disruption in autism (e.g., Hood et al., 1998; Pelphrey et al., 2002). Infants as young as 7 weeks attend to the eyes of faces during social interaction (Haith et al., 1977) and express a preference for faces with open eyes by 3 months of age (Bakti et al., 2000). These findings suggest a rudimentary appreciation of the significance of eyes vis-à-vis their social value. Eye contact and gaze following are not only key components of social perception, but they modulate other cognitive processes during infancy and later development. For instance, Hood and colleagues (2003) asked both adults and 6- to 7-year-old children to memorize faces from photographs that were making direct eye contact with the viewer or gazing in other directions. During a test phase in which the faces were presented with closed eyes, both adults and children were better able to identify individuals who displayed direct gaze in the original pictures. This study illustrates the importance of gaze in basic face perception and, possibly, the beginnings of social interaction. Eye contact in the context of joint attention has also been linked prospectively to ToM development. Charman and colleagues (2000) tested 20-month-old infants on various joint attention tasks relating to interaction with both adults and a moving toy. At 3.5 years of age, children who had previously demonstrated more advanced joint attention abilities scored higher on a standard false belief task, underscoring an early link between basic building blocks of social cognition and later ToM.

Perceiving emotions

In addition to parsing the social world according to the analysis of agents, their actions, goals, and intentions, individuals use emotion as an important source of information about others. Newborn infants show sensitivity to and synchrony with others' emotions. Field and Walden (1982a) observed that neonates cry when other infants cry and discriminate between happy and sad facial expressions. Neonates (as young as 36 h) also display some discrimination and imitation of facial expression. Field and colleagues (1982) observed the facial expressions of newborns in response to models' discrete emotional expressions. Independent observers were able to determine the model's posed emotion from the facial display of the infant, who imitated that expression. Similarly, Meltzoff and Moore (1983) observed young infants' imitations of modeled facial expressions and suggested that this early form of imitation might provide a basis for emotion perception. In addition, Nelson and colleagues (1979) found evidence for early discrimination of fearful and happy faces. Further studies confirm that infants can reliably distinguish facial expressions in the second half of their first year of life (Caron et al., 1988; Nelson, 1987).

The development of facial expression recognition and understanding of the situational determinants of emotion continues to develop throughout early childhood. For example, 3-year-olds are able to match drawings of people with the same facial expression to each other (Walden and Field, 1982) and produce a facial expression to match a verbal label that can be easily guessed by adults (Field and Walden, 1982b). Children as young as 2 years are able to assign a verbal emotion label to an emotional expression. At 3 years of age, they are consistently able to attach the correct emotional facial expression to a faceless puppet that re-enacts an emotional story (Denham, 1986). Finally, Camras and Allison (1985) reported that preschool children are able to match emotions with stories both verbally, by naming an emotion, and non-verbally, by pointing to facial expression in a photograph, with great (over 80%) accuracy. This skill continues to improve well into middle childhood. All of these studies of emotion perception report variability related to specific emotions, with happiness being the easiest and disgust and surprise being the most difficult to identify, enact, and match.

Aside from discriminating emotions conveyed by facial expressions, infants use this information to guide their own actions, a skill referred to as "social referencing" (Feinman, 1992). Sorce and colleagues (1985) reported that 1-year-old infants used such emotional information to resolve perceived ambiguities in the environment. In a now classic experiment, infants looked to their mother when prompted by the experimenter to cross to the deep side of a visual cliff. If the mother posed a happy facial expression, infants were more likely to cross than if the mother posed a fearful or angry facial expression. Finally social referencing to emotional cues can occur in both the visual and auditory domains. For instance, Mumme and others (1996) reported that fearful maternal vocalizations were sufficient to regulate 1-year-old infants' behavior without the accompanying facial expression. Strikingly, 5-month-old infants have been found to respond to negative speech intonations even outside their native language and to use this emotional information to direct their behavior (Fernald, 1993).

By 14 months, infants can combine information about a person's gaze and emotion both to infer the person's goal and to direct their own actions. For instance, infants observed an actor looking, smiling, and positively vocalizing toward one of two objects and later observed the actor holding that same object or the second object. Infants looked longer in trials where the adult held the second object, suggesting surprise that the actor's gaze and expressed positive regard toward a particular object did not influence their later action (Phillips et al., 2002). Finally, Moses and colleagues (2001) showed that infants are able to assimilate knowledge of gaze direction and the referent of an emotional message. When an actor made a negatively valenced noise, 12- and 18-month-olds looked up from the novel object in their hand to follow the gaze of the actor. Later, the infants avoided the object associated with the negative feelings of the actor.

Representing other people's beliefs

Theory of mind (ToM) refers to our abilities to make inferences about the beliefs of others and to understand that individuals act according to those beliefs (Wimmer and Perner, 1983). The prerequisite for ToM is the awareness that others' thoughts, beliefs, and motivation differ from our own and can explain their behavior (Frith and Frith, 1999). Within developmental psychology, children are typically credited with possessing a mature ToM when they understand that a person's beliefs motivate his or her actions (even when those beliefs conflict with reality) (Wellman et al., 2001). While sometimes discussed as an all-or-nothing construct, ToM follows a developmental course over the first 5 years that includes apparent transitions and intermediate steps. Most accounts of the development of ToM highlight the transition between the ages of 3 and 4 years, observed by the passing of tests involving object transfer problems, with most typically developing children achieving full competency in ToM by age 5 years.

The "False Belief" task has been the most widely used test of children's ToM (Wimmer and Perner, 1983). In the now classic version of this task (the "object transfer" problem), the child listens to a story in which a character's belief about the location of an object is falsified when it is moved unbeknownst to the character. Children who state that the character believes that the object is in its original location pass the test. Dozens of versions of the standard False Belief problem have now been used, and while the precise age of success varies between children and between task versions (Wellman et al., 2001), children younger than 3 or 4 years generally fail, while older children generally pass.

With slight modifications, younger children can demonstrate a grasp of ToM problems if not required to give a verbal explanation (Clements and Perner, 1994; Garnham and Ruffman, 2001). Moreover, Onishi and Baillargeon (2005) have shown that infants as young as 15 months may possess some false belief understanding. Infants viewed an actor playing with a toy and then depositing it into a green box. As a partition shielded the actor's view, the toy was moved from the green box to a neighboring yellow box. When the scene once again became visible to the actor, infants looked longer when the actor reached for the toy in the yellow box (an unexpected event), than when the actor reached to the green box. The authors interpreted this as evidence that 15-month-old infants possess a rudimentary ToM.

Brain mechanisms for social cognition

Our abilities to understand and anticipate the actions, intentions, emotions, thoughts, and beliefs of others have long been the subject of psychological and philosophical inquiry. More recently, cognitive neuroscientists have begun to identify the neural circuitry that supports these social abilities. In the first extensive review of the literature on non-human primates, Brothers (1990) proposed that a range of social functions engages three neuroanatomical

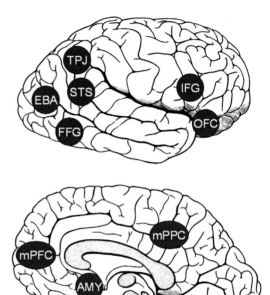

Figure 5.1. Brain regions that have been implicated in various aspects of social cognition. AMY, amygdala; EBA, extrastriate body area; FFG, fusiform gyrus; IFG, inferior frontal gyrus; mPFC, medial prefrontal cortex; mPPC, medial parietal cortex; OFC, orbital frontal cortex; STS, superior temporal sulcus region; TPJ, temporal parietal junction.

regions: the superior temporal sulcus (STS), the amygdala (AMY), and the orbitofrontal cortices (OFC). Over the past decade, non-invasive functional neuroimaging in adults has dramatically expanded our knowledge of the key cortical and subcortical regions implicated in social-cognitive processes, some of which are illustrated in Figure 5.1. Before focusing on the development of these regions, we discuss below some of what neuroimaging studies of adult humans have told us about the function of each region implicated in social cognition.

Social perception: biological motion, goal-directed action, gaze, and facial expressions

Located in the ventral occipitotemporal cortex, the lateral fusiform gyrus (FFG) containing the "fusiform face area" (FFA) is implicated in face perception and recognition (e.g., Kanwisher et al., 1997; Puce et al., 1996). The extrastriate body area (EBA) has been implicated in the visual perception of human bodies (e.g., Downing et al., 2001). The STS region, particularly the posterior STS, has been implicated in the interpretation of the actions and social intentions of others through the analysis of biological motion cues, including eye, hand, and other body movements (e.g., Bonda et al., 1996; Pelphrey et al., 2003a). Also related to the perception of biological motion, the "mirror neuron system" in humans is involved in both the execution of a motor action and the observation of a motor action performed by another person (e.g., Buccino et al., 2001; Rizzolatti et al., 1996). Finally, the AMY is highly interconnected with other cortical and subcortical brain structures and implicated in determining the emotional states of others through analysis of facial expressions (e.g., Morris et al., 1996). Beyond its role in the perception of facial emotion, the AMY plays a central and complex role in multiple aspects of emotion (e.g., Davis and Whalen, 2001; Kluver and Bucy, 1997; LeDoux, 2000).

Theory of mind and self cognition

Four additional cortical regions have been reliably implicated in certain advanced aspects of social cognition including ToM and self cognition. The temporo-parietal junction (TPJ) appears, in adults, to be particularly selective for reasoning about beliefs (e.g., Saxe and Kanwisher, 2003). Further, the medial prefrontal cortex (MPFC) is implicated in reasoning about others' beliefs, self-reflection (e.g., Kelley et al., 2002; Northoff et al., 2006; Saxe and Powell, 2006), and autobiographical memory (Shannon and Buckner, 2004). The MPFC has been shown to be activated by a wide range of tasks that require inferences concerning intentions and mental states (e.g., Castelli et al., 2002; Frith and Frith, 1999), the attribution of emotion to self and others (e.g., Ochsner et al., 2004), self-reflection generally (e.g., Kelley et al., 2002), and representations of

semantic knowledge about the psychological aspects of others (e.g., Mitchell et al., 2005). Additionally, the precuneus or posterior cingulate in the medial parietal cortex (MPPC) is more active during self-knowledge retrieval than during other types of social or semantic tasks (e.g., D'Argembeau et al., 2005).

Functional neuroimaging of the developing social brain

Mapping the development of brain mechanisms for social perception

While cognitive neuroscientists have generated a wealth of information regarding the brain regions involved in social perception and social cognition in the mature adult brain, only within the last few years has research begun to address the brain mechanisms involved in these domains in children using fMRI. These investigations have focused on the perception of faces, whole-body biological motion, and direction of gaze. Table 5.1 provides a summary of the major findings to date. It is noteworthy that no study in this table employed a longitudinal design and no study using fMRI included children younger than 6 years, reflecting the challenges involved in scanning young children (see Chapter 17).

The earliest developmental studies of brain regions involved in social cognition examined the role of the lateral FFG in face processing using fMRI and event-related potentials (ERPs). In an elegant series of studies, Taylor and colleagues (1999) found that the specialization of the N170 response to faces, thought to reflect face-specific structural encoding, continues to develop from early childhood into young adulthood. Notably, the response to eyes appears to mature more rapidly than responses to other facial features (Taylor et al., 2001), underscoring the precedence of gaze perception in face processing. In an fMRI study of face perception, Aylward and her colleagues (2005) found that 12- to 14-year-old children exhibited a greater blood-oxygen-level-dependent (BOLD) response in the FFG when viewing faces than did younger children (8–10 years old).

More recent studies have continued to examine the development of face-selective activity in the FFG in children of various ages. Scherf and colleagues (2007) compared the activity evoked by pictures of places, faces, and objects in 6- to 14-year-old children. They were unable to identify face-selective activation in the FFG of the younger children, but were able to localize the FFA in adolescents and adults. In contrast, the localization of brain activity evoked by pictures of places and objects was equivalent in adults and children. Subsequently, Golijeh and colleagues (2007) reported that children as young as 7 years had face-selective activity localized to the FFG, but the volume of activation in the FFA and parahippocampal gyrus to faces and places, respectively, increased across the 7- to 11-years-old period and correlated with improved recognition memory for the respective stimulus class. In contrast, when children viewed faces, a face-sensitive region of the posterior STS was equally active across the observed age range. The picture emerging from this line of research is one of different developmental pathways for the FFA versus the face-sensitive area of the posterior STS. The FFA appears to follow a more protracted developmental course compared to the STS region. The posterior STS region is thought to be involved in representing dynamic aspects of faces including emotional expressions and direction of eye gaze (Hoffman and Haxby, 2000; LaBar et al., 2002; McCarthy, 1999; Mosconi et al., 2005), while the FFA has been implicated in the structural encoding of a face as such (McCarthy, 1999) and in the recognition of identity (Hoffman and Haxby, 2000).

Our laboratory has examined the development of brain regions involved in representing the actions and intentions of others. In one study (Pelphrey et al., 2003a), we examined the degree to which the STS activation is selective for biological motion in children aged 7–10 years. We compared four different motion conditions: a walking man, a walking robot, a moving disjointed mechanical figure with the same components as the robot, and a grandfather clock with a swinging pendulum. The walking man and robot represent biological movement, while the mechanical figure and swinging pendulum represent non-biological motion. A network of brain regions was identified that exhibited greater responses to biological (robot and human walking) than to non-biological motion (disjointed mechanical figure and grandfather clock). Included in the posterior STS and portions of the purported human mirror neuron system were the inferior frontal gyri, the precentral gyri, and middle and superior frontal gyri (Carter and Pelphrey, 2007). Additionally, increasing specificity for biological motion with age was seen in the right posterior STS. The magnitude of the difference between the response to biological and nonbiological motion was positively correlated with age in the right posterior STS region ($r = 0.64$, $p < 0.05$).

Moving beyond the representation of whole-body biological motion to the question of how the developing brain represents the intentions of actions, we used an incongruent versus congruent gaze paradigm, illustrated in Figure 5.2, with children aged 7–10 years. A flashing checkerboard was shown on one side of an animated face, which either turned to look at (congruent) or away from (incongruent) the flashing object. Based on our prior findings in adults (Pelphrey et al., 2003b), we hypothesized that STS activity would differentiate congruent from incongruent trials, reflecting the ability of typically developing children to link the perception of the gaze shift with its mentalistic significance. Similar to adults, children showed greater activation of the posterior STS to congruent, relative to incongruent, eye gaze, reflecting sensitivity to intent (Mosconi et al., 2005). These findings suggest that the neural circuitry underlying the processing of eye gaze and the detection of intentions from gaze in children is functional by middle childhood and similar to that of adults.

Development of brain circuitry for emotion understanding

Several fMRI studies to date have focused on the biological basis of emotion processing, with

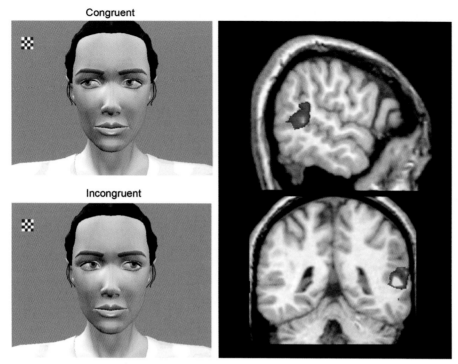

Figure 5.2. Brain activation in response to gaze shifts. Experiment to determine brain activation in response to expected and unexpected gaze shifts on the part of another person (left panel) and corresponding brain activation to biological motion (observed human movements; right panel). Incongruent trials evoked greater right hemisphere STS activity than did congruent trials, demonstrating the sensitivity of the STS region to the intentions conveyed by eye-gaze shifts. Reprinted with permission from Pelphrey and Morris (2006), courtesy of Blackwell Publishing.

a particular concentration on the AMY's response to emotional facial expressions. Baird and colleagues (1999) demonstrated AMY activation to fearful faces in adolescents, aged 12–17 years. Subsequently, it was reported that adults demonstrated greater AMY activation to fearful than to neutral facial expressions, whereas 11-year-old children showed greater AMY activation in response to neutral faces (Thomas et al., 2001). The authors argued that the neutral faces were experienced as more ambiguous than fearful facial expressions, thereby accounting for the increased AMY activation in response to the neutral faces. In addition, Killgore and others (2001) reported sex differences in AMY development in children and adolescents. Whereas the left AMY responded to fearful facial expressions in all children,

activity decreased over the adolescent period in females, but not in males. In addition, Lobaugh and colleagues (2006) found that, similar to adults (Phan et al., 2002), different patterns of activation for the processing of fearful, disgusted, and sad facial expressions were present in 10-year-old children, even in the absence of conscious processing. These circuits included the AMY, parahippocampal gyrus, insula and cingulate gyrus, the fusiform, and superior temporal gyri. Adolescents did, however, show more activation in the "fear circuit" comprising the AMY, orbitofrontal cortices, and the anterior cingulate cortex in response to fearful faces than did adults. Finally, Guyer and colleagues (2008) found that AMY and fusiform activation to fearful expressions was greater in adolescents than adults.

Furthermore, Guyer et al. observed greater functional connectivity between the AMY and hippocampus in adults compared to adolescents and speculated that this effect may indicate greater learning or habituation to fearful faces in adults. Eye movement data collected outside the scanner indicated that this difference was not attributable to differential scanpaths when viewing a fearful face.

Brain mechanisms for the development of theory of mind

Only a few studies have addressed the development of brain mechanisms supporting the emergence of ToM. Studying children, aged 9–16 years, Moriguchi et al. (2007) reported significant activation in the STS and MPFC during a ToM task. A significant positive correlation between age and activation in the dorsal MPFC and a significant negative correlation between age and the ventral MPFC were seen. No age-related changes were found in the STS. The authors argued that ToM-related activation in the MPFC shifts from ventral to dorsal during late childhood and adolescence and that this effect may be related to the maturation of the prefrontal cortex.

Pfeifer and colleagues (2007) conducted the first developmental study addressing the role of the MPFC in self and other reflection. In a clever design, children, aged 9–10, and adults were shown short phrases taken from standard self-esteem scales (e.g., I am popular) during fMRI scanning. Subjects were asked to rate how much the phrase described them and how much the phrase described the fictional character Harry Potter. Both children and adults activated the MPFC more during self, as compared to other, knowledge retrieval. Children activated the MPFC to a higher degree than adults while reflecting upon the self, possibly due to a lack of automaticity in the children's self-knowledge retrieval process.

When considering the behavioral literature on the development of social cognition relative to the fMRI studies of the brain mechanisms supporting social cognitive processes (particularly those

related to ToM) in children, one notices an obvious disconnect between the ages of the children studied in the two lines of research. Most of the behavioral findings focus on transitions occurring from birth to age 5 years. In contrast, the fMRI studies generally begin at age 6 years. This disconnect reflects methodological hurdles. Continued development of more "infant-friendly" neuroimaging techniques such as near infrared spectroscopy (NIRS) and advances in the use of fMRI for younger children will allow researchers to begin to directly study changes in brain function occurring before, during, and after key developmental transitions in social cognitive abilities. Currently, the study of resting levels of functional connectivity between social brain structures (identified in studies of adults; Greicius et al., 2003) in infant and toddlers is a promising approach. These measurements can be taken while children sleep inside of the scanner.

Age-related differences in findings reported in behavioral versus imaging studies also raise important theoretical issues. Several of the neuroimaging studies reviewed here have exposed progressive changes across childhood and adolescence in the functioning of social brain structures, with a trend toward increasing specialization of function for various social cognitive abilities with age. These neurofunctional changes are occurring later than many of the well-documented developmental transitions in social cognitive abilities. The changes in social brain function occurring in children from age 6 years onwards cannot logically be the neural basis for earlier occurring behavioral changes.

What might account for the differences in age-related changes related to ToM reported in behavioral versus neuroimaging studies? The developmental transitions reported in behavioral studies at younger ages are likely to be related to as yet undiscovered earlier changes in the structures that comprise the social brain. The structures involved in the acquisition of social cognition abilities may differ from those that ultimately come to serve and elaborate those functions. For instance, early in development, a nascent ability could be subserved via one set of mechanisms. Then, with further brain maturation, the growing ability might come to depend on another set of mechanisms. For example, we might expect to observe subcortical to cortical shifts in the localization of function. We might also predict that functions become increasingly shifted toward heteromodal, tertiary association cortices. Importantly, this possibility highlights the potential role of the child's own activity in shaping the development of specialization in particular brain structures. On the whole, developmental psychologists tend to focus on the earliest emergence of particular cognitive abilities. Later, more subtle developmental changes (e.g., integration with language abilities, and integration with executive function systems involved in planning and action monitoring) will be important to understand in relationship to neuroimaging findings. Neuroimaging studies might actually serve to highlight important transition points for novel behavioral studies.

To illustrate this last point, Saxe and colleagues (personal communication) presented 6- to 11-year-old children with short stories and illustrations during fMRI scanning. Within each story were subsections describing physical facts, the story character's appearance and social relations, or the character's mental states. Similar to adults, the right TPJ was most active while listening to descriptions of others' mental states. Further, in the youngest children, the right TPJ was recruited equally for any information about a person, while in the older children, as in adults, the right TPJ was recruited only for descriptions of mental states. Developmental studies of ToM have historically focused on children between ages 3 and 5, when they first pass the paradigmatic False Belief task. These new results, by contrast, suggest that the neural organization underlying ToM is still changing 5 years later. Future work will be necessary to determine the possible behavioral correlates of this late neural specialization. It is also conceivable that the neural substrates change in the absence of noticeable behavioral change, or that behaviors simply become more automatic (but do not change qualitatively) with associated neural changes.

Functional connectivity as a mechanism of development

Models of the social brain frequently emphasize the unique contributions of specific neuroanatomical regions. While this analytical perspective is helpful in providing a framework for organizing our emerging understanding of social brain development, it fails to capture the complexity of interactions among these and other brain regions. These are better understood as nodes in a network of regions subserving different aspects of social cognition. To date, these components have been studied extensively in adults and occasionally in older children and adolescents, but their integration has been relatively neglected in both age groups. More critically, the role of changes in functional connectivity over development as a mechanism underlying ontogenetic changes in social cognition has only begun to be examined (Guyer et al., 2008). Recent developments in techniques for studying functional connectivity (e.g., Just et al., 2004; Meyer-Lindenberg et al., 2005) now allow neuroscientists to move toward brain-based, mechanistic theories of social cognitive development.

In Figure 5.3, we propose a preliminary model for thinking about the development of social cognition and its neural basis from an "interacting, developing brain systems" perspective. The aspects of social cognition include those most frequently studied in children and/or adults using fMRI. As illustrated by the left-to-right stair-step-like organization, we think of each component as an increasingly sophisticated aspect of social cognition. The horizontal arrows originating from each construct represent the continued development of each along its own trajectory from infancy through adolescence that serves as a building block for, and is influenced by, more sophisticated aspects of social cognition. The model predicts that increasingly sophisticated forms of social cognition arise as a result of the enhanced efficiency or connectivity within brain networks involved in various components of social cognition. Contributing

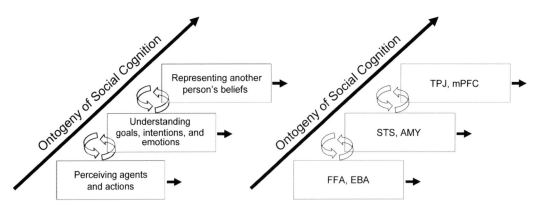

Figure 5.3. Model of the development of social cognition as driven by increases in functional connectivity. The rectangles on the left represent aspects of social cognition most frequently studied to date using functional magnetic resonance imaging (fMRI). The rectangles on the right indicate the brain regions onto which the constructs on the left loosely map. The stairstep organization reflects an increasing developmental sophistication, with the lower boxes providing the supporting building blocks for those above. The horizontal and circular arrows originating from each box indicate continued development from infancy through adolescence, whereby the processes and their neural substrates both support the development of higher-level processes (bottom-up) while being influenced by them (top-down). AMY, amygdala; EBA, extrastriate body area; FFA, fusiform face area; mPFC, medial prefrontal cortex; STS, superior temporal sulcus; TPJ, temporo-parietal junction.

neurobiological mechanisms may include myelination and synaptic pruning.

Conclusion: the value of a developmental perspective on the social brain

At present, little is known about the neural correlates of social cognition in children or about the changes in brain function that underlie normative development in this domain. Fundamental questions concerning brain maturation in relationship to changes in social cognition remain unanswered. But the future is bright. Developmental fMRI studies will allow the field to construct normative developmental curves for the functioning of circuits supporting different aspects of social cognition. The availability of these normative data will facilitate efforts to characterize atypical developmental pathways and may improve our ability to develop more effective interventions for social deficits.

Functional brain correlates or neurobiological markers of social cognition may prove useful in the early identification of children at risk for difficulties in social information processing. The STS region has emerged as a region that is quite promising in this regard (Pelphrey et al., 2005). To the extent that this research can elucidate developmental trajectories of the neural circuitry supporting pivotal, early social cognitive abilities, it can inform the design of more effective programs for the identification and remediation of children at risk for difficulties in these areas. Further, functional neuroimaging techniques may provide a means for assessing the efficacy of treatment and reveal whether behavioral improvements correspond to compensatory changes in brain function or the normalization of developmental pathways.

Finally, by defining brain phenotypes based on neurofunctional activation patterns, fMRI studies of children hold potential for dissecting the heterogeneity present in neurodevelopmental disorders such as autism. Early and longitudinal study will be critical in defining brain phenotypes because developmental trajectories of brain functioning are expected to be more informative than the analysis of brain phenotypes in adults. As a case in point, imaging studies have provided noteworthy findings regarding correlations between structural developmental trajectories and level of general intelligence in older children and adolescents. Shaw and colleagues (2006) reported that the pattern of age-related change in cortical thickness (rather than that at any one time point) was closely related to level of intelligence. Knowledge of the development of the "social brain" may aid researchers in their search for genetic and possible environmental factors related to suboptimal social cognitive development.

ACKNOWLEDGMENTS

Kevin Pelphrey is supported by a Career Development Award from the National Institutes of Health, NIMH Grant MH071284 and by an award from the John Merck Scholars Fund. We gratefully acknowledge our collaborators, especially Gregory McCarthy, Elizabeth Carter, James Morris, and Truett Allison.

REFERENCES

Adolphs, R. (2001) The neurobiology of social cognition. *Curr Opin Neurobiol*, **2**(1), 231–9.

Argyle, M. and Cook, M. (1976) *Gaze and Mutual Gaze.* Cambridge, England: Cambridge University Press, Cambridge, 1976.

Aylward, E. H., Park, J. E., Field, K. M., et al. (2005) Brain activation during face perception: evidence of a developmental change. *J Cogn Neurosci*, **17**(2), 308–19.

Baird, A. A., Gruber, S. A., Fein, D. A., et al. (1999) Functional magnetic resonance imaging of facial affect recognition in children and adolescents. *J Am Acad Child Adolesc Psychiatry*, **38**(2), 195–9.

Bakti, A., Baron-Cohen, S., Wheelwright, A., et al. (2000) Is there an innate gaze module? Evidence from human neonates. *Infant Behav Dev*, **23**, 223–9.

Bonda, E., Petrides, M., Ostry, D., et al. (1996) Specific involvement of human parietal systems and the amygdala in the perception of biological motion. *J Neurosci*, **16**(11), 3737–44.

Brooks, R. and Meltzoff, A. N. (2002) The importance of eyes: how infants interpret adult looking behavior. *Dev Psychol*, **38**(6), 958–66.

Brothers, L. (1990) The neural basis of primate social communication. *Motiv Emot*, **14**(2), 81–91.

Buccino, G., Binkofski, F., Fink, G. R., et al. (2001) Action observation activates premotor and parietal areas in a somatotopic manner: an fMRI study. *Eur J Neurosci*, **13**(2), 400–4.

Camras, L. A. and Allison, K. (1985) Children's understanding of emotional facial expressions and verbal labels. *J Nonverbal Behav*, **9**(2), 84–94.

Caron, A. J., Butler, S. C., and Brooks, R. (2002) Gaze following at 12 and 14 months: do the eyes matter? *Br J Dev Psychol*, **20**(2), 225–39.

Caron, A. J., Caron, R. F., and MacLean, D. J. (1988) Infant discrimination of naturalistic emotional expressions: the role of face and voice. *Child Dev*, **59**(3), 604–16.

Carpenter, M., Akhtar, N., and Tomasello, M. (1998) Fourteen-through 18-month-old infants differentially imitate intentional and accidental actions. *Infant Behav Dev*, **21**(2), 315–30.

Carter, E. J. and Pelphrey, K. A. (2007) School-aged children exhibit domain-specific responses to biological motion. *Soc Neurosci*, **1**(3–4), 396–411.

Castelli, F., Frith, C., Happe, F., et al. (2002) Autism, Asperger syndrome and brain mechanisms for the attribution of mental states to animated shapes. *Brain*, **125**(8), 1839–49.

Charman, T., Baron-Cohen, S., Swettenham, J., et al. (2000) Testing joint attention, imitation, and play as infancy precursors to language and theory of mind. *Cogn Dev*, **15**(4), 481–98.

Clements, W. A. and Perner, J. (1994) Implicit understanding of belief. *Cogn Dev*, **9**(4), 377–95.

D'Argembeau, A., Collette, F., Van der Linden, M., et al. (2005) Self-referential reflective activity and its relationship with rest: a PET study. *NeuroImage*, **25**(2), 616–24.

Darwin, C. (1872) *The Expression of the Emotions in Man and Animals.* London: J. Murray.

Davis, M. and Whalen, P. J. (2001) The amygdala: vigilance and emotion. *Mol Psychiatry*, **6**(1), 13–34.

Denham, S. A. (1986) Social cognition, prosocial behavior, and emotion in preschoolers: contextual validation. *Child Dev*, **57**(1), 194–201.

Dennet, D. C. (1987) *The Intentional Stance.* Cambridge, MA: MIT Press.

Doherty, M. J. and Anderson, J. R. (1999) A new look at gaze: preschool children's understanding of eye-direction. *Cogn Dev*, **14**(4), 549–71.

Downing, P. E., Jiang, Y., Shuman, M., et al. (2001) A cortical area selective for visual processing of the human body. *Science*, **293**(5539), 2470–73.

Feinman, S. (1992) *Social Referencing and the Social Construction of Reality in Infancy*. New York: Plenum.

Fernald, A. (1993) Approval and disapproval: infant responsiveness to vocal affect in familiar and unfamiliar languages. *Child Dev*, **64**, 657–74.

Field, T. M. and Walden, T. A. (1982a) Production and perception of facial expression in infancy and early childhood. In: Reese H. W., Lipsitt, L. P. (Eds.) *Advances in Child Development and Behavior*. Vol. **16**. New York: Academic Press, pp. 169–211.

Field, T. M. and Walden, T. A. (1982b) Production and discrimination of facial expressions by preschool children. *Child Dev*, **53**(5), 1299–311.

Field, T. M., Woodson, R., Greenberg, D., et al. (1982) Discrimination and imitation of facial expression by neonates. *Science*, **218**(4568), 179–81.

Frith, C. D. and Frith, U. (1999) Interacting minds – a biological basis. *Science*, **286**, 1692–5.

Garnham, W. A. and Ruffman, T. (2001) Doesn't see, doesn't know: is anticipatory looking really related to understanding or belief? *Dev Sci*, **4**(1), 94–100.

Gergely, G. and Csibra, G. (1997) Teleological reasoning in infancy: The infant's naïve theory of rational action. A reply to Premack and Premack. *Cognition*, **63**(2), 227–33.

Golijeh, G., Ghahremani, D. G., Whitfield-Gabrieli, S., et al. (2007) Differential development of high level visual cortex correlates with category-specific recognition memory. *Nature Neurosci*, **10**, 512–22.

Greicius, M. D., Krasnow, B., Reiss, A. L., et al. (2003) Functional connectivity in the resting brain: a network analysis of the default mode hypothesis. *Proc Natl Acad Sci*, **100**(1), 253–8.

Guyer, A. E., Monk, C. S., McClure, E. B., et al. (2008) Developmental differences in amygdala response to fearful facial expressions. *J Cogn Neurosci*, **20**(9), 1565–82.

Haith, M. M., Bergman, T., and Moore, M. J. (1977) Eye contact and face scanning in early infancy. *Science*, **198** (4319), 853–5.

Hoffman, E. A. and Haxby, J. V. (2000) Distinct representations of eye gaze and identity in the distributed human neural system for face perception. *Nature Neurosci*, **3**(1), 80–4.

Hood, B. M., MaCrae, C. N., Cole-Davies, V., et al. (2003) Eye remember you: the effects of gaze direction on face recognition in children and adults. *Dev Sci*, **6**(1), 67–71.

Hood, B. M., Willen, J. D., and Driver, J. (1998) Adult's eyes trigger shifts of visual attention in human infants. *Psychol Sci*, **9**(2), 131–4.

Johansson, G. (1973) Visual perception of biological motion and a model for its analysis. *Percept Psychophysics*, **14**, 201–11.

Johnson, S. C., Booth, A., and O'Hearn, K. (2001) Inferring the goals of a nonhuman agent. *Cogni Dev*, **16**, 637–56.

Just, M. A., Cherkassky, V. L., Keller, T. A., et al. (2004) Cortical activation and synchronization during sentence comprehension in high functioning autism: evidence for underconnectivity. *Brain*, **127**(8), 1811–21.

Kanwisher, N., McDermott, J., and Chun, M. M. (1997) The fusiform face area: a module in human extrastriate cortex specialized for face perception. *J Neurosci*, **17**(11), 4302–11.

Kelley, W. M., MaCrae, C. N., Wyland, C. L., et al. (2002) Finding the self? An event-related fMRI study. *J Cogn Neurosci*, **5**(14), 785–94.

Killgore, W. D. and Yurgelun-Todd, D. A. (2004) Sex-related developmental differences in the lateralized activation of the prefrontal cortex and amygdala during perception of facial affect. *Percept Mot Skills*, **99**(2), 371–91.

Killgore, W. D., Oki, M., and Yurgelun-Todd, D. A. (2001) Sex specific developmental changes in amygdala responses to affective faces. *NeuroReport*, **12**(2), 427–33.

Kleinke, C. L. (1986) Gaze and eye contact: a research review. *Psychol Bull*, **100**, 78–100.

Kluver, H. and Bucy, P. C. (1997) Preliminary analysis of functions of the temporal lobes in monkeys. *J Neuropsychiatry Clin Neurosci*, **9**(4), 606–20.

Kunda, Z. (1999) *Social Cognition: Making Sense of People*. Cambridge, MA: MIT Press.

LaBar, K. S., Crupian, M. J., Voyvodic, J. T., et al. (2002) Dynamic perception of facial affect and identity in the human brain. *Cerebr Cortex*, **13**(10), 1023–33.

LeDoux, J. E. (2000) Emotion circuits in the brain. *Annu Rev Neurosci*, **23**, 155–84.

Lobaugh, N. J., Gibson, E., and Taylor, M. J. (2006) Children recruit distinct neural systems for implicit emotional face processing. *NeuroReport*, **17**(2), 215–19.

McCarthy, G. (1999) Physiological studies of face processing in humans. In: Gazzaniga, M. S. (ed.) *The New Cognitive Neurosciences*. Cambridge, MA: MIT Press, pp. 393–410.

McClure, E. B., Monk, C. S., Nelson, E. E., et al. (2004) A developmental examination of gender differences in brain engagement during evaluation of threat. *Biol Psychiatry*, **11**(1), 1047–55.

Meltzoff, A. N. (1995) Understanding the intentions of others: re-enactment of intended acts by 18-month-old children. *Dev Psychol*, **31**(5), 838–50.

Meltzoff, A. N. and Moore, M. K. (1983) Newborn infants imitate adult facial gestures. *Child Dev*, **54**(3), 702–9.

Meyer-Lindenberg, A., Hariri, A. R., Munoz, K. E., et al. (2005) Neural correlates of genetically abnormal social cognition in Williams syndrome. *Nature Neurosci*, **8**(8), 991–3.

Mitchell, J. P., Banaji, M. R., and Macrae, C. N. (2005) General and specific contributions of the medial prefrontal cortex to knowledge about mental states. *NeuroImage*, **28**(4), 757–62.

Monk, C. S., McClure, E. B., Nelson, E. E., et al. (2003) Adolescent immaturity to attention-related brain engagement to emotional facial expression. *NeuroImage*, **20**(1), 420–8.

Moriguchi, Y., Ohnishi, T., Mori, T., et al. (2007) Changes of brain activity in the neural substrates for theory of mind during childhood and adolescence. *Psychiatry Clin Neurosci*, **61**(4), 355–63.

Morris, J. S., Frith, C. D., Perrett, D. I., et al. (1996) A differential neural response in the human amygdala to fearful and happy expressions. *Nature*, **383**(6603), 812–15.

Mosconi, M. W., Mack, P. B., McCarthy, G., et al. (2005) Taking an "intentional stance" on eye-gaze shifts: a functional neuroimaging study of social perception in children. *NeuroImage*, **27**(1), 247–52.

Moses, L. J., Baldwin, D. A., Rosicky, J. G., et al. (2001) Evidence for referential understanding in the emotions domain at twelve and eighteen months. *Child Dev*, **72**(3), 718–35.

Mumme, D. L., Fernald, A., and Herrera, C. (1996) Infants' responses to facial and vocal emotional signals in a social referencing paradigm. *Child Dev*, **67**(6), 3219–37.

Nelson, C. A. (1987) The recognition of facial expression in the first two years of life: mechanisms of development. *Child Dev*, **58**(4), 889–909.

Nelson, C. A., Morse, P. A., and Leavitt, L. A. (1979) Recognition of facial expressions by seven-month-old infants. *Child Dev*, **50**(4), 1239–42.

Nelson, E. E., Leibenluft, E., McClure, E. B., et al. (2005) The social re-orientation of adolescence: a neuroscience perspective on the process and its relation to psychopathology. *Psychol Med*, **35**(2), 163–74.

Nelson, E. E., McClure, E. B., Monk, C. S., et al. (2003) Developmental differences in neuronal engagement during implicit encoding of emotional faces: an event-related fMRI study. *J Child Psychol Psychiatry*, **44**(7), 1015–24.

Northoff, G., Heinzel, A., de Greck, M., et al. (2006) Self-referential processing in our brain: a meta-analysis of imaging studies on the self. *NeuroImage*, **31**(1), 440–57.

Ochsner, K. N., Knierim, K., Ludlow, D. H., et al. (2004) Reflecting upon feelings: an fMRI study of neural systems supporting the attribution of emotion to self and other. *J Cogn Neurosci*, **16**(10), 1746–72.

Ohnishi, T., Moriguchi, Y., Matsuda, H., et al. (2004) The neural network for the mirror system and mentalizing in normally developed children: an fMRI study. *NeuroReport*, **15**(9), 1483–7.

Onishi, K. H. and Baillargeon, R. (2005) Do 15-month-old infants understand false belief? *Science*, **308**(5719), 255–8.

Pelphrey, K. A. and Morris, J. P. (2006) Brain mechanisms for interpreting the actions of others from biological-motion cues. *Curr Direct Psychol Sci*, **15**(3), 136–40.

Pelphrey, K. A., Adolphs, R., and Morris, J. P. (2004) Neuroanatomical substrates of social cognition dysfunction in autism. *Mental Retard Dev Disabil Res Rev*, **10**(4), 259–71.

Pelphrey, K. A., Mitchell, T. V., McKeown, M. J., et al. (2003a) Brain activity evoked by the perception of human walking: controlling for meaningful coherent motion. *J Neurosci*, **23**(17), 6819–25.

Pelphrey, K. A., Morris, J. P., and McCarthy, G. (2005) Neural basis of eye gaze processing deficits in autism. *Brain*, **128**(5), 1038–48.

Pelphrey, K. A., Sasson, N. J., Reznick, J. S., et al. (2002) Visual scanning of faces in autism. *J Autism Dev Disord*, **32**(4), 249–61.

Pelphrey, K. A., Singerman, J. D., Allison, T., et al. (2003b) Brain activation evoked by perception of gaze shifts: the influence of context. *Neuropsychologia*, **41**(2), 156–70.

Pfeifer, J. H., Lieberman, M. D., and Dapretto, M. (2007) "I know you are but what am I?!": Neural basis of self- and social knowledge retrieval in children and adults. *J Cogn Neurosci*, **19**(8), 1323–37.

Phan, K. L., Wager, T., Taylor, S. F., et al. (2002) Functional neuroanatomy of emotion: a meta-analysis of emotion activation studies in PET and fMRI. *NeuroImage*, **16**(2), 331–48.

Phillips, A. T., Wellman, H. M., and Spelke, E. S. (2002) Infants' ability to connect gaze and emotional expression to intentional action. *Cognition*, **85**(1), 53–78.

Premack, D. and Woodruff, G. (1978) Does the chimpanzee have a theory of mind? *Behav Brain Sci*, **3**, 615–36.

Puce, A., Allison, T., Asgari, M., et al. (1996) Differential sensitivity of human visual cortex to faces, letterstrings, and textures: a functional magnetic resonance imaging study. *J Neurosci*, **16**(16), 5205–15.

Rizzolatti, G., Fadiga, L., Gallese, V., et al. (1996) Premotor cortex and the recognition of motor actions. *Cogn Brain Res*, **3**(2), 131–41.

Saxe, R. and Kanwisher, N. (2003) People thinking about people: the role of the temporo-parietal junction in "theory of mind". *NeuroImage*, **19**(4), 1835–42.

Saxe, R. and Powell, L. J. (2006) It's the thought that counts: specific brain regions for one component of theory of mind. *Psychol Sci*, **17**(8), 692–9.

Scherf, S. K., Behrmann, M., Humphreys, K., et al. (2007) Visual category-selectivity for faces, places, and objects emerges along different developmental trajectories. *Dev Sci*, **10**(4), F15–F30.

Shannon, B. J. and Buckner, R. L. (2004) Functional-anatomic correlates of memory retrieval that suggest nontraditional processing roles for multiple distinct regions within posterior parietal cortex. *J Neurosci*, **24**(45), 10084–92.

Shaw, P., Greenstein, D., Lerch, J., et al. (2006) Intellectual ability and cortical development in children and adolescents. *Nature*, **440**(7084), 676–9.

Siegler, R. S. (2000) The rebirth of children's learning. *Child Dev*, **71**, 26–35.

Sommerville, J. A., Woodward, A. L., and Needham, A. (2005) Action experience alters 3-month-old infants' perception of others' actions. *Cognition*, **1**, B1–B11.

Sorce, J. F., Emde, R. N., Campos, J. J., et al. (1985) Maternal emotional signaling: its effect on the visual cliff behavior of 1-year-olds. *Dev Psychol*, **21**(1), 195–200.

Taylor, M. J., Edmonds, G. E., McCarthy, G., et al. (2001) Eyes first! Eye processing develops before face processing in children. *NeuroReport*, **12**(8), 1671–6.

Taylor, M. J., McCarthy, G., Saliba, E., et al. (1999) ERP evidence for developmental changes in processing of faces. *Clin Neurophysiol*, **110**(5), 910–15.

Thomas, K. M., Drevets, W. C., Whalen, P. J., et al. (2001) Amygdala response to facial expression in children and adults. *Biol Psychiatry*, **49**(4), 309–16.

Tzourio-Mazoyer, N., De Schonen, S., Crivello, F., et al. (2002) Neural correlates of woman face processing by 2-month-old infants. *NeuroImage*, **15**(2), 454–61.

Walden, T. A. and Field, T. M. (1982) Discrimination of facial expressions by preschool children. *Child Dev*, **53**(5), 1312–19.

Wellman, H. M., Cross, D., and Watson, J. (2001) Meta-analysis of theory of mind development: the truth about false belief. *Child Dev*, **72**(3), 655–84.

Wimmer, H. and Perner, J. (1983) Beliefs about beliefs: representation and constraining function of wrong beliefs in young children's understanding of deception. *Cognition*, **13**(1), 103–28.

Woodward, A. L. (1998) Infants selectively encode the goal object of an actor's reach. *Cognition*, **69**, 1–34.

6

Language and the developing brain: insights from neuroimaging

Kristin McNealy, Mirella Dapretto and Susan Bookheimer

Introduction

Language is part of our biological heritage, an achievement that is no doubt dependent upon the unique characteristics of the human brain. By studying the neural basis of language in the developing brain, important insights may be gained about the complex relationship between brain and language, which may ultimately shed light on the neural mechanisms that allowed for the emergence of language. At a more applied level, a better understanding of the neural events that accompany language processing in the typically developing brain is also critical to identify brain dysfunction in a variety of developmental disorders characterized by linguistic and/or communicative impairments.

The rapid and continuous developments in non-invasive imaging techniques and data analysis methods witnessed over the last decade have greatly expanded our ability to study brain–language relationships in vivo, thus allowing for significant strides in delineating the functional representation of language in the adult brain (see Bookheimer, 2002; Friederici, 2002; Poeppel and Hickok, 2004; and Price, 2000 for reviews). While we still know relatively much less about the neural correlates of language processing in the developing brain, in recent years there has been a surge in the number of developmental studies focusing on language, thanks in part to the increased use of functional

magnetic resonance imaging (fMRI). While the main goal of this chapter is to review this growing neuroimaging literature, we will begin by briefly summarizing language-related research findings from studies using "older" electrophysiological measures (i.e., event-related potentials, ERPs) as well as a relatively "young" technique (i.e., near infrared spectroscopy, NIRS). Whenever possible, we will separately discuss studies focusing on distinct linguistic aspects such as phonological, semantic, and syntactic processes, as well as studies focusing on extralinguistic processes such as prosody (i.e., intonation and vocal stress) and higher-level linguistic functions such as discourse and pragmatics (i.e., language use in a communicative context). We will then discuss some methodological considerations and highlight avenues for future research in this field.

Evidence from event-related potentials

The practical difficulties of using fMRI with infants and very young children have severely constrained the use of this technique in these populations. However, the neural underpinnings of early language development have been examined using electroencephalographic techniques (i.e., ERPs). As these studies have recently been comprehensively reviewed by Friederici (2005, 2006), here we

Neuroimaging in Developmental Clinical Neuroscience, eds. Judith M. Rumsey and Monique Ernst. Published by Cambridge University Press. © Cambridge University Press 2009.

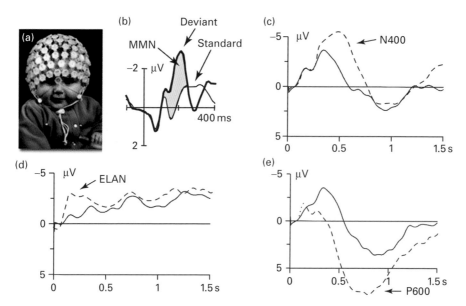

Figure 6.1a–e. Event-related potentials (ERPs) associated with language processing. (a) Five-month-old infant wearing a 128-channel Hydrocel Geodesic Sensor net, seated in the Geodesic photogrammetry system for registration of 3D electrode positions. (Photo courtesy of Electrical Geodesics, Inc.) (b) Graph showing the mismatch negativity (MMN) response that is related to the discrimination of the mismatch between deviant and standard stimuli in the auditory oddball paradigm. (c) Graph showing the N400 component in response to hearing sentences containing semantic violations. (d and e) Graph showing the ELAN (d) and P600 (e) components in response to hearing sentences containing syntactic violations. Figures adapted with permission from (a) Electrical Geodesics, Inc., (b) Kujala and Näätänen (2001), and (c–e) Friederici (2002).

will only present a brief overview of recent research findings in infants and young children. Event-related potentials provide an index of brain activity in response to a stimulus with precise temporal resolution (see Figure 6.1). In this technique, the electrical cortical responses to many presentations of a particular type of stimulus are averaged together, allowing researchers to separate an ERP waveform from background noise. An ERP waveform contains different component peaks that vary in polarity, latency and scalp distribution, and, through extensive language research in adults, these waveform components have been associated with various linguistic processes. Recent developmental language research has focused on answering the following questions: do infants and children display the same language-related ERP components as adults display? If so, when do they appear

and how might they be different in young language learners than in experienced adults? This line of research expands upon what has been learned about the developmental timecourse of linguistic skills through behavioral studies and provides valuable information via electrophysiological markers of the neural processes that underlie language acquisition.

Early phonological processing

Event-related potential research focusing on language processing at the phonological level has shown that very young infants are able to discriminate language-relevant sounds and patterns (Dehaene-Lambertz and Gliga, 2004; Friederici, 2006). One popular ERP paradigm, the auditory oddball paradigm, involves recording from electrodes during ongoing auditory stimulation in which one phoneme is repeated many times (standard stimulus) and another phoneme is intermittently interspersed (deviant stimulus). In adults, listening to a deviant phoneme produces what is called a mismatch negativity (MMN) response, which is a negative peak at about 100–250 ms after the onset of the deviant phoneme (see Dehaene-Lambertz and Gliga, 2004, for a review). This paradigm has been used to assess whether infants are able to discriminate between a range of acoustically and phonetically different stimuli, such as phonemes, syllables, consonants, and vowels in both native and non-native languages. Newborns show a mismatch response to different vowels, and infants as young as 2 or 3 months show a mismatch response to different syllables, albeit with different peak onsets, amplitudes, durations, directionality, and scalp distributions than adults (see Cheour et al., 2000; Dehaene-Lambertz and Gliga, 2004, for reviews). The directionality of the mismatch response in infants has been found to vary; it is positive in some children (Dehaene-Lambertz and Dehaene, 1994; Friederici et al., 2002; Leppänen et al., 1999; Pihko et al., 1999; Weber et al., 2004) and negative in others (Cheour et al., 1998; Kushnerenko et al., 2001). Although the variability in these findings is not yet well understood, there is initial evidence that differences in the neural underpinnings of phonetic processing may be related not only to age but also to linguistic experience. While there is likely a general age-related progression toward resemblance of the mismatch response seen in adults, the individual linguistic skills of infants may also influence the electrophysiological response to phonemic processing. For example, in 11-month-olds, differences in the timing, polarity, and scalp distribution of the mismatch response to non-native speech contrasts (indexing the detection of phonetic differences for foreign speech stimuli) were shown to be related to later productive vocabulary scores assessed at timepoints between 18 and 30 months of age (Riviera-Gaxiola et al., 2005a). Findings of differential responses to native and non-native contrasts is evidence of the emergence of language-specific

phonemic discrimination between 6 and 12 months, whereby infants' ability to discriminate between non-native contrasts diminishes as the processing of the patterns of their native language is strengthened (Cheour et al., 1998; Kuhl et al., 2006; Riviera-Gaxiola et al., 2005b). Decreased ability to discriminate non-native contrasts behaviorally (Kuhl et al., 2006) and changes in the timing, polarity, and scalp distribution of mismatch responses to native and non-native contrasts may reflect a continuous process of neural commitment by which the brain's networks become attuned to the patterns characteristic of one's native language (for a review, see Kuhl, 2004).

As infants learn to recognize the specific sounds present in their native language, they also continue to learn its rhythmical properties. Similar auditory oddball paradigms have demonstrated that infants begin to have knowledge of the stress patterns in their native language as early as 5 months of age, which represents an important advancement toward determining word boundaries within continuous speech (Weber et al., 2004). In addition, 8-month-old infants have been shown to display an ERP component that indexes detection of prosodic phrase boundaries, called a closure positive shift, that is similar to, albeit slower than, what has been found in adults (Pannekamp et al., 2006). These developments mark the beginning steps in infants' ability to recognize words as discrete units, which paves the way for the emergence of semantic processing.

Event-related potential studies have also examined the neural bases of semantic processing as children's understanding of words emerges. In adults, semantic processing has been associated with an N400 effect, a negativity peaking at around 400 ms, with a greater negativity for pseudowords than for words (for a review, see Kutas and Federmeier, 2000). At the single-word level, a comparison of the ERP response to known versus unknown, or familiar versus unfamiliar, words provides an electrophysiological marker of word recognition in preverbal children. In children between 11 and 20 months old, there is a negativity between 200 and 500 ms that is greater for familiar words than for unfamiliar words, with a transition from a bilateral distribution to left hemisphere dominance with age (Kooijman et al., 2005; Mills et al., 1997, 2004; Thierry et al., 2003). In addition to this earlier-occurring familiarity effect, a more adult-like N400 effect begins to appear between 14 and 19 months of age. For example, when presented with a picture of an object at the same time as an auditory stimulus that either matches the object's name or not, 14- and 19-month-olds show an N400 effect for picture–word pairs that do not match (Friedrich and Friederici, 2004, 2005a, 2005b). The effect occurs later and lasts longer than in adults, which has been interpreted as reflecting slower lexical-semantic processing in children.

At the sentence level, the N400 effect is observed as a larger negative amplitude for sentences containing words that are not appropriate for the context and is taken to index semantic integration difficulty in adults and in children as young as 19 months old (Friedrich and Friederici, 2005c). This effect has been demonstrated with various paradigms in children across a wide age range through adolescence and into adulthood (Atchley et al., 2006; Friedrich and Friederici, 2005b, 2005c; Hahne et al., 2004; Holcomb et al., 1992; Silva-Pereyra et al., 2005a, 2005b). There appears to be an age-related decrease in the duration, amplitude and width of the N400 effect as speakers expand their lexicon. Overall, semantic processing that is highly similar to that of adults develops between 2.5 and 3 years of age (Friederici, 2006).

Early syntactic processing

There are only a few ERP studies in children that have focused specifically on syntactic processing. In adults, violations of syntactic rules typically engender two waveform components that are thought to underlie different aspects of syntactic analysis (see Friederici, 2002, for a review). A left anterior negativity (LAN, or a very early-occurring ELAN) between 150 and 350 ms is thought to reflect initial structure building and later morpho-syntactic

processing. A P600 component, a positive-going peak with a centro-parietal distribution at about 600 ms, is thought to be involved with syntactic reanalysis or repair (i.e., the process whereby a grammatically correct representation of an ungrammatical sentence is constructed). Studies have reported a P600 effect, but not a LAN component, in response to syntactic violations in 2-, 3-, and 4-year-olds, although the P600 occurs later and lasts longer than it does in adults (Oberecker and Friederici, 2006; Silva-Pereyra et al., 2005b) until about 8 years of age when the P600 component latency, amplitude and scalp location do not differ from adults (Atchley et al., 2006). One study found a delayed ELAN-like component in addition to the P600 in children age two and a half (Oberecker et al., 2005). These studies posit that the LAN emerges later than the P600 because it likely reflects automatic processes that are not yet developed in young children, whereas the P600 reflects late controlled processes.

Summary of ERP findings in infancy and early childhood

To summarize, the findings of ERP studies in infants and young children report a large degree of similarity to what is observed in adults during phonological, semantic, and syntactic processing, suggesting that, while the neural mechanisms of language processing may undergo considerable "fine tuning" as a function of age and linguistic experience, the "blue prints" appear to be established quite early in life.

Evidence from near infrared spectroscopy

An imaging technique that researchers have recently begun to employ in young infants is near infrared spectroscopy (NIRS), which measures changes in the cortical concentration of hemoglobin that are associated with neural activity (see Figure 6.2 and Obrig and Villringer, 2003 for a description of this technique and its use in

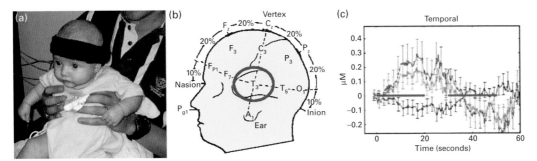

Figure 6.2. Near-infrared-spectroscopy (NIRS) detected changes in infant brain activity. (a) Infant wearing a headband containing the NIRS instrument. (b) Diagram showing the region of interest in the temporal lobe. (c) An illustration of the hemodynamic response function in temporal cortex across the 60-s blocks of stimulus presentation. Audiovisual segments are indicated by the solid red bar; visual segments are indicated by the solid pink bar. The y-axis indicates relative changes in concentration of oxyhemoglobin (red), deoxyhemoglobin (blue), and total hemoglobin (green). Figure adapted with permission from Bortfeld et al. (2007).

neuroimaging). While this technique does not provide the temporal resolution of ERPs or the spatial resolution of fMRI, the procedure is silent and can be used to ascertain information about the hemodynamic response accompanying brain activity in infants, who are not easily scanned with fMRI. An initial study in sleeping infants less than 1 week old revealed greater activity in the left temporal cortex for regular speech than for speech played backward (Pena et al., 2003). Despite the fact that the infants were asleep, the differential responses observed suggest that passive auditory processing in the neonate brain is already capable of distinguishing speech. A recent study in older, awake 6- to 9-month-olds also found left temporal activity when comparing the response to the presentation of multimodal audiovisual stimuli versus visual stimuli alone (although data were not collected from the right hemisphere; Bortfeld et al., 2007). The right hemisphere has been shown to be involved in speech processing as well, as one study reported greater activity in a right temporo-parietal region when infants were presented with speech with a normal intonational contour compared to prosodically flattened speech (Homae et al., 2006). The results from two other recent studies suggest unique roles for each hemisphere

that are refined as infants and children gain linguistic experience. Infants showed an early bilateral activity in response to speech sounds with a progression toward left hemisphere dominance over time, as data collected from 3- to 28-month-olds during an auditory oddball paradigm with a Japanese durational vowel contrast demonstrated a left hemisphere dominance beginning at 13 months of age (Minagawa-Kawai et al., 2007). When 4-year-olds listened to normal, hummed, or prosodically flat sentences, right fronto-temporal regions displayed greater activity to hummed sentences containing prosody in isolation, whereas left hemisphere fronto-temporal regions were dominant when the full content of linguistic and prosodic information was available (Wartenburger et al., 2007). Taken together, these data suggest an early mechanism of auditory processing that is bilaterally distributed, with an experience-dependent specialization of the left and right hemispheres for linguistic information and prosodic cues, respectively, that aid in comprehending the full meaning of speech. While only a few such studies have been conducted so far, this technology represents a promising avenue of research to track changes in the neural correlates of language acquisition in young infants and children.

Evidence from functional magnetic resonance imaging

Despite several logistical challenges involved in keeping an infant comfortable and still in the very noisy and unfriendly MRI scanner, researchers have recently begun using fMRI in infants as young as 8 weeks of age. While fMRI lacks the same precise temporal resolution of ERPs, these studies have allowed researchers to assess with good spatial resolution the initial localization of auditory language processing. In a comparison of speech to a silent resting baseline, or to speech played in reverse, researchers measured relative differences in the blood-oxygenation-level-dependent (BOLD) signal associated with the hemodynamic response accompanying brain activity. In 3-month-old infants, an adult-like pattern of response was found in bilateral superior temporal regions with greater activation in the left hemisphere in response to sentences (Dehaene-Lambertz et al., 2002). The left angular gyrus and precuneus may play a role in processing speech sounds in particular, as they displayed greater activity to forward speech than backward speech. Additionally, for infants who were awake during the scan, activity in the right frontal cortex while listening to forward speech was taken to reflect the allocation of attentional resources during language processing. In a later study, the timecourse of activation in language-related regions of interest in the left hemisphere was examined by presenting awake infants with one of several short sentences from a children's story every 14 s. The fastest BOLD response was observed in Heschl's gyrus, with progressively slower responses moving anteriorly along the bilateral superior temporal sulci and posteriorly in the left hemisphere toward Wernicke's area (Dehaene-Lambertz et al., 2006). Interestingly, activity was also observed in the left inferior frontal gyrus (IFG), suggesting that this region may be important for perceptual learning of complex sound production in addition to facilitating motor production after learning has taken place. When the timecourse of response to sentences in infants was compared to that in adults, the same temporal progression was observed (see Figure 6.3). These results have been interpreted as evidence that a hierarchical organization for language processing in the brain is already present in the first months of life.

The scanning of such young infants, while considerably challenging, has a relatively high success rate, with useable data being acquired from approximately half of participating infants. Typically, the infants are swaddled, have headphones placed over their ears that minimize the scanner noise and deliver auditory stimuli, and, if they are awake, are distracted with interesting visual stimuli such as faces and toys (as is commonly done in ERP studies with infants as well). In addition, a pediatrician is often also present in the scanner room to monitor the comfort and wakefulness of each infant during the scanning session, which can last up to 30 min.

The majority of fMRI studies of language processing, however, have been conducted in older children, with the youngest participants being around age 5 when children are better able to hold still for short periods of time. Scanning sessions are typically kept under 45 min in total, with each particular activation task lasting only 5–10 min. Many fMRI studies in typically developing children have focused on assessing the extent of language lateralization and on comparing the observed pattern of activity with that observed in adults. Canonical language networks in adults have been identified as left hemisphere dominant and include bilateral primary auditory cortices and superior temporal gyri, extending into middle temporal, inferior parietal and inferior frontal regions (for reviews, see Bookheimer, 2002; Friederici, 2002; Poeppel and Hickok, 2004). In general, the left hemisphere has been associated with the identification of phonetic, lexical, and syntactic elements, their sequential processing, and the formation of relationships between linguistic elements. The right hemisphere has been associated with keeping track of the topic of conversation, narrative comprehension, and integrating prosodic information to ascertain a speaker's

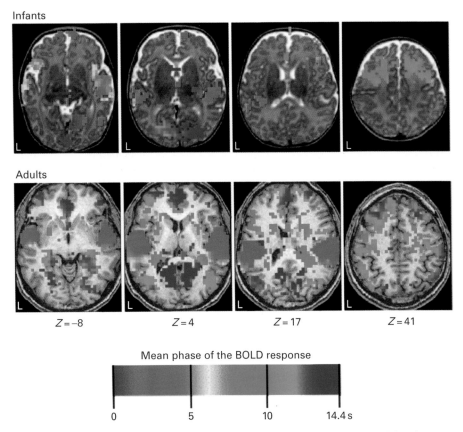

Figure 6.3. Comparison of cerebral responses to a single sentence in infants and adults. Shown are axial slices placed at similar locations in infant (upper tier) and adult (lower tier) standard brains. The same temporal progression is observed in adults and infants. Regions in shades of blue may correspond to a "resting state" network. Figure reprinted with permission from Dehaene-Lambertz et al. (2006). Copyright 2006 National Academy of Sciences, U.S.A.

full meaning. Does the developing brain recruit the same neural networks for processing speech as the adult brain? How might one characterize differences in the activation patterns during language processing between children and adults? Are there developmental changes in the degree to which each hemisphere is recruited to perform particular tasks? Using the adult data as a neurodevelopmental endpoint, a variety of fMRI paradigms have been used to explore how the neural correlates of phonological and semantic processing, as well as higher-level linguistic functions, change throughout childhood and adolescence.

Phonological processing

Several studies have attempted to isolate the neural correlates of phonological processing in children. These studies have used paradigms that require participants to make rhyming and spelling judgments on words presented in either the visual or auditory modality. A comparison of the activation in visual phonological (rhyming) and orthographic (spelling) tasks in children between the ages of 9 and 15 years revealed greater activation in bilateral IFG (Brodmann's area (BA) 45 and 47) for the rhyming task, which likely reflects both phonological and semantic processing (Bitan et al., 2007). A subsequent study using dynamic causal modeling to assess functional connectivity reported that, during the rhyming task, activation in the IFG exerted a top-down modulatory influence on the amount of activation in temporal cortices, which are thought to subserve phonological processing, as is the case in adults (Bitan et al., 2006). The strength of this modulation, however, was weaker in children, suggesting that mechanisms of cognitive control are still developing throughout late childhood and early adolescence. It appears that children and adults may recruit a highly similar neural network during phonological processing, but the extent to which each region is activated changes during development. Indeed, another study that used a similar paradigm focusing on auditory phonological representations revealed significantly overlapping clusters of activation in bilateral fronto-temporal regions (including posterior superior temporal gyrus (STG) and IFG) between children and adults; however, greater activity was observed in adults in the angular gyrus, a parietal region thought to be involved in the mapping of phonological and orthographic representations (Booth et al., 2004). This developmental difference perhaps suggests that representational systems become more interactive and convergent with age and linguistic experience.

Semantic processing

The majority of developmental neuroimaging studies of language have focused on semantic processing, using paradigms that require word generation or semantic relatedness/category judgments. In adults, semantic processing at the word level has been shown to engage the middle temporal gyrus (MTG), STG, inferior parietal regions, and IFG (in particular, BA 47) in the left hemisphere as well as the right STG, with increasing frontal involvement when strategic or memory aspects are needed (Bookheimer, 2002). In children, semantic association has been demonstrated to activate those same regions, with the addition of right IFG (Chou

et al., 2006). Observed age-related increases in left MTG and right IFG activity have been attributed to the development of more efficient access to semantic representations and broader semantic searches, respectively.

In developmental studies, activation differences between children and adults have been reported in both magnitude and extent, in addition to differences in the distribution of activation across regions within the language network. Developmental differences have been reported as greater overall activation in children than in adults, including activity in the right IFG (Gaillard et al., 2000), greater activation observed in adults than in children, including activity in the left IFG and MFG (Gaillard et al., 2003; Schapiro et al., 2004; Wood et al., 2004), and an increase in the degree of left lateralization as a function of age (Holland et al., 2001; Szaflarski et al., 2006a). Some studies, however, report no age-related differences in laterality (Balsamo et al., 2006; Gaillard et al., 2003; Wilke et al., 2005; Wood et al., 2004). A clear picture of how word meanings are represented in the developing brain and how this representation maps onto findings in the adult literature has yet to emerge.

The high degree of variability in functional activation patterns in children due to maturational and experiential factors represents a significant challenge in developmental neuroimaging (Berl et al., 2006). While reported findings across studies have been highly variable, many of the inconsistencies may be accounted for by the different ages and skill levels of children being studied and by variations in task design. In order to ameliorate the effects of high variability in activity in children, activation paradigms should attempt to control for age-related differences in performance (e.g., older children being able to retrieve a larger number of exemplars for a given category) and/or task difficulty, two variables known to have an impact on the pattern of cortical activity during fMRI (e.g., Raichle et al., 1994). Indeed, Ahmad and colleagues (Ahmad et al., 2003) found no reliable correlations between age and regional asymmetry indices during an auditory comprehension task where 15 children listened to stories that were selected to match their educational level, though this finding may reflect the relatively narrow age range of the subjects (5–7 years of age).

In particular, comparisons between children and adults should take accuracy and reaction time differences into account, as task performance has been shown to impact the extent to which various brain regions are activated. For example, when 15 9- to 12-year-olds were asked to make semantic relatedness judgments in the scanner, high performers displayed more activity in posterior temporal regions, while low performers engaged more anterior frontal regions (Blumenfeld et al., 2006). Using a word-generation task (in response to visually presented words), Schlaggar and colleagues were amongst the first to tease apart performance-related versus age-related differences between adults and school-age children in the functional networks underlying single-word processing (Schlaggar et al., 2002). Independent of task performance, children showed greater activity in left extrastriate cortex whereas adults showed greater activity in the left inferior frontal region. These results were later confirmed in a large-scale study of lexical access in 95 individuals in whom age-related increases in activity were observed in left frontal and parietal regions and age-related decreases were seen in early processing regions including extrastriate cortices (Brown et al., 2005).

Longitudinal studies may provide additional power for teasing apart the various factors impacting developmental differences in language networks. One such study that followed 30 children, aged 5–7, each year for 5 years reported age-related increases in activity in inferior and middle frontal, middle temporal and inferior parietal gyri in the left hemisphere and right inferior temporal gyrus, as well as age-related decreases in left posterior insula, extrastriate cortex, superior frontal gyrus, thalamus, and right anterior cingulate gyrus during covert verb generation to auditorily presented nouns (Szaflarski et al., 2006b). These findings indicate that true developmental changes in the neural architecture subserving language processing are still occurring during middle to late childhood.

Syntactic processing

As very few ERP studies and, to our knowledge, no fMRI studies of syntactic processing have been conducted in children, the neural representation of syntax in the developing brain is a knowledge gap that needs to be addressed in future research. Regions such as the left anterior STG and portions of the IFG (BA 44, in particular), which have been associated with syntactic processing in adults, may be shown to subserve a similar function in children (Cooke et al., 2006; Dapretto and Bookheimer, 1999; Davis et al., 2004). Although developmental fMRI studies have yet to concentrate specifically on the neural correlates of syntactic processing, a component of the activity in the broad fronto-temporal-parietal language networks observed during sentence-level or narrative tasks (see next section below) is likely attributable to syntactic processing and the interaction of multiple linguistic representations.

Higher-level linguistic processing

The findings of several studies that examined higher-level linguistic functions, such as discourse comprehension and the integration of prosodic information, have pointed to a unique contribution of the right hemisphere for performance of these tasks. In a large-scale study of 313 children ages 5–18, listening to stories was associated with significant activity in a number of separate task-related networks, including a right-lateralized region in the posterior STG, a region that is thought to be of particular importance for integration of various linguistic cues during narrative comprehension (Karunanayaka et al., 2007; Schmithorst et al., 2006). As is the case with phonemic and semantic processing, the neural underpinnings of narrative comprehension continue to change throughout development, as age-related increases in activity were also observed in bilateral STG and left IFG, with age-related decreases in activity observed in the left angular gyrus.

The extent to which activity during higher-level linguistic processing is lateralized to one hemisphere or the other is not static, but rather may shift dynamically depending on the nature of the task being performed. In a study using a discourse-monitoring task, activity in both children and adults was strongly left-lateralized when the appraisal of discourse coherence involved detecting a break in the logic of a conversation; however, activity was remarkably bilateral when this assessment rested on detecting a break in the conversation topic (Dapretto et al., 2005). In another study, when children between ages 8 and 15 listened to potentially ironic comments, the presence of prosodic cues as an indication of irony or sincerity was associated with activity in bilateral STG, MTG, and IFG, and the presence of contextual cues in the scenarios was associated with additional activation in the right temporal pole (Wang et al., 2006).

Researchers have also focused on ascertaining whether there is a right-lateralized involvement of the STG for linguistic and emotional prosody processing in children, as has been suggested by studies in adults (Meyer et al., 2004; Plante et al., 2002). In another large-scale study of children ranging from 5 to 18 years of age, listening to low-pass-filtered sentences (in which word meanings were obscured but prosodic contours were preserved) to detect a match to a previously heard target sentence resulted in bilateral activation in fronto-temporal regions, with a right-lateralized pattern in STG and MTG (Plante et al., 2006a). Interestingly, while activity increased as a function of age bilaterally in the STG, BOLD signal in only the left, and not right, STG was predictive of correct task performance. Further analyses led the researchers to speculate that the right hemisphere may be preferentially recruited for the processing of sentence-level intonation, whereas the left hemisphere may be recruited more for the processing of stress cues related to word segmentation. Another study with over 200 participating children conducted by the same research group used several tasks (story comprehension, prosodic processing, picture identification, and verb generation) and analyzed the data in regions of interest (ROI) in right and left frontal and temporal cortices while taking into account age, sex, task, and performance (Plante et al., 2006b). Their analysis revealed left lateralization in all tasks except for the prosody task, on which there were no significant differences in laterality in the frontal or temporal ROIs. In addition, there was a tendency for small increases in activation with age in most tasks, with larger age-related increases observed in the left frontal ROI during verb generation and in the left and right temporal ROIs for the story processing task. Taken together, these findings then suggest that hemispheric specialization for language processing does not uniformly change with development; rather, the two cerebral hemispheres may differentially contribute to language processing as a function of both age and linguistic task.

Summary

Table 6.1 lists some of the studies published using different methodologies at different age ranges. In line with behavioral data and the typical time-table of early language development, the ERP studies suggest a progression of differential neural responses with age. Namely, responses to phonological manipulations are apparent in the youngest cohorts (i.e., in the first year of life), whereas word-related responses tend to emerge in the second year, and responses to syntactic anomalies appear no earlier than the beginning of the third year. With regard to the fMRI studies, mainly conducted in children older than five who have mastered basic linguistic skills and are competent language speakers, the data suggest a general trend toward increasing functional specialization within language networks as a function of age and/or linguistic competence, likely reflecting experiential factors as well as structural maturation, both of which may contribute to increased automaticity of language processing.

Table 6.1 Summary of published developmental imaging studies

Age	Phonological processing	Semantic processing	Syntactic processing	Higher-level linguistic processing
Birth – 12 months	**ERP – auditory oddball paradigm** (Cheour et al., 1998, 2000; Dehaene-Lambertz and Dehaene, 1994; Friederici et al., 2002; Kuhl et al., 2006; Kushnerenko et al., 2001; Leppänen et al., 1999; Pannekamp et al., 2006; Pihko et al., 1999; Riviera-Gaxiola et al., 2005a, 2005b; Weber et al., 2004) **NIRS – auditory oddball paradigm** (Minagawa-Kawai et al., 2007)	**ERP – listening to words** (Friedrich and Friederici, 2004, 2005a, 2005b; Kooijman et al., 2005; Thierry et al., 2003)		**NIRS – listening to speech** (Bortfeld et al., 2007; Homae et al., 2006; Pena et al., 2003) **fMRI – listening to speech** (Dehaene-Lambertz et al., 2002, 2006)
13 months – 4 years	**NIRS – auditory oddball paradigm** (Minagawa-Kawai et al., 2007)	**ERP – listening to words** (Mills et al., 1997, 2004; Friedrich and Friederici, 2004, 2005a, 2005b) **ERP – listening to sentences with semantic anomalies** (Friedrich and Friederici 2005b, 2005c; Silva-Pereyra et al., 2005a, 2005b)	**ERP – listening to sentences with syntactic anomalies** (Oberecker et al., 2005; Oberecker and Friederici 2006; Silva-Pereyra et al., 2005b)	**NIRS – listening to speech** (Wartenburger et al., 2007)
5 years and older		**ERP – listening to sentences with semantic anomalies** (Hahne et al., 2004; Holcomb et al., 1992) **fMRI – semantic relatedness judgment task** (Balsamo et al., 2006)		**fMRI – listening to sentences or discourse** (Ahmad et al., 2003; Karunanayaka et al., 2007; Plante et al., 2006a, 2006b; Schmithorst et al., 2006; Wilke et al., 2005)

Table 6.1 (*cont.*)

Age	Phonological processing	Semantic processing	Syntactic processing	Higher-level linguistic processing
		fMRI – word generation task (Plante et al., 2006b; Schapiro et al., 2004; Szaflarski et al., 2006a, 2006b; Wilke et al., 2005)		fMRI – listening to sentences or discourse (Dapretto et al., 2005; Wang et al., 2006)
7 years and older	fMRI – rhyming task (Bitan et al., 2006, 2007; Booth et al., 2004)	ERP – listening to sentences with semantic anomalies (Atchley et al., 2006) fMRI – spelling task (Bitan et al., 2006, 2007; Booth et al., 2004) fMRI – semantic relatedness judgment task (Blumenfeld et al., 2006; Chou et al., 2006) fMRI – word generation task (Brown et al., 2005; Gaillard et al., 2000, 2003; Holland et al., 2001; Schlaggar et al., 2002; Wood et al., 2004)	ERP – listening to sentences with syntactic anomalies (Atchley et al., 2006)	

Methodological considerations and future directions

The majority of the fMRI studies described above have used blocked designs, in which BOLD signal is compared between alternating blocks of an experimental task and a baseline control task. In many cases, the control task is silent rest, and in others it is a task chosen to be as similar to the experimental task as possible without engaging the key cognitive dimension of interest. It is important to note, however, that language-mediated cognitive processing during silent rest blocks may vary between age groups, thereby impacting the ability to make inferences about developmental changes in the experimental task alone. To ameliorate this potential confound, future fMRI studies should increasingly utilize fast event-related designs (e.g., Buckner et al., 1998; Burock et al., 1998; Dale and Buckner, 1997) in which experimental and control stimuli are randomly intermixed, together with brief "null" events (e.g., a blank screen with a fixation cross). Fast event-related designs present an additional advantage over traditional blocked designs in that they allow the researcher to control for potential differences in performance across different age groups. For instance, in event-related designs, one can compare brain activity between groups only for trials showing correct responses, or for trials where responses occur within a specified time window, thus avoiding potential confounds due to differences in accuracy and response times, two variables known to impact the magnitude and extent of brain activity (e.g., Petersen et al., 1998; Raichle et al., 1994).

Another methodological consideration for the design of language studies that employ auditory stimuli is interference from the scanner noise. In general, researchers have been satisfied by the use of insulated headphones that minimize background scanner noise and by the use of subtraction in the statistical analysis of the data, whereby effects due to the scanner noise are "cancelled out" when two activation conditions are directly compared (since the scanner noise impacts both conditions equally). However, concerns about auditory interference from the scanner noise can also be addressed by the use of behaviorally interleaved acquisition sequences, which capitalize on the fact that the hemodynamic response as measured with fMRI is delayed (by 4–6 s) with respect to stimulus presentation. Accordingly, the gradients can be turned on and off so that auditory stimuli can be presented without the scanner noise during the "off" periods and the brain's delayed hemodynamic responses captured during the "on" periods (e.g., Eden et al., 1999).

An additional methodological consideration generally relevant to studies comparing brain activity in children and adults concerns the need to control for group differences in the amount of head motion occurring during data acquisition. While it is common practice to correct for head motion on a frame-by-frame basis as part of the initial stages of image analysis, there is likely a certain amount of residual head motion that goes uncorrected (particularly out-of-plane motion), which may impact the results of between-group comparisons. Accordingly, in the presence of significant group differences in head movement, motion parameters can and should be entered as covariates in the statistical analyses. The use of vacuum cushions to limit head movement during data acquisition can significantly reduce head movement, particularly in young children, as can the use of operant conditioning techniques used to train children to lie still for increasingly longer periods of time in a mock scanner.

To better characterize the neural foundations of language in the developing brain, future developmental imaging studies should directly relate brain activity to behavioral measures – collected outside the scanner – that tap into different linguistic skills. For instance, researchers may wish to examine how the patterns of brain activity during language tasks might be related to a child's phonological processing skills, vocabulary size, experience with foreign languages, and so on. This correlational approach has been successfully used with structural MRI data to better delineate how anatomical changes might be related to linguistic skills. For instance, building on evidence of continued increases in gray matter density in regions subserving linguistic processing throughout childhood and adolescence (for a review of developmental structural MRI studies, see Chapter 1), one recent study demonstrated that higher vocabulary scores (on a subtest of the Wechsler Intelligence Scale for Children) are positively correlated with gray matter density in bilateral supramarginal gyri (Lee et al., 2007). Also, in children ages 5–11 who were assessed over a 2-year interval, gray matter thickening in the left IFG was associated with improving phonological processing skills (Lu et al., 2007). Moreover, gray matter density decreases in the left dorsolateral frontal and lateral parietal regions have been found to correlate with increased vocabulary and performance on a verbal learning task in adolescents (Sowell et al., 2001, 2004).

These interesting findings highlight the need to relate developmental changes in the functional organization of language – as assessed with fMRI – not just to behavioral indices of linguistic abilities, but also to concomitant structural changes in brain anatomy. The complex relationship between structural and functional brain development is presently not well understood. Relating developmental changes in brain structure to age-related changes in brain activity and linguistic competence is likely to become a significant focus of future research. Together with detailed structural analyses examining changes in cortical thickness and the development of white matter tracts connecting language-relevant cortices, the increasing use of sophisticated data-analytic techniques to examine changes in functional connectivity within language networks (e.g., dynamic causal modeling, Friston et al., 2003) will significantly aid in the elucidation of the neural underpinnings of language in the developing brain.

In addition, to further characterize the neural correlates of different linguistic functions, future developmental fMRI research should also examine the neural mechanisms underlying the actual process of learning a novel language. While several studies in adults have begun to examine language learning by exposing subjects to artificial grammars while they are being scanned (McNealy et al., 2006; Opitz and Friederici, 2003, 2004; Thiel et al., 2003), to our knowledge no published study to date has directly examined the neural correlates of language learning in children. Charting developmental changes in the neural architecture subserving language learning may shed light on the long-standing question of why infants and young children are better language learners than adults. Importantly, this line of research may also have important applied implications, by informing our understanding of what goes awry in developmental language disorders as well as aiding the creation of optimal interventions for children diagnosed with developmental disorders impacting language functions (e.g., dyslexia, autism). Admittedly, there is still much to be learned about the neural underpinnings of language in the developing brain. However, armed with powerful new tools and insights from different fields and disciplines (see Kuhl, 2007), we are now closer than ever to understanding the neurobiological basis of this quintessential human function.

REFERENCES

Ahmad, Z., Balsamo, L. M., Sachs, B. C., et al. (2003) Auditory comprehension of language in young children: neural networks identified with fMRI. *Neurology*, **60**, 1598–605.

Atchley, R. A., Rice, M. L., Betz, S. K., et al. (2006) A comparison of semantic and syntactic event related potentials generated by children and adults. *Brain Lang*, **99**, 236–46.

Balsamo, L. M., Xu, B., and Gaillard, W. D. (2006) Language lateralization and the role of the fusiform gyrus in semantic processing in young children. *NeuroImage*, **31**, 1306–14.

Berl, M., Vaidya, C. J., and Gaillard, W. D. (2006) Functional imaging of developmental and adaptive changes in neurocognition. *NeuroImage*, **30**, 679–91.

Bitan, T., Burman, D. D., Chou, T. L., et al. (2007) The interaction between orthographic and phonological information in children. *Hum Brain Mapp*, **28**(9), 880–91.

Bitan, T., Burman, D. D., Lu, D., et al. (2006) Weaker top-down modulation from the left inferior frontal gyrus in children. *NeuroImage*, **33**(3), 991–8.

Blumenfeld, H. K., Booth, J. R., and Burman, D. D. (2006) Differential prefrontal-temporal neural correlates of semantic processing in children. *Brain Lang*, **99**(3), 226–35.

Bookheimer, S. (2002) Functional MRI of language: new approaches to understanding the cortical organization of semantic processing. *Annu Rev Neurosci*, **25**, 151–88.

Booth, J. R., Burman, D. D., Meyer, J. R., et al. (2004) Development of brain mechanisms for processing orthographic and phonologic representations. *J Cogn Neurosci*, **16**, 1234–49.

Bortfeld, H., Wruck, E., and Boas, D. A. (2007) Assessing infants' cortical response to speech using near-infrared spectroscopy. *NeuroImage*, **34**(1), 407–15.

Brown, T. T., Lugar, H. M., Coalson, R. S., et al. (2005) Developmental changes in human cerebral functional organization for word generation. *Cereb Cortex*, **15**(3), 275–90.

Buckner, R. L., Goodman, J., Burock, M., et al. (1998) Functional-anatomic correlates of object priming in humans revealed by rapid presentation event-related fMRI. *Neuron*, **20**, 285–96.

Burock, M. A., Buckner, R. L., Woldorff, M. G., et al. (1998) Randomized event-related experimental designs allow for extremely rapid presentation rates using functional MRI. *NeuroReport*, **9**, 3735–9.

Cheour, M., Ceponiene, R., Lehtokoski, A., et al. (1998) Development of language-specific phoneme representations in the infant brain. *Nat Neurosci*, **1**, 351–3.

Cheour, M., Leppanen, P., and Kraus, N. (2000) Mismatch negativity (MMN) as a tool for investigating auditory discrimination and sensory memory in infants and children. *Clin Neurophysiol*, **111**, 4–16.

Chou, T. L., Booth, J. R., Burman, D. D., et al. (2006) Developmental changes in the neural correlates of semantic processing. *NeuroImage*, **29**, 1141–9.

Cooke, A., Grossman, M., DeVita, C., et al. (2006) Large-scale neural network for sentence processing. *Brain Lang*, **96**(1), 14–36.

Dale, A. M. and Buckner, R. L. (1997) Selective averaging of rapidly presented individual trials using fMRI. *Hum Brain Mapp*, **5**, 329–40.

Dapretto, M. and Bookheimer, S. (1999) Form and content: dissociating syntax and semantics in sentence comprehension. *Neuron*, **24**, 427–32.

Dapretto, M., Lee, S. S., and Caplan, R. (2005) A functional magnetic resonance imaging study of discourse coherence in typically developing children. *NeuroReport*, **16**, 1661–5.

Davis, M. H., Meunier, F., and Marslen-Wilson, W. D. (2004) Neural responses to morphological, syntactic, and semantic properties of single words: an fMRI study. *Brain Lang*, **89**(3), 439–49.

Dehaene-Lambertz, G. and Dehaene, S. (1994) Speed and cerebral correlates of syllable discrimination in infants. *Nature*, **370**, 292–5.

Dehaene-Lambertz, G. and Gliga, T. (2004) Common neural basis for phoneme processing in infants and adults. *J Cogn Neurosci*, **16**(8), 1375–87.

Dehaene-Lambertz, G., Dehaene, S., and Hertz-Pannier, L. (2002) Functional neuroimaging of speech perception in infants. *Science*, **298**, 2013–15.

Dehaene-Lambertz, G., Hertz-Pannier, L., Dubois, J., et al. (2006) Functional organization of perisylvian activation during presentation of sentences in preverbal infants. *Proc Natl Acad Sci*, **103**, 14240–5.

Eden, G. F., Joseph, J. E., Brown, H. E., et al. (1999) Utilizing hemodynamic delay and dispersion to detect fMRI signal change without auditory interference: the Behavioral Interleaved Gradients Technique. *Magn Reson Med*, **41**(1), 13–20.

Friederici, A. D. (2002) Towards a neural basis of auditory sentence processing. *Trends Cogn Sci*, **6**, 78–84.

Friederici, A. D. (2005) Neurophysiological markers of early language acquisition: from syllables to sentences. *Trends Cogn Sci*, **9**(10), 481–8.

Friederici, A. D. (2006) The neural basis of language development and its impairment. *Neuron*, **52**, 941–52.

Friedrich, M. and Friederici, A. D. (2004) N400-like semantic incongruity effect in 19-month-olds: processing known words in picture contexts. *J Cogn Neurosci*, **16**, 1465–77.

Friedrich, M. and Friederici, A. D. (2005a) Phonotactic knowledge and lexical-semantic processing in one-year-olds: brain responses to words and nonsense words in picture contexts. *J Cogn Neurosci*, **17**, 1785–802.

Friedrich, M. and Friederici, A. D. (2005b) Lexical priming and semantic integration reflected in the ERP of 14-month-olds. *NeuroReport*, **16**, 653–6.

Friedrich, M. and Friederici, A. D. (2005c) Semantic sentence processing reflected in the event-related potentials of one- and two-year-old children. *NeuroReport*, **16**, 1801–4.

Friederici, A. D., Friedrich, M., and Weber, C. (2002) Neural manifestation of cognitive and precognitive mismatch detection in early infancy. *NeuroReport*, **13**, 1251–4.

Friston, K. J., Harrison, L., and Penny, W. (2003) Dynamic causal modelling. *NeuroImage*, **19**, 1273–302.

Gaillard, W. D., Hertz-Pannier, L., Mott, S. H., et al. (2000) Functional anatomy of cognitive development: fMRI of verbal fluency in children and adults. *Neurology*, **54**, 180–5.

Gaillard, W. D., Sachs, B. C., Whitnah, J. R., et al. (2003) Developmental aspects of language processing: fMRI of verbal fluency in children and adults. *Hum Brain Mapp*, **18**, 176–85.

Hahne, A., Eckstein, K., and Friederici, A. D. (2004) Brain signatures of syntactic and semantic processes during children's language development. *J Cogn Neurosci*, **16**, 1302–18.

Holcomb, P. J., Coffey, S. A., and Neville, H. J. (1992) Visual and auditory sentence processing: a developmental analysis using event related brain potentials. *Dev Neuropsychol*, **8**, 203–41.

Holland, S. K., Plante, E., Weber Byars, A., et al. (2001) Normal fMRI brain activation patterns in children performing a verb generation task. *NeuroImage*, **14**, 837–43.

Homae, F., Watanabe, H., Nakano, T., et al. (2006) The right hemisphere of sleeping infant perceives sentential prosody. *Neurosci Res*, **54**, 276–80.

Karunanayaka, P. R., Holland, S. K., Schmithorst, V. J., et al. (2007) Age-related connectivity changes in fMRI data from children listening to stories. *NeuroImage*, **34**(1), 349–60.

Kooijman, V., Hagoort, P., and Cutler, A. (2005) Electrophysiological evidence for prelinguistic infants' word recognition in continuous speech. *Brain Res Cogn Brain Res*, **24**, 109–16.

Kuhl, P. K. (2004) Early language acquisition: cracking the speech code. *Nat Rev Neurosci*, **5**, 831–43.

Kuhl, P. K. (2007) Is speech learning "gated" by the social brain? *Dev Sci*, **10**(1), 110–20.

Kuhl, P. K., Stevens, E., Hayashi, A., et al. (2006) Infants show a facilitation effect for native language phonetic perception between 6 and 12 months. *Dev Sci*, **9**(2), F13–F21.

Kujala, T. and Näätänen, R. (2001) The mismatch negativity in evaluating central auditory dysfunction in dyslexia. *Neurosci Biobehav Rev*, **25**, 535–43.

Kushnerrenko, E., Cheour, M., Ceponiene, R., et al. (2001) Central auditory processing of durational changes in complex speech patterns by newborns: an event-related brain potential study. *Dev Neuropsychol*, **19**, 83–97.

Kutas, M. and Federmeier, K. D. (2000) Electrophysiology reveals semantic memory use in language comprehension. *Trends Cogn Sci*, **4**, 463–70.

Lee, H., Devlin, J. T., Shakeshaft, C., et al. (2007) Anatomical traces of vocabulary acquisition in the adolescent brain. *J Neurosci*, **27**(5), 1184–9.

Leppänen, P. H. T., Pihko, E., Eklund, K. M., et al. (1999) Cortical responses of infants with and without a genetic risk for dyslexia: II. Group effects. *NeuroReport*, **10**, 969–73.

Lu, L. H., Leonard, C. M., Thompson, P. M., et al. (2007) Normal developmental changes in inferior frontal gray matter are associated with improvement in phonological processing: a longitudinal MRI analysis. *Cereb Cortex*, **17**(5), 1092–9.

McNealy, K., Mazziotta, J. C., and Dapretto, M. (2006) Cracking the language code: neural mechanisms underlying speech parsing. *J Neurosci*, **26**(29), 7629–39.

Meyer, M., Steinhauer, K., Alter, K., et al. (2004) Brain activity varies with modulation of dynamic pitch variance in sentence melody. *Brain Lang*, **89**, 277–89.

Mills, D. L., Coffey-Corina, S., and Neville, H. J. (1997) Language comprehension and cerebral specialization from 13 to 20 months. *Dev Neuropsychol*, **13**, 397–445.

Mills, D. L., Prat, C., Zangl, R., et al. (2004) Language experience and the organization of brain activity to phonetically similar words: ERP evidence from 14- and 20-month-olds. *J Cogn Neurosci*, **16**, 1452–64.

Minagawa-Kawai, Y., Mori, K., Naoi, N., et al. (2007) Neural attunement processes in infants during the acquisition of a language-specific phonemic contrast. *J Neurosci*, **27**(2), 315–21.

Oberecker, R. and Friederici, A. D. (2006) Syntactic ERP components in 24-month-olds' sentence comprehension. *NeuroReport*, **17**, 1017–21.

Oberecker, R., Friedrich, M., and Friederici, A. D. (2005) Neural correlates of syntactic processing in two-year-olds. *J Cogn Neurosci*, **17**(10), 1667–78.

Obrig, H. and Villringer, A. (2003) Beyond the visible: imaging the human brain with light. *J Cereb Blood Flow Metab*, **23**, 1–18.

Opitz, B. and Friederici, A. D. (2003) Interactions of the hippocampal system and the prefrontal cortex in learning language-like rules. *NeuroImage*, **19**, 1730–7.

Opitz, B. and Friederici, A. D. (2004) Brain correlates of language learning: the neuronal dissociation of rule-based versus similarity-based learning. *J Neurosci*, **24**, 8436–40.

Pannekamp, A., Weber, C., and Friederici, A. D. (2006) Prosodic processing at the sentence level in infants. *NeuroReport*, **17**, 675–8.

Pena, M., Maki, A., Kovacic, D., et al. (2003) Sounds and silence: an optical topography study of language recognition at birth. *Proc Natl Acad Sci*, **1000**, 11702–5.

Petersen, S. E., van Mier, H., Fiez, J. A., et al. (1998) The effects of practice on the functional anatomy of task performance. *Proc Natl Acad Sci USA*, **95**, 853–60.

Pihko, E., Leppänen, P. H. T., Eklund, K. M., et al. (1999) Cortical responses of infants with and without a genetic risk for dyslexia: I. Age effects. *NeuroReport*, **10**, 901–5.

Plante, E., Creusere, M., and Sabin, C. (2002) Dissociating sentential prosody from sentence processing: activation interacts with task demands. *NeuroImage*, **17**, 401–10.

Plante, E., Holland, S. K., and Schmithorst, V. J. (2006a) Prosodic processing by children: an fMRI study. *Brain Lang*, **97**, 332–42.

Plante, E., Schmithorst, V. J., Holland, S. K., et al. (2006b) Sex differences in the activation of language cortex during childhood. *Neuropsychologia*, **44**, 1210–21.

Poeppel, D. and Hickok, G. (2004) Towards a new functional neuroanatomy of speech perception. *Cognition*, **92**, 1–12.

Price, C. J. (2000) The anatomy of language: contributions from functional neuroimaging. *J Anat*, **197**, 335–59.

Raichle, M. E., Fiez, J. A., Videen, T. O., et al. (1994) Practice-related changes in human brain functional anatomy during nonmotor learning. *Cereb Cortex*, **4**, 8–26.

Rivera-Gaxiola, M., Klarman, L., Garcia-Sierra, A., et al. (2005a) Neural patterns to speech and vocabulary growth in American infants. *NeuroReport*, **16**, 495–8.

Rivera-Gaxiola, M., Silva-Pereyra, J., and Kuhl, P. K. (2005b) Brain potentials to native and non-native speech contrasts in 7- and 11-month-old American infants. *Dev Sci*, **8**(2), 162–72.

Schapiro, M. B., Schmithorst, V. J., Wilke, M., et al. (2004) BOLD fMRI signal increases with age in selected brain regions in children. *NeuroReport*, **15**, 2575–8.

Schlaggar, B. L., Brown, T. T., Lugar, H. M., et al. (2002) Functional neuroanatomical differences between adults and school-age children in the processing of single words. *Science*, **296**, 1476–9.

Schmithorst, V. J., Holland, S. K., and Plante, E. (2006) Cognitive modules utilized for narrative comprehension in children: a functional magnetic resonance imaging study. *NeuroImage*, **29**, 254–66.

Silva-Pereyra, J., Klarman, L., Lin, L. J., et al. (2005a) Sentence processing in 30-month-old children: an event-related potential study. *NeuroReport*, **16**, 645–8.

Silva-Pereyra, J., Rivera-Gaxiola, M., and Kuhl, P. K. (2005b) An event-related brain potential study of sentence comprehension in preschoolers: semantic and morphosyntactic processing. *Brain Res*, **23**, 247–58.

Sowell, E. R., Delis, D., Stiles, J., et al. (2001) Improved memory functioning and frontal lobe maturation between childhood and adolescence: a structural MRI study. *J Int Neuropsychol Soc*, **7**(3), 312–22.

Sowell, E. R., Thompson, P. M., Leonard, C. M., et al. (2004) Longitudinal mapping of cortical thickness and brain growth in normal children. *J Neurosci*, **23**(38), 8223–31.

Szaflarski, J. P., Holland, S. K., Schmithorst, V. J., et al. (2006a) fMRI study of language lateralization in children and adults. *Hum Brain Mapp*, **27**, 202–12.

Szaflarski, J. P., Schmithorst, V. J., Altaye, M., et al. (2006b) A longitudinal functional magnetic resonance imaging study of language development in children 5 to 11 years old. *Ann Neurol*, **59**, 796–807.

Thiel, C. M., Shanks, D. R., Henson, R. N. A., et al. (2003) Neuronal correlates of familiarity-driven decisions in artificial grammar learning. *NeuroReport*, **14**, 131–6.

Thierry, G., Vihman, M., and Roberts, M. (2003) Familiar words capture the attention of 11-month-olds in less than 250 ms. *NeuroReport*, **14**, 2307–10.

Wang, A. T., Lee, S. S., Sigman, M., et al. (2006) Neural basis of irony comprehension in children with autism: the role of prosody and context. *Brain*, **129**, 932–43.

Wartenburger, I., Steinbrink, J., Telkemeyer, S., et al. (2007) The processing of prosody: evidence of interhemispheric specialization at the age of four. *NeuroImage*, **34**, 416–25.

Weber, C., Hahne, A., Friedrich, M., et al. (2004) Discrimination of word stress in early infant perception: electrophysiological evidence. *Cogn Brain Res*, **18**, 149–61.

Wilke, M., Lidzba, K., Staudt, M., et al. (2005) Comprehensive language mapping in children, using functional magnetic resonance imaging: what's missing counts. *NeuroReport*, **16**, 915–19.

Wood, A. G., Harvey, A. S., Wellard, R. M., et al. (2004) Language cortex activation in normal children. *Neurology*, **63**(6), 1035–44.

Atypical processes in developmental neuropsychiatric disorders

Introduction to Section 2

Neuroimaging has been used extensively to study a variety of disorders in adults, but the study of disorders in children and adolescents remains less developed overall. Developmental studies hold promise for understanding which and when critical neural processes deviate from their expected normal trajectories. Such studies present opportunities for identifying neural risk factors that may be apparent before the onset of symptoms and for identifying neural correlates of illness unconfounded by treatment or the impact of disease itself. Neural processes can be tracked as treatment is implemented to determine correlates of symptom reduction and the direct effects of drugs or other interventions. Tracking the longitudinal course of disease using neuroimaging may also identify the neural correlates of age-related improvement or remission versus persistence of symptoms and impairments and distinguish primary from secondary, compensatory pathophysiology.

In this section, Epstein analyzes the extensive neuroimaging literature on attention deficit hyperactivity disorder (ADHD) (Chapter 7). Zilbovicius, Boddaert and Chabane review the burgeoning neuroimaging literature on autism spectrum disorders (Chapter 8). Keshavan, Diwadkar, Prasad and Stanley describe developmental aspects of schizophrenia (Chapter 9). Kalmar, Shah and Blumberg then discuss cortico-limbic brain circuitry development in bipolar disorder (Chapter 10). Pine assesses developmental findings in anxiety and depressive disorders (Chapter 11). Marsh, Gorman, Royal and Peterson address developmental findings in Tourette syndrome and obsessive-compulsive disorder, two related childhood syndromes (Chapter 12). Using the case of fragile X syndrome, Rivera and Reiss demonstrate how neuroimaging can further the understanding of complex relationships among genes, brain and behavior (Chapter 13). Bjork describes how alcohol exposure affects the developing brain, both through in utero exposures and through abuse of alcohol during adolescence (Chapter 14). Lastly, Frank and Kaye (Chapter 15) discuss the use of neuroimaging for understanding the eating disorders of anorexia and bulimia nervosa.

From this section, the reader will quickly gain an overview of recent research developments and pathophysiological theories of specific disorders. Also to be gained is knowledge of the various methodological approaches to studying these disorders and the integration of knowledge across modalities and methods.

7

A pathophysiology of attention deficit/hyperactivity disorder: clues from neuroimaging

Jeffery N. Epstein

Introduction

This chapter focuses on the growing structural and functional neuroimaging literature that investigates the pathophysiology of attention deficit/hyperactivity disorder (ADHD). Functional imaging studies are reviewed according to cognitive constructs (i.e., response inhibition, working memory, attention, resting state). An integration of the morphometric and functional literature is provided along with an identification of methodological issues and future directions for the field.

As the ADHD neuroimaging literature is quite substantial, this chapter focuses on studies that have compared children, adolescents or adults with ADHD nearly exclusively to normal controls. Research comparing subtypes of ADHD or examining comorbidities within ADHD is sparse. A number of neuroimaging studies of treatment response have been published. These are covered in Chapter 21 and will not be considered here.

Clinical phenomenology and epidemiology

Attention-deficit/hyperactivity disorder is defined by core symptoms of inattention, hyperactivity, and impulsivity, an onset by age 7, and impairment in more than one setting. *The Diagnostic and Statistical Manual of Mental Disorders* 4th edition (American Psychiatric Association, 1994) specifies three sub-types: (1) predominantly inattentive, (2) predominantly hyperactive/impulsive, and (3) combined types. A recent national study of 8- to 15- year-old school children found an 8.6% prevalence rate of ADHD (Froehlich et al., 2007). Developmentally, overt hyperactivity and impulsivity appear to wane with age (Biederman et al., 2000), while symptoms of inattention are highly persistent into adulthood. A national sample of adults revealed a 4.4% prevalence rate among adults aged 18–44 (Kessler et al., 2006).

Children with ADHD are likely to underachieve academically and experience social and disciplinary problems. Adolescents with ADHD are more likely to have problems with driving (e.g., traffic violations, accidents), substance use, high school graduation, delinquency, teenage pregnancy, and injuries than adolescents without ADHD. Adults with ADHD evidence high rates of occupational failure and underachievement, substance abuse and dependence, and unsuccessful relationships. Across all age ranges, patients with ADHD have high rates of comorbid mental disorders (see Barkley, 2006 for a review of impairments across the lifespan).

Neurobiology of attention deficit/ hyperactivity disorder

Twin studies, family studies, adoption studies, and molecular genetic research have all recognized a

Neuroimaging in Developmental Clinical Neuroscience, eds. Judith M. Rumsey and Monique Ernst. Published by Cambridge University Press. © Cambridge University Press 2009.

strong genetic component to ADHD. More recently, molecular research has implicated specific dopaminergic genes, particularly DRD4 and DAT1, as well as COMT (Faraone et al., 2001; Maher et al., 2002), as being related to ADHD. A dopamine deficit hypothesis has been proposed (Solanto, 2002) and supported by the efficacy of psychostimulants (MTA Cooperative Group, 1999), which exert their action in large part via increasing dopamine availability.

Neuropsychological research has identified deficits in a variety of executive functions, including attention, working memory and motoric/response inhibition, pointing to fronto-striatal neurocircuitry dysfunction, in both pediatric and adult samples (Hervey et al., 2004; Willcutt et al., 2005). Intolerance to delay (e.g., Sonuga-Barke, 2005), excessive response variability (e.g., Hervey et al., 2006), and a poor sense of timing (e.g., Bauermeister et al., 2005) have also been reported in ADHD. Over the past two decades, multiple imaging technologies have allowed a more direct investigation of the neural correlates of ADHD-related deficits.

Structural imaging

Approximately 20–25 volumetric magnetic resonance imaging (MRI) studies comparing ADHD subjects to normal controls, most with relatively small sample sizes (modal $n = 15$), have been published. All but two have studied pediatric samples. A recent comprehensive meta-analysis of 21 published pediatric studies (Valera et al., 2007), which together include 565 ADHD and 583 normal subjects, has integrated this literature and calculated between-group effect sizes using a standardized mean difference (SMD).

Taken together, these studies suggest reductions in whole-brain volumes resulting from perturbations in both white matter (fiber tracts) and gray matter (cell bodies). The corpus callosum (CC) is the fiber tract most consistently reported to be altered in ADHD. The posterior CC region (splenium), which contains fibers from the occipital and parietal cortices, emerges as the CC region with the largest between-group difference. The caudate nucleus has been found to show volumetric reductions and altered asymmetries, although results across studies have been inconsistent. Based on the meta-analysis, reduced size of the right caudate nucleus appears to be the most reliable caudate finding. The most reliable and largest effect between-group difference is a reduction in the size of the posterior inferior cerebellar vermis (lobules VIII–X). Finally, the volume of the frontal cortex, particularly the prefrontal region and deep white matter, is smaller among children with ADHD. These findings implicate fronto-striato-cerebellar networks in the pathophysiology of ADHD.

Few studies have focused on adolescents (e.g., Mataro et al., 1997), and only two morphometric studies with adults have been conducted (Hesslinger et al., 2002; Seidman et al., 2006). Both adult studies reported smaller frontal lobe volumes despite different parcellation methods. Notable age-related differences between the pediatric and adult literature include the lack of findings in the caudate and cerebellum in the adult studies. However, the small number of adult studies makes comparisons of the two literatures difficult. Also, differences in measurement approaches across these literatures may be particularly relevant for the cerebellum, where more localized measures have yielded more findings in the pediatric studies.

Structural imaging studies thus suggest both enduring and transient volumetric brain abnormalities, potentially reflecting genetic, environmental, treatment-related, or compensatory influences. Longitudinal studies of individuals whose symptoms persist compared to those whose symptoms remit may be highly informative.

Functional imaging

Brain function has been compared between patients with ADHD and normal controls using measures of regional cerebral blood flow (rCBF) or glucose metabolism, both proxy measures of neuronal activity level, under resting conditions and during cognitive task performance. A number of studies have employed positron emission tomography (PET) or single photon emission computed tomography (SPECT) to measure blood flow and glucose metabolic rates as well as to assay specific neurochemical systems. These studies are limited in children because of the involvement of radiation exposure. Thus, extensive research in pediatric samples has developed using functional MRI (fMRI) paired with cognitive paradigms. Consistent with areas of neuropsychological deficits, fMRI studies of ADHD have primarily employed response inhibition, working memory, and attention paradigms. Response inhibition is often operationally defined as the ability to suppress a prepotent response. Working memory refers to the processes by which information is temporarily stored and manipulated to permit higher-order cognitive processing. Attention is a heterogeneous process: sustained attention involves the continuous monitoring of a stream of stimuli. Selective attention involves the allocation of attentional resources to specific aspects of a set of stimuli. Finally, executive attention involves the resolution of conflict among responses.

Response inhibition

In tasks designed to assess response inhibition, the participant is asked to respond to a defined set of non-target stimuli and to inhibit responding to target stimuli. Based on their preponderant and frequent presentation, non-target stimuli provoke prepotent responding. Within this context, sporadic and unpredictable presentations of target stimuli test participants' ability to inhibit their prepotent response bias. The two most commonly used response inhibition paradigms are the go/no-go (GNG) task and the stop-signal task (SST). While both have many versions and parametric manipulations, the general task design for each is as follows. The GNG requires a motoric response to all non-target stimuli, usually letters of the alphabet. One letter of the alphabet is then designated as

a target (e.g., the letter "X") to which the participant is not to respond.

Similarly, the SST requires participants to make one response when they see one stimulus (e.g., the letter "X") and another response when shown a competing stimulus (e.g., the letter "O"). Sporadically throughout the task, an auditory tone is presented after the stimulus presentation. Participants are told to withhold responding when they hear the tone. The length of time between the stimulus presentation and the tone (i.e., stop signal delay) is varied according to individual performance.

Using errors of commission on the GNG task and reaction times on the SST, children, adolescents, and adults with ADHD demonstrate behavioral deficits on these response inhibition outcomes (see Hervey et al., 2004; Willcutt et al., 2005 for reviews). These effects are some of the largest seen on any neuropsychological measure in the ADHD literature.

As illustrated in Figure 7.1, studies in cognitive neuroscience using fMRI indicate that response inhibition-related activation is localized to right lateralized prefrontal regions such as the inferior and middle frontal gyri (Brodmann's area (BA) 44/45; BA 9/46; Buchsbaum et al., 2005). Dorsolateral prefrontal cortex (DLPFC) is speculated to be engaged when there is a more deliberate response selection process, as opposed to a prepotent response (Rowe et al., 2002) which recruits more ventral regions. Parietal activation is often seen in tandem with prefrontal activation. A prefrontal-parietal network may be necessary for more deliberate or "controlled" inhibitions (Garavan et al., 2002).

Attention deficit/hyperactivity disorder functional neuroimaging results

Twelve ADHD functional neuroimaging studies using response inhibition tasks (nine using GNG and three using SST) have been published (see Table 7.1). Despite similarities, designs (block versus event-related), presentation rates, and stimuli have varied. As shown in Figure 7.1, most studies reported less frontal lobe activation for ADHD subjects relative to controls. Abnormally low

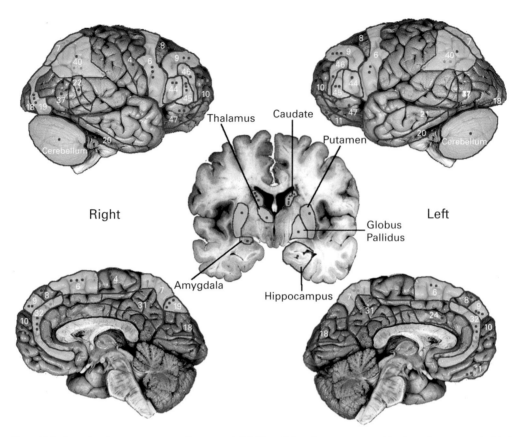

Figure 7.1. Functional neuroanatomy of *response inhibition*. Colored areas represent brain regions purported to be responsible for performance of response inhibition tasks. Brain regions are either labeled with the respective names or are numbered by Brodmann's area. Studies that compared attention deficit/hyperactivity disorder (ADHD) versus normal controls are included with areas of over- (green circles) or under-activation (red circles) illustrated. Studies reporting x, y, z coordinates of activated regions (local maxima) are included (Booth et al., 2005; Durston et al., 2003, 2006; Epstein et al., 2007; Pliszka et al., 2006; Rubia et al., 1999, 2005; Schulz et al., 2004, 2005; Smith et al., 2006; Tamm et al., 2004).

activation appears to be distributed throughout the frontal cortex including orbital, inferior, middle, and superior frontal gyri. There does appear to be a pattern of right lateralized activation deficits, though many studies also localize activation deficits to the left frontal lobe. It should be noted that two studies have reported greater frontal activation in children and adolescents with ADHD (Durston et al., 2003; Shulz et al., 2004). These two studies are unique in terms of the young age of the participants (Durston et al., 2003; mean age = 8.5) and the fast stimulus presentation rate (Shultz et al., 2004).

Within the striatum, reduced activation is seen most often in the caudate nucleus (Booth et al., 2005; Durston et al., 2003; Epstein et al., 2007; Rubia et al., 1999; Vaidya et al., 1998). Notably, all studies have reported less caudate activation among patients with ADHD, with inconsistent lateralization of these deficits. Other brain regions that consistently displayed activation abnormalities were the dorsal anterior cingulate cortex (dACC) and the parietal lobes, particularly BA 40, which demonstrated both less and greater activation among patients with ADHD (Booth et al., 2005; Durston

Table 7.1 Positron emission tomography (PET) and functional magnetic resonance imaging (fMRI) imaging studies comparing patients with attention deficit/hyperactivity disorder (ADHD) to controls

Study	Approximate age group	Method[a]	Design[b]	Task	Contrast/measure
Response inhibition tasks					
Vaidya et al. (1998)	Children	fMRI	Block	Go/no-go	No-go vs. go
Rubia et al. (1999)	Adolescents	fMRI	Block	Stop-signal test	Stop vs. control
Durston et al. (2003)	Children	fMRI	Event	Go/no-go	Correct No-go vs. go
Schulz et al. (2004)	Adolescents	fMRI	Event	Go/no-go	Correct No-go vs. go
Tamm et al. (2004)	Adolescents	fMRI	Event	Go/no-go	No-go vs. go
Booth et al. (2005)	Children	fMRI	Block	Go/no-go	No-go vs. go
Schulz et al. (2005)	Adolescents	fMRI	Event	Go/no-go	Correct No-go vs. go
Rubia et al. (2005)	Adolescents	fMRI	Event	Stop-signal test	Success vs. unsuccessful inhibition
Smith et al. (2006)	Children and adolescents	fMRI	Event	Go/no-go	Correct no-go vs. oddball
Durston et al. (2006)	Children	fMRI	Event	Go/no-go	Correct no-go vs. go
Pliszka et al. (2006)	Children and adolescents	fMRI	Event	Stop-signal test	Success vs. unsuccessful inhibition
Epstein et al. (2007)	Adolescents	fMRI	Event	Go/no-go	Correct no-go vs. go
Epstein et al. (2007)	Adults	fMRI	Event	Go/no-go	Correct no-go vs. go
Working memory tasks					
Schweitzer et al. (2000)	Adults	^{15}O-PET	Block	PASAT	PASAT-number generate
Schweitzer et al. (2004)	Adults	^{15}O-PET	Block	PASAT	PASAT-number generate
Valera et al. (2005)	Adults	fMRI	Block	n-back	2-Back – vigilance
Attention tasks					
Zametkin et al. (1990)	Adults	^{18}F-PET	Continuous	Auditory CPT	CPT
Zametkin et al. (1993)	Adults	^{18}F-PET	Continuous	Auditory CPT	CPT
Ernst et al. (1994)	Adolescents	PET	Continuous	Auditory CPT	CPT
Ernst et al. (1997)	Adolescents	^{18}F-PET	Continuous	Auditory CPT	CPT
Bush et al. (1999)	Adults	fMRI	Block	Counting Stroop	Interference vs. neutral
Shafritz et al. (2004)	Adolescents	fMRI	Event	Selective attention task	Selective attend vs. direction
Zang et al. (2005)	Children	fMRI	Event	Stroop-like task	Interference vs. neutral
Konrad et al. (2006)	Children	fMRI	Event	Attention Network Test	Incongruent vs. congruent
Vaidya et al. (2005)	Children	fMRI	Event	Flanker task	Incongruent vs. neutral
Booth et al. (2005)	Children	fMRI	Block	Visual search	9 Stimuli-1 Stimuli
Adler et al. (2005)	Adolescents	fMRI	Block	CPT-identical pairs	Vigilance vs. control
Smith et al. (2006)	Children and adolescents	fMRI	Event	Motor Stroop	Incongruent vs. incongruent
Tamm et al. (2006)	Adolescents	fMRI	Event	Oddball task	Oddball vs. standard

117

Table 7.1 (*cont.*)

Study	Approximate age group	Method[a]	Design[b]	Task	Contrast/measure
Miscellaneous tasks					
Rubia et al. (1999)	Adolescents and adults	fMRI	Block	Motor timing/delay task	Delayed vs. short event rate
Ernst et al. (2003)	Adults	15O-PET	Block	Gambling task	Gambling vs. control task
Mostofsky et al. (2006)	Children	fMRI	Block	Sequential finger tapping	Finger tapping vs. rest
Silk et al. (2005)	Adolescents	fMRI	Event	Mental rotation	Rotation vs. control
Smith et al. (2006)	Children and adolescents	fMRI	Event	Switch task	Switch vs. repeats
Scheres et al. (2007)	Adolescents	fMRI	Event	Monetary incentive delay task	Reward vs. non-reward
Resting state studies					
Ernst et al. (1998)	Adults	18F-DOPA PET	Resting state	N/A	DOPA accumulation
Ernst et al. (1999)	Adolescents	18F-DOPA PET	Resting state	N/A	DOPA accumulation
Teicher et al. (2000)	Children	fMRI relaxometry	Resting state	N/A	Indirect measure of steady-state blood flow/volume
Cao et al. (2006)	Children	fMRI	Resting state	N/A	Regional homogeneity
Zang et al. (2007)	Children	fMRI	Resting state	N/A	Amplitude of low-frequency fluctuation
Tian et al. (2006)	Adolescents	fMRI	Resting state	N/A	Low-frequency fluctuation

Notes:

[a]fMRI, functional MRI using the blood-oxygen-level-dependent (BOLD) response as a measure of blood flow/volume.

15O-PET, Oxygen-15-labeled water used with positron emission tomography (PET) to measure regional cerebral blood flow.

18F-PET, Fluorine-18-labeled fluorodeoxyglucose used with PET to measure regional cerebral glucose utilization.

18F-DOPA PET, Fluorine-18-labeled fluorodopa used to measure DOPA accumulation.

[b]Block designs present control vs. experimental condition stimuli in separate blocks.

Event-related designs present control and experimental stimuli/items interspersed with each other. For data analysis, trials are then sorted by type of item and/or accuracy of performance for statistical comparison.

Continuous designs present a single task continuously to achieve a steady state and require no contrast condition because the imaging technique yields a quantitative measure of some physiological variable, e.g., cerebral metabolic rate for glucose, in absolute units.

et al., 2006; Epstein et al., 2007; Rubia et al., 1999; Schulz et al., 2004). Interestingly, ADHD-related activation abnormalities in both the dACC and parietal lobes were generally congruent with the direction of deficits (abnormally high or low activation) observed in the frontal lobes across studies.

Overall, patients with ADHD seem to underactivate fronto-striatal regions during response inhibition (see Figure 7.1). Within these regions, functional deficits of left caudate and the inferior frontal gyrus (IFG) (BA 44, 45, and 46) emerge as the most reliable findings. In addition, ADHD subjects tend to overactivate the superior parietal lobe (BA 40). This region is particularly involved in visuospatial attention processes.

Developmental considerations

The samples used in studies of response inhibition are children, with most subjects ranging in age from 10 to 15 years. Within this limited age range, no relationships have been noted between age and activation patterns (Durston et al., 2003, 2006; Rubia et al., 2005). Epstein et al. (2007) used an innovative research design in which adolescents with ADHD and their parents who were also diagnosed with ADHD were imaged while performing a GNG task. Activation patterns were compared to those of age-matched normal parent-adolescent dyad controls. Similar activation abnormalities were observed between adolescent and adult ADHD samples, particularly in the right IFG and caudate regions where both parents and adolescents with ADHD showed reductions in activation relative to controls. Thus, ADHD-related functional activation deficits during response inhibition may not change with age.

Working memory

Children with ADHD demonstrate significant deficits compared to normal controls on a variety of working memory (WM) tasks (for reviews see Willcutt et al., 2005; Martinussen et al., 2005), with larger between-group effect sizes for spatial than for verbal WM tasks (Martinussen et al., 2005).

Neuroimaging studies of WM in ADHD have used the n-back task and Paced Auditory Serial Addition Task (PASAT). The n-back task presents a series of letters or numbers individually. The participant must decide whether the target stimulus matches or does not match the stimulus that was presented one, two, or even three stimuli back in the presentation stream. The PASAT presents a stimulus stream of numbers. The participant must add each number to the previous number and state the correct sum of the last two presented numbers. Both the n-back task and PASAT require the storage *and* manipulation of information.

As shown in Figure 7.2, a prefrontal network is consistently implicated in working memory (Cabeza et al., 2000). Areas in the middle frontal gyrus (MFG, BA 9 and 46) and inferior frontal gyrus (IFG, BA 44) are most consistently activated during working memory performance. In addition, parietal regions (BA 7 and 40) are frequently activated during verbal/numeric tasks. Parietal activation tends to be left-lateralized, which suggests both a relationship to linguistic operations and coordination with left frontal IFG activation (i.e., Broca's area). Some have postulated that left parietal activation reflects the phonological store, while left frontal activation represents the rehearsal process (Paulesu et al., 1993).

Attention deficit/hyperactivity disorder functional neuroimaging results

Three functional imaging studies have been conducted using WM paradigms, all with adult ADHD subjects (Schweitzer et al., 2000, 2004; Valera et al., 2005). Schweitzer et al. (2000) did not directly compare activation patterns across groups. Between the Schweitzer et al. (2004) and Valera et al. (2005) studies, it is difficult to find consistencies in the results. This may be due to the different imaging methods (^{15}O-PET versus fMRI) and different behavioral paradigms (PASAT versus n-back). Using ^{15}O-PET, Schweitzer et al. (2004) found less activation among the adults with ADHD in left-sided circuitry

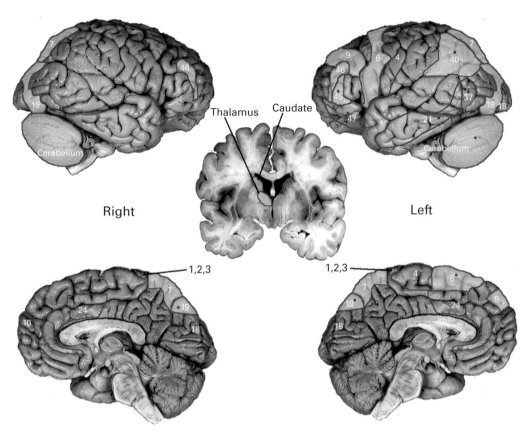

Figure 7.2. Functional neuroanatomy of *working memory.* Colored areas represent brain regions purported to be responsible for performance of working memory tasks. Brain regions are either labeled with the respective names or are numbered by Brodmann's area. Studies that compared ADHD versus normal controls are included with areas of over- (green circles) or under-activation (red circles) illustrated. Studies reporting x, y, z coordinates of activated regions (local maxima) are included (Schweitzer et al., 2004; Valera et al., 2005).

including frontal, temporal, and occipital areas. Greater activation was observed in the right midbrain, right caudate, anterior cerebellum, and left MFG. However, Valera et al. (2005) used fMRI paired with an n-back task and reported less cerebellar and occipital lobe activation in adults with ADHD compared to controls. Hence, in the only area where both studies report differences in activation (i.e., cerebellum), the directions of the findings conflict. Further interpretation of these findings awaits a larger corpus of neuroimaging studies of WM in ADHD.

Attention

Sustained attention tasks, such as Continuous Performance Tasks (CPT), require the participant to identify targets in a stream of stimuli. Selective attention tasks, e.g., visual search, typically require the participant to search within a field of stimuli for a target object. These tasks become progressively more difficult as the features of the non-target stimuli (e.g., shape, color, movement) become more similar to those of the target stimuli.

Tasks that assess executive attention present stimuli that pose some form of response conflict. For at least some of the stimuli, an initial response set must be overcome to produce the correct response. Two such widely used tasks are the Stroop and the flanker tasks. The Stroop task has been adapted for imaging environments as a Counting Stroop (Bush et al., 1998). Number words (e.g., "one") are printed multiple times. The respondent must respond according to the number of times the word is printed on the screen and overcome the conflict to respond to the printed word. The flanker task presents a central target arrow pointing either right or left. The target arrow is flanked by arrows on either side that are either pointed the same way (congruent) or in the opposite direction as the target arrow (incongruent). The participant must resolve the conflict presented by the flanking stimuli on incongruent trials to indicate the direction of the central target stimulus.

Higher-order attentional functioning, as shown in Figure 7.3, involves a fronto-parietal association network that may be mediated by the thalamus (Coull et al., 1996; Sarter et al., 2001). Basic arousal systems involve large parts of the brain, including the ventral stream (occipital and temporal and pre-frontal cortices) and dorsal stream (occipital and parietal and prefrontal cortices). The parietal cortex in the dorsal stream plays a role in spatial repre-sentation and the location of attended objects in space. The temporal cortex in the ventral stream plays a role in object identity and the coding of complex objects. The end-point of these attention streams, the prefrontal cortex, is involved in both selective and sustained attention and likely plays a role in the motor-intentional aspects of attention.

Attention deficit/hyperactivity disorder functional neuroimaging results

Figure 7.3 illustrates between-group imaging results across studies. Position emission tomography stud-ies of sustained attention using an auditory CPT (Ernst et al., 1994, 1997; Zametkin et al., 1990, 1993)

have reported a lower cerebral metabolic rate for glucose (CMRglc) in the frontal lobes among adolescents and adults with ADHD compared with their controls. However, a gender effect was reported by Ernst et al. (1994) suggesting that girls, but not boys, demonstrated abnormally low CMRglc. A follow-up study of a larger sample of adolescent girls (Ernst et al., 1997) revealed no differ-ences in CMRglc. However, regional CMRglc was less in the left parietal lobe and striatum, specifically putamen, compared to normals. Discrepancies across these PET studies are possibly explained by small samples with differing characteristics (e.g., sexual maturation levels).

Using an oddball task and fMRI, Tamm et al. (2006) found no activation differences between ADHD adolescents and their controls in the frontal lobes. Rather, between-group differences were noted in the mid-cingulate cortex, parietal lobe bilaterally and right precuneus. In a study focused on comorbid ADHD, Adler et al. (2005) compared adolescents with bipolar disorder to adolescents with bipolar disorder and ADHD using a CPT with fMRI. Subjects with comorbid bipolar + ADHD had less activation in ventrolateral prefrontal cortex (BA 10) and ACC and greater activation in the posterior parietal cortex and middle temporal gyrus than did adolescents with bipolar disorder alone. The authors concluded that having comorbid ADHD may be related to preferentially recruiting portions of the parietal and temporal cortex to perform attentional functions.

Two studies have examined selective attention. Shafritz et al. (2004) used a selective attention task that required participants to attend either aurally or visually to combined auditory/visual presentations. Booth et al. (2005) used a visual search paradigm. Shafritz et al. noted less activation among patients with ADHD in the basal ganglia and temporal lobe, while Booth et al. noted activation differences, some with less activation and some with more, in frontal, parietal, occipital, cingulate, thalamic, and striatal regions. Less activation in the basal ganglia was reported in both studies, although the

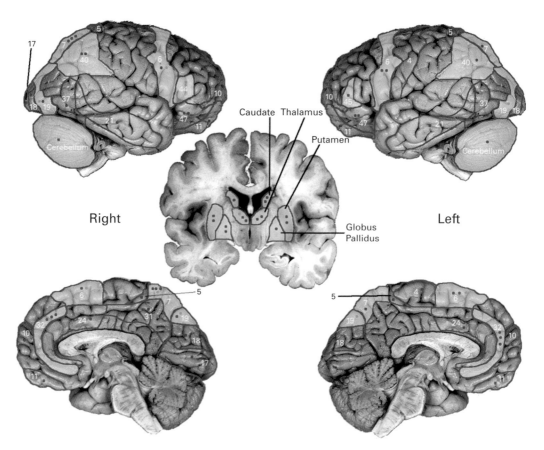

Figure 7.3. Functional neuroanatomy of *attention*. Colored areas represent brain regions purported to be responsible for performance of attention tasks. Brain regions are either labeled with the respective names or are numbered by Brodmann's area. Studies that compared ADHD versus normal controls are included with areas of over- (green circles) or under-activation (red circles) illustrated. Studies reporting *x, y, z* coordinates of activated regions (local maxima) are included (Booth et al., 2005; Bush et al., 1999; Konrad et al., 2006; Shafritz et al., 2004; Smith et al., 2006; Tamm et al., 2006; Vaidya et al., 2005; Zang et al., 2005).

structures involved differed–ventral striatum in Shafritz et al. (2004) and globus pallidus in Booth et al. (2005).

Studies of executive attention involving conflict have used the Stroop (three studies), flanker task (one study), and the Attention Network Test (ANT) Conflict condition, a task involving valid and invalid cues (one study). Consistent findings include left-sided underactivation of the frontal lobes during conflict (Konrad et al., 2006; Vaidya et al., 2005; Zang et al., 2005), but with variation in

the locus. In addition, the dACC appears to be underactivated during the Stroop for patients with ADHD (Bush et al., 1999; Zang et al., 2005). Finally, two studies found underactivation in the putamen (Konrad et al., 2006; Zang et al., 2005). Notably, in addition to putamen underactivation among children with ADHD, Zang et al. (2005) found under-activation in bilateral cerebellum, caudate, and globus pallidus. The study by Bush et al. (1999) also suggested the presence of underactivation of frontal, parietal, and occipital cortices, as well as overactivation of IFG, bilateral insula, and bilateral striatum. However, between-group statistical comparisons were not reported for these regions.

Taken together, findings converge upon a pattern of fronto-parietal-striatal underactivation for patients with ADHD when performing attention tasks. Frontal underactivation appears to be localized to the dACC and caudal regions of the superior and middle frontal gyri. Striatal underactivation is rather pervasive and has included areas of the caudate, putamen, and globus pallidus. Functional deficits of these dopaminergic structures are consistent with findings of perturbed dopamine transmission evidenced by PET studies (Ernst et al., 1998; Volkow et al., 2005). Together, these findings support the dopaminergic hypothesis of ADHD pathophysiology.

Developmental considerations

Since few attentional paradigms have been used across children and adults, it is difficult to determine whether developmental trends exist. The exceptions are the Zametkin et al. (1990, 1993) fluorodeoxyglucose PET studies which used a CPT with adults and adolescents, and the Bush et al. (1999) and Zang et al. (2005) fMRI studies which used Stroop-like tasks with adults and children, respectively. Zametkin et al. (1990, 1993) found reductions of global glucose metabolism in adults, but not in adolescents, although both groups showed some frontal regions with reductions that reached significance. Both Bush et al. (1999) and Zang et al. (2005) found underactivation in dACC. The Zang et al. and Bush et al. findings differed in terms of striatal and frontal activation. Whereas Zang et al. found striatal and frontal underactivation for the children with ADHD during the Stroop, Bush et al. presented some data (but without between-group statistical comparisons) suggestive of overactivation in these regions for adults with ADHD. Key differences in task design (block versus event) and task parameters (Counting Stroop versus Stroop-like directional task) further limit comparisons.

Studies using other activation tasks

In addition to deficits in response inhibition, WM, and attention, there is evidence that ADHD is associated with motor overflow (Mostofsky et al., 2006), a poor sense of timing (Barkley et al., 1997; Bauermeister et al., 2005), and an impaired ability to delay gratification (Sonuga-Barke, 2005), the neural correlates of which have also been investigated (though to a lesser extent) using fMRI. To examine the neural correlates of anomalous motor function (e.g., excessive overflow movements) in ADHD, Mostofsky et al. (2006) used a sequential finger tapping task in children with and without ADHD. These investigators reported less activation among patients with ADHD in the right superior parietal region and primary motor cortex, although between-group differences in the primary motor cortex were not due to magnitude of activation, but rather to the spatial extent of the activation, suggesting a possible ADHD-related under-recruitment in primary motor cortex for coordination of finger tapping.

Silk et al. (2005) studied adolescents with ADHD and controls during a mental rotation task. Such mental rotation tasks have been shown to activate the superior parietal and middle frontal areas, which contribute to the coding for spatio-visual attention processes (Booth et al., 2000; Parsons, 2003). Less activation among children with ADHD was observed in bilateral caudate, left lateralized inferior and superior frontal gyri, visual association cortex, and temporal and parietal regions. Also observed was greater activation in temporal, parietal, and cingulate regions. These results were interpreted as evidence of widespread fronto-striatal-parietal dysfunction in adolescents with ADHD.

Rubia et al. (1999) used a motor delay task requiring the timing of a motor response to a cue, a task which may target ADHD-related time estimation deficits reported by several investigators (e.g., Barkley et al., 1997). Less activation in the anterior and posterior cingulate and greater activation in the supplementary motor cortex were seen in adolescents and young adults with ADHD compared to normal controls. Smith et al. (2006) assessed children's and adolescents' performance on a task that required them to switch response sets throughout testing. Two large areas of underactivation were noted for subjects with ADHD compared to controls: (1) left IFG extending to the superior temporal lobe and insula and (2) right IFG extending to the superior temporal and inferior parietal gyri.

To examine the neural correlates of reward processing (e.g., delay aversion, risk taking), Scheres et al. (2007) used a monetary incentive task that consistently engages the striatum. Consistent with predictions, these investigators found less ventral striatal activation during reward anticipation for adolescents with ADHD compared to controls. Finally, Ernst et al. (2003) used ^{15}O-PET and a gambling task with adults with and without ADHD. This study probed decision-making rather than reward anticipation. The gambling task produced activation in prefrontal and insular regions; however, patients with ADHD exhibited smaller areas of activation in these regions, as well as less intense activation in the hippocampus and insula relative to normal controls. Adults with ADHD activated the ACC more than normal controls. Such differences in patterns of activation may suggest either reliance on different strategies or an under-recruitment of task-specific regions (i.e., insula).

Resting state studies

Resting state studies include no cognitive challenge and may help to elucidate whether activation differences result from differences present in the processes under study or in baseline resting activity. Early resting state SPECT research offers limited conclusions due to the many methodological concerns with these early studies (for a review of these concerns see Bush et al., 2005; Castellanos, 2002).

Teicher et al. (2000) examined blood volumes in the striatum (putamen and caudate) using fMRI relaxometry. They reported that boys with ADHD seemed to have decreased blood flow in the putamen relative to controls, presumably reflecting lower metabolic demand, during rest. Several resting state fMRI studies have recently been completed using a variety of methods and dependent measures (Cao et al., 2006; Tian et al., 2006; Zang et al., 2007). Several brain regions including IFG, ACC, caudate, cerebellum, and motor cortex appear to have atypical patterns of activity during rest among patients with ADHD (Cao et al., 2006; Zang et al., 2005, 2007). Tian et al. (2006) reported greater functional connectivity between dACC and bilateral thalamus, bilateral cerebellum, bilateral insula, and bilateral brainstem (pons), which was interpreted as abnormal autonomic system functional control.

Summary and integration of results

Until recently, hypotheses concerning the pathophysiology of ADHD were mainly focused on frontostriatal circuits (Castellanos, 1997). The recent explosion of neuroimaging studies of ADHD has expanded this view to include the involvement of the parietal lobes, cingulate cortex, and cerebellum. The parietal lobe appears to play an active role in response inhibition, WM, and attention. Parietal lobe findings seem to converge on BA 7 and 40, which include the superior parietal lobe, precuneus, and supramarginal gyrus. Given the co-occurrence of ADHD-related frontal and parietal activation abnormalities and the known involvement of frontoparietal networks in attentional processes (Coull et al., 1996), altered functional connectivity between these two areas may be an important aspect of ADHD. Cingulate abnormalities have been reported primarily in the dACC region and are seen concurrently with other frontal findings, suggesting here again perturbed functional connectivity between these regions.

Finally, morphometric and functional imaging studies demonstrate smaller volumes and abnormal activation patterns in the cerebellum, especially in the vermis (Schulz et al., 2004; Schweitzer et al., 2000, 2004; Valera et al., 2005). Cerebellar findings may be specific to certain task requirements, as cerebellar abnormalities were detected more frequently with

WM tasks than with response inhibition and attention tasks. Cerebellar abnormalities also were seen in all of the resting state fMRI studies, suggesting atypical cerebellar activity during rest (Cao et al., 2006; Tian et al., 2006; Zang et al., 2007).

The networks potentially involved in the pathophysiology of ADHD have thus been expanded to include fronto-striatal-cerebellar, fronto-parietal-thalamic, and fronto-cingulate networks. Dysfunction in any or all of these networks could be caused by any one or a combination of: (1) a structural anomaly; (2) altered connectivity; and (3) a neurochemical perturbation.

Structural anomaly

The one region common to all of the cited networks is the frontal cortex. The frontal lobes have been found to be smaller in size (Valera et al., 2007), to exhibit cortical thinning (Makris et al., 2007; Shaw et al., 2006), and to show correlations between their volume and neuropsychological performance in patients with ADHD (Casey et al., 1997; Hill et al., 2003). Other regions evidencing structural abnormalities may be the striatum, including caudate and putamen, and cerebellum. A recent study found that hyperintensities in subcortical areas may be more prevalent in children with ADHD, suggesting the possibility that injury to striatal regions may be involved in ADHD (Amat et al., 2006).

Altered connectivity

An initial diffusion tensor imaging (DTI) study (Ashtari et al., 2005) found lower fractional anisotropy (suggestive of altered anatomical connectivity) in many of the tracts connecting regions implicated in the pathophysiology of ADHD (e.g., frontal lobes, striatum, cerebellum). In addition, the observed correspondences between abnormal activation patterns in regions such as frontal and parietal areas (as reviewed earlier in this chapter) suggest that altered connectivity between regions may be involved in the pathophysiology of ADHD.

Neurochemical perturbation

Dopamine neurotransmission is a primary candidate for a role in the pathogenesis of ADHD and is strongly supported by the therapeutic efficacy of stimulants in patients with ADHD (MTA Cooperative Group, 1999). However, it is highly likely that other neurotransmitter systems are involved as well. For example, noradrenergic transmission, given its role in attention and the beneficial effects of noradrenergic agents (e.g., clonidine, atomoxetine) in ADHD, is also likely to be involved. Similarly, the serotonergic system, based on its role in impulsivity, also needs to be considered. Attention deficit/hyperactivity disorder is a complex disorder with a heterogeneous presentation that can be accounted for by a range of pathophysiological mechanisms. Solutions to the puzzle of the neural mechanisms underlying ADHD will come from the integration of research findings across multiple domains, including genetics, neuroimaging, and animal models.

The integration of the ADHD imaging research literature presents several challenges. The literature is heterogeneous in terms of technologies (i.e., fMRI, SPECT, PET), experimental designs, and patient populations. Imaging techniques probe different biological substrates (e.g., blood-oxygen-level-dependent response versus cerebral glucose metabolism) and present distinct spatial and temporal resolutions. Experimental designs include both blocked and event-related designs and a wide variety of behavioral paradigms and task parameters, which target different neural systems. A further complication is the variability in data analytic techniques, e.g., use of regions of interest versus whole brain analyses, statistics applied (e.g., t tests versus ANOVAs). Patient samples vary in their comorbidities, treatment histories, severity, and ability to remain still in the scanner. Particularly with respect to ADHD, significant numbers of subjects are lost to data analysis due to head motion (Durston et al., 2003; Epstein et al., 2007; Pliszka et al., 2006; Tian et al., 2006), thus biasing findings toward less severely hyperactive subjects. Finally, differences in ADHD versus control groups may be contaminated by differences in effort expended to remain still in the scanner.

Future directions

The existing imaging literature has provided preliminary clues regarding the pathophysiology of ADHD. Indeed, both structural and functional studies suggest fronto-striatal-cerebellar abnormalities with associated dysfunction in parietal and cingulate regions. Much can be accomplished with future imaging studies to further define this pathophysiology. It is not surprising that the structural imaging literature has demonstrated the most consistent results, as the methods are generally more uniform than those used in the functional imaging literature. One future goal for functional imaging studies should be to standardize methods and to utilize behavioral paradigms that have been validated in the larger cognitive neuroscience literature.

Secondly, the majority of imaging research has been pediatric in nature. Despite the chronicity of ADHD, only two structural imaging studies have been published using adult samples and the functional studies conducted with adults are mostly PET and SPECT studies with only a couple of fMRI studies. Given this dearth of adult studies, it is difficult to draw conclusions as to the developmental course of ADHD-related brain abnormalities. While morphometric findings have been reported developmentally (Castellanos et al., 2002), no study has reported on repeated assessment of ADHD-related functional impairments over time. Longitudinal assessment of ADHD-related functional impairments may shed new light on the neural correlates of persistence versus remission of symptoms and help define which functional signatures reflect primary pathophysiology versus compensatory activity.

Thirdly, future research should focus on integrating findings across modalities. As noted, structural and functional imaging has implicated several areas of dysfunction, but the mechanism by which these structural and/or functional abnormalities produce ADHD behavior is unknown. For example,

it is unclear whether the functional abnormalities being observed are the result of top-down or bottom-up deficits. Combining the spatial resolution of fMRI with the temporal resolution of magnetoelectro-encephalography or EEG affords an opportunity to better understand the circuitry involved in ADHD-related brain abnormalities. Along these same lines, technologies such as DTI and magnetic resonance spectroscopy should be utilized to better understand connectivity and brain chemistry.

Finally, given the strong heritability of ADHD, a relatively large number of genetic studies have focused on identifiying ADHD-related genes. Increasingly, genetic researchers are interested in using endophenotypes to define disorders rather than the heterogeneous clinical behavioral criteria (Doyle et al., 2005). Therefore, effort should be directed at defining endophenotypes. Much work needs to be done to establish whether the neuroimaging findings can meet the criteria that define an endophenotype.

ACKNOWLEDGMENT

Work on this chapter was supported by NIH grant #MH64478.

REFERENCES

Adler, C. M., Del Bello, M. P., Mills, N. P., et al. (2005) Comorbid ADHD is associated with altered patterns of neuronal activation in adolescents with bipolar disorder performing a simple attention task. *Bipolar Disord*, **7**, 577–88.

Amat, J. A., Bronen, R. A., Saluja, S., et al. (2006) Increased number of subcortical hyperintensities on MRI in children and adolescents with Tourette's Syndrome, Obsessive-Compulsive Disorder, and Attention Deficit Hyperactivity Disorder. *Am J Psychiatry*, **163**, 1106–8.

American Psychiatric Association (1994) *Diagnostic and Statistical Manual of Mental Disorders*, 4th edn. Arlington, VA: American Psychiatric Association.

Ashtari, M., Kumra, S., Bhaskar, S. L., et al. (2005) Attention-deficit/hyperactivity disorder: a preliminary diffusion tensor imaging study. *Biol Psychiatry*, **57**, 448–55.

Barkley, R. A. (2006) *Attention-Deficit Hyperactivity Disorder: A Handbook for Diagnosis and Treatment.* New York: The Guilford Press.

Barkley, R. A., Koplowicz, S., Anderson, T., et al. (1997) Sense of time in children with ADHD: effects of duration, distraction, and stimulant medication. *J Int Neuropsychol Soc*, **3**, 359–69.

Bauermeister, J. J., Barkley, R. A., Martinez, J. V., et al. (2005) Time estimation and performance on reproduction tasks in subtypes of children with attention deficit hyperactivity disorder. *J Clin Child Adolesc Psychol*, **34**, 151–62.

Biederman, J., Mick, E., and Faraone, S. V. (2000) Age-dependent decline of symptoms of attention deficit hyperactivity disorder: impact of remission definition and symptom type. *Am J Psychiatry*, **157**, 816–18.

Booth, J. R., Burman, D. D., Meyer, J. R., et al. (2005) Larger deficits in brain networks for response inhibition than for visual selective attention in attention deficit hyperactivity disorder (ADHD). *J Child Psychol Psychiatry*, **46**, 94–111.

Booth, J. R., Macwhinney, B., Thulborn, K. R., et al. (2000) Developmental and lesion effects in brain activation during sentence comprehension and mental rotation. *Dev Neuropsych*, **18**, 139–69.

Buchsbaum, B. R., Greer, S., Chang, W. L., et al. (2005) Meta-analysis of neuroimaging studies of the Wisconsin card-sorting task and component processes. *Hum Brain Mapp*, **25**, 35–45.

Bush, G., Frazier, J. A., Rauch, S. L., et al. (1999) Anterior cingulate cortex dysfunction in attention-deficit/hyperactivity disorder revealed by fMRI and counting stroop. *Biol Psychiatry*, **45**, 1542–52.

Bush, G., Valera, E. M., and Seidman, L. J. (2005) Functional neuroimaging of attention-deficit/hyperactivity disorder: a review and suggested future directions. *Biol Psychiatry*, **57**, 1273–84.

Bush, G., Whalen, P. J., Rosen, B. R., et al. (1998) The counting Stroop: an interference task specialized for functional neuroimaging – validation study with functional MRI. *Hum Brain Mapp*, **6**, 270–82.

Cabeza, R. and Nyberg, L. (2000) Imaging cognition II: An empirical review of 275 PET and fMRI studies. *J Cogn Neurosci*, **12**, 1–47.

Cao, Q., Zang, Y., Sun, L., et al. (2006) Abnormal neural activity in children with attention deficit hyperactivity disorder: a resting-state functional magnetic resonance imaging study. *NeuroReport*, **17**, 1033–6.

Casey, B. J., Castellanos, F. X., Giedd, J. N., et al. (1997) Implication of right frontostriatal circuitry in response inhibition and attention-deficit/hyperactivity disorder. *J Am Acad Child Adolesc Psychiatry*, **36**, 374–83.

Castellanos, F. X. (1997) Toward a pathophysiology of attention-deficit/hyperactivity disorder. *Clin Pediat*, **36**, 381–93.

Castellanos, F. X. (2002) Proceed, with caution: SPECT cerebral blood flow studies of children and adolescents with attention deficit hyperactivity disorder [comment]. *J Nucl Med*, **43**, 1630–3.

Castellanos, F. X., Lee, P. P., Sharp, W., et al. (2002) Developmental trajectories of brain volume abnormalities in children and adolescents with attention-deficit/hyperactivity disorder. *J Am Med Assoc*, **288**(14), 1740–8.

Coull, J. T., Frith, C. D., Frackowiak, R. S., et al. (1996) A fronto-parietal network for rapid visual information processing: a PET study of sustained attention and working memory. *Neuropsychologia*, **34**, 1085–95.

Doyle, A. E., Willcutt, E. G., Seidman, L. J., et al. (2005) Attention-deficit/hyperactivity disorder endophenotypes. *Biol Psychiatry*, **57**, 1324–35.

Durston, S., Mulder, M., Casey, B. J., et al. (2006) Activation in ventral prefrontal cortex is sensitive to genetic vulnerability for attention-deficit hyperactivity disorder. *Biol Psychiatry*, **60**, 1062–70.

Durston, S., Tottenham, N. T., Thomas, K. M., et al. (2003) Differential patterns of striatal activation in young children with and without ADHD. *Biol Psychiatry*, **53**, 871–8.

Epstein, J. N., Casey, B. J., Tonev, S. T., et al. (2007) ADHD- and medication-related brain activation differences in concordantly affected parent-child dyads with ADHD. *J Child Psychol Psychiatry*, **48**, 899–913.

Ernst, M., Cohen, R. M., Liebenauer, L. L., et al. (1997) Cerebral glucose metabolism in adolescent girls with attention-deficit/hyperactivity disorder. *J Am Acad Child Adolesc Psychiatry*, **36**, 1399–406.

Ernst, M., Kimes, A. S., London, E. D., et al. (2003) Neural substrates of decision making in adults with attention deficit hyperactivity disorder. *Am J Psychiatry*, **160**, 1061–70.

Ernst, M., Liebenauer, L. L., King, A. C., et al. (1994) Reduced brain metabolism in hyperactive girls. *J Am Acad Child Adolesc Psychiatry*, **33**, 858–68.

Ernst, M., Zametkin, A. J., Matochik, J. A., et al. (1998) DOPA decarboxylase activity in attention deficit hyperactivity disorder adults. A [fluorine-18]fluorodopa positron emission tomographic study. *J Neurosci*, **18**, 5901–7.

Ernst, M., Zametkin, A. J., Matochik, J. A., et al. (1999) High midbrain [^{18}F]dopa accumulation in children with attention deficit hyperactivity disorder. *Am J Psychiatry*, **156**, 1209–15.

Faraone, S. V., Doyle, A. E., Mick, E., et al. (2001) Meta-analysis of the association between the 7-repeat allele of the dopamine D(4) receptor gene and attention deficit hyperactivity disorder. *Am J Psychiatry*, **158**, 1052–7.

Froehlich, T. E., Lanphear, B. P., Epstein, J. N., et al. (2007) Prevalence and treatment of attention-deficit/hyperactivity disorder in a national sample of U.S. children. *Arch Pediatr Adolesc Med*, **161**(9), 857–64.

Garavan, H., Ross, T. J., Murphy, K., et al. (2002) Dissociable executive functions in the dynamic control of behavior: inhibition, error detection, and correction. *NeuroImage*, **17**, 1820–9.

Hervey, A. S., Epstein, J. N., and Curry, J. F. (2004) The neuropsychology of adults with attention deficit hyperactivity disorder: a meta-analytic review. *Neuropsychology*, **18**, 485–503.

Hervey, A., Epstein, J. N., Curry, J. F., et al. (2006) Reaction time distribution analysis of neuropsychological performance in an ADHD sample. *Child Neuropsychol*, **12**, 125–40.

Hesslinger, B., Tebartz Van Elst, L., Thiel, T., et al. (2002) Frontoorbital volume reductions in adult patients with attention deficit hyperactivity disorder. *Neurosci Lett*, **328**, 319–21.

Hill, D. E., Yeo, R. A., Campbell, R. A., et al. (2003) Magnetic resonance imaging correlates of attention-deficit/hyperactivity disorder in children. *Neuropsychology*, **17**, 496–506.

Kessler, R. C., Adler, L., Barkley, R., et al. (2006) The prevalence and correlates of adult ADHD in the United States: results from the National Comorbidity Survey Replication. *Am J Psychiatry*, **163**, 716–23.

Konrad, K., Neufang, S., Hanisch, C., et al. (2006) Dysfunctional attentional networks in children with attention deficit/hyperactivity disorder: evidence from an event-related functional magnetic resonance imaging study. *Biol Psychiatry*, **59**, 643–51.

Maher, B. S., Marazita, M. L., Ferrell, R. E., et al. (2002) Dopamine system genes and attention deficit hyperactivity disorder: a meta-analysis. *Psychiatr Genet*, **12**, 207–15.

Makris, N., Biederman, J., Valera, E. M., et al. (2007) Cortical thinning of the attention and executive function networks in adults with attention-deficit/hyperactivity disorder. *Cerebr Cortex*, **17**(6), 1364–75.

Martinussen, R., Hayden, J., Hogg-Johnson, S., et al. (2005) A meta-analysis of working memory impairments in children with attention-deficit/hyperactivity disorder. *J Am Acad Child Adolesc Psychiatry*, **44**, 377–84.

Mataro, M., Garcia-Sanchez, C., Junque, C., et al. (1997) Magnetic resonance imaging measurement of the caudate nucleus in adolescents with attention-deficit hyperactivity disorder and its relationship with neuropsychological and behavioral measures. *Arch Neurol*, **54**, 963–8.

Mostofsky, S. H., Rimrodt, S. L., Schafer, J. G. B., et al. (2006) Atypical motor and sensory cortex activation in attention-deficit/hyperactivity disorder: a functional magnetic resonance imaging study of simple sequential finger tapping. *Biol Psychiatry*, **59**, 48–56.

MTA Cooperative Group (1999) A 14-month randomized clinical trial of treatment strategies for attention-deficit/hyperactivity disorder. *Arch Gen Psychiatry*, **56**, 1073–86.

Parsons, L. M. (2003) Superior parietal cortices and varieties of mental rotation. *Trends Cogn Neurosci*, **7**, 515–17.

Paulesu, E., Frith, C. D., and Frackowiak, R. S. J. (1993) The neural correlates of the verbal component of working memory. *Nature*, **362**, 342–5.

Pliszka, S. R., Glahn, D. C., Semrud-Clikeman, M., et al. (2006) Neuroimaging of inhibitory control areas in children with attention deficit hyperactivity disorder who were treatment naïve or in long-term treatment. *Am J Psychiatry*, **163**, 1052–60.

Rowe, J., Friston, K., Frackowiak, R., et al. (2002) Attention to action: specific modulation of corticocortical interactions in humans. *NeuroImage*, **17**, 988–98.

Rubia, K., Overmeyer, S., Taylor, E., et al. (1999) Hypofrontality in attention deficit hyperactivity disorder during higher-order motor control: a study with functional MRI. *Am J Psychiatry*, **156**, 891–6.

Rubia, K., Smith, A. B., Brammer, M. J., et al. (2005) Abnormal brain activation during inhibition and error detection in medication-naive adolescents with ADHD. *Am J Psychiatry*, **162**, 1067–75.

Sarter, M., Givens, B., Bruno, J. P., et al. (2001) The cognitive neuroscience of sustained attention: where top-down meets bottom-up. *Brain Res Brain Res Rev*, **35**, 146–60.

Scheres, A., Milham, M. P., Knutson, B., et al. (2007) Ventral striatal hyporesponsiveness during reward anticipation in attention-deficit/hyperactivity disorder. *Biol Psychiatry*, **61**, 720–4.

Schulz, K. P., Fan, J., Tang, C. Y., et al. (2004) Response inhibition in adolescents diagnosed with attention deficit hyperactivity disorder during childhood: an event-related fMRI study. *Am J Psychiatry*, **161**, 1650–7.

Schulz, K. P., Newcorn, J. H., Fan, J., et al. (2005) Brain activation gradients in ventrolateral prefrontal cortex related to persistence of ADHD in adolescent boys. *J Am Acad Child Adolesc Psychiatry*, **44**, 47–54.

Schweitzer, J. B., Faber, T. L., Grafton, S. T., et al. (2000) Alterations in the functional anatomy of working memory in adult attention deficit hyperactivity disorder. *Am J Psychiatry*, **157**, 278–80.

Schweitzer, J. B., Lee, D. O., Hanford, R. B., et al. (2004) Effect of methylphenidate on executive functioning in adults with attention-deficit/hyperactivity disorder: normalization of behavior but not related brain activity. *Biol Psychiatry*, **56**, 597–606.

Seidman, L. J., Valera, E. M., Makris, N., et al. (2006) Dorsolateral prefrontal and anterior cingulate cortex volumetric abnormalities in adults with attention-deficit/hyperactivity disorder identified by magnetic resonance imaging. *Biol Psychiatry*, **60**, 1071–80.

Shafritz, K. M., Marchione, K. E., Gore, J. C., et al. (2004) The effects of methylphenidate on neural systems of attention in attention deficit hyperactivity disorder. *Am J Psychiatry*, **161**, 1990–7.

Shaw, P., Lerch, J., Greenstein, D., et al. (2006) Longitudinal mapping of cortical thickness and clinical outcome in children and adolescents with attention-deficit/ hyperactivity disorder. *Arch Gen Psychiatry*, **63**, 540–9.

Silk, T., Vance, A., Rinehart, N., et al. (2005) Fronto-parietal activation in attention-deficit hyperactivity disorder, combined type: functional magnetic resonance imaging study. *Br J Psychiatry*, **187**, 282–3.

Smith, A. B., Taylor, E., Brammer, M., et al. (2006) Task-specific hypoactivation in prefrontal and temporoparietal brain regions during motor inhibition and task switching in medication-naive children and adolescents with attention deficit hyperactivity disorder. [see comment]. *Am J Psychiatry*, **163**, 1044–51.

Solanto, M. V. (2002) Dopamine dysfunction in AD/HD: integrating clinical and basic neuroscience research. *Behav Brain Res*, **130**, 65–71.

Sonuga-Barke, E. J. (2005) Causal models of attention-deficit/hyperactivity disorder: from common simple deficits to multiple developmental pathways. *Biol Psychiatry*, **57**, 1231–8.

Tamm, L., Menon, V., and Reiss, A. L. (2006) Parietal attentional system aberrations during target detection in adolescents with attention deficit hyperactivity disorder: event-related fMRI evidence [see comment]. *Am J Psychiatry*, **163**, 1033–43.

Tamm, L., Menon, V., Ringel, J., et al. (2004) Event-related fMRI evidence of frontotemporal involvement in aberrant response inhibition and task switching in attention deficit/hyperactivity disorder. *J Am Acad Child Adolesc Psychiatry*, **43**, 1430–40.

Teicher, M. H., Anderson, C. M., Polcari, A., et al. (2000) Functional deficits in basal ganglia of children with attention-deficit/hyperactivity disorder shown with functional magnetic resonance imaging relaxometry. *Nat Med*, **6**, 470–3.

Tian, L., Jiang, T., Wang, Y., et al. (2006) Altered resting-state functional connectivity patterns of anterior cingulate cortex in adolescents with attention deficit hyperactivity disorder. *Neurosci Lett*, **400**, 39–43.

Vaidya, C. J., Austin, G., Kirkorian, G., et al. (1998) Selective effects of methylphenidate in attention deficit hyperactivity disorder: a functional magnetic resonance study. *Proc Nat Acad Sci USA*, **95**, 1444–9.

Vaidya, C. J., Bunge, S. A., Dudukovic, N. M., et al. (2005) Altered neural substrates of cognitive control in childhood ADHD: evidence from functional magnetic resonance imaging. *Am J Psychiatry*, **162**, 1605–13.

Valera, E. M., Faraone, S. V., Biederman, J., et al. (2005) Functional neuroanatomy of working memory in adults with attention deficit/hyperactivity disorder. *Biol Psychiatry*, **57**, 439–47.

Valera, E. M., Faraone, S. V., Murray, K. E., et al. (2007) Meta-analysis of structural imaging findings in attention-deficit/hyperactivity disorder. *Biol Psychiatry*, **61**, 1361–9.

Volkow, N. D., Wang, G., Fowler, J. S., et al. (2005) Imaging the effects of methylphenidate on brain dopamine: new model on its therapeutic actions for attention-deficit/hyperactivity disorder. *Biol Psychiatry*, **57**, 1410–15.

Willcutt, E. G., Doyle, A. E., Nigg, J. T., et al. (2005) Validity of the executive function theory of attention-deficit/ hyperactivity disorder: a meta-analytic review. *Biol Psychiatry*, **57**, 1336–46.

Zametkin, A. J., Liebenauer, L. L., Fitzgerald, G. A., et al. (1993) Brain metabolism in teenagers with attention-deficit hyperactivity disorder [see comments]. *Arch Gen Psychiatry*, **50**, 333–40.

Zametkin, A. J., Nordahl, T. E., Gross, M., et al. (1990) Cerebral glucose metabolism in adults with hyperactivity of childhood onset. *N Engl J Med*, **323**, 1361–6.

Zang, Y., He, Y., Zhu, C. Z., et al. (2007) Altered baseline brain activity in children with ADHD revealed by resting-state functional MRI. *Brain Dev*, **29**, 83–91.

Zang, Y.-F., Jin, Z., Weng, X.-C., et al. (2005) Functional MRI in attention-deficit hyperactivity disorder: evidence for hypofrontality [see comment]. *Brain Dev*, **27**, 544–50.

8

Brain imaging of autism spectrum disorders

Monica Zilbovicius, Nathalie Boddaert and Nadia Chabane

Introduction

Autistic disorder ("autism") is defined by an onset of symptoms by 3 years of age and a triad of impairments that include deficits in social interaction and in communication and restricted, repetitive and stereotyped patterns of behavior (American Psychiatric Association, 1994). It is classified as one of several pervasive developmental disorders (PDDs), all of which share the triad of symptoms outlined above. These include Asperger's disorder, which is distinguished from autistic disorder by a lack of general developmental language delay, and the less specific PDD-not otherwise specified (PDD-NOS). While most cases of the above disorders are idiopathic, a minority are associated with known infectious, metabolic and genetic disease. Approximately one-third of idiopathic cases are associated with the development of seizures (Tuchman, 2003). For purposes of this chapter, autism spectrum disorders (ASD) includes idiopathic autistic disorder, Asperger's, and PDD-NOS.

Autism spectrum disorders are highly heterogeneous in presentation, with levels of intellectual functioning ranging from severe mental retardation to above average and language ranging from absent (mutism) to fluent, grammatical speech that may be characterized by extralinguistic disturbances, e.g., dysprosodic speech and pragmatic deficits.

Further phenotypic complexity is added by uneven cognitive profiles, variable course (e.g., regression in language in some), and sensory-perceptual disturbances.

This chapter reviews what has grown to be an extensive neuroimaging literature on ASD, encompassing a variety of structural and functional imaging modalities. Improved technology and advances in social neuroscience have stimulated a recent upsurge in research directed toward understanding what are considered to be the core social deficits of these disorders. Covered are the main anatomical and functional brain imaging investigations over the past 20 years. Included are structural imaging studies, resting state measurements of regional cerebral glucose metabolism (rCMRglu) or cerebral blood flow (rCBF), and studies performed during brain activation paradigms. Some caveats include the fact that much of the literature involves small sample sizes, focuses predominantly on higher functioning subjects (thus limiting generalization to more severe forms of ASD), and is complicated by performance deficits in patients that may confound the interpretation of functional imaging findings. Positron emission tomography (PET) receptor occupancy studies are not covered, but see the review by Rumsey and Ernst (2000). Finally, we highlight some promising findings and avenues for future research.

Neuroimaging in Developmental Clinical Neuroscience, eds. Judith M. Rumsey and Monique Ernst. Published by Cambridge University Press. © Cambridge University Press 2009.

Anatomical imaging studies

Structural MRI

Since the first structural imaging studies of autism were published in the late 1980s, about 300 studies have appeared in the literature, many of which report non-replicated findings. The main brain structures implicated by these studies have included the cerebellum, amygdala, hippocampus, corpus callosum, and cingulate. There were, until recently, few MRI data on neocortical involvement in autism.

The cerebellum is one of the most studied structures in autism. In 1988, a quantitative MRI study showed evidence of hypoplasia of the vermian lobules VI and VII in a group of patients with autism (Courchesne et al., 1988). This vermian hypoplasia was not replicated by other researchers and appears not to be specifically linked to autism, but rather related to mental retardation (Piven et al., 1992).

Concerning the amygdala, some studies have shown increased volume (Howard et al., 2000); some, decreased volume (Aylward et al., 1999); and others, no significant abnormalities (Haznedar et al., 2000). Likewise, hippocampal findings have been inconsistent. Some studies have revealed no hippocampal size anomalies (Howard et al., 2000; Piven et al., 1998; Saitoh et al., 1995), whereas others have reported decreased volumes (Aylward et al., 1999) or increased volumes that were proportional to increases in whole-brain volumes (Sparks et al., 2002). Concerning the cingulate, Haznedar et al. (2000) has reported that individuals with autism display a decreased volume.

The possible failures in replicating localized brain anomalies in autism using quantitative MRI may be attributed to methodological design limitations (e.g., IQ heterogeneity, age range of autistic subjects, inclusion of epileptic autistic subjects). Moreover, classic MRI morphometric techniques are based upon region of interest metrics, which are inherently subjective and operator-dependent.

Greater consistency has been seen across studies of the corpus callosum. Several MRI studies have reported decreases in callosal volume. Egaas et al. (1995) found that the area of the caudal third of the corpus callosum was reduced in subjects with autism compared to healthy controls, a result confirmed by subsequent studies (Manes et al., 1999; Piven et al., 1997). After controlling for total brain volume, gender, and performance IQ, Piven et al. (1997) reported a significantly smaller size of the body and posterior subregions of the corpus callosum in autism. Manes et al. (1999) found a significantly smaller corpus callosum, with the most marked decreases in size localized to the body, in a sample of autistic individuals with moderate and severe mental retardation compared to mentally retarded non-autistic controls. Hardan et al. (2000) found that areas of the anterior subregions were smaller in a group of autistic subjects who were not mentally retarded.

A recent convergence of findings of increased total brain volume in autism has also been seen across MRI studies and further supported by postmortem studies and documentation of increased head size associated with autism (for review see Lainhart, 2006). While the finding of brain volume enlargement in autism now appears to be well established, some controversies exist about the timing of this enlargement. For example, while cross-sectional studies have suggested that enlargement is present only in the first few years of life (Courchesne et al., 2001, 2003) or in pre-schoolers (Sparks et al., 2002), some studies have detected increased brain volume in adolescent and adult samples (Hardan et al., 2001; Piven et al., 1992, 1995, 1996). Others have reported enlargement in 8- to 12-year-olds, but not in 13- to 46-year-olds (Aylward et al., 2002). Similarly, the pattern of enlargement across gray and white matter tissue compartments throughout the brain has not been well characterized (Lainhart 2006). It seems premature to conclude, from cross-sectional studies, that these effects are limited to early ages in autistic individuals. Because all published results of age-related changes in total brain volume in autism are from cross-sectional studies, longitudinal investigations are needed to confirm and extend them.

Recently, quantitative structural imaging studies have benefited greatly from both new technologies for data acquisition and new approaches to image analysis. These upgraded methods are better suited to the study of complex neocortical anatomy and are beginning to yield promising results.

Using parametric mesh-based analytic techniques to create a three-dimensional model of the cerebral cortex and detailed maps of 22 major sulci in stereotaxic space, Levitt et al. (2003a) showed significant differences in cortical sulcal patterns in children with autism, localized mainly to the frontal and temporal lobes.

Voxel-based morphometry (VBM), a fully automated technique, provides voxel-wise measures of regional gray and white matter in the whole brain independent of a priori constraints, enabling the detection of subtle group differences in gray and white matter. A pioneering study of 15 high-functioning adults with autism using VBM (Abell et al., 1999) reported fronto-temporal gray matter abnormalities, including decreases in the right paracingulate sulcus and left inferior frontal gyrus and increases in the amygdala/peri-amygdaloid cortex, middle temporal gyrus, and inferior temporal gyrus. Using VBM procedures optimized to improve spatial normalization and segmentation, Boddaert et al. (2004b) identified bilaterally significant decreases of gray matter localized to the superior temporal sulcus (STS) in 21 children with autism (mean age 9.3 ± 2.2 years), relative to 12 age-matched healthy controls. This localization is remarkably consistent with findings of bilateral temporal hypoperfusion found in autistic children by two independent functional studies, one using PET (Zilbovicius et al., 2000) and one using single photon emission computed tomography (SPECT) (Ohnishi et al., 2000) (see Figure 8.1). McAlonan et al. (2005) have mapped regional gray matter differences across the whole brain and reported reductions within fronto-striatal and parietal networks and in ventral and superior temporal cortices in 17 children with autism (all with IQ >80), relative to age-matched controls. Using an automated analysis of the cerebral cortical thickness to generate cross-subject statistics in a coordinate system based on cortical anatomy, Hadjikhani et al. (2006) found, in a group of 14 high-functioning ASD adults, local decreases in gray matter in the inferior frontal gyrus, inferior parietal lobule, and the STS, regions implicated in social cognition. In addition, cortical thinning in these regions was correlated with ASD symptom severity. Recent neuroimaging findings involving regions implicated in social perception and cognition are highlighted in Table 8.1.

Diffusion tensor imaging

Diffusion tensor imaging (DTI) is a non-invasive method for mapping the diffusion properties of tissue water (see Kim, Chapter 19). Diffusion tensor imaging measures are sensitive to subtle differences in the architecture of white matter at the microstructural level. The diffusion tensor defines the magnitude, anisotropy (the property of being directionally dependent, in that case imposed by white matter tracts and fibers), as opposed to isotropy (homogeneity in all directions), and orientation of anisotropic water diffusion in biological tissues.

To date, only three studies examining DTI in ASD have been published. Barnea-Goraly et al. (2004) used DTI to investigate white matter structure in seven male children and adolescents with autism and nine age-, gender-, and IQ-matched controls. Reduced fractional anisotropy (FA) (a normalized measure of anisotropy) was observed in white matter adjacent to the ventromedial prefrontal cortices and in the anterior cingulate gyri, as well as in the temporo-parietal junctions in the autistic group. Additional clusters of reduced FA values were seen adjacent to the superior temporal sulcus bilaterally, in the temporal lobes approaching the amygdala bilaterally, in occipitotemporal tracts, and in the corpus callosum. These investigators suggested that the disrupted development of white matter tracts in these regions might contribute to impaired social cognition in autism. Keller et al. (2007) used DTI to examine age-related differences in the organization of white matter in 34 participants with autism and 31 controls between the ages of

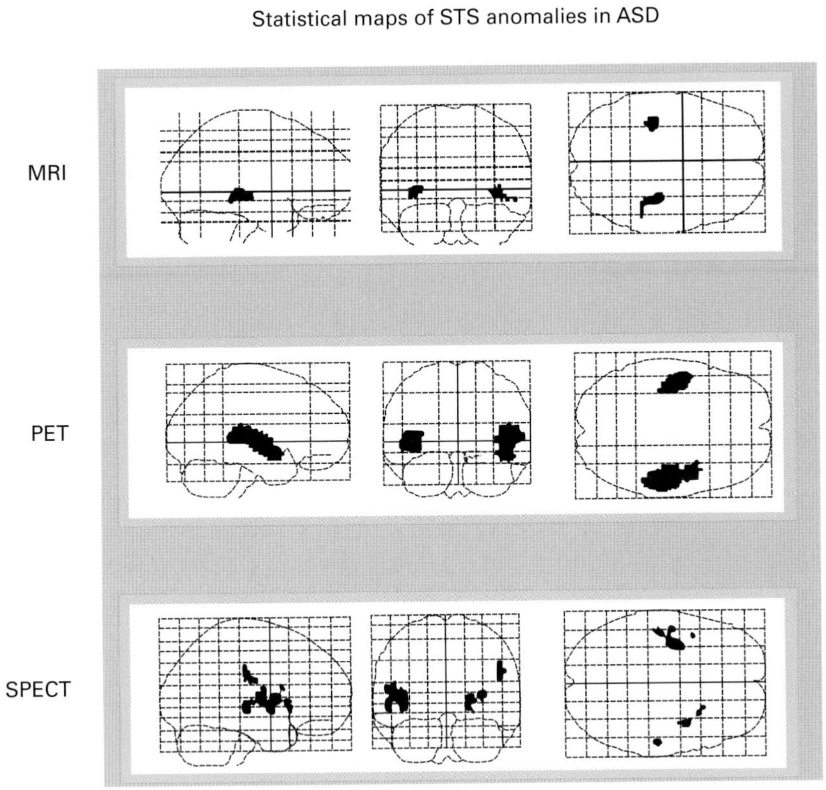

Statistical maps of STS anomalies in ASD

Figure 8.1. Convergent anatomical and resting state functional temporal lobe anomalies in children with autism spectrum disorders (ASD). MRI: SPM glass brain represents the superior temporal sulcus (STS) regions where a significant decrease of the gray matter concentration was found in ASD children, relative to controls (Boddaert et al., 2004b). PET and SPECT: The same regions had a significant decrease of cerebral blood flow as measured with PET (Zilbovicius et al., 2000) and SPECT (Ohnishi et al., 2000) relative to controls.

10 and 35 years. Participants with autism had lower FA in areas within and near the corpus callosum and in the right retrolenticular portion of the internal capsule. Only one area, in the posterior limb of the right internal capsule, showed an interaction between age and group, such that FA increased with age in autism, but decreased with age in controls. The findings suggest that reductions in the structural integrity of white matter in autism persist into adulthood. Alexander et al. (2007) performed DTI in 43 subjects with high-functioning autism and 34 controls matched for age, handedness, IQ, and head size. Diffusion tensor imaging and volumetric measurements of the total corpus callosum and

subregions (genu, body, and splenium) revealed significant differences in volume and fractional anisotropy between groups. The autism group exhibited significantly smaller total and regional corpus callosum volumes (genu, body, and splenium) and significantly lower mean FA values in the total corpus callosum, the genu and splenium.

Magnetic resonance spectroscopy

Magnetic resonance spectroscopy (MRS) is a non-invasive method for investigating cellular neuro-chemistry in vivo. It provides spectra that can be

Table 8.1 Recent neuroimaging findings involving regions implicated in social perception and cognition

Brain imaging method	Study	Autistic patients	Controls	Results
Structural MRI				
VBM	Boddaert et al. (2004b)	21 children LFA (9 ± 2 years) (IQ 55 ± 15)	10 normal children (10 ± 3 years) (IQ 98)	↓Gray matter STS bilaterally
VBM	McAlonan et al. (2005)	17 children with HFA (12 ± 2 years) (IQ 101 ± 10)	17 normal children (11 ± 1 years) (IQ 11 ± 14)	↓GM fronto-parietal and STS bilaterally
Sulcus cartography	Levitt et al. (2003)	21 children HFA (10 ± 3 years) (IQ 100)	20 normal children (11 ± 3 years) (IQ 117)	Abnormal placement of STS, SFS, and IFS
Rest functional				
SPECT	Ohnishi et al. (2000)	23 children LFA (2.6–13 years) (IQ 48 ± 19.5)	26 children MR (3–12 years) (IQ 57 ± 17)	↓CBF insula, STS, IFG, and MFG
PET $H_2{}^{15}O$	Zilbovicius et al. (2000)	32 children LFA (5–13 years) (IQ 44.5 ± 27)	10 children MR (5–13 years) (IQ 50 ± 17)	↓CBF STG, and STS
Activation				
PET $H_2{}^{15}O$, auditory – verbal task – repetition	Muller et al. (1999)	5 adults HFA (18–31 years) (IQ 77)	5 normal adults (23–30 years) (IQ normal)	Reversed hemispheric dominance during verbal auditory stimulation
PET $H_2{}^{15}O$, listening to speech-like sounds	Boddaert et al. (2003)	6 adults with autism (19 ± 4 years) (IQ 64 ± 5)	5 normal adults (21 ± 3 years) (IQ normal)	↓STG left activation and ↑STG right activation during passive listening to speech-like sounds
fMRI, theory of mind	Baron-Cohen et al. (1999)	6 adults with HFA or AS (26 ± 2 years) (IQ 108 ± 10)	5 normal adults (25 ± 2 years) (IQ 110 ± 8)	↓Activation IFG, amygdala and insula
PET $H_2{}^{15}O$, theory of mind	Castelli et al. (2002)	10 adults with HFA or AS (33 ± 7 years) (IQ normal)	10 normal adults (25 ± 4 years) (IQ normal)	↓STS, temporal poles and PFC activation during "ToM" animated sequences of geometric figures
fMRI, voice	Gervais et al. (2004)	5 adults with HFA (25 ± 5 years) (IQ normal)	8 normal adults (27 ± 2 years) (IQ normal)	No STS activation during listening to voice
fMRI, face	Schultz et al. (2000)	14 adults with HFA or AS (23 ± 12 years) (IQ 109 ± 19)	28 normal adults (21 ± 8 years) (IQ 109 ± 16)	↓FFA activation, ↑ITG right activation during face discrimination

fMRI, face	Pierce et al. (2004)	8 adults with autism (16–42 years) (IQ 80 ± 18)	10 normal adults (16–40 years) (IQ normal)	Normal FFA activation during discrimination of familiar faces
fMRI, face and eye tracking	Dalton et al. (2005)	14 subjects with autism (15 ± 4 years) (IQ 94 ± 19)	28 normal adults (21 ± 8 years) (IQ normal)	↓ FFA activation ↑ amygdala activation during face and emotion discrimination; correlated with eye fixation
fMRI, eye gaze	Pelphrey et al. (2005)	10 adults with autism (19–50 years) (IQ 107 ± 16)	9 normal adults (15–32 years) (IQ 118 ± 9)	↓ STS activation during eye-gaze perception
fMRI, imitation	Dappreto et al. (2006)	9 children with autism (12 ± 2 years)	10 normal children (12 ± 2 years)	↓ Activation of frontal MSN during imitation

Notes:

ASD, autism spectrum disorders; CBF, cerebral blood flow; fMRI, functional magnetic resonance imaging; FFA, fusiform face area; GM, gray matter; HFA, high-functioning ASD – IQ generally normal; IFG, inferior frontal gyrus; IFS, inferior frontal sulcus; IQ normal, IQ > 80; ITG, inferior temporal gyrus; LFA, low-functioning ASD – IQ cutoff < 70; MFG, middle frontal gyrus; MR, mental retardation, IQ < 70; MNS, mirror neuron system; PET, positron emission tomography; PFC, prefrontal cortex; SFS, superior frontal sulcus; SPECT, single photon emission computer tomography; STG, superior temporal gyrus; STS, superior temporal sulcus; ToM, theory of mind; VBM, voxel-based morphometry.

used to measure *N*-acetylaspartate (NAA), choline-containing compounds (Cho), and creatine and phosphocreatine (Cr) (see Silveri, et al., Chapter 18). The NAA signal, the most prominent spectral peak, is present at high concentrations in gray matter and neurons, and its synthesis may be related to mitochondrial function; therefore, NAA is often used to assess neuronal density. The Cho signal may indicate glial cell density and also increases in intensity with increased membrane synthesis and turnover. The Cr signal is thought to reflect glial or overall (neurons plus glia) cellular density.

Findings from single-voxel MRS studies have shown reduced NAA concentrations (suggestive of reduced neuronal density or integrity) in frontal (Chugani et al., 1999), lateral temporal (Hisaoka et al., 2001), and medial temporal (Otsuka et al., 1999) lobes, and the cerebellar hemispheres (Chugani et al., 1999; Otsuka et al., 1999) among children with autism. In contrast, Murphy et al. (2002) reported increased NAA concentration in the medial prefrontal lobe among adults with Asperger's that correlated with their obsessive behavior. Findings from multiple-voxel MRS studies have indicated reduction of NAA concentration in the left frontal white matter, bilateral cingulate, right thalamus, and right superior temporal gyrus among young children with ASD (Friedman et al., 2003). Two recent studies reported decreased NAA concentrations of gray matter in children with ASD (DeVito et al., 2007; Friedman et al., 2006). Furthermore, altered Cho metabolism in the anterior cingulate, caudate nucleus, thalamus, and temporal lobe have also been reported in ASD subjects (Friedman et al., 2003; Levitt et al., 2003b), and Sokol et al. (2002) reported that the ratios of Cho to Cr in the left hippocampus region were increased among children with autism. In conclusion, findings from MRS studies investigating individuals with ASD, although inconsistent, suggest neuronal developmental problems and/or altered membrane metabolism in various brain regions in individuals with ASD.

Functional brain imaging

Resting state studies

Positron emission tomography (PET) and single photon emission tomography (SPECT) measures of regional cerebral glucose utilization (rCMRglu) and/or regional cerebral blood flow (rCBF) reflect changes in synaptic activity. Neurons increase their utilization of glucose in direct proportion to their activity. In addition, under normal conditions, the rCMRglu is coupled to rCBF.

The first functional imaging studies of autism, conducted in the 1980s, focused on adult subjects and many (though not all) reported normal resting cerebral glucose utilization (CMRglu) rates, as measured with PET and [18F]-fluorodeoxyglucose (FDG) (Rumsey and Ernst, 2000). The first PET study of children with autism ($n = 18$), ages 2–18, was reported by DeVolder et al. in 1988, but results were compared to adult controls and a small number of children with varied brain pathology. This study reported normal rates and regional distribution of brain glucose metabolism.

In our laboratory, early studies of rCBF with SPECT in children with idiopathic autism found no evidence of localized brain cortical dysfunction (Zilbovicius et al., 1992). The lack of SPECT-detected focal rCBF abnormalities in autistic children was also confirmed by Chiron et al. (1995).

However, these negative results may be explained by some methodological limitations. These studies all were performed with low-spatial-resolution cameras (20 mm) and analyzed using a region-of-interest approach that allowed analysis of only large cerebral regions. All were performed in the resting state. Thus, these approaches may have had limited sensitivity for identifying abnormalities involving small or widely distributed brain regions and those involving white matter and/or functional connectivity.

More recent PET and SPECT studies using high-resolution cameras (about 5 mm of resolution) and voxel-based whole-brain analysis (Statistical Parametric Mapping software) have described localized bilateral temporal hypoperfusion in children with autism. Using SPECT, Ohnishi et al. (2000) detected a significant hypoperfusion located bilaterally in the superior temporal gyrus and in the left frontal region in a sample of 23 autistic children (mean age 6 years, no persistent seizures), compared with 26 control children. Using PET, Zilbovicius et al. (2000) detected significant hypoperfusion bilaterally in the superior temporal gyrus and in the right right STS of 21 autistic children (mean age 8.5 years, without any history of seizures), relative to 10 control children. In both studies, children with idiopathic mental retardation were matched for age and developmental quotients with the group, such that the findings could not be attributed to mental retardation. In addition, an individual-level analysis that compared each autistic child to the control group revealed significant temporal hypoperfusion in 16 of 21 autistic children (77%) (Zilbovicius et al., 2000). Moreover, a replication study performed in an independent group of 12 autistic children confirmed both the group and individual results (Zilbovicius et al., 2000). Thus, the finding of bitemporal hypoperfusion has been reported in three independent samples of autistic children, providing the first robust evidence for temporal lobe dysfunction in school-aged children with autism.

A correlational analysis was performed in order to investigate a putative relationship between rCBF and the clinical profile of 45 autistic children. Autistic behavior was evaluated with the Autism Diagnostic Interview-Revised (ADI-R; Lord et al., 1994). A significant negative correlation was observed between rCBF and the ADI-R score in the left superior temporal gyrus. The higher the ADI-R score (indicating greater symptom severity), the lower the left temporal rCBF (Meresse et al., 2005).

The STS abnormalities found in ASD children are highly consistent with the social deficits seen in autism, as the STS is known to be a critical region for social cognition and is implicated in several aspects of social interaction, including auditory and visual social perception (eye gaze, gestures, facial displays of emotions and voice perception)

and more complex social cognition (theory of mind, mentalizing) (for review see Zilbovicius et al., 2006).

Activation studies

Activation studies measure local changes of CBF or blood oxygenation, reflecting synaptic activity, during sensory, cognitive, or motor paradigms. The PET, SPECT and functional MRI (fMRI) studies described below suggest that autistic subjects display brain activation patterns while performing tasks that differ from those seen in healthy controls.

Language stimulation studies

Muller et al. (1999) studied five high-functioning autistic males and five normal adults using PET during verbal and non-verbal auditory stimulation (five conditions: resting, listening to a sequence of simple tones, listening to simple sentences, repeating simple sentences, and generating sentences from a stimulus sentence and word prompt). Results indicated a significant reversal of the more typical left hemisphere language dominance seen in controls among the autistic participants during the sentence-listening condition.

More recently, an auditory activation PET study was performed in 5 autistic adults and 12 controls during passive listening to speech-like stimuli (with an acoustic structure similar to consonant-vowel-consonant, although normal volunteers never recognized them as speech sounds). Compared to controls, autistic patients showed a significantly greater activation of the right posterior superior temporal gyrus and a significantly lower activation of the left superior temporal gyrus, indicating abnormal auditory processing (Boddaert et al., 2003). Applying the same auditory paradigm to children (passive listening to speech-like stimuli), Boddaert et al. (2004a, b) detected a significantly lower activation of the left posterior superior temporal gyrus in the 11 autistic children compared to 6 age- and IQ-matched controls.

These three PET activation studies suggest that autism is associated with an abnormal pattern of activation of the left temporal cortex in response to both verbal and speech-like auditory stimulation. Because the left temporal region is implicated in the brain's organization for language, this abnormal left hemisphere activation may contribute to language impairments, as well as to inadequate behavioral responses to sounds.

Social perception and mentalizing studies

Face perception

Although not part of current diagnostic criteria, much evidence suggests that persons with ASDs process faces differently than do normal controls and are impaired in their ability to recognize faces (Schultz, 2005). For example, persons with an ASD show less of an inversion effect for faces, i.e., their performance is not much degraded for the recognition of upside-down faces versus upright face (Teunisse and de Gelder, 2003).

Functional neuroimaging studies performed in normal volunteers have revealed the presence of an area in the right fusiform gyrus that is more strongly activated during face perception than during the perceptual processing of any other class of visual stimuli (Haxby et al., 2000; Kanwisher et al., 1997). This area is known as the fusiform face area (FFA). At least five fMRI studies have shown that older children, adolescents and adults with ASD have reduced levels of activity to images of the human face in this specialized face region of the right hemisphere (for review see Schultz, 2005).

Schultz et al. (2000) were the first to use fMRI to study face processing in autistic persons. They reported that a group of 14 high-functioning individuals with autism or Asperger's syndrome showed significantly less activation of the middle aspect of the right fusiform gyrus (FFA) compared to controls. Hypoactivation of the FFA was replicated in a series of functional studies. Critchley et al. (2000) studied nine high-functioning autistic adults and nine age-matched controls with a task involving explicit (conscious) and implicit (subconscious) processing of emotional facial expressions. In the explicit task, subjects were asked to judge the facial expression of each stimulus (happy/angry or neutral). In the implicit task, subjects attended to and judged the gender of each face (male or female). Thus, both conditions examined the processing of happy and angry faces relative to neutral faces. Autistic subjects made more errors than controls during the explicit processing of facial expressions, potentially contributing to differences in brain activity described below. Autistic subjects differed significantly from controls in their activation of the cerebellum and the mesolimbic and temporal lobe cortical regions when observing facial expressions (explicitly and implicitly). Notably, they failed to activate the FFA when explicitly appraising expressions. Pierce et al. (2001) used an active perceptual task involving gender discrimination of neutral faces in a sample of six adults with autism and found reduced FFA activation. Hubl et al. (2003) also showed FFA hypoactivation in seven adult males with autism using both a gender discrimination and a neutral-versus-expressive discrimination task. Hall et al. (2003) used PET in a group of eight high-functioning males with autism as compared to eight healthy male controls during an emotion-recognition task and showed hypoactivation of the FFA.

In contrast to these studies, two recent studies of autism failed to find FFA hypoactivation. Using familiar and unfamiliar faces in a further fMRI study of eight adult males with autism and 10 healthy control males, Pierce et al. (2004) found significantly greater FFA activation to familiar, personally meaningful faces (friends, family) as compared to unfamiliar faces in an autism sample. Hadjikhani et al.'s (2004) study of 11 adults with ASD and 10 normal controls used a passive viewing task with facial and objective stimuli that contained a red fixation cross that was continuously present in the center. The participants' task was to visually fixate on the cross. Individuals with ASD activated the FFA and other brain areas normally involved in face processing when they viewed faces as compared to non-face stimuli (Hadjikhani et al., 2004). These results suggest that attentional mechanisms may modulate FFA engagement in ASD.

Dalton et al. (2005) used eye-gaze tracking while measuring functional activity during two facial discrimination tasks (Study I: emotion discrimination of faces; and Study II: face recognition of familiar and unfamiliar faces) in individuals with autism (Study I, $n = 14$; Study II, $n = 16$) and normal controls. In both studies, the control group showed significantly greater activation than the autistic group in response to face stimuli in the fusiform gyrus bilaterally, but the autistic group showed greater activation of the amygdala (left in Study I and right in Study II). In addition, in the autistic group, activation of the fusiform gyrus and amygdala was strongly and positively correlated with the amount of time spent fixating on the eye region. These results suggest that diminished gaze fixation may account for the fusiform hypoactivation reported in autism. The positive correlation between amygdala activation and gaze fixation suggests that gaze fixation is associated with a greater emotional response, such as anxiety, in autism.

Voice perception

In the auditory domain, voice is at the epicenter of human social interactions and may be thought of as the auditory counterpart to faces in the visual domain. Like face, each voice contains in its acoustic structure information about the speaker's identity and emotional state, which are readily perceived by healthy individuals. A recent fMRI study identified, in normal adults, voice-selective areas located bilaterally along the upper bank of the STS (Belin et al., 2000). Individuals with autism have difficulties in voice perception, such as lack of a preference for their mother's voice (Klin, 1991) and impairment in the extraction of mental states from voices (Rutherford et al., 2002). This suggests that abnormal cortical voice processing is a feature of autism.

To test this hypothesis, Gervais et al. (2004) used fMRI to study brain activation during voice processing in adults with autism. In normal controls, listening to voice, compared to non-voice (environmental) sounds, significantly activated a "voice-selective area" located bilaterally along the upper bank of

the STS in both group and individual control subject analyses. In accord with the hypothesis, voice perception in the autistic group failed to activate any brain region relative to non-voice perception; cortical activation was equivalent for voice and non-voice stimuli as compared to silence. In addition, in contrast to the controls (all of whom activated the voice-selective area), only a single autistic subject activated the voice-selective area. Together with reports of failures to normally activate the FFA, this finding suggests a more pervasive difficulty processing salient social stimuli in either the visual or auditory modality.

Mentalizing and theory of mind

Theory of mind, or mentalizing, refers to the ability to think about oneself and the minds of other people and provides a basis for predicting other people's behavior. Thinking about what others think, as opposed to thinking about the external physical world, is essential for engaging in complex social activity. Deficits in theory of mind are thought to contribute to impaired social communication in autism (Frith and Happe, 2005).

Fletcher et al. (1995) and Happe et al. (1996) performed ^{15}O-PET to determine the neural correlates of the ability to recognize the mental states of others (theory of mind). Normal volunteers and five patients with Asperger's syndrome were compared while silently reading stories and answering a question about each. "Theory of mind" passages required the attribution of complex mental states, e.g., "white lie" (an unimportant lie); stories dealing with physical events required the integration of information and inferences; and passages of unlinked sentences required neither. On the theory of mind tasks, both groups activated similar regions in the temporal and parietal lobes thought to be involved in language processing and narrative comprehension. However, differences were seen in the localization of a left medial frontal region uniquely activated by the theory of mind task (Brodmann's area 8 in controls; Brodmann's area 9/10 in the Asperger's group) (Happe et al., 1996), raising the possibility of an aberrant organization of the brain system that underlies the understanding of others' minds in ASD.

Baron-Cohen et al. (1999) used fMRI with a facial perception task in which subjects viewed photographs of eyes and were asked to indicate whether each stimulus was a man or a woman and the person's mental state, e.g., concerned, unconcerned ("theory of mind" task). In normal subjects, the theory of mind task activated two main brain systems: (1) fronto-temporal neocortical regions, comprising left dorsolateral prefrontal cortex, the left medial frontal cortex, supplementary motor area, and bilateral temporo-parietal regions (middle and superior temporal, angular, and supramarginal gyri) and (2) subcortical regions, including the left amygdala, left hippocampal gyrus, bilateral insulae, and left striatum. The autism group activated the frontal components less extensively than did the controls and failed to activate the amygdala, suggesting amygdala involvement in impaired mental state processing.

Using PET, Castelli et al. (2002) studied brain activation patterns elicited by the viewing of animated geometric shapes that moved and interacted in a manner suggesting human actions and intentions. Three conditions, each using two triangles, were compared: random movement, goal-directed movement (e.g., chasing, fighting), and interactive movements (e.g., coaxing, tricking) that elicited descriptions in terms of intentions/mental states that viewers attributed to the triangles (mentalizing) in both groups. Ten adults with high-functioning autism or Asperger's syndrome and 10 normal volunteers were scanned. While viewing animations that elicited mentalizing, in contrast to randomly moving shapes, the normal group showed increased activation in a previously identified "mentalizing network" (medial prefrontal cortex, STS at the temporo-parietal junction, and temporal poles). In contrast, the ASD group, which gave fewer and less accurate descriptions of the latter two conditions, showed less activation in these regions, but similar activation of extrastriate cortex. This region showed reduced functional connectivity with the STS, an area associated with the processing of biological motion as well as with mentalizing, in the ASD group.

More recently, Pelphrey et al. (2005) reported an abnormal STS activation in autistic adults during an eye-gaze perception task. On congruent trials, subjects watched as a virtual actor looked toward a flickering checkerboard that appeared in the actor's visual field, confirming the subject's expectation regarding what the actor "ought to do" in this context. On incongruent trials, the virtual actor looked toward empty space, violating the subject's expectation. In controls, incongruent trials evoked more activity in the STS and other brain regions linked to social cognition. The same brain regions were activated during the observation of gaze shifts in subjects with autism, but did not differentiate between congruent and incongruent trials, indicating that activity in these regions was not modulated by the context of the perceived gaze shift. The authors suggested that this lack of modulation contributes to the eye-gaze-processing deficits associated with autism. The differences in activation of the superior temporal sulcus across several tasks and studies are illustrated in Figure 8.2.

The mirror neuron system

The mirror neuron system (MNS) supports imitation and goal understanding in healthy adults. Recently, it has been proposed that a deficit in this system may contribute to poor imitation and be a cause of social disabilities in children with ASD.

Based on work in both humans and monkeys, the MNS has been found to be composed of a network of areas, including the pars opercularis of the inferior frontal gyrus and its adjacent ventral area (inferior frontal cortex), the inferior parietal lobule, and the STS, which are activated during the observation and imitation of an action (Rizzolatti and Craighero, 2004). In this way, the MNS allows matching between the actions of the self and of others and supports inference of the goals and intentions of other people (Hamilton and Grafton, 2006). Thus, at least two categories of behaviors, imitation and action understanding, are supported by the MNS.

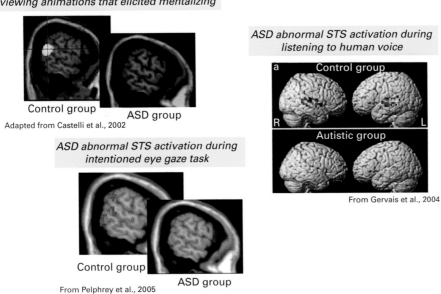

STS and abnormal social cognition in ASD

ASD abnormal STS activation during viewing animations that elicited mentalizing

Control group   ASD group

Adapted from Castelli et al., 2002

ASD abnormal STS activation during listening to human voice

a   Control group

R   L

Autistic group

From Gervais et al., 2004

ASD abnormal STS activation during intentioned eye gaze task

Control group   ASD group

From Pelphrey et al., 2005

Figure 8.2. Superior temporal sulcus (STS) and abnormal social cognition in ASD. Brain imaging studies have shown reduced (relative to controls) or absent STS activation in subjects with ASD during tasks involving social cognition. Castelli et al. (2002) used animated geometric figures that moved (1) randomly, (2) in a goal-directed fashion (chasing, fighting), and (3) interactively with implied intentions (coaxing, tricking), a condition that elicited descriptions in terms of mental states. Using functional magnetic resonance imaging (fMRI), Gervais et al. (2004) compared vocal to non-vocal sounds. Pelphrey et al. (2005) compared congruent versus incongruent gaze. On congruent trials, a virtual actor looked toward a checkerboard that appeared in her visual field, confirming the subject's expectation regarding what the actor "ought to do" in this context. On incongruent trials, the virtual actor looked toward empty space, violating the subject's expectation. In normal subjects, incongruent trials evoked more activity in the STS and other brain regions linked to social cognition, indicating a strong effect of intention.

It has been suggested that children with autism might have an abnormal MNS that impairs imitation, the development of theory of mind skills, and social cognition (Williams et al., 2001). The MNS hypothesis of autism has been tested repeatedly in recent years, using various techniques and approaches.

A morphometric study has demonstrated cortical thinning in regions belonging to the MNS in ASD, compared with typically developing controls (Hadjikhani et al., 2006). Furthermore, there is evidence of disordered functional connectivity between visual and inferior frontal mirror neuron areas and between frontal and parietal areas in patients with ASD (Villalobos et al., 2005).

Electroencephalographic mu power is reduced, or suppressed, in typically developing individuals both when they perform actions and when they view others performing actions. Oberman et al. (2005) reported that ten high-functioning subjects with ASD showed mu suppression when performing hand movements, but not when viewing videos of others' hand movements, while controls showed mu suppression under both circumstances.

Using fMRI, Dapretto et al. (2006) studied high-functioning children with autism and matched controls while they imitated and observed emotional facial expressions. Although both groups performed the tasks equally well, children with autism showed no activation of the inferior frontal gyrus (pars opercularis). Notably, activity in this area was inversely related to symptom severity in the social domain.

Also using fMRI, Hadjikhani et al. (2007) studied ten ASD subjects and seven healthy controls during passive viewing of non-emotional faces. Autism spectrum disorder subjects activated the FFA and inferior occipital gyrus, but showed hypoactivation in the broader network of regions involved in face processing, including the right amygdala, inferior frontal cortex, STS, and face-related somatosensory and premotor cortex. The ASD group showed functional correlations between some areas belonging to the MNS (inferior frontal cortex, STS) and other face-processing areas, with the severity of their social symptoms correlating with right inferior frontal cortical thickness and functional activation. These results suggest involvement of the MNS in face-processing deficits in ASD. However, other recent behavioral evidence of the ability to imitate the goals of observed hand actions (Hamilton et al., 2007) argues against a general imitation deficit and global MNS deficit in children with ASD and in favor of multiple brain systems for different types of imitation and action understanding.

Studies of exceptional abilities

A ^{15}O-PET study of an autistic adult with unusual calendar calculating abilities (associating dates to the corresponding days of the week) (savant) (Boddaert et al., 2005) explored regions activated during a calendar task. When asked the day of the week on which a certain date fell (relative to a word-repetition control task), the subject activated regions that are usually implicated in verbal memory – left hippocampus and left fronto-temporal regions, suggesting that this prodigious skill was sustained by memory processing.

Using a task on which autistic subjects generally perform unusually well (embedded figures test), Ring et al. (1999) tested visual search. Normal controls activated prefrontal cortical areas that were not recruited in the group with autism. In contrast, subjects with autism demonstrated greater activation of ventral occipitotemporal regions, suggesting the use of different cognitive strategies. The pattern of activation seen in controls suggested a greater reliance on working memory systems, while that seen in the autistic group suggested a reliance on visual systems for object feature analysis.

Conclusions and future directions

In this review, we have attempted to provide an overview of past decades of neuroimaging research in the field of ASDs. We focused primarily on the main structural and functional studies that have produced current hypotheses about the neurobiology underlying this complex disorder. The history of imaging research in autism has included changing foci of particular interest, e.g., from cerebellum to frontal lobes, now temporal lobes, mirror neuron and other systems believed to subserve social cognition. These shifts are in part due to improvements in brain imaging technologies that allow for a better understanding of the neural basis of social and language functions in the normal brain. While the newer findings are intriguing, we are still far from understanding the neural basis of autism.

The most compelling structural findings suggest an early overgrowth of the brain, the nature, timing, and developmental trajectory of which require longitudinal investigation. A number of possible mechanisms proposed to explain this apparent overgrowth include increased neurogenesis, decreased neuronal cell death, increased production of non-neuronal brain tissues (i.e., glia), decreased synaptic pruning, and abnormalities of myelin (Courchesne and Pierce, 2005). At this point, there is no firm pathological evidence to support any of these suggested hypotheses. Understanding this phenomenon should advance our understanding of the pathogenesis of this disorder.

The neurobiology of social behavior represents another particularly promising avenue of inquiry. Indeed, social impairments distinguish autism from other developmental disorders. A number of brain imaging studies in ASD have found anatomical and functional abnormalities in regions belonging to social-cognitive neural networks (STS, FFA, amygdala, MNS, frontal and parietal cortex) (see Table 8.1). All these regions were found to be abnormally activated in ASD during tasks of social cognition and/or to be anatomically abnormal in individuals with ASD. These data provide a new outlook on our understanding in ASD, arguing for a deficit in the perception and processing of socially relevant stimuli. This view may help to stimulate new therapeutic approaches focused on face, eye gaze, and voice perception in order to approximate normal or more socially effective strategies for processing these stimuli.

Promising avenues for future research include the integration of findings across imaging modalities with those from genetic studies. This may provide insights into the nature of risk genes and the timing of their expression during development, thereby also providing useful clues to relevant environmental factors.

Acknowledgments

This work was supported by France Foundation and Orange Foundation. The authors thank Professor Francis Brunelle and Professor Yves Samson for their helpful advice.

REFERENCES

Abell, F., Krams, M., Ashburner, J., et al. (1999) The neuroanatomy of autism: a voxel-based whole brain analysis of structural scans. *NeuroReport*, **10**, 1647–51.

Alexander, A. L., Lee, J. E., Lazar, M., et al. (2007) Diffusion tensor imaging of the corpus callosum in autism. *NeuroImage*, **34**, 61–73.

American Psychiatric Association (1994) *Diagnostic and Statistical Manual of Mental Disorders*, 4th edn. Washington, D. C.: American Psychiatric Association.

Aylward, E. H., Minshew, N. J., Field, K., et al. (2002) Effects of age on brain volume and head circumference in autism. *Neurology*, **59**, 158–9.

Aylward, E. H., Minshew, N. J., Goldstein, G., et al. (1999) MRI volumes of amygdala and hippocampus in non-mentally retarded autistic adolescents and adults. *Neurology*, **53**, 2145–50.

Barnea-Goraly, N., Kwon, H., Menon, V., et al. (2004) White matter structure in autism: preliminary evidence from diffusion tensor imaging. *Biol Psychiatry*, **55**, 323–6.

Baron-Cohen, S., Ring, H. A., Wheelwright, S., et al. (1999) Social intelligence in the normal and autistic brain: an fMRI study. *Eur J Neurosci*, **11**, 1891–8.

Belin, P., Zatorre, R. J., Lafaille, P., et al. (2000) Voice-selective areas in human auditory cortex. *Nature*, **403**, 309–12.

Boddaert, N., Barthelemy, C., Poline, J. B., et al. (2005) Autism: functional brain mapping of exceptional calendar capacity. *Br J Psychiatry*, **187**, 83–6.

Boddaert, N., Belin, P., Chabane, N., et al. (2003) Perception of complex sounds: abnormal pattern of cortical activation in autism. *Am J Psychiatry*, **160**, 2057–60.

Boddaert, N., Chabane, N., Belin, P., et al. (2004a) Perception of complex sounds in autism: abnormal auditory cortical processing in children. *Am J Psychiatry*, **161**, 2117–20.

Boddaert, N., Chabane, N., Gervais, H., et al. (2004b) Superior temporal sulcus anatomical abnormalities in childhood autism: a voxel-based morphometry MRI study. *NeuroImage*, **23**, 364–9.

Castelli, F., Frith, C., Happe, F., et al. (2002) Autism, Asperger syndrome and brain mechanisms for the attribution of mental states to animated shapes. *Brain*, **125**, 1839–49.

Chiron, C., Leboyer, M., Leon, F., et al. (1995) SPECT of the brain in childhood autism: evidence for a lack of normal hemispheric asymmetry. *Dev Med Child Neurol*, **37**, 849–60.

Chugani, D. C., Sundram, B. S., Behen, M., et al. (1999) Evidence of altered energy metabolism in autistic children. *Prog Neuropsychopharmacol Biol Psychiatry*, **23**, 635–41.

Courchesne, E. and Pierce, K. (2005) Brain overgrowth in autism during a critical time in development: implications for frontal pyramidal neuron and interneuron development and connectivity. *Int J Dev Neurosci*, **23**, 153–70.

Courchesne, E., Carper, R., and Akshoomoff, N. (2003) Evidence of brain overgrowth in the first year of life in autism. *JAMA*, **16**, 337–44.

Courchesne, E., Karns, C. M., Davis, H. R., et al. (2001) Unusual brain growth patterns in early life in patients with autistic disorder: an MRI study. *Neurology*, **57**, 245–54.

Courchesne, E., Yeung-Courchesne, R., Press, G. A., et al. (1988) Hypoplasia of cerebellar vermal lobules VI and VII in autism. *N Engl J Med*, **318**, 1349–54.

Critchley, H. D., Daly, E. M., Bullmore, E. T., et al. (2000) The functional neuroanatomy of social behaviour: changes in cerebral blood flow when people with autistic disorder process facial expressions. *Brain*, **123**, 2203–12.

Dalton, K. M., Nacewicz, B. M., Johnstone, T., et al. (2005) Gaze fixation and the neural circuitry of face processing in autism. *Nat Neurosci*, **8**, 519–26.

Dapretto, M., Davies, M. S., Pfeifer, J. H., et al. (2006) Understanding emotions in others: mirror neuron dysfunction in children with autism spectrum disorders. *Nat Neurosci*, **9**, 28–30.

DeVito, T. J., Drost, D. J., Neufeld, R. W., et al. (2007) Evidence for cortical dysfunction in autism: a proton magnetic resonance spectroscopic imaging study. *Biol Psychiatry*, **61**, 465–73.

DeVolder, A. G., Bol, A., Michel, C., et al. (1988) Cerebral glucose metabolism in autistic children. Study and positron emission tomography. *Acta Neurol Belg*, **88**, 75–90.

Egaas, B., Courchesne, E., and Saitoh, O. (1995) Reduced size of corpus callosum in autism. *Arch Neurol*, **52**, 794–801.

Fletcher, P. C., Happe, F., Frith, U., et al. (1995) Other minds in the brain: a functional imaging study of "theory of mind" in story comprehension. *Cognition*, **57**, 109–28.

Friedman, S. D., Shaw, D. W., Artru, A. A., et al. (2003) Regional brain chemical alterations in young children with autism spectrum disorder. *Neurology*, **60**, 100–7.

Friedman, S. D., Shaw, D. W., Artru, A. A., et al. (2006) Gray and white matter brain chemistry in young children with autism. *Arch Gen Psychiatry*, **63**, 786–94.

Frith, U. and Happe, F. (2005) Autism spectrum disorder. *Curr Biol*, **15**, 786–90.

Gervais, H., Belin, P., Boddaert, N., et al. (2004) Abnormal cortical voice processing in autism. *Nat Neurosci*, **7**, 801–2.

Hadjikhani, N., Joseph, R. M., Snyder, J., et al. (2004) Activation of the fusiform gyrus when individuals with autism spectrum disorder view faces. *NeuroImage*, **22**, 1141–50.

Hadjikhani, N., Joseph, R. M., Snyder, J., et al. (2006) Anatomical differences in the mirror neuron system and social cognition network in autism. *Cereb Cortex*, **16**, 1276–82.

Hadjikhani, N., Joseph, R. M., Snyder, J., et al. (2007) Abnormal activation of the social brain during face perception in autism. *Hum Brain Mapp*, **28**, 441–9.

Hall, G. B., Szechtman, H., and Nahmias, C. (2003) Enhanced salience and emotion recognition in autism: a PET study. *Am J Psychiatry*, **160**, 1439–41.

Hamilton, A. F. and Grafton, S. T. (2006) Goal representation in human anterior intraparietal sulcus. *J Neurosci*, **25**, 1133–7.

Hamilton, A. F., Brindley, R. M., and Frith, U. (2007) Imitation and action understanding in autistic spectrum disorders: how valid is the hypothesis of a deficit in the mirror neuron system? *Neuropsychologia*, **9**, 1859–68.

Happe, F., Ehlers, S., Fletcher, P., et al. (1996) 'Theory of mind' in the brain. Evidence from a PET scan study of Asperger syndrome. *NeuroReport*, **8**, 197–201.

Hardan, A. Y., Minshew, N. J., and Keshavan, M. S. (2000) Corpus callosum size in autism. *Neurology*, **55**, 1033–36.

Hardan, A. Y., Minshew, N. J., Mallikarjuhn, M., et al. (2001) Brain volume in autism. *J Child Neurol*, **16**, 421–4.

Haxby, J. V., Hoffman, E. A., and Gobbini, M. I. (2000) The distributed human neural system for face perception. *Trends Cogn Sci*, **4**, 223–33.

Haznedar, M. M., Buchsbaum, M. S., Wei, T. C., et al. (2000) Limbic circuitry in patients with autism spectrum disorders studied with positron emission tomography and magnetic resonance imaging. *Am J Psychiatry*, **157**, 1994–2001.

Hisaoka, S., Harada, M., Nishitani, H., et al. (2001) Regional magnetic resonance spectroscopy of the brain in autistic individuals. *Neuroradiology*, **43**, 496–8.

Howard, M. A., Cowell, P. E., Boucher, J., et al. (2000) Convergent neuroanatomical and behavioural evidence of an amygdala hypothesis of autism. *NeuroReport*, **11**, 2931–5.

Hubl, D., Bolte, S., Feineis-Matthews, S., et al. (2003) Functional imbalance of visual pathways indicates alternative face processing strategies in autism. *Neurology*, **61**, 1232–7.

Kanwisher, N., McDermott, J., and Chun, M. M. (1997) The fusiform face area: a module in human extrastriate cortex specialized for face perception. *J Neurosci*, **17**, 4302–11.

Keller, T. A., Kana, R. K., and Just, M. A. (2007) A developmental study of the structural integrity of white matter in autism. *NeuroReport*, **18**, 23–7.

Klin, A. (1991) Young autistic children's listening preferences in regard to speech: a possible characterization of the symptom of social withdrawal. *J Autism Dev Disord*, **21**, 29–42.

Lainhart, J. E. (2006) Advances in autism neuroimaging research for the clinician and geneticist. *Am J Med Genet C Semin Med Genet*, **15**, 33–9.

Levitt, J. G., Blanton, R. E., Smalley, S., et al. (2003a) Cortical sulcal maps in autism. *Cereb Cortex*, **13**, 728–35.

Levitt, J. G., O'Neill, J., Blanton, R. E., et al. (2003) Proton magnetic resonance spectroscopic imaging of the brain in childhood autism. *Biol Psychiatry*, **54**, 1355–66.

Lord, C., Rutter, M., and Le Couteur, A. (1994) Autism Diagnostic Interview-Revised: a revised version of a diagnostic interview for caregivers of individuals with possible pervasive developmental disorders. *J Autism Dev Disord*, **24**, 659–85.

McAlonan, G. M., Cheung, V., Cheung, C., et al. (2005) Mapping the brain in autism. A voxel-based MRI study of volumetric differences and intercorrelations in autism. *Brain*, **128**, 268–76.

Manes, F., Piven, J., Vrancic, D., et al. (1999) An MRI study of the corpus callosum and cerebellum in mentally retarded autistic individuals. *J Neuropsychiatry Clin Neurosci*, **11**, 470–4.

Meresse, I., Zilbovicius, M., Boddaert, N., et al. (2005) Autism severity and temporal lobe functional abnormalities. *Ann Neurol*, **58**, 466–9.

Muller, R. A., Behen, M. E., Rothermel, R. D., et al. (1999) Brain mapping of language and auditory perception in high-functioning autistic adults: a PET study. *J Autism Dev Disord*, **29**, 19–31.

Murphy, D. G., Critchley, H. D., Schmitz, N., et al. (2002) Asperger syndrome: a proton magnetic resonance spectroscopy study of brain. *Arch Gen Psychiatry*, **59**, 885–91.

Oberman, L. M., Hubbard, E. M., McCleery, J. P., et al. (2005) EEG evidence for mirror neuron dysfunction in autism spectrum disorders. *Brain Res Cogn Brain Res*, **24**, 190–8.

Ohnishi, T., Matsuda, H., Hashimoto, T., et al. (2000) Abnormal regional cerebral blood flow in childhood autism. *Brain*, **123**, 1838–44.

Otsuka, H., Harada, M., Mori, K., et al. (1999) Brain metabolites in the hippocampus-amygdala region and cerebellum in autism: an ^{1}H-MR spectroscopy study. *Neuroradiology*, **41**, 517–19.

Pelphrey, K. A., Morris, J. P., and McCarthy, G. (2005) Neural basis of eye gaze processing deficits in autism. *Brain*, **128**, 1038–48.

Pierce, K., Haist, F., Sedaghat, F., et al. (2004) The brain response to personally familiar faces in autism: findings of fusiform activity and beyond. *Brain*, **127**, 2703–16.

Pierce, K., Muller, R. A., Ambrose, J., et al. (2001) Face processing occurs outside the fusiform "face area" in autism: evidence from functional MRI. *Brain*, **124**, 2059–73.

Piven, J., Arndt, S., Bailey, J., et al. (1995) An MRI study of brain size in autism. *Am J Psychiatry*, **152**, 1145–9.

Piven, J., Arndt, S., Bailey, J., et al. (1996) Regional brain enlargement in autism: a magnetic resonance imaging study. *J Am Acad Child Adolesc Psychiatry*, **35**, 530–6.

Piven, J., Bailey, J., Ranson, B. J., et al. (1997) An MRI study of the corpus callosum in autism. *Am J Psychiatry*, **154**, 1051–6.

Piven, J., Bailey, J., Ranson, B. J., et al. (1998) No difference in hippocampus volume detected on magnetic resonance imaging in autistic individuals. *J Autism Dev Disord*, **28**, 105–10.

Piven, J., Nehme, E., Simon, J., et al. (1992) Magnetic resonance imaging in autism: measurement of the cerebellum, pons, and fourth ventricle. *Biol Psychiatry*, **31**, 491–504.

Ring, H. A., Baron-Cohen, S., Wheelwright, S., et al. (1999) Cerebral correlates of preserved cognitive skills in autism: a functional MRI study of embedded figures task performance. *Brain*, **122**, 1305–15.

Rizzolatti, G., and Craighero, L. (2004) The mirror-neuron system. *Annu Rev Neurosci*, **27**, 169–92.

Rumsey, J. M., and Ernst, M. (2000) Functional neuroimaging of autistic disorders. *Ment Retard Dev Disabil Res Rev*, **6**, 171–9.

Rutherford, M. D., Baron-Cohen, S., and Wheelwright, S. (2002) Reading the mind in the voice: a study with normal adults and adults with Asperger syndrome and high functioning autism. *J Autism Dev Disord*, **32**, 189–94.

Saitoh, O., Courchesne, E., Egaas, B., et al. (1995) Cross-sectional area of the posterior hippocampus in autistic patients with cerebellar and corpus callosum abnormalities. *Neurology*, **45**, 317–24.

Schultz, R. T. (2005) Developmental deficits in social perception in autism: the role of the amygdala and fusiform face area. *Int J Dev Neurosci*, **23**, 125–41.

Schultz, R. T., Gauthier, I., Klin, A., et al. (2000) Abnormal ventral temporal cortical activity during face discrimination among individuals with autism and Asperger syndrome. *Arch Gen Psychiatry*, **57**, 331–40.

Sokol, D. K., Dunn, D. W., Edwards-Brown, M., et al. (2002) Hydrogen proton magnetic resonance spectroscopy in autism: preliminary evidence of elevated choline/ creatine ratio. *J Child Neurol*, **17**, 245–9.

Sparks, B. F., Friedman, S. D., Shaw, D. W., et al. (2002) Brain structural abnormalities in young children with autism spectrum disorder. *Neurology*, **59**, 184–92.

Teunisse, J. P. and de Gelder, B. (2003) Face processing in adolescents with autistic disorder: the inversion and composite effects. *Brain Cogn*, **52**, 285–94.

Tuchman, R. (2003) Autism. *Neurol Clin*, **21**, 915–32, viii.

Villalobos, M. E., Mizuno, A., Dahl, B. C., et al. (2005) Reduced functional connectivity between V1 and inferior frontal cortex associated with visuomotor performance in autism. *NeuroImage*, **25**, 916–25.

Williams, J. H., Whiten, A., Suddendorf, T., et al. (2001) Imitation, mirror neurons and autism. *Neurosci Biobehav Rev*, **25**, 287–95.

Zilbovicius, M., Boddaert, N., Belin, P., et al. (2000) Temporal lobe dysfunction in childhood autism: a PET study; positron emission tomography. *Am J Psychiatry*, **157**, 1988–93.

Zilbovicius, M., Meresse, I., Chabane, N., et al. (2006) Autism, the superior temporal sulcus and social perception. *Trends Neurosci*, **29**, 359–66.

Zilbovicius, M., Garreau, B., Tzourio, N., et al. (1992) Regional cerebral blood flow in childhood autism: a SPECT study. *Am J Psychiatry*, **149**, 924–30.

9

Neuroimaging of schizophrenia and its development

Matcheri S. Keshavan, Vaibhav A. Diwadkar, Konasale Prasad and Jeffrey A. Stanley

Introduction

Schizophrenia is a severe and chronic mental disorder with a life time prevalence of about 1%. Onset is typically in adolescence or early adulthood, and very rare before age 11; onset before age 13 is called very early onset schizophrenia, while onset before age 18 is called early-onset schizophrenia. The characteristic symptoms of schizophrenia include false beliefs (delusions), altered perceptions (hallucinations), disordered thinking, disorganized behavior (collectively, "positive symptoms") and deficits in motivation, affect, and socialization ("negative symptoms"). Diagnosis by the *Diagnostic and Statistical Manual of Mental Disorders*, 4th edn (DSM-IV; American Psychiatric Association, 1994) requires the presence of at least two of these symptoms (one in case of bizarre delusions or hallucinations), along with a decline in functioning lasting at least 6 months, after ensuring that these symptoms cannot be better explained by another medical disease, substance use or another psychiatric disorder.

The early course of schizophrenia is characterized by a sequence of phases. The *premorbid phase* is associated with cognitive and social difficulties, dating back to early childhood in many cases. Initial symptoms are often mild, with mood, thought and personality changes and sub-threshold psychotic symptoms beginning insidiously over time. This *prodromal phase* is only retrospectively diagnosed, though operational criteria are being developed to define this phase prospectively. The *psychotic phase* is heralded by florid positive symptoms described above, followed by a *transitional phase* lasting months to years and characterized by a tendency to frequent relapse before stabilization. The *stable phase* is characterized by persistent negative symptoms, cognitive deficits, and remissions and exacerbations of psychotic symptoms.

While the clinical presentations of adolescent and adult-onset schizophrenia are very similar, there are differences as well. Early-onset patients are more often male, tend to have more prominent impairments in cognition, structural brain anomalies and negative symptoms, are more often diagnosed as having an undifferentiated type of illness (fewer delusions and hallucinations), and tend to respond less well to treatment. Early-onset patients often fail to achieve expected levels of academic and interpersonal achievement and show evidence of greater familial risk.

Impairments in emotion and cognition in recent years have emerged as central features of schizophrenia. Deficits are consistently seen in **p**sychomotor **s**peed, **m**emory (working memory, visual and verbal memory), **a**ttention, **r**easoning and **t**act (i.e., aspects of reasoning related to social cognition and interaction) (note the mnemonic SMART). These deficits are strongly predictive of impaired

Neuroimaging in Developmental Clinical Neuroscience, eds. Judith M. Rumsey and Monique Ernst. Published by Cambridge University Press. © Cambridge University Press 2009.

functional outcome. Cognitive deficits are seen during the first psychotic episode, as well as during the premorbid and prodromal phases, and persist during the long-term course of the illness. Large-scale population cohort studies have indicated impaired educational performance and avoidant social behavior dating back to early childhood (Cannon et al., 2002). This suggests that cognitive impairments are invariant, trait-like features of schizophrenia, providing a functional context for the observed neuroanatomical, neurochemical, and neurophysiological alterations, discussed below.

Neurobiology of schizophrenia

Despite over a century of research, only a limited understanding of the pathogenesis of schizophrenia and related psychotic disorders exists. Early studies largely relied on postmortem brain examinations and were confounded by the effects of aging, illness chronicity, and medications. Studies of individuals in the early phases of schizophrenia, especially those in the first episode, avoid such difficulties. During the past three decades, these studies have greatly helped elucidate the course and nature of the neurobiological changes early in the illness.

In recent years, schizophrenia has been increasingly viewed as a disorder of brain development. Brain development begins in intrauterine life and continues into early adulthood. *Early brain development* (pre- and perinatal periods) is characterized by neurogenesis and neuronal migration, followed by an abundant proliferation of synapses continuing into childhood. This is followed, during *late brain development*, by programmed elimination or pruning of redundant synapses (late childhood and adolescence) and myelination (continuing into adulthood). Further synaptic and neuronal loss occurs later in life as a result of *aging*.

Derailments in the developmental and degenerative processes outlined above have led to three pathophysiological models (Figure 9.1), proposed to explain brain mechanisms underlying schizophrenia.

The so-called *early developmental* (or "doomed from the womb") model, suggests that abnormalities in brain development around or before birth mediate the failure of brain functions in early adulthood. This idea is supported by an array of findings, such as increased rates of birth complications, minor physical anomalies, neurological soft signs and subtle behavioral abnormalities, in children who later developed schizophrenia (Cannon et al., 2000, 2002; Dalman et al., 1999). It has been suggested that neuronal migration, which occurs mainly during the second trimester of fetal development, may be defective (Akbarian et al., 1993). Animal models have suggested that neonatal lesions of the hippocampus may lead to a reduction in dopamine transporter (DAT) messenger RNA (mRNA) (Lipska et al., 2003). Other models have suggested that a developmental lesion in the hippocampus leads to its abnormal gating of limbic system activity, resulting in valence-based (i.e., amygdala related), as opposed to goal-directed (i.e., prefrontal related), response strategies (Grace, 2000).

The fact that the onset of the salient symptoms of schizophrenia does not begin typically until adolescence or early adulthood points to a possible *late developmental* derailment around or prior to the onset of psychosis. If the process of programmed pruning is excessive, a pronounced loss of synapses, perhaps of the glutamatergic system, may result, leading to the emergence of the illness. In support of this view are postmortem observations showing reductions in dendrite density in cortical brain regions in schizophrenia (Glantz and Lewis, 2000).

Finally, the observation that at least a subgroup of patients deteriorate over the first few years of the illness has led to the view that there may be a *post-illness-onset* degenerative process, perhaps involving loss of neuronal or glial elements (Lieberman et al., 2001). These models that cite developmental versus degenerative-related deficits in schizophrenia are not necessarily mutually exclusive. Some investigators have suggested that a sequential combination of these processes may occur. Environmental factors such as illicit drug use and psychosocial

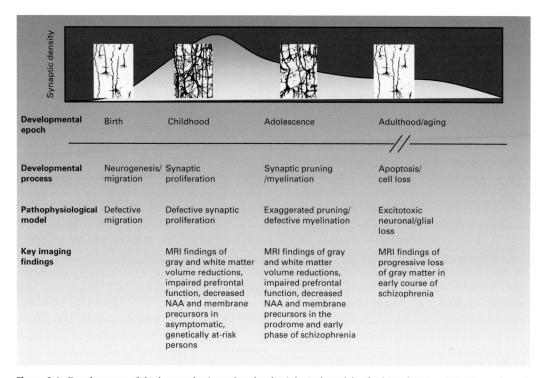

Figure 9.1. Development of the human brain and pathophysiological models of schizophrenia. NAA, *N*-acetylaspartate.

stress may also be potential secondary triggers accompanying illness onset and course. Neurobiological studies of early schizophrenia, especially the neuroimaging studies, have the potential to examine predictions generated by these seemingly contrasting models.

What neuroimaging can tell us about brain development

Structural magnetic resonance imaging (sMRI) studies have been used to map the anatomical trajectory of normal brain development (see Chapter 1). While total brain size is about 90% of its adult size by age 6 years, the gray and white matter proportions dynamically change throughout adolescence. Gray matter volumes increase during early childhood followed by substantive reductions through adolescence (an inverted U-shaped curve); these changes

parallel maturation of cognitive and affective measures. Structural MRI studies show that cortical gray matter loss occurs earliest in the primary sensorimotor areas and latest in the dorsolateral prefrontal and lateral temporal cortices, consistent with primate postmortem studies showing that the prefrontal cortex is among the last brain region to mature (Huttenlocher, 1979). Diffusion tensor imaging (DTI) studies (see Chapter 19) suggest that regional development of prefrontal connectivity parallels cognitive development.

Positron emission tomography (PET) and blood-oxygenation-level-dependent- (BOLD-) based functional MRI (fMRI) provide the best balance of spatial and temporal resolution for studying the neural correlates of cognitive development. Cognitive control, the capacity to exercise control over thoughts and actions, including the ability to inhibit inappropriate actions, matures during the first two decades of life (see Chapter 2). Functional

MRI studies examining cognitive control using tasks such as the Stroop and go/no-go suggest that, compared to adults, children recruit more diffuse and larger regions of the brain when performing such tasks. During development, activation of task-relevant brain regions becomes more fine-tuned with greater inter-regional coherence in brain activity suggesting increased collaboration as a function of normal development (Casey et al., 2005).

Magnetic resonance spectroscopy (MRS) allows non-invasive longitudinal evaluation of neuro-chemical brain maturation. Magnetic resonance spectroscopy generates a "spectrum" consisting of bio-chemical peaks of different resonance frequencies (or chemical shifts) with intensities proportional to the biochemical concentration. In vivo MRS provides information on the viability of neurons, axons, and astrocytes, as well as on energy status and membrane constituents (Stanley et al., 2000). In vivo *proton* ^{1}H spectroscopy has been the most widely used approach to MRS. *N*-Acetylaspartate (NAA), the most abundant signal, is primarily stored in neurons. ^{1}H spectroscopy can detect and monitor metabolite changes as a function of normal brain development. The NAA levels are low early in postnatal brain development, increase, and then plateau as the brain reaches maturation (Van Der Knaap et al., 1992).

In vivo *phosphorus* ^{31}P spectroscopy offers information on metabolites that are part of the anabolic and catabolic pathway of membrane phospholipids (MPLs) by quantifying the phosphomonoester (freely mobile precursors of MPLs, which are termed *free*-PME) and phosphodiester (freely mobile breakdown products of MPLs or free-PDE) (Stanley and Pettegrew, 2001; Stanley et al., 2000). Early in postnatal brain development, levels of PMEs are high whereas breakdown products, PDEs, are low; this is followed by dramatic decreases in precursor levels and increases in breakdown products which then plateau as the brain reaches maturation. The dramatic changes in MPL metabolites, which form neuronal membranes, may reflect the pruning of dendrites and synaptic connections during adolescence.

Neuroimaging studies in schizophrenia

Alterations in brain structure and connectivity

Over the past three decades, advances in sMRI have led to the identification of a number of brain structural abnormalities in schizophrenia that have confirmed earlier postmortem findings. Systematic reviews and meta-analyses of sMRI studies in schizophrenia (mostly using region of interest, or ROI, approaches) indicate that the whole-brain and gray-matter volume is reduced and ventricular volume is increased (Steen et al., 2006; Ward et al., 1996; Woodruff et al., 1995; Wright et al., 2000). Larger reductions are seen in temporal lobe structures, in particular the hippocampus, amygdala, and the superior temporal gyri (STG) (Lawrie and Abukmeil, 1998; Nelson et al., 1998), the prefrontal cortex, and the thalamus (Konick and Friedman, 2001). Automated regional parcellation and voxel-based morphometry (VBM) techniques have helped to validate these ROI-based findings. Voxel-based morphometry studies consistently find gray matter density reductions in medial temporal lobes and the STG (Honea et al., 2005). Superior temporal gyri volumes correlate with positive symptoms, while medial temporal lobe (MTL) reductions correlate with memory impairment (Antonova et al., 2004). Studies in childhood-onset schizophrenia patients have also demonstrated that these impairments are progressive, observed in structures including the hippocampus (Nugent et al., 2007), frontal and cingulate cortices (Vidal et al., 2006), and specific to adolescents with a diagnosis of schizophrenia (Gogtay et al., 2004).

Brain structural alterations in schizophrenia are present in first-episode schizophrenia patients. Recent systematic meta-analyses of first-episode schizophrenia patients versus controls (Steen et al., 2006) have shown whole-brain and hippocampal volume reductions. Brain structural changes appear to persist during the course of the schizophrenic illness, although it remains unclear whether they progress. Whitworth et al. (2005) found no evidence for progression, while other studies suggest ongoing changes in the brains of schizophrenic patients during the early course of illness (DeLisi et al., 2004; Ho et al., 2003). Taken together, it appears that brain structural alterations are persistent and may be trait-related.

Structural imaging data have been reported in offspring, siblings, and parents, as well as unaffected co-twins, of schizophrenia patients. Regions-of-interest-based studies of young relatives at risk (offspring and siblings) show amygdala and hippocampal volume reductions in relatives compared to controls (Keshavan et al., 1997; Lawrie et al., 1999). Cross-sectional VBM studies have also found reduced gray matter density in prefrontal cortex in unaffected relatives or offspring of schizophrenia patients who are known to be at higher risk for schizophrenia by comparison to healthy controls (Diwadkar et al., 2006; Job et al., 2003). Follow-up of unaffected relatives in one of these studies has shown that individuals at high genetic risk for developing psychosis have amygdala-hippocampal and thalamic volume reductions (Job et al., 2005). These findings support the view that the genetic risk for schizophrenia may underlie premorbid brain structural alterations.

Another approach to identifying at-risk individuals is to operationally define the prodrome based on subthreshold delusions, hallucinations or thought disorder, or functional declines that are not yet severe enough to meet full criteria for schizophrenia. Using such criteria, Pantelis et al. (2003) collected MRI scans from individuals with prodromal symptoms, about a third of whom developed psychosis. Those who later became psychotic (after 12 months of follow-up) had progressive loss of gray matter in the parahippocampal, fusiform, orbitofrontal and cerebellar cortices, and the cingulate cortices. This suggests that an active disease process may be taking place in the brain during the prodromal phase of transition to psychosis.

Magnetic resonance imaging methods have been employed to prospectively examine brain developmental trajectories in childhood-onset schizophrenia (Thompson et al., 2001). The investigators observed a subtle 1%–2% decrease per year of gray matter volume in parietal cortices in healthy children from ages 13 to 18, whereas a relatively more rapid loss was observed in the superior frontal, temporal, and parietal cortices reaching almost 3%–4% loss per year in children with schizophrenia studied over this same age range. All of this reduction may not be accounted for by neuronal loss, but may also result from decreases in myelination, lipid metabolism, and axonal fibers (Toga et al., 2006). Such studies point to the progressive structural brain changes that occur following illness onset, at least in a subgroup of patients with schizophrenia.

Studies of neural connectivity are made possible by DTI, which measures the orientation of water diffusion along the axis of tissue elements, such as axons. Fractional anisotropy (FA) measures the degree to which diffusion within a voxel is anisotropic (non-random) and is affected by the degree of local orientational coherence and the structural integrity of white matter.

Several DTI studies of adult patients have documented reduced FA in white matter tracts in schizophrenia, including the corpus callosum, the arcuate and the uncinate fasciculi, although findings have been somewhat inconsistent (Kanaan et al., 2005). Kumra et al. (2004) investigated early-onset schizophrenia and found reduced FA in frontal white matter bilaterally and in right occipital white matter. However, most studies have used relatively small sample sizes (range $n = 5$ to $n = 30$ patients) and widely varying methodologies (Kanaan et al., 2005).

In summary, there is robust evidence for global alterations in brain structure as well as prominent reductions in regional brain volumes and connectivity, especially in medial and superior temporal and prefrontal cortices. These abnormalities appear to be present in asymptomatic individuals with a genetic predisposition to the disorder, may evolve during the prodromal phase of the illness, may progress early in the illness, and may be more prominent in early-onset forms of the disorder.

Alterations in brain function

Non-invasive fMRI studies are eminently well suited for the study of functional brain abnormalities in schizophrenia and have produced a large literature in adolescent and adult patients. We focus on three domains/systems: working memory, cognitive control, and emotion processing, all of which are central to normal development.

Prefrontal function is of particular relevance to the developmental pathophysiology of schizophrenia (Lewis, 1997). Working memory tasks such as the n-back (Braver et al., 1997) have been used to study both schizophrenia patients and their adolescent and adult relatives. Controlling for performance differences, patients show more prefrontal activity than controls (Thermenos et al., 2005), suggesting an inefficient frontal response. Other studies of working memory however suggest that patients do not increase activation in key regions such as dorsolateral prefrontal cortex in response to increases in working memory demands to the same extent as is observed in controls (Tan et al., 2005). Our own data (Keedy et al., 2006) (Figure 9.2) show deficient prefrontal activation with an oculomotor delayed-response task used to examine spatial working memory. Similar findings have been seen in a large multi-site study of first-episode schizophrenia (Schneider et al., 2007) and in siblings (Callicott et al., 2003) and offspring (Keshavan et al., 2002) as well. Furthermore, using fMRI with non-working-memory tasks such as sentence completion, decreased fronto-cerebellar functional connectivity (to be distinguished from anatomical connectivity observed in DTI studies) has been reported in genetically at-risk individuals (Whalley et al., 2005), suggesting that part of the developmental deficit in the illness may indeed be functional disconnectivity (Friston, 2005). A recent meta-analysis of fMRI studies of subjects defined to be in the prodromal phase of the illness showed abnormalities in prefrontal cognitive functions that were less severe than, but qualitatively similar to, those observed in the first episode of illness (Fusar-Poli et al., 2007). In sum, impaired prefrontal

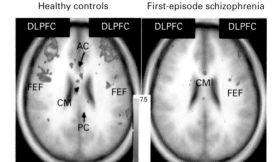

Healthy controls   First-episode schizophrenia

Figure 9.2. Activation maps for oculomotor delayed-response task versus visually guided saccade task contrast in an axial slice overlaid on a group anatomical average. Left: healthy controls; right: first-episode schizophrenia. Significant voxels are color coded to reflect t statistics. AC, anterior cingulate; CM, cingulate motor area; DLPFC, dorsolateral prefrontal cortex; FEF, frontal eye fields; PC, posterior cingulate (adapted from Keedy et al., 2006). These results suggest deficits in neural regions supporting spatial working memory.

function with fMRI studies may be seen in the premorbid and prodromal phases of the illness, as well as after the illness is established.

The anterior cingulate cortex is central to cognitive control and self-monitoring (Kerns et al., 2004). Using conflict-inducing tasks such as the Stroop to examine cingulate function, schizophrenia patients show both reduced error monitoring (i.e., reduced awareness of performance errors) and reduced anterior cingulate activity compared to controls (Carter et al., 2001; Kerns et al., 2005). Furthermore, there is evidence that cingulate abnormalities may affect the ability of these structures to exercise control of regions such as the prefrontal and temporal cortices in memory, leading to a relative lack of prefrontal-temporal functional connectivity in schizophrenia (Fletcher et al., 1999).

The amygdala-hippocampal complex is involved in the appraisal of emotion and affect/mood (amygdala) and episodic and associative memory (hippocampus) (Eichenbaum, 2004; LeDoux, 2002).

Social cognition deficits manifested in difficulty in interpreting facial expressions are associated with amygdala impairment (Adolphs et al., 2002). Since social deficits are common in schizophrenia, several fMRI studies have assessed amygdala function in this illness (Aleman and Kahn, 2005). In general, patients perform poorly on affect labeling tasks (Hempel et al., 2003) and correspondingly have reduced responsivity of the amygdala (Gur et al., 2002; Takahashi et al., 2004). Further studies are needed to carefully investigate the relationship between these deficits and social functioning and the emergence of these deficits in at-risk populations.

Plasticity or change in cortical organization and function in association with changes in behavior is a key element of normal development and function (Edelman, 1993; Innocenti et al., 1995). Studies that assess neural changes or lack of changes in relationship to behavior in schizophrenia may help uncover vital elements of dysfunction (Stephan et al., 2006), as well as provide insights into the functioning of the schizophrenia brain "in action." Functional MRI studies examining relationships between gene expression and cortical function may also help explore new frontiers (Hariri and Weinberger, 2003), although the physiological pathways from genes to the BOLD response are complex.

Neurochemical brain alterations

The majority of MRS studies in schizophrenia have focused on ^{1}H MRS, an in vivo neuroimaging approach discussed earlier. A recent meta-analysis of in vivo ^{1}H spectroscopy studies in schizophrenia showed reduced NAA (a marker of neuronal integrity), primarily in the prefrontal cortex and hippocampus, that was indistinguishable between first-episode and chronic schizophrenia patients (Steen et al., 2005). However, not all studies of first-episode schizophrenia patients show NAA reductions; inconsistencies between studies may be due to variations in methodologies (Stanley et al., 2000). In our study, adolescent-onset, antipsychotic-naive, first-episode

patients showed lower NAA levels compared to age-matched controls, while adult-onset patients and adult healthy comparison subjects did not differ (Stanley et al., 2007). ^{1}H spectroscopy studies of children and adolescents with schizophrenia show reduced NAA (expressed as a ratio to the creatine signal and often used as an internal standard) in the left prefrontal white matter (Brooks et al., 1998), medial prefrontal cortex (Thomas et al., 1998), and in the prefrontal cortex bilaterally (Bertolino et al., 1998). A recent study (O'Neill et al., 2004) reported no prefrontal NAA abnormalities in a small sample ($n = 11$, 7 male) of children and adolescents with schizophrenia. In the thalamus, there was a diagnosis-by-gender interaction, whereby NAA was lower in male, but not in female, patients. Bertolino et al. (1998) observed bilateral reduction of NAA/creatine ratios in the prefrontal cortex in childhood-onset schizophrenia patients, as well as in schizophreniform and chronic schizophrenia patients (Bertolino et al., 1996). Collectively, these observations suggest greater prefrontal NAA deficits in schizophrenia patients with a younger age of onset compared to those with an older onset age. Considering the evidence of greater premorbid deficits in early-onset subjects with schizophrenia noted above, these results further support the presence of greater neurobiological abnormalities in patients with an earlier age of onset (Stanley et al., 2007). The lower NAA may reflect impaired neuronal integrity perhaps related to excessive synaptic pruning (Pettegrew et al., 1991) or a reduction in neuronal or glial elements and/or underdeveloped myelin.

Reductions in NAA may be present prior to the illness onset. We (Keshavan et al., 1997) observed a trend for NAA/choline ratio reductions in the anterior cingulate region in adolescent offspring at genetic risk for schizophrenia. Interestingly, reduced NAA/choline in the anterior cingulate predicted conversion to psychosis in individuals who met criteria for the prodromal phase of schizophrenia (Jessen et al., 2006).

A relatively small number of studies have used ^{31}P MRS. Pettegrew and colleagues (1991) initially observed decreased PMEs and increased PDE levels (membrane precursors and breakdown products, as discussed earlier) in the combined right and left prefrontal region of first-episode, antipsychotic-naive schizophrenia subjects compared with controls, suggesting decreased synthesis and increased breakdown products of MPLs. This was the first study supporting Feinberg's hypothesis of abnormal neurodevelopment in schizophrenia due to an exaggeration of normal preadolescent synaptic pruning in the prefrontal region (Pettegrew et al., 1991). Subsequent studies of first-episode schizophrenia patients demonstrated similar MPL alterations in the left prefrontal and in the right and left temporal lobe (Fukuzako et al., 1999). More recent studies using multi-voxel chemical shift imaging have shown prefrontal PDE reductions (Volz et al., 2000) and prefrontal reductions in both PMEs and PDEs in first-episode patients compared to controls (Smesny et al., 2007). At low field strength such as 1.5 Tesla, the PME and PDE peaks in many cases also include the contribution of a broad underlying signal of larger and less mobile molecules with PDE and PME moieties (Stanley and Pettegrew, 2001), which may complicate the interpretation of results. Overall, there is compelling evidence of diminished mass or content of MPL (i.e., neuropil) at the early stage of schizophrenia that is consistent with postmortem studies (Selemon and Goldman-Rakic, 1999). Our work showing prefrontal PME reductions in adolescent offspring at genetic risk for schizophrenia suggests these alterations may precede symptomatic manifestations (Keshavan et al., 2003).

In summary, MRS studies have revealed reductions in neuronal and membrane integrity in the early stages of schizophrenia and in individuals at risk for the disorder in brain regions where structural and functional alterations are also observed with sMRI and fMRI. Future prospective high-field ^{31}P and ^{1}H spectroscopy studies of children and adolescents at risk for and those already presenting with schizophrenia may identify the point in time at which regional deviations from the normal course of development occur.

Methodological issues in neuroimaging research

Certain design and methodological issues may particularly impact the interpretation of neuroimaging findings in schizophrenia. First, matching patient's education/IQ levels to controls is problematic because of the effect of schizophrenia on intellectual function; matching for parental education is generally preferred. Secondly, possible effects of antipsychotic drugs (Keshavan et al., 1994) as well as alcohol abuse on MRI brain morphology (Mathalon et al., 2003) need to be considered in interpreting the findings. Finally, studies seeking to identify systematic differences in brain structure across disease populations are complicated by the enormous variability of brain structure between individuals, as well as across developmental periods. Recent approaches to image analysis, such as cortical pattern matching (see Chapter 1), can be used to compare and pool morphometric data over time and across subjects. Brain structural differences, not otherwise seen in individual scans, can then be identified between groups. Studies of early-onset schizophrenia using cortical pattern matching have thus observed differences between patients and controls in cortical thickness and gray matter density (Narr et al., 2005). Probabilistic brain atlases incorporating information on growth rates and brain development can be used to identify specific deviations in development from longitudinal neuroimaging datasets.

The interpretation of extant developmental fMRI studies is also rendered difficult by competing conceptual themes. For example, is less fMRI activation of a structure better because it reflects efficient processing and a healthier brain? Or is more activation better because it reflects more effective recruitment and therefore more normal cortical organization? The answers to such questions are complex and depend (among other factors) on whether the experimental paradigms used are demanding or involve obligatory processing or whether parametric manipulations are being used in an experiment (Cohen et al., 1997). The answers to such questions will drive the interpretation of fMRI results and related mechanisms of disease (Wishart et al., 2002).

Technical challenges continue to pose limitations for spectroscopy. An example is the difficulty of precisely separating glutamate and glutamine neurotransmitter systems, which are important because of their implication in development (Rothman et al., 1999). The term "Glx," which is commonly used in the literature, is somewhat problematic due to multiple definitions (i.e., either defined as glutamate plus glutamine or as the total signal integrated over the glutamate and glutamine spectral region, including gamma-aminobutyric acid (GABA) and signals from macromolecules), leading to differences in results depending on the definition employed in a study. Higher field magnets (3 T or higher) allow for greater spectral separation of these metabolites.

Does neuroimaging help in diagnosis or outcome prediction?

The potential diagnostic value of neuroimaging studies in schizophrenia is uncertain for several reasons. First, the range of values observed among patients overlaps with those of healthy controls. In addition, brain structure changes in affective disorder appear to be less marked, but qualitatively similar to schizophrenia (Bearden et al., 2001; Hoge et al., 1999; McDonald et al., 2004; Strakowski et al., 2000), suggesting structural alterations may be diagnostically non-specific. A second confound arises from poorly defined clinical boundaries among psychotic disorders and within the subtypes of schizophrenia. Thirdly, rapid advances in molecular genetics and neurobiology point to the heterogeneity that may underlie variability in the imaging findings observed across the studies. Finally, schizophrenia lacks construct validity, meaning that there is no reference "gold" standard for the diagnosis against which to validate neuroimaging as a diagnostic tool.

Prognostication refers to the ability to predict qualitative (which outcomes), quantitative (how likely), and the temporal (over what period) dimensions of outcome. Interestingly, reductions in the volume of the STG at the beginning of a psychotic illness may predict who will eventually specifically develop schizophrenia as opposed to affective illness (Hirayasu et al., 1998). In recent-onset psychotic disorders other than schizophrenia, Sylvian fissure and third ventricular volumes predicted negative symptoms and unemployment (Van Os et al., 1995). Other studies have reported ventricular enlargement (Lieberman et al., 2001) and reduced cerebellar volume (Wassink et al., 1999) in persons with poor-outcome schizophrenia. Left dorsolateral prefrontal cortex volume has been shown to predict functional outcome (social and employment subscales of Strauss-Carpenter scale) at 1 year but not at 2 years (Prasad et al., 2005). Further prospective studies on sufficiently powered samples are required for neuroimaging to meet evidence-based medicine standards as a prognosticating tool.

Implications of neuroimaging for future research

In the spirit of the current focus on translational neuroscience, neuroimaging studies in schizophrenia are increasingly focusing on the application of these techniques for elucidating pathophysiology, developing approaches for early detection and prediction of illness, and individualizing and monitoring treatment. We discuss these trends in research briefly below.

Neuroimaging studies are valuable for unraveling the etiopathology of schizophrenia. Brain structural and functional alterations have been proposed as potential endophenotypes, i.e., intermediate phenotypes on the causal pathway from the genotype to the phenotype. Structural brain changes appear to meet criteria for endophenotypes (Gottesman and Gould, 2003): they are robustly associated with schizophrenia and are primarily state independent, i.e., present during both periods of illness and wellness, but may cut across the diagnostic boundaries of major psychotic illnesses. Brain morphometric measures are highly heritable; they co-segregate with the broadly defined neurocognitive and behavioral phenotypes within families and are present in unaffected family members more frequently than in the general population. Neuroimaging studies have also begun to elucidate genes that may underlie specific neuroanatomical and functional abnormalities in schizophrenia, as well as specific etiological factors such as exposure to viral infections that may confer risk to the disease (Prasad et al., 2007). Several ongoing studies are investigating individuals at high genetic risk for schizophrenia with repeated, longitudinal neuroimaging assessments to characterize the etiological antecedents of neural changes and whether they predict an imminent onset of psychosis (Keshavan et al., 2005).

Future studies using brain imaging techniques are also likely to be of help in quantitatively monitoring and predicting therapeutic response to pharmacological and psychosocial treatment. Positron emission tomography can be used to measure receptor occupancy of central neurotransmitter systems targeted by psychopharmacological treatments. For example, using $[^{11}C]$raclopride, it has been shown that the therapeutic response to antipsychotics is associated with 60%–70% occupancy of D2 receptors, while occupancy of 80% or higher leads to significant extrapyramidal side-effects (Kapur et al., 1998), leading to proposed algorithms for deciding how best to utilize PET neuroimaging for both disease-related and drug-related questions (Kapur, 2001). Neurochemical, functional, and structural imaging have the potential to reveal underlying systems and molecular pathology. An integrated approach using the information from molecular genetic studies, functional genomics, and proteomics with hypothesis-driven neuroimaging studies may pave the way for designing novel approaches to treatment.

Conclusions

The past three decades of research in brain imaging of schizophrenia have yielded important clues regarding the pathophysiology of schizophrenia. Compelling data point to brain structural, functional, and neurochemical alterations in fronto-limbic and related brain regions and their connecting circuitries. These alterations appear to begin early in development and evolve during the course of the illness, suggesting a sequential derailment of developmental processes. While several conceptual and methodological challenges remain, the future holds promise for further elucidating the neurobiology of schizophrenia at all stages of development. Application of the accumulating knowledge base from neuroimaging studies of schizophrenia for better diagnosis, prognostication, and etiopathological research may be expected in the near future.

REFERENCES

Adolphs, R., Baron-Cohen, S., and Tranel, D. (2002) Impaired recognition of social emotions following amygdala damage. *J Cogn Neurosci*, **14**, 1264–74.

Akbarian, S., Vinuela, A., Kim, J. J., et al. (1993) Distorted distribution of nicotinamide-adenine dinucleotide phosphate-diaphorase neurons in temporal lobe of schizophrenics implies anomalous cortical development. *Arch Gen Psychiatry*, **50**, 178–87.

Aleman, A. and Kahn, R. S. (2005) Strange feelings: do amygdala abnormalities dysregulate the emotional brain in schizophrenia? *Prog Neurobiol*, **77**, 283–98.

American Psychiatric Association (1994) *Diagnostic and Statistical Manual of Mental Disorders*, 4th edn. Arlington, VA: American Psychiatric Association.

Antonova, E., Sharma, T., Morris, R., et al. (2004) The relationship between brain structure and neurocognition in schizophrenia: a selective review. *Schizophr Res*, **70**, 117–45.

Bearden, C. E., Hoffman, K. M., and Cannon, T. D. (2001) The neuropsychology and neuroanatomy of bipolar affective disorder: a critical review. *Bipolar Disord*, **3**, 106–50; discussion 151–3.

Bertolino, A., Kumra, S., Callicott, J. H., et al. (1998) Common pattern of cortical pathology in childhood-onset and adult-onset schizophrenia as identified by proton magnetic resonance spectroscopic imaging. *Am J Psychiatry*, **155**, 1376–83.

Bertolino, A., Nawroz, S., Mattay, V. S., et al. (1996) Regionally specific pattern of neurochemical pathology in schizophrenia as assessed by multislice proton magnetic resonance spectroscopic imaging. *Am J Psychiatry*, **153**, 1554–63.

Braver, T. S., Cohen, J. D., Nystrom, L. E., et al. (1997) A parametric study of prefrontal cortex involvement in human working memory. *NeuroImage*, **5**, 49–62.

Brooks, W. M., Hodde-Vargas, J., Vargas, L. A., et al. (1998) Frontal lobe of children with schizophrenia spectrum disorders: a proton magnetic resonance spectroscopic study. *Biol Psychiatry*, **43**, 263–9.

Callicott, J. H., Egan, M. F., Mattay, V. S., et al. (2003) Abnormal fMRI response of the dorsolateral prefrontal cortex in cognitively intact siblings of patients with schizophrenia. *Am J Psychiatry*, **160**, 709–19.

Cannon, M., Caspi, A., Moffitt, T. E., et al. (2002) Evidence for early-childhood, pan-developmental impairment specific to schizophreniform disorder: results from a longitudinal birth cohort. *Arch Gen Psychiatry*, **59**, 449–56.

Cannon, T. D., Rosso, I. M., Hollister, J. M., et al. (2000) A prospective cohort study of genetic and perinatal influences in the etiology of schizophrenia. *Schizophr Bull*, **26**, 351–66.

Carter, C. S., MacDonald, A. W., 3rd, Ross, L. L., et al. (2001) Anterior cingulate cortex activity and impaired self-monitoring of performance in patients with schizophrenia: an event-related fMRI study. *Am J Psychiatry*, **158**, 1423–8.

Casey, B. J., Galvan, A., and Hare, T. A. (2005) Changes in cerebral functional organization during cognitive development. *Curr Opin Neurobiol*, **15**, 239–44.

Cohen, J. D., Perlstein, W. M., Braver, T. S., et al. (1997) Temporal dynamics of brain activation during a working memory task. *Nature*, **386**, 604–8.

Dalman, C., Allebeck, P., Cullberg, J., et al. (1999) Obstetric complications and the risk of schizophrenia: a longitudinal study of a national birth cohort. *Arch Gen Psychiatry*, **56**, 234–40.

DeLisi, L. E., Sakuma, M., Maurizio, A. M., et al. (2004) Cerebral ventricular change over the first 10 years after the onset of schizophrenia. *Psychiatry Res*, **130**, 57–70.

Diwadkar, V. A., Montrose, D. M., Dworakowski, D., et al. (2006) Genetically predisposed offspring with schizotypal features: an ultra high-risk group for schizophrenia? *Prog Neuropsychopharmacol Biol Psychiatry*, **30**, 230–8.

Edelman, G. M. (1993) Neural Darwinism: selection and reentrant signaling in higher brain function. *Neuron*, **10**, 115–25.

Eichenbaum, H. (2004) Hippocampus: cognitive processes and neural representations that underlie declarative memory. *Neuron*, **44**, 109–20.

Fletcher, P., McKenna, P. J., Friston, K. J., et al. (1999) Abnormal cingulate modulation of fronto-temporal connectivity in schizophrenia. *NeuroImage*, **9**, 337–42.

Friston, K. (2005) Disconnection and cognitive dysmetria in schizophrenia. *Am J Psychiatry*, **162**, 429–32.

Fukuzako, H., Fukuzako, T., Hashiguchi, T., et al. (1999) Changes in levels of phosphorus metabolites in temporal lobes of drug-naive schizophrenic patients. *Am J Psychiatry*, **156**, 1205–8.

Fusar-Poli, P., Perez, J., Broome, M., et al. (2007) Neurofunctional correlates of vulnerability to psychosis: a systematic review and meta-analysis. *Neurosci Biobehav Rev*, **31**, 465–84.

Glantz, L. A. and Lewis, D. A. (2000) Decreased dendritic spine density on prefrontal cortical pyramidal neurons in schizophrenia. *Arch Gen Psychiatry*, **57**, 65–73.

Gogtay, N., Sporn, A., Clasen, L. S., et al. (2004) Comparison of progressive cortical gray matter loss in childhood-onset schizophrenia with that in childhood-onset atypical psychoses. *Arch Gen Psychiatry*, **61**, 17–22.

Gottesman, I. I. and Gould, T. D. (2003) The endophenotype concept in psychiatry: etymology and strategic intentions. *Am J Psychiatry*, **160**, 636–45.

Grace, A. A. (2000) Gating of information flow within the limbic system and the pathophysiology of schizophrenia. *Brain Res Brain Res Rev*, **31**, 330–41.

Gur, R. E., McGrath, C., Chan, R. M., et al. (2002) An fMRI study of facial emotion processing in patients with schizophrenia. *Am J Psychiatry*, **159**, 1992–9.

Hariri, A. R. and Weinberger, D. R. (2003) Imaging genomics. *Br Med Bull*, **65**, 259–70.

Hempel, A., Hempel, E., Schonknecht, P., et al. (2003) Impairment in basal limbic function in schizophrenia during affect recognition. *Psychiatry Res*, **122**, 115–24.

Hirayasu, Y., Shenton, M. E., Salisbury, D. F., et al. (1998) Lower left temporal lobe MRI volumes in patients with first-episode schizophrenia compared with psychotic patients with first-episode affective disorder and normal subjects. *Am J Psychiatry*, **155**, 1384–91.

Ho, B. C., Andreasen, N. C., Nopoulos, P., et al. (2003) Progressive structural brain abnormalities and their relationship to clinical outcome: a longitudinal magnetic resonance imaging study early in schizophrenia. *Arch Gen Psychiatry*, **60**, 585–94.

Hoge, E. A., Friedman, L., and Schulz, S. C. (1999) Meta-analysis of brain size in bipolar disorder. *Schizophr Res*, **37**, 177–81.

Honea, R., Crow, T. J., Passingham, D., et al. (2005) Regional deficits in brain volume in schizophrenia: a meta-analysis of voxel-based morphometry studies. *Am J Psychiatry*, **162**, 2233–45.

Huttenlocher, P. R. (1979) Synaptic density in human frontal cortex. Developmental changes and effects of aging. *Brain Res*, **163**, 195.

Innocenti, G. M., Aggoun-Zouaoui, D., and Lehmann, P. (1995) Cellular aspects of callosal connections and their development. *Neuropsychologia*, **33**, 961–87.

Jessen, F., Scherk, H., Traber, F., et al. (2006) Proton magnetic resonance spectroscopy in subjects at risk for schizophrenia. *Schizophr Res*, **87**, 81–8.

Job, D. E., Whalley, H. C., Johnstone, E. C., et al. (2005) Grey matter changes over time in high risk subjects developing schizophrenia. *NeuroImage*, **25**, 1023–30.

Job, D. E., Whalley, H. C., McConnell, S., et al. (2003) Voxel-based morphometry of grey matter densities in subjects at high risk of schizophrenia. *Schizophr Res*, **64**, 1–13.

Kanaan, R. A., Kim, J. S., Kaufmann, W. E., et al. (2005) Diffusion tensor imaging in schizophrenia. *Biol Psychiatry*, **58**, 921–9.

Kapur, S. (2001) Neuroimaging and drug development: an algorithm for decision making. *J Clin Pharmacol Suppl*, 64S–71S.

Kapur, S., Zipursky, R. B., Remington, G., et al. (1998) 5-HT2 and D2 receptor occupancy of olanzapine in schizophrenia: a PET investigation. *Am J Psychiatry*, **155**, 921–8.

Keedy, S. K., Ebens, C. L., Keshavan, M. S., et al. (2006) Functional magnetic resonance imaging studies of eye movements in first episode schizophrenia: smooth pursuit, visually guided saccades and the oculomotor delayed response task. *Psychiatry Res*, **146**, 199–211.

Kerns, J. G., Cohen, J. D., MacDonald, A. W., 3rd, et al. (2004) Anterior cingulate conflict monitoring and adjustments in control. *Science*, **303**, 1023–6.

Kerns, J. G., Cohen, J. D., MacDonald, A. W., 3rd, et al. (2005) Decreased conflict- and error-related activity in the anterior cingulate cortex in subjects with schizophrenia. *Am J Psychiatry*, **162**, 1833–9.

Keshavan, M. S., Bagwell, W. W., Haas, G. L., et al. (1994) Changes in caudate volume with neuroleptic treatment. *Lancet*, **344**, 1434.

Keshavan, M. S., Diwadkar, V. A., Montrose, D. M., et al. (2005) Premorbid indicators of risk for schizophrenia: a selective review and update. *Schizophr Res*, **79**, 45–57.

Keshavan, M. S., Diwadkar, V. A., Spencer, S. M., et al. (2002) A preliminary functional magnetic resonance imaging study in offspring of schizophrenic parents. *Prog Psychopharmacol Biol Psychiatry*, **26**, 1143–9.

Keshavan, M. S., Montrose, D. M., Pierri, J. N., et al. (1997) Magnetic resonance imaging and spectroscopy in offspring at risk for schizophrenia: preliminary studies. *Prog Neuropsychopharmacol Biol Psychiatry*, **21**, 1285–95.

Keshavan, M. S., Stanley, J. A., Montrose, D. M., et al. (2003) Prefrontal membrane phospholipid metabolism of child and adolescent offspring at risk for schizophrenia or schizoaffective disorder: an in vivo ^{31}P MRS study. *Mol Psychiatry*, **8**, 316–23, 251.

Konick, L. C. and Friedman, L. (2001) Meta-analysis of thalamic size in schizophrenia. *Biol Psychiatry*, **49**, 28–38.

Kumra, S., Ashtari, M., McMeniman, M., et al. (2004) Reduced frontal white matter integrity in early-onset schizophrenia: a preliminary study. *Biol Psychiatry*, **55**, 1138–45.

Lawrie, S. M. and Abukmeil, S. S. (1998) Brain abnormality in schizophrenia. A systematic and quantitative review of volumetric magnetic resonance imaging studies. *Br J Psychiatry*, **172**, 110–20.

Lawrie, S. M., Whalley, H., Kestelman, J. N., et al. (1999) Magnetic resonance imaging of brain in people at high risk of developing schizophrenia. *Lancet*, **353**, 30–3.

LeDoux, J. (2002) The emotional brain, fear, and the amygdala. *Cell Mol Neurobiol*, **23**, 727–38.

Lewis, D. A. (1997) Development of the prefrontal cortex during adolescence: insights into vulnerable neural circuits in schizophrenia. *Neuropsychopharmacology*, **16**, 385–98.

Lieberman, J., Chakos, M., Wu, H., et al. (2001) Longitudinal study of brain morphology in first episode schizophrenia. *Biol Psychiatry*, **49**, 487–99.

Lipska, B. K., Lerman, D. N., Khaing, Z. Z., et al. (2003) The neonatal ventral hippocampal lesion model of schizophrenia: effects on dopamine and GABA mRNA markers in the rat midbrain. *Eur J Neurosci*, **18**, 3097–104.

Mathalon, D. H., Pfefferbaum, A., Lim, K. O., et al. (2003) Compounded brain volume deficits in schizophrenia – alcoholism comorbidity. *Arch Gen Psychiatry*, **60**, 245–52.

McDonald, C., Zanelli, J., Rabe-Hesketh, S., et al. (2004) Meta-analysis of magnetic resonance imaging brain morphometry studies in bipolar disorder. *Biol Psychiatry*, **56**, 411–17.

Narr, K. L., Bilder, R. M., Toga, A. W., et al. (2005) Mapping cortical thickness and gray matter concentration in first episode schizophrenia. *Cereb Cortex*, **15**, 708–19.

Nelson, M. D., Saykin, A. J., Flashman, L. A., et al. (1998) Hippocampal volume reduction in schizophrenia as assessed by magnetic resonance imaging: a meta-analytic study. *Arch Gen Psychiatry*, **55**, 433–40.

Nugent, T. F., 3rd, Herman, D. H., Ordonez, A., et al. (2007) Dynamic mapping of hippocampal development in childhood onset schizophrenia. *Schizophr Res*, **90**, 62–70.

O'Neill, J., Levitt, J., Caplan, R., et al. (2004) ^{1}H MRSI evidence of metabolic abnormalities in childhood-onset schizophrenia. *NeuroImage*, **21**, 1781–9.

Pantelis, C., Velakoulis, D., McGorry, P. D., et al. (2003) Neuroanatomical abnormalities before and after onset of psychosis: a cross-sectional and longitudinal MRI comparison. *Lancet*, **361**, 281–8.

Pettegrew, J. W., Keshavan, M. S., Panchalingam, K., et al. (1991) Alterations in brain high-energy phosphate and membrane phospholipid metabolism in first-episode, drug-naive schizophrenics. A pilot study of the dorsal prefrontal cortex by in vivo phosphorus 31 nuclear magnetic resonance spectroscopy. *Arch Gen Psychiatry*, **48**, 563–8.

Prasad, K. M., Sahni, S. D., Rohm, B. R., et al. (2005) Dorsolateral prefrontal cortex morphology and short-term outcome in first-episode schizophrenia. *Psychiatry Res*, **140**, 147–55.

Prasad, K. M., Shirts, B. H., Yolken, R. H., et al. (2007) Brain morphological changes associated with exposure to HSV1 in first-episode schizophrenia. *Mol Psychiatry*, **12**, 105–13, 1.

Rothman, D. L., Sibson, N. R., Hyder, F., et al. (1999) In vivo nuclear magnetic resonance spectroscopy studies of the relationship between the glutamate-glutamine neurotransmitter cycle and functional neuroenergetics. *Philos Trans R Soc Lond B Biol Sci*, **354**, 1165–77.

Schneider, F., Habel, U., Reske, M., et al. (2007) Neural correlates of working memory dysfunction in first-episode schizophrenia patients: an fMRI multi-center study. *Schizophr Res*, **89**, 198–210.

Selemon, L. D. and Goldman-Rakic, P. S. (1999) The reduced neuropil hypothesis: a circuit based model of schizophrenia. *Biol Psychiatry*, **45**, 17–25.

Smesny, S., Rosburg, T., Nenadic, I., et al. (2007) Metabolic mapping using 2D ^{31}P-MR spectroscopy reveals frontal and thalamic metabolic abnormalities in schizophrenia. *NeuroImage*, **35**, 729–37.

Stanley, J. A. and Pettegrew, J. W. (2001) A post-processing method to segregate and quantify the broad components underlying the phosphodiester spectral region of *in vivo* ^{31}P brain spectra. *Magnet Reson Med*, **45**, 390–6.

Stanley, J. A., Pettegrew, J. W., and Keshavan, M. S. (2000) Magnetic resonance spectroscopy in schizophrenia: methodological issues and findings-part I. *Biol Psychiatry*, **48**, 357–68.

Stanley, J. A., Vemulapalli, M., Nutche, J., et al. (2007) Reduced *N*-acetyl-aspartate levels in schizophrenia patients with a younger onset age: a single-voxel (1)H spectroscopy study. *Schizophr Res*, **93**, 23–32.

Steen, R. G., Hamer, R. M., and Lieberman, J. A. (2005) Measurement of brain metabolites by ^{1}H magnetic resonance spectroscopy in patients with schizophrenia: a systematic review and meta-analysis. *Neuropsychopharmacology*, **30**, 1949–62.

Steen, R. G., Mull, C., McClure, R., et al. (2006) Brain volume in first-episode schizophrenia: systematic review and meta-analysis of magnetic resonance imaging studies. *Br J Psychiatry*, **188**, 510–18.

Stephan, K. E., Baldeweg, T., and Friston, K. J. (2006) Synaptic plasticity and dysconnection in schizophrenia. *Biol Psychiatry*, **59**(10), 929–39.

Strakowski, S. M., Del Bello, M. P., Adler, C., et al. (2000) Neuroimaging in bipolar disorder. *Bipolar Disord*, **2**, 148–64.

Takahashi, H., Koeda, M., Oda, K., et al. (2004) An fMRI study of differential neural response to affective pictures in schizophrenia. *NeuroImage*, **22**, 1247–54.

Tan, H. Y., Choo, W. C., Fones, C. S., et al. (2005) fMRI study of maintenance and manipulation processes within working memory in first-episode schizophrenia. *Am J Psychiatry*, **162**, 1849–58.

Thermenos, H. W., Goldstein, J. M., Buka, S. L., et al. (2005) The effect of working memory performance on functional MRI in schizophrenia. *Schizophr Res*, **74**, 179–94.

Thomas, M. A., Ke, Y., Levitt, J., et al. (1998) Preliminary study of frontal lobe ^{1}H MR spectroscopy in childhood-onset schizophrenia. *J Magn Reson Imaging*, **8**, 841–6.

Thompson, P. M., Vidal, C., Giedd, J. N., et al. (2001) Mapping adolescent brain change reveals dynamic wave of accelerated gray matter loss in very early-onset schizophrenia. *Proc Natl Acad Sci U S A*, **98**, 11650–5.

Toga, A. W., Thompson, P. M., and Sowell, E. R. (2006) Mapping brain maturation. *Trends Neurosci*, **29**, 148–59.

Van Der Knaap, M. S., Van Der Grond, J., Luyten, P. R., et al. (1992) ^{1}H and ^{31}P magnetic resonance spectroscopy of the brain in degenerative cerebral disorders. *Ann Neurol*, **31**, 202–11.

Van Os, J., Fahy, T. A., Jones, P., et al. (1995) Increased intracerebral cerebrospinal fluid spaces predict unemployment and negative symptoms in psychotic illness: a prospective study. *Br J Psychiatry*, **166**, 750–8.

Vidal, C. N., Rapoport, J. L., Hayashi, K. M., et al. (2006) Dynamically spreading frontal and cingulate deficits mapped in adolescents with schizophrenia. *Arch Gen Psychiatry*, **63**, 25–34.

Volz, H. R., Riehemann, S., Maurer, I., et al. (2000) Reduced phosphodiesters and high-energy phosphates in the frontal lobe of schizophrenic patients: a (31)P chemical shift spectroscopic-imaging study. *Biol Psychiatry*, **47**, 954–61.

Ward, K. E., Friedman, L., Wise, A., et al. (1996) Meta-analysis of brain and cranial size in schizophrenia. *Schizophr Res*, **22**, 197–213.

Wassink, T. H., Andreasen, N. C., Nopoulos, P., et al. (1999) Cerebellar morphology as a predictor of symptom and psychosocial outcome in schizophrenia. *Biol Psychiatry*, **45**, 41–8.

Whalley, H. C., Simonotto, E., Marshall, I., et al. (2005) Functional disconnectivity in subjects at high genetic risk of schizophrenia. *Brain*, **128**(pt 9), 2097–108.

Whitworth, A. B., Kemmler, G., Honeder, M., et al. (2005) Longitudinal volumetric MRI study in first- and multiple-episode male schizophrenia patients. *Psychiatry Res*, **140**, 225–37.

Wishart, H. A., Saykin, A. J., and McAllister, T. W. (2002) Functional magnetic resonance imaging: emerging clinical applications. *Curr Psychiatry Rep*, **4**, 338–45.

Woodruff, P. W., McManus, I. C., and David, A. S. (1995) Meta-analysis of corpus callosum size in schizophrenia. *J Neurol Neurosurg Psychiatry*, **58**, 457–61.

Wright, I. C., Rabe-Hesketh, S., Woodruff, P. W., et al. (2000) Meta-analysis of regional brain volumes in schizophrenia. *Am J Psychiatry*, **157**, 16–25.

10

Cortico-limbic development in bipolar disorder: a neuroimaging view

Jessica H. Kalmar, Maulik P. Shah and Hilary P. Blumberg

Introduction

Bipolar disorder (BD) is a mood disorder characterized by the dramatic affective changes of the acute "highs" of manias and "lows" of depressions. The neural underpinning of emotional and motivated behavior, cortico-limbic circuitry that includes the amygdala, hippocampus, ventral striatum, and ventral prefrontal cortex (VPFC), is therefore implicated in the disorder. Structural and functional neuroimaging studies of adults with BD have provided evidence for abnormal morphology and functioning of the involved cortico-limbic structures. Proton magnetic resonance spectroscopy (^{1}H MRS) studies have provided evidence of cellular and metabolic factors that may contribute to these regional brain abnormalities in adults with BD.

Neuroimaging research of pediatric BD is a growing field. Adolescence is increasingly recognized as a developmental period when symptoms of BD emerge. During healthy adolescent neurodevelopment, maturation of the VPFC enables higher-order regulation of mesial temporal and ventral striatum responses in emotional and motivated behavior. In this chapter, we will offer a model of how this neurodevelopmental trajectory may contribute to a subcortical-to-cortical pattern for the emergence of cortico-limbic brain disturbances in BD. We will then present neuropsychological and structural, functional and spectroscopic neuroimaging evidence

to support the involvement of the components of cortico-limbic circuitry in this model. Pediatric research will be discussed against the background of adult literature. Details of the pediatric imaging studies are provided in Tables 10.1–10.3. We will review promising new findings suggesting that existing treatments may help reverse cortico-limbic structural and functional abnormalities in BD. Finally, we will briefly present emerging translational research approaches, such as imaging genetics, which may provide insights into the pathophysiology of BD and point to novel strategies for early detection and treatment.

Clinical features of bipolar disorder

Sustained changes in emotional and motivational states are the core characteristics of the manic and depressive episodes of BD. These changes are accompanied by symptoms that reflect disrupted functioning of lower- and higher-order cortico-limbic structures. The former includes dysregulation of biological rhythms and appetitive drives manifested by shifts in sleep, eating, and sexual behavior. Changes reflecting higher-order structures may include impairment within multiple cognitive domains, such as attention and memory.

Depressive episodes bring sadness and an inability to derive pleasure from activities previously enjoyed,

Neuroimaging in Developmental Clinical Neuroscience, eds. Judith M. Rumsey and Monique Ernst. Published by Cambridge University Press. © Cambridge University Press 2009.

Table 10.1 Structural neuroimaging studies of children and adolescents with BD

Study	Sample	Mean age ± SD, in years	BD sample characteristics	MRI parameters	Main findings in BD compared to HC
Blumberg et al. (2003a)	23 HC 14 BD	14.4 ± 3.5 15.7 ± 4.0	**Medications** Unmedicated = 6, Lithium = 3, Anticonvulsants = 5, Antipsychotics = 2, Antidepressants = 3, Stimulants = 1, Clonidine = 1 **Comorbidities** 2 ADHD, 2 ODD, 1 PTSD, 1 OCD, 1 Social phobia, 1 Coordination disorder, 2 LD NOS	1.5 T MRI scan ROIs: amygdala, hippocampus	Bilateral ↓ amygdala volume Bilateral ↓ hippocampus volume
Blumberg et al. (2005b)	8 HC 10 BD	At Time 1 15.3 ± 2.8 At Time 2 17.4 ± 2.7 At Time 1 15.0 ± 4.0 At Time 2 17.5 ± 3.9	**Medications (scan 1, scan 2)** Unmedicated = 5, 7; Lithium = 2, 1; Anticonvulsants = 3, 2; Antipsychotics = 2, 2; Antidepressants = 2, 2; Stimulants = 1, 2; Atomoxetine = 0, 1; Clonidine = 1, 0 **Comorbidities** 1 ADHD, 2 ODD, 2 Anxiety disorders, 1 Substance abuse and alcohol dependence, 1 Coordination disorder, 2 LD NOS	1.5 T MRI scans were performed at two time points, approximately 2 years apart, in a within-subject longitudinal design ROI: amygdala	↓ Amygdala volume that remained stable over a 2-year time period
Blumberg et al. (2006)	23 HC 14 BD	14.7 ± 3.6 15.7 ± 4.0	**Medications** Unmedicated = 6, Lithium = 3, Anticonvulsants = 5, Antipsychotics = 3, Antidepressants = 3, Clonidine = 1, Benzodiazepine = 1 **Comorbidities** 2 ADHD, 2 ODD, 1 PTSD, 1 OCD, 1 Social phobia, 1 Coordination disorder, 2 LD NOS	1.5 T MRI scan ROI: VPFC	Non-significant decreases detected in VPFC volume

Study	Sample	Age	Medications / Comorbidities	Imaging	Findings
Chang et al. (2005a)	20 HC 20 BD	14.1 ± 2.8 14.6 ± 2.8	**Medications** Unmedicated = 4, Current medications not specified **Comorbidities** 17 ADHD, 12 ODD, 7 Anxiety disorders	3 T MRI scan ROIs: 4 lobes, PFC, 8 subregions of PFC	No differences detected in volumes of ROIs, including orbitofrontal cortex
Chang et al. (2005b)	20 HC 20 BD	14.1 ± 2.8 14.6 ± 2.8	**Medications** Current medications not specified **Comorbidities** 16 ADHD, 11 ODD, 7 Anxiety disorders	3 T MRI scan ROIs: amygdala, hippocampus, caudate, and thalamus	Bilateral ↓ amygdala volume No differences detected in volumes of other structures
Chen et al. (2004)	21 HC 16 BD	17 ± 4 16 ± 3	**Medications** Unmedicated = 2, Lithium = 10, Anticonvulsants = 8, Antipsychotics = 1, Antidepressants = 4, Stimulants = 2, Anxiolytics/sedatives = 1 **Comorbidities** 5 ADHD, 1 ODD, 1 CD	1.5 T MRI scan ROIs: amygdala, hippocampus, temporal lobe	Trend towards ↓ left amygdala volume No differences detected in volumes of other subcortical and cortical structures
Del Bello et al. (2004)	20 HC 23 BD	17.2 ± 1.9 16.3 ± 2.4	**Medications** Mood stabilizers = 20, Antipsychotics = 15, Antidepressants = 4, Stimulants = 5 **Comorbidities** 10 ADHD, 7 Substance use disorders	1.5 T MRI scan ROIs: amygdala, caudate, putamen, globus pallidus, thalamus	↓ Amygdala volume ↑ Putamen volume No differences detected in volumes of other structures
Dickstein et al. (2005)	20 HC 20 BD	13.3 ± 2.3 13.4 ± 2.5	**Medications** Unmedicated = 1, Lithium = 10, Anticonvulsants = 19, Antipsychotics = 13, Antidepressants = 7, Stimulants = 4, Anxiolytics/sedatives = 4 **Comorbidities** 12 ADHD, 14 Anxiety disorders	1.5 T MRI scan VBM ROIs: amygdala, hippocampus, accumbens, dorsolateral prefrontal cortex, orbitofrontal cortex	↓ Left amygdala and left accumbens volume No differences detected in volumes of hippocampus or orbitofrontal cortex

Table 10.1 (*cont.*)

Study	Sample	Mean age ± SD, in years	BD sample characteristics	MRI parameters	Main findings in BD compared to HC
Frazier et al. (2005a)	15 HC, 32 BD	11.2 ± 3.0, 11.2 ± 2.8	**Medications** Lithium and anticonvulsants = 14, Antipsychotics = 26, Antidepressants = 13, Stimulants = 8. **Comorbidities** 20 ADHD, 20 Anxiety disorders	1.5 T MRI scan. ROIs: parcellation of cortical gyri	No difference detected in VPFC volume
Frazier et al. (2005b)	20 HC, 43 BD	11.0 ± 2.6, 11.3 ± 2.7	**Medications** Lithium = 11, Anticonvulsants = 18, Antipsychotics = 33, Stimulants = 9, Other = 8. **Comorbidities** 22 ADHD, 29 ODD	1.5 T MRI scan. ROIs: amygdala, hippocampus, thalamus	↓ Hippocampus volume. No differences detected in volumes of amygdala or thalamus
Kaur et al. (2005)	21 HC, 16 BD	16.9 ± 3.8, 15.5 ± 3.4	**Medications** Unmedicated = 2, Lithium = 10, Anticonvulsants = 8, Antipsychotics = 1, Antidepressants = 4, Stimulants = 2, Anxiolytics/sedatives = 1. **Comorbidities** 5 ADHD, 1 ODD, 1 CD	1.5 T MRI scan. ROI: cingulate cortex	↓ Anterior and posterior cingulate cortex volume
Sanches et al. (2005a)	21 HC, 15 BD	16.9 ± 3.8, 15.9 ± 3.2	**Medications** Lithium = 10, Anticonvulsants = 8 **Comorbidities** 5 ADHD, 1 ODD	1.5 T MRI scan. ROIs: caudate, putamen	No differences detected in volumes of caudate or putamen
Sanches et al. (2005b)	21 HC, 15 BD	16.9 ± 3.8, 15.5 ± 3.5	**Medications** Lithium = 10, Anticonvulsants = 7 **Comorbidities** 5 ADHD, 1 ODD, 1 CD	1.5 T MRI scan. ROI: subgenual prefrontal cortex	No differences detected in volume of subgenual prefrontal cortex

Wilke et al. (2004)	52 HC 10 BD	14.5 ± 1.3 14.5 ± 1.8	**Medications** Unmedicated = 10 **Comorbidities** 2 ADHD, 2 PTSD	3.0 T MRI scan Automated methods, including VBM ROIs: numerous regions studied	↑ Striatal and thalamic volume ↓ Amygdala, subgenual anterior cingulate and orbitofrontal cortex volume

Notes:

Abbreviations: ADHD, attention deficit/hyperactivity disorder; BD, bipolar disorder; CD, conduct disorder; HC, healthy control; LD NOS, learning disability not otherwise specified; MRI, magnetic resonance imaging; OCD, obsessive compulsive disorder; ODD, oppositional defiant disorder; PFC, prefrontal cortex; PTSD, post-traumatic stress disorder; ROI, region of interest; SD, standard deviation; VBM, voxel-based morphometry; VPFC, ventral prefrontal cortex.

Table 10.2 Functional neuroimaging studies of children and adolescents with BD

Study	Sample	Mean age ± SD, in years	BD sample characteristics	Activation task	Main findings in BD compared to HC
Blumberg et al. (2003c)	10 HC 10 BD	14.6 ± 2.8 13.6 ± 2.8	**Medications** Unmedicated = 3, Lithium = 4, Anticonvulsants = 3, Antipsychotics = 2, Antidepressants = 3, Stimulants = 1, Clonidine = 1 **Comorbidities** 2 ADHD, 2 ODD, 4 Anxiety disorders, 2 Substance or alcohol abuse, 1 LD NOS	Stroop Color Word Test	↑ Striatal and thalamic activation Ventral striatal activation correlated with mood ratings No significant difference in VPFC activation
Chang et al. (2004)	10 HC 12 BD	14.4 ± 3.2 14.7 ± 3.0	**Medications** Unmedicated = 1, Lithium = 5, Anticonvulsants = 8, Antipsychotics = 4, Antidepressants = 8 **Comorbidities** 11 ADHD, 7 ODD, 4 Anxiety disorders	Visuospatial n-back Processing of non-facial emotional stimuli (IAPS)	↑ Striatal, thalamic, and prefrontal activation ↑ Striatal, thalamic, and prefrontal activation during processing of positively valenced pictures
Leibenluft et al. (2007)	17 HC 26 BD	14.6 ± 1.8 13.6 ± 2.6	**Medications** Unmedicated = 13, Lithium = 6, Anticonvulsants = 11, Antipsychotics = 11, Antidepressants = 4, Stimulants = 4, Other = 2 **Comorbidities** 15 ADHD, 10 ODD or CD, 14 Anxiety disorders	Stop-signal task	↓ Striatal and VPFC activation during commission errors
Pavuluri et al. (2007)	10 HC 10 BD	14.3 ± 2.4 14.9 ± 1.8	**Medications** Unmedicated = 10 **Comorbidities** Only ADHD, rates not specified	Processing of emotional face stimuli	During processing of angry and happy faces: ↑ amygdala activation ↓ VPFC activation

Rich et al. (2006)	21 HC	14.5 ± 2.5	**Medications**	Attending to emotional and non-emotional aspects of neutral face stimuli	↑ Amygdala, ventral striatal activation when rating face hostility or fear of face
	22 BD	14.2 ± 3.1	Unmedicated = 4, Lithium = 6, Mood stabilizers = 14, Antipsychotics = 10, Antidepressants = 7, Stimulants = 5, Anxiolytics/sedatives = 4		↑ VPFC activation when rating face hostility
			Comorbidities		
			9 ADHD, 8 Anxiety disorders		

Notes:
Abbreviations: ADHD, attention deficit/hyperactivity disorder; BD, bipolar disorder; CD, conduct disorder; HC, healthy control; IAPS, international affective picture system; LD NOS, learning disability not otherwise specified; ODD, oppositional defiant disorder; SD, standard deviation; VPFC, ventral prefrontal cortex.

Table 10.3 Proton magnetic resonance spectroscopy studies of children and adolescents with BD

Study	Sample	Mean age ± SD, in years	BD sample characteristics	MR parameters	Main findings in BD compared to HC
Castillo et al. (2000)	10 HC 10 BD	Not specified 8 (SD not specified)	**Medications** All unmedicated for 1 week prior to scan **Comorbidities** 88% BD participants had comorbidities	1.5 T Proton MRS of voxels placed in temporal and frontal lobes	↑ Glx/Cr in frontal lobe and basal ganglia No differences detected in NAA or Cho
Chang et al. (2003)	11 HC 15 BD	12.6 ± 2.9 12.6 ± 2.9	**Medications** Unmedicated = 1, Current medications not specified **Comorbidities** 13 ADHD, 8 ODD, 4 Anxiety disorders	3.0 T Proton MRS of voxels placed in DLPFC	↓ NAA/Cr in right DLPFC No differences detected in Cho or mI
Davanzo et al. (2001)	11 HC 11 BD	Not specified 11.5 ± 3.3	**Medications** Unmedicated =2, Anticonvulsants = 2, Antipsychotics = 5, Stimulants = 4 **Comorbidities** 5 ADHD, 4 ODD	1.5 T Proton MRS of voxels placed in ACC, scans conducted before and after 1 week of lithium treatment, HCs only scanned at baseline	Baseline ↑ mI/Cr in ACC reduced following lithium treatment Largest mI decreases associated with greatest amelioration of manic symptomatology No differences detected in Glx/Cr, Cho/Cr or NAA/Cr

Study	Subjects	Age	Medications/Comorbidities	MRS method	Findings
Davanzo et al. (2003)	13 HC 10 BD 10 IED	11.7 ± 3.6 9.8 ± 2.0 9.6 ± 3.0	**Medications of BD** Unmedicated = 5, Anticonvulsants = 2, Antipsychotics = 2, Antidepressants = 1, Stimulants = 2, Guanfacine = 1 **Medications of IED** Unmedicated = 3, Anticonvulsants = 2, Stimulants = 4, Clonidine = 1, Guanfacine = 2 **Comorbidities of BD** 8 ADHD, 9 ODD, 7 CD, 1 Anxiety disorder **Comorbidities of IED** 7 ADHD, 9 ODD, 2 CD, 3 Anxiety disorders	1.5 T Proton MRS of voxels placed in ACC and occipital cortex	↑ mI/Cr and mI in ACC in BD compared to IED and HC mI levels in ACC were correlated with mania ratings
Del Bello et al. (2006)	10 HC 11 BD remitters 8 BD non-remitters	15 ± 2 14 ± 2 15 ± 2	**Medications** Unmedicated = 19, excluding study med **Comorbidities** 8 ADHD	1.5 T Proton MRS of voxels placed in VPFC, scans conducted at baseline and days 7 and 28 of olanzapine treatment, HCs underwent 3 scans but not treatment	↑ NAA in VPFC following 4 weeks of treatment in BD remitters compared to non-remitters ↑ Cho in VPFC at baseline in BD remitters compared to non-remitters
Galleli et al. (2005)	26 HC 32 BD	14.2 ± 2.8 14.1 ± 3.0	**Medications** Unmedicated = 4, Lithium = 5, Anticonvulsants = 8, Antipsychotics = 13, Stimulants = 8 **Comorbidities** 29 ADHD, 18 ODD, 11 Anxiety disorders	3.0 T Proton MRS of voxels placed in DLPFC	No differences detected in NAA/Cr

Table 10.3 (*cont.*)

Study	Sample	Mean age ± SD, in years	BD sample characteristics	MR parameters	Main findings in BD compared to HC
Moore et al. (2006)	7 HC, 8 BD	Age range: 6–13 for all subjects	**Medications** Unmedicated = 6, Current medications not specified **Comorbidities** 8 ADHD, 8 ODD, 1 CD, 2 Anxiety disorders	1.5 T Proton MRS of voxel placed in ACC	No differences detected in mI levels
Patel et al. (2006)	28 BD	15.5 ± 1.5	**Medications** Unmedicated = 28, excluding study med **Comorbidities** 7 ADHD	1.5 T Proton MRS of voxels placed in prefrontal cortices, scans conducted at baseline and days 7 and 42 of lithium treatment	No differences detected in mI levels between baseline and after lithium treatment
Sassi et al. (2005)	18 HC, 14 BD	17.3 ± 3.7, 15.5 ± 3.0	**Medications** Unmedicated = 2, Lithium = 8, Anticonvulsants = 8, Antidepressants = 4, Stimulants = 2, Benzodiazepine = 1 **Comorbidities** 5 ADHD, 1 ODD, 1 CD	1.5 T Proton MRS of voxel placed in left DLPFC	↓ NAA in left DLPFC

Notes:

Abbreviations: ACC, anterior cingulate cortex; ADHD, attention deficit/hyperactivity disorder; BD, bipolar disorder; CD, conduct disorder; Cho, choline; Cr, creatine; DLPFC, dorsolateral prefrontal cortex; Glx, glutamate/glutamine; HC, healthy control; IED, intermittent explosive disorder; mI, *myo*-inositol; MR, magnetic resonance; MRS, magnetic resonance spectroscopy; NAA, *N*-acetylaspartate; ODD, oppositional defiant disorder; SD, standard deviation; VPFC, ventral prefrontal cortex.

i.e., anhedonia. Energy and sex drive are low, while sleep and carbohydrate craving are increased. Reduced mental flexibility and motivation lead to constricted and isolative behavior.

Manic episodes, the hallmark of BD, are characterized by elevated mood states, which can be either euphoric or irritable and must persist for at least one week in order to satisfy diagnostic criteria. The episodes are frequently heralded by decreases in sleep to a few hours per night accompanied by feelings of high energy. Drive towards hedonic, risky behaviors is high and can have severe consequences. Speech is fast, distractibility is severe, and flow of ideas is rapid. Thoughts are often grandiose and may reach psychotic levels. Mixed episodes have features of both depression and mania, producing high-energy dysphoric states, and these are the episodes associated with the highest suicide risk. In adults, acute episodes tend to be distinct and are demarcated by euthymic periods of variable length when adaptive functioning can return to normal levels. For those with the rapid-cycling subtype, episodes occur at least four times a year. Rapid-cycling and mixed episodes are associated with treatment resistance and poor prognosis.

Pediatric presentations of BD may be more likely than adult presentations to include irritability, chronic symptoms, rapid-cycling and mixed symptoms, imparting challenges for the delineation of BD diagnostic boundaries in youth. Narrow diagnostic criteria for pediatric BD require that manic episodes meet adult criteria for duration and are characterized by euphoria and grandiosity. Broader criteria include children with chronic, rapid mood swings that may manifest predominantly with irritability and impulsivity (Leibenluft et al., 2003). Establishing reliable diagnostic criteria for pediatric BD is complicated by similarities in clinical presentation with other childhood disorders, such as the impulsivity and distractibility of attention deficit/hyperactivity disorder (ADHD) or the depressive episodes of major depressive disorder (MDD). Researchers are investigating possible neurobehavioral and neurobiological markers to distinguish pediatric BD from other psychiatric disorders.

A neurodevelopmental cortico-limbic model of bipolar disorder

Behavioral neurology provided the first evidence to implicate cortico-limbic circuitry in the emotional and motivational dysregulation of BD. Since at least the late 1800s, VPFC lesions have been reported to produce symptoms similar to those of BD, including caustic euphoric states, inappropriate jocularity, uncharacteristic impulsive behaviors, and depression (Jastrowitz, 1888; Oppenheim, 1889). Whereas deficits in VPFC functioning are implicated in BD, excesses in mesial temporal lobe activity are indicated by observations that complex partial seizure disorders with mesial temporal lobe foci are often associated with mood abnormalities that may display a bipolar course (Flor-Henry, 1969). Anticonvulsants that blunt mesial temporal activity excesses are currently among the first-line treatments for BD (Bowden et al., 2000). This literature, combined with evidence for significant inhibitory connections between VPFC and amygdala (Amaral and Price, 1984), suggests that mood symptomatology in BD could be produced by VPFC deficits, amygdala excesses or abnormal connections between these structures. Ventral prefrontal cortex and amygdala also have major connectivity with the hippocampus and ventral striatum (Amaral and Price, 1984; Morecraft et al., 1992). Abnormalities in these structures could contribute to the mnemonic dysfunction and dysregulation of hedonic drives and motivated behavior observed in BD.

Dynamic developmental changes occur in the cortico-limbic structures implicated in BD from childhood through young adulthood. The trajectory of these developmental changes may help explain the emergence of specific regional brain abnormalities and associated behavioral changes during certain developmental epochs. In particular, the adolescent peak for acute BD episodes coincides with VPFC structural and functional maturation and its regulation of complex emotional and motivational behaviors. Thus, abnormalities in VPFC neurodevelopment during adolescence might contribute to the behavioral expression of BD in adolescence and young adulthood. The developmental model proposed herein suggests that mesial temporal lobe abnormalities are present by puberty, whereas divergence in VPFC structure and function between individuals with and without BD may still be progressing over the course of adolescence. Ventral prefrontal cortex disturbances may not be expressed fully until the region passes through its course of programmed development in late adolescence/early adulthood, when characteristics of the prototypic adult phenotype emerge (Blumberg et al., 2004).

Studies in both non-human primates and humans demonstrate greater reliance on subcortical structures prior to puberty and more focused postpubertal recruitment of prefrontal structures in the adaptive regulation of behavior across multiple functional domains. Non-human primates increasingly recruit VPFC relative to amygdala as more mature, motivated social behavior emerges over adolescence (Machado and Bachevalier, 2003). Human functional neuroimaging studies provide support for increased prefrontal recruitment during adolescence for inhibition of prepotent responses (Blumberg et al., 2003c; Rubia et al., 2000). Furthermore, over adolescence, the inhibition of prepotent responses to rewarding stimuli is associated with replacement of striatal and widespread cortical engagement with more focal and robust VPFC activation (Blumberg et al., 2003c, 2004; Galvan et al., 2006). In human emotional behavior, a similar progression over adolescence to greater prefrontal, relative to subcortical, reliance has been found, e.g., prefrontal recruitment, relative to that of amygdala, increases over adolescence during emotional face processing (Killgore et al., 2001). Underlying these functional maturational changes are changes in gray and white matter structure that are reviewed in detail in Chapter 1 of this text on neurodevelopment.

Neuroimaging evidence presented below suggests that abnormalities in subcortical components of cortico-limbic circuitry in BD are present by adolescence, whereas VPFC abnormalities may progress during adolescence, consistent with an interaction between illness processes and the neurodevelopmental changes that occur in VPFC during adolescence.

As the VPFC is still changing dynamically during adolescence, the expression of VPFC abnormalities may continue to evolve over this developmental period. This raises the important question of whether it might be possible to intervene in this process to halt the progression of brain abnormalities in the disorder. Hope for this possibility lies in evidence that medications that treat BD have neurotrophic and neuroprotective effects, and in early evidence suggesting that they may have the potential to reverse cortico-limbic structural abnormalities. Studies using longitudinal within-subject designs and systematic medication assignment may reveal important insights.

Amygdala

Deficits in emotional processing, present in pediatric BD samples, suggest that amygdala dysfunction may be an early feature of the disorder and/or reflect vulnerability to the illness. Prepubescent children and adolescents with BD have difficulty identifying facial emotions (McClure et al., 2003). The amygdala contains populations of cells that respond to faces, particularly to facial emotion (Fried et al., 2002), and receives input from posterior association cortices that process face stimuli (Amaral and Price, 1984). Adults with BD also exhibit deficits in processing facial affect (Addington and Addington, 1998) and recognizing negative facial emotion (Lembke and Ketter, 2002), and are biased toward positive emotional stimuli (Elliott et al., 2002; Murphy et al., 1999). Thus, abnormal processing of both negative and positive emotional stimuli is present in BD from the prepubescent period through adulthood.

Morphometric studies of the amygdala have yielded the most consistent neuroimaging findings in pediatric BD: reduction of amygdala volumes in adolescents with BD (Blumberg et al., 2005b, 2003a; Chang et al., 2005b; Chen et al., 2004; Del Bello et al., 2004; Dickstein et al., 2005; Wilke et al., 2004). While most of these studies are cross-sectional, a longitudinal, within-subject study of adolescents with BD found smaller amygdala volumes in patients relative

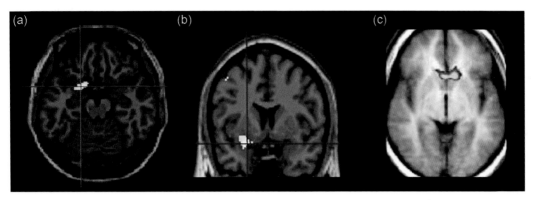

Figure 10.1a–c. Increased amygdala and ventral striatal activation in bipolar adolescents. The functional magnetic resonance imaging (fMRI) images illustrate increased neural activation in adolescents with bipolar disorder, relative to healthy controls, in amygdala (a) and ventral striatum (b) when viewing emotional face stimuli (Rich et al., 2006). Shown in (c) are regions in ventral striatum where increased activation in bipolar subjects, relative to controls, during a Stroop task correlated positively with ratings of mood symptoms (Blumberg et al., 2003c). Reprinted with permission from the *American Journal of Psychiatry.*

to controls that remained over a 2-year period (Blumberg et al., 2005b). Thus, amygdala structural abnormalities may appear early, manifest by adolescence, and remain stable during adolescence. Taken together with findings of abnormalities in pediatric BD samples in the functions subserved by the amygdala, such as emotional face processing, this suggests that the amygdala plays a role early in the disorder.

In contrast, structural magnetic resonance imaging (MRI) studies in adults with BD have yielded inconsistent results, with reports of adults with BD displaying decreased, increased or no difference in amygdala volumes relative to healthy comparison groups (reviewed in Blumberg et al., 2003a). The increased consistency of amygdala volumetric findings in adolescents with BD suggests that adolescents with BD may represent a more homogeneous subgroup, perhaps a separable phenotype with early disorder onset, than adults with the disorder. However, it is also possible that while amygdala volume reductions appear early in the course of BD, other factors may lead to increases in amygdala volume over time in subsets of patients. Finally, inconsistent findings in amygdala volumetric studies in adults with BD may relate to sampling differences, as studies describing

increases in amygdala volume tended to primarily include males or those with a history of psychosis.

In line with structural neuroimaging reports, amygdala abnormalities have also been observed in functional MRI (fMRI) studies of adolescents with BD. As shown in Figure 10.1, fMRI scanning performed while participants process emotional face stimuli demonstrated elevations in amygdala activation in adolescents with BD across differing mood states, consistent with findings in adults with BD (Pavuluri et al., 2007; Rich et al., 2006). Rich et al. (2006) studied adolescents with BD and found that amygdala hyper-responsivity was elicited when attention was directed to the emotional components of face stimuli or the participants' responses to the faces (e.g., subjects asked to rate hostility on the face or their own fear). In fMRI studies in which non-facial emotional stimuli were presented, differences in amygdala activation were not detected between adolescents with and without BD (Chang et al., 2004). The combination of the amgydala's role in responding to emotional faces and the abnormalities revealed in imaging research suggests that the amygdala may be responsible for salient features of BD that appear early and persist into adulthood.

Hippocampus

Evidence of deficits in verbal memory and learning (Glahn et al., 2005; McClure et al., 2005a; Pavuluri et al., 2006), as well as in visuospatial memory (Dickstein et al., 2004; McClure et al., 2005a), has begun to accumulate in studies of children and adolescents with BD, suggesting early mesial temporal lobe dysfunction. Consistent with findings in adults, verbal learning and memory dysfunction has been detected in euthymic medicated and in acutely ill unmedicated children and adolescents with BD, suggesting that these impairments represent early trait features of the disorder (Pavuluri et al., 2006). Moreover, memory deficits have been demonstrated in unaffected first-degree relatives of individuals with BD (Ferrier et al., 2004; Keri et al., 2001). Taken together, these data suggest that memory deficits, similar to emotional processing dysfunction, are reflective of mesial temporal abnormalities, may be an early disorder feature, and/or reflect vulnerability to the illness.

Hippocampal volume deficits may be one of the early brain abnormalities in BD, as indicated by Frazier et al.'s report of decreased hippocampal volumes in prepubertal children with BD compared to healthy controls (Frazier et al., 2005b). This finding was more pronounced in girls, pointing to possible sex differences in hippocampal volumes in BD. Blumberg et al. (2003a) noted hippocampal volume deficits in adolescents with BD but did not observe statistically significant hippocampal volume reductions in adults with the disorder. Other studies of hippocampal volumes in adults with BD have yielded variable findings (reviewed in Blumberg et al., 2003a). A report of hippocampal volume decreases in the ill member of monozygotic twin pairs discordant for BD suggests that hippocampal volume reductions may relate to the disease process (Noga et al., 2001). Variable findings in adults may reflect the impact of stress and acute mood episodes, as these factors have been shown to lead to reductions in hippocampus volume in MDD research (Sheline, 2000).

The hippocampus has not been specifically evaluated with fMRI or ^{1}H MRS in pediatric BD.

However, adult BD studies (Atmaca et al., 2006; Blasi et al., 2004), including one of an unmedicated BD sample of patients who were experiencing their first manic episode at the time of study (Atmaca et al., 2006), consistently report decreased hippocampal N-acetylaspartate/creatine ratios (NAA/Cr). The levels of NAA within the right hippocampus were shown to correlate inversely with illness duration (Deicken et al., 2003). This supports the presence of impaired neuronal functioning in the hippocampus early in the course of BD and is consistent with structural MRI data suggesting that illness duration may have a deleterious effect on the structural integrity of the hippocampus.

Striatum

The ventral striatum is of particular interest in BD owing to its role in reward processing and motivated behavior and its possible contribution to the impulsive and risky behaviors of pediatric BD. Ernst et al. (2004) conducted one of the earliest studies to specifically investigate the regulation of reward-related behaviors in a pediatric BD sample and reported deficits in the ability to respond properly to feedback and exhibit confidence in favorable outcomes. Evidence of impulsivity in pediatric BD is indicated by poor inhibition of prepotent responses on a stop-signal task (McClure et al., 2005b). In a further effort to study reward mechanisms in pediatric BD, an affect-modulated startle paradigm was implemented, but failed to show differences in performance between youth with and without BD (Rich et al., 2005). The authors highlighted some methodological challenges inherent in studying reward-related processing in pediatric populations, including the choice of age-appropriate stimuli that are sufficiently aversive or rewarding to evoke valid and reliable responses in a laboratory setting. The development of alternate paradigms is a goal for further research in this area.

Early ventral striatum volume abnormalities are indicated in BD based on Dickstein et al.'s (2005) report using voxel-based morphometry (VBM) to demonstrate smaller left nucleus accumbens volumes in children with the disorder when compared to controls. As accumbens boundaries are difficult to delineate manually on MRI, the striatum has been studied more broadly in region-of-interest (ROI) analyses, with variable findings. Similar to studies of adults, studies of adolescents with BD have reported larger striatal volumes (Del Bello et al., 2004; Wilke et al., 2004) or failed to find any volumetric differences compared with healthy control groups (Sanches et al., 2005a). Most patients studied have been medicated, an exception being 10 patients studied by Wilke et al. (2004). Hwang et al. (2006) noted abnormalities in striatal shape that were found only in persons with BD who were medication-naive. Taken together with reports of changes in basal ganglia morphology in association with psychotropic medications (Chakos et al., 1994), these data suggest the need to consider medication exposure in future studies of striatal volumes in BD.

Functional MRI evidence also suggests that abnormalities in the striatum may be present in adolescents with BD, as they display abnormal striatal activation relative to healthy controls on tasks that require the inhibition of prepotent responses (Blumberg et al., 2003c; Leibenluft et al., 2007). In one report, the degree of ventral striatum activation was positively associated with mood symptoms (see Figure 10.1) (Blumberg et al., 2003c). Consistent with findings in adults with BD, excessive ventral striatal activation was also demonstrated in response to faces and positively valenced pictures in adolescents with BD (Chang et al., 2004; Rich et al., 2006). These findings provide preliminary indications that abnormal striatal response may contribute to impulsive and affective symptomatology early in the disorder and into adulthood.

While ^{1}H MRS has not been applied to study the striatum specifically in pediatric BD, it has been used to investigate the basal ganglia more generally. Castillo et al. (2000) reported increased glutamate/glutamine (Glx) to Cr ratios in the basal ganglia in unmedicated children with BD, but did not detect differences in NAA or choline (Cho). In contrast, in adults, although there have been several reports of

negative findings (Blasi et al., 2004; Ohara et al., 1998), basal ganglia Cho/Cr and Cho have been reported to be increased in euthymic and depressed adults with BD relative to healthy controls in some studies (Hamakawa et al., 1998; Kato et al., 1996), suggesting membrane dysfunction in this region in adults with BD. Ohara et al. (1998) reported an inverse association between basal ganglia NAA/Cr and age. Taken together, these findings suggest that there are early excesses in basal ganglia glutamatergic activity that may have excitotoxic ramifications, such that membrane instability and decreased neural density progress into adulthood.

Ventral prefrontal cortex

A unifying construct for the behavioral dysregulation of acute episodes of BD is the inability to respond adaptively to changes in the emotional and motivational valence of environmental stimuli, reflective of VPFC dysfunction. Pediatric BD samples have exhibited these deficits on measures of reversal learning (Gorrindo et al., 2005), response flexibility (McClure et al., 2005b), set-shifting (Dickstein et al., 2004, 2007), and the ability to adapt to changing response contingencies (Rich et al., 2005, 2007), as has been demonstrated in adults.

The neurodevelopmental model described above posits that prefrontal abnormalities continue to progress over the course of adolescence in BD. Ventral prefrontal cortex volume findings in adolescents with BD have been variable and include reports of decreases in sub- and pre-genual anterior cingulate and orbitofrontal cortices (Kaur et al., 2005; Wilke et al., 2004) and of failures to detect differences (Adler et al., 2007; Chang et al., 2005a; Dickstein et al., 2005; Frazier et al., 2005a; Sanches et al., 2005b). In contrast, studies of adults with BD have consistently yielded reports of abnormalities in multiple VPFC subregions including decreases in gray matter volume in subgenual anterior cingulate cortex (ACC), orbitofrontal cortex, and inferior frontal cortex (Blumberg et al., 2006; Drevets et al., 1997; Lopez-Larson et al., 2002; Lyoo et al., 2004). The VPFC volume decreases reported by Blumberg et al. (2006) were statistically significant in young adults, but not in adolescents, with BD. Based on these cross-sectional results, the authors suggest that VPFC volume abnormalities may progress over the course of adolescence and are less likely to emerge as significantly divergent until late adolescence or early adulthood.

While fMRI evidence is beginning to accumulate that subcortical functional abnormalities in BD are apparent by adolescence and persist into adulthood, differences in VPFC recruitment between adolescents with and without BD may progress during adolescence and be more difficult to detect prior to late adolescence/early adulthood. Whereas VPFC activation deficits during Stroop performance have been observed repeatedly in adults with BD across acute mood states and during euthymia (Blumberg et al., 2003b; Kronhaus et al., 2006; Malhi et al., 2005), these deficits were not detected during a study of Stroop performance in adolescents with BD (Blumberg et al., 2003c). In the latter study, age-related increases in VPFC activation were observed in healthy adolescents but not in adolescents with BD, suggesting abnormalities in VPFC functional maturation over adolescence in BD. Some recent studies have reported VPFC dysfunction in adolescents with BD, including VPFC deficits during commission errors on a stop-signal task (Leibenluft et al., 2007), as well as both increases and decreases in VPFC activation relative to controls during the performance of emotional-face-processing tasks (Pavuluri et al., 2007; Rich et al., 2006). It is possible that these study designs were more sensitive to VPFC dysfunction that emerges earlier in adolescence, highlighting the importance of interpreting fMRI findings in light of the specific behavioral task employed and the need for further research to develop paradigms that are more sensitive and specific to neural circuitry dysfunction in pediatric BD.

Emerging ^{1}H MRS studies of adolescents with BD provide early molecular and cellular indicators of developing frontal pathology, although not necessarily limited to the VPFC. Abnormalities in the phosphatidylinositol signaling pathway in the ACC

in children and adolescents with BD are implicated by elevations in *myo*-inositol (mI) that appear to be associated with acute mood states. Increased mI in the ACC has been demonstrated in children and adolescents experiencing manic episodes, whose mI levels were correlated with mania severity scores (Davanzo et al., 2003). Euthymic adolescents and adults with BD generally did not differ from healthy controls on measures of mI (Chang et al., 2003; Moore et al., 2006; Winsberg et al., 2000). These findings are some of the earliest brain findings in BD and differentiated manic children from those with intermittent explosive disorder (Davanzo et al., 2003).

Castillo et al. (2000) reported bilateral Glx/Cr elevations in the frontal lobes of adolescents with BD. This raises the possibility that glutamatergic neurotoxicity may be a mechanism that contributes to frontal pathology in the disorder during adolescence. However, these findings may not be specific to BD. For example, increased ACC Glx/Cr has been reported in ADHD (Courvoisie et al., 2004; MacMaster et al., 2003; Moore et al., 2006). Decreased NAA measures have also been reported in the frontal cortex in adolescents (Chang et al., 2003; Sassi et al., 2005) and unmedicated adults (Winsberg et al., 2000) with BD, further suggesting that neuronal integrity is disrupted by adolescence and persists into adulthood in BD. One study of adolescents with BD did not detect decreases in frontal NAA (Gallelli et al., 2005), although participants in this study were medicated. Thus, the ^{1}H MRS data provide in vivo evidence for frontal abnormalities in children and adolescents with BD in glutamatergic and phosphatidylinositol signaling pathways, as well as in cell membrane stability and neuronal integrity.

Treatment

Preclinical studies have demonstrated up-regulation of neurotrophic and neuroprotective factors in mammalian frontal cortex in association with mood-stabilizing medication (Manji et al., 2000), suggesting that medication treatment has the potential to reverse or protect against structural brain abnormalities. Moore et al. (2000) reported an increase in cortical gray matter after 4 weeks of lithium treatment in adults with BD. Subsequently, several studies noted that research participants with BD taking mood-stabilizing medications had VPFC volumes that were significantly larger than the VPFC volumes of unmedicated participants and were similar to VPFC volumes of individuals without BD (Blumberg et al., 2006; Drevets, 2001; Sassi et al., 2004). Similarly, Chang et al. (2005b) noted that adolescents with BD who were taking mood-stabilizing medications had amygdala volumes that were more similar to those of adolescents without BD. These data provide preliminary evidence that mood-stabilizing medications may yield beneficial effects for cortico-limbic structures in BD; however, within-subject longitudinal designs with systematic treatment assignments are needed in order to generate more conclusive findings.

Preliminary functional neuroimaging evidence suggests that medication treatments for BD normalize cortico-limbic functioning as well. Adults with BD taking mood-stabilizing medications at the time of scanning showed resting activity or task-related activation in amygdala (Drevets et al., 2002), ventral striatum (Caligiuri et al., 2003; Goodwin et al., 1997), and VPFC (Blumberg et al., 2005a; Goodwin et al., 1997; Kruger et al., 2006) that was more similar to that in healthy controls than in unmedicated BD subjects. In one report, amygdala activation in response to emotional faces was elevated in unmedicated BD adults, while adults with BD who were taking anticonvulsants had amygdala activation similar to that of controls. Treatment with lithium carbonate was associated with increases in VPFC activation to healthy control levels (Blumberg et al., 2005a). This suggests that efficacious treatment for BD may help to normalize functioning within cortico-limbic neural circuitry, with different treatment subclasses affecting different nodes within the circuitry. Future study with systematic medication assignment and investigation for possible effects of specific treatment subclasses

on particular regions within cortico-limbic circuitry is warranted. These data raise the important question of whether early intervention might help to reverse or prevent progression of functional brain abnormalities in the disorder.

Of the imaging modalities reviewed herein, MRS is the only one that has been implemented using within-subject longitudinal designs and systematic treatment assignments in pediatric BD. Consistent with the theory that inositol depletion underlies lithium's mood-stabilizing effects, pretreatment elevations in ACC mI/Cr were reduced within 1 week of lithium treatment in children and adolescents with BD (Davanzo et al., 2001). The largest decreases in mI/Cr ratios were associated with the greatest amelioration of manic symptoms. While a subsequent study of lithium treatment in adolescents failed to find this reduction (Patel et al., 2006), Davanzo et al.'s findings are consistent with a report by Moore et al. (1999) of decreased frontal mI observed on serial scans of adults within a few days of initiating lithium treatment. The Davanzo and Moore studies differed in the observed time-course of clinical effects. While Davanzo et al. reported clinical improvement after 1 week of lithium treatment, Moore et al. observed a several-week lag in clinical improvement, thought to reflect time for downstream signaling events and induction of gene expression (Moore et al., 1999). Similar mI changes have been observed following valproate treatment in adults (Silverstone et al., 2002), suggesting that different subclasses of pharmacotherapies efficacious in the treatment of BD may potentiate similar changes in mI.

A study of olanzapine monotherapy for adolescents in their first hospitalization for mania reported increased VPFC NAA after 4 weeks of treatment in participants whose mania remitted (Del Bello et al., 2006). Increased pretreatment Cho also distinguished those participants whose mania did and did not remit. These findings are consistent with neurotrophic and membrane-stabilizing effects of BD pharmacotherapies including recent preclinical evidence for up-regulation of neurotrophic factors by olanzapine (Parikh et al., 2004).

Conclusions and future directions

Considerable neuroimaging data provide evidence for the involvement of a cortico-limbic neural system that subserves the regulation of emotion and motivation in the pathophysiology of BD. Indications of amygdala, hippocampus, ventral striatum, and VPFC contributions to BD are consistent with regional brain abnormalities associated with BD-type emotional dysregulation described in the behavioral neurology literature. The affected brain regions also play important roles in the functional domains implicated in BD by the neuropsychological literature, including the processing of emotional stimuli, mnemonic encoding, and adaptive responding to shifting reinforcement contingencies.

The model presented herein is a working model that attempts to consider the dimension of development in understanding the neuropathophysiology of BD. In particular, it is hoped that it will generate questions about how neurodevelopmental changes during specific epochs may interact with the neural and behavioral expression of BD. Preliminary evidence suggests that the developmental epochs during which specific regional brain abnormalities and associated behavioral dysfunction emerge in BD parallel the programmed timing of healthy regional neurodevelopment. Subcortical abnormalities appear to be present early in the disorder. For example, volumetric and functional amygdala abnormalities and disturbances in amygdala-related functions such as the recognition and processing of emotional face stimuli have been demonstrated in children and adolescents with BD. Findings in some structural and functional neuroimaging studies, as well as behavioral studies, of BD suggest the emergence of VPFC abnormalities in adolescence. These VPFC abnormalities may continue to progress over adolescence, and their full expression may not be apparent until adulthood. Magnetic resonance spectroscopy studies provide evidence for frontal metabolic and cellular changes in children and adolescents that may be the precursors to structural and functional VPFC deficits reported consistently in adults with BD.

Identification of trait markers for pediatric BD is necessary in order to develop effective strategies to differentiate BD from clinically overlapping disorders (e.g., ADHD and MDD) and thus prevent the consequences of misdiagnosis and improve outcomes for children suffering emotional and behavioral dysregulation. Subcortical abnormalities detected in children and adolescents with BD have also been detected in children and adolescents with other disorders of emotion and impulse regulation, including MDD and ADHD. The common subcortical abnormalities across these disorders in youth may contribute to the symptom overlap and difficulties in distinguishing them clinically in pediatric populations. The timing and regional distribution of the specific cortical abnormalities that may emerge in these disorders over adolescence may provide clues to help dissociate them. Future efforts to parse early neural and behavioral signs of fronto-cortical abnormalities specific to these disorders would be important in understanding their pathophysiology and could be especially valuable in identifying trait markers for the disorders that could be targeted by treatment.

The field of pediatric BD neuroimaging research is still relatively new. Data inconsistencies may relate to subject sampling and imaging methodological issues, making comparisons across studies difficult. Participant characteristics that vary across studies and may affect results include sex, age of disease onset, diagnostic subtype, mood state, number and frequency of acute mood episodes, psychiatric comorbidity, and exposure to stressors, medications, or substances of abuse. Medications that treat mood disorders may increase regional brain volumes through the stimulation of neurotrophic mechanisms. With regard to methodology, differences across studies include scanner characteristics, acquisition parameters, the utilization of manual or automated methods for morphometric measures, neuroanatomical landmarks used for manual delineation, and the behavioral task employed during fMRI. Ongoing studies with larger sample sizes and in-depth phenotypic characterization hold great promise. Many of the age-dependent associations discussed are based on cross-sectional data, and these data are especially limited for prepubescent samples. Caution is therefore needed in drawing conclusions about developmental factors. Ongoing research using within-subject longitudinal designs to study prepubertal children and adolescents will be instrumental in furthering our understanding of developmental influences in BD. Moreover, investigations of unaffected first-degree relatives of persons suffering from BD may shed light on abnormalities associated with vulnerability to developing the disorder and lead to improvements in early identification and treatment interventions.

Future directions include investigations of the molecular mechanisms that contribute to cortico-limbic abnormalities in BD. Preliminary cross-sectional studies of current treatments for BD suggest that these treatments may reverse structural and functional cortico-limbic abnormalities in the disorder. Future MRS investigations hold promise, as MRS studies have provided potential early markers of abnormalities in glutamatergic and second-messenger signaling pathways, as well as of early disturbances in neuronal integrity, which may be normalized via treatment with mood-stabilizing agents. Another promising emerging line of research in pediatric BD is imaging genetics. The study of genetic factors may be especially fruitful in researching pediatric BD, as studies suggest there is a high familial loading in pediatric-onset BD. Genetic variations that have been identified as having effects on the cortico-limbic regions involved in BD, such as variation in the gene encoding brain-derived neurotrophic growth factor (Geller et al., 2004), have also been associated with pediatric BD. Studies are underway to understand the relationships between such genes, the development of cortico-limbic circuitry, and BD. Molecular mechanisms shown to underlie the developmental pathophysiology of BD could be targeted by early treatment interventions and may lead to prevention of the disorder in the future.

REFERENCES

Addington, J., Addington, D. (1998) Facial affect recognition and information processing in schizophrenia and bipolar disorder. *Schizophr Res*, **32**, 171–81.

Adler, C. M., Del Bello, M. P., Jarvis, K., et al. (2007) Voxel-based study of structural changes in first-episode patients with bipolar disorder. *Biol Psychiatry*, **61**, 776–81.

Amaral, D. G. and Price, J. L. (1984) Amygdalo-cortical projections in the monkey (*Macaca fascicularis*). *J Comp Neurol*, **230**, 465–96.

Atmaca, M., Yildirim, H., Ozdemir, H., et al. (2006) Hippocampal ^{1}H MRS in first-episode bipolar I patients. *Prog Neuropsychopharmacol Biol Psychiatry*, **30**, 1235–9.

Blasi, G., Bertolino, A., Brudaglio, F., et al. (2004) Hippocampal neurochemical pathology in patients at first episode of affective psychosis: a proton magnetic resonance spectroscopic imaging study. *Psychiatry Res*, **131**, 95–105.

Blumberg, H. P., Donegan, N. H., Sanislow, C. A., et al. (2005a) Preliminary evidence for medication effects on functional abnormalities in the amygdala and anterior cingulate in bipolar disorder. *Psychopharmacology (Berl)*, **183**, 308–13.

Blumberg, H. P., Fredericks, C., Wang, F., et al. (2005b) Preliminary evidence for persistent abnormalities in amygdala volumes in adolescents and young adults with bipolar disorder. *Bipolar Disord*, **7**, 570–6.

Blumberg, H. P., Kaufman, J., Martin, A., et al. (2003a) Amygdala and hippocampal volumes in adolescents and adults with bipolar disorder. *Arch Gen Psychiatry*, **60**, 1201–8.

Blumberg, H. P., Kaufman, J., Martin, A., et al. (2004) Significance of adolescent neurodevelopment for the neural circuitry of bipolar disorder. *Ann N Y Acad Sci*, **1021**, 376–83.

Blumberg, H. P., Krystal, J. H., Bansal, R., et al. (2006) Age, rapid-cycling, and pharmacotherapy effects on ventral prefrontal cortex in bipolar disorder: a cross-sectional study. *Biol Psychiatry*, **59**, 611–18.

Blumberg, H. P., Leung, H. C., Skudlarski, P., et al. (2003b) A functional magnetic resonance imaging study of bipolar disorder: state- and trait-related dysfunction in ventral prefrontal cortices. *Arch Gen Psychiatry*, **60**, 601–9.

Blumberg, H. P., Martin, A., Kaufman, J., et al. (2003c) Frontostriatal abnormalities in adolescents with bipolar disorder: preliminary observations from functional MRI. *Am J Psychiatry*, **160**, 1345–7.

Bowden, C. L., Calabrese, J. R., Mcelroy, S. L., et al. (2000) A randomized, placebo-controlled 12-month trial of divalproex and lithium in treatment of outpatients with bipolar I disorder. Divalproex Maintenance Study Group. *Arch Gen Psychiatry*, **57**, 481–9.

Caligiuri, M. P., Brown, G. G., Meloy, M. J., et al. (2003) An fMRI study of affective state and medication on cortical and subcortical brain regions during motor performance in bipolar disorder. *Psychiatry Res*, **123**, 171–82.

Castillo, M., Kwock, L., Courvoisie, H., et al. (2000) Proton MR spectroscopy in children with bipolar affective disorder: preliminary observations. *AJNR Am J Neuroradiol*, **21**, 832–8.

Chakos, M. H., Lieberman, J. A., Bilder, R. M., et al. (1994) Increase in caudate nuclei volumes of first-episode schizophrenic patients taking antipsychotic drugs. *Am J Psychiatry*, **151**, 1430–6.

Chang, K., Adleman, N., Dienes, K., et al. (2003) Decreased *N*-acetylaspartate in children with familial bipolar disorder. *Biol Psychiatry*, **53**, 1059–65.

Chang, K., Adleman, N. E., Dienes, K., et al. (2004) Anomalous prefrontal-subcortical activation in familial pediatric bipolar disorder: a functional magnetic resonance imaging investigation. *Arch Gen Psychiatry*, **61**, 781–92.

Chang, K., Barnea-Goraly, N., Karchemskiy, A., et al. (2005a) Cortical magnetic resonance imaging findings in familial pediatric bipolar disorder. *Biol Psychiatry*, **58**, 197–203.

Chang, K., Karchemskiy, A., Barnea-Goraly, N., et al. (2005b) Reduced amygdalar gray matter volume in familial pediatric bipolar disorder. *J Am Acad Child Adolesc Psychiatry*, **44**, 565–73.

Chen, B. K., Sassi, R., Axelson, D., et al. (2004) Cross-sectional study of abnormal amygdala development in adolescents and young adults with bipolar disorder. *Biol Psychiatry*, **56**, 399–405.

Courvoisie, H., Hooper, S. R., Fine, C., et al. (2004) Neurometabolic functioning and neuropsychological correlates in children with ADHD-H: preliminary findings. *J Neuropsychiatry Clin Neurosci*, **16**, 63–9.

Davanzo, P., Thomas, M. A., Yue, K., et al. (2001) Decreased anterior cingulate *myo*-inositol/creatine spectroscopy resonance with lithium treatment in children with bipolar disorder. *Neuropsychopharmacology*, **24**, 359–69.

Davanzo, P., Yue, K., Thomas, M. A., et al. (2003) Proton magnetic resonance spectroscopy of bipolar disorder versus intermittent explosive disorder in children and adolescents. *Am J Psychiatry*, **160**, 1442–52.

Deicken, R. F., Pegues, M. P., Anzalone, S., et al. (2003) Lower concentration of hippocampal *N*-acetylaspartate in familial bipolar I disorder. *Am J Psychiatry*, **160**, 873–82.

Del Bello, M. P., Cecil, K. M., Adler, C. M., et al. (2006) Neurochemical effects of olanzapine in first-

hospitalization manic adolescents: a proton magnetic resonance spectroscopy study. *Neuropsychopharmacology*, **31**, 1264–73.

Del Bello, M. P., Zimmerman, M. E., Mills, N. P., et al. (2004) Magnetic resonance imaging analysis of amygdala and other subcortical brain regions in adolescents with bipolar disorder. *Bipolar Disord*, **6**, 43–52.

Dickstein, D. P., Milham, M. P., Nugent, A. C., et al. (2005) Frontotemporal alterations in pediatric bipolar disorder: results of a voxel-based morphometry study. *Arch Gen Psychiatry*, **62**, 734–41.

Dickstein, D. P., Nelson, E. E., McClure, E. B., et al. (2007) Cognitive flexibility in phenotypes of bipolar disorder. *J Am Acad Child Adolesc Psychiatry*, **46**, 341–55.

Dickstein, D. P., Treland, J. E., Snow, J., et al. (2004) Neuropsychological performance in pediatric bipolar disorder. *Biol Psychiatry*, **55**, 32–9.

Drevets, W. C. (2001) Neuroimaging and neuropathological studies of depression: implications for the cognitive-emotional features of mood disorders. *Curr Opin Neurobiol*, **11**, 240–9.

Drevets, W. C., Price, J. L., Bardgett, M. E., et al. (2002) Glucose metabolism in the amygdala in depression: relationship to diagnostic subtype and plasma cortisol levels. *Pharmacol Biochem Behav*, **71**, 431–47.

Drevets, W. C., Price, J. L., Simpson, J. R., Jr., et al. (1997) Subgenual prefrontal cortex abnormalities in mood disorders. *Nature*, **386**, 824–7.

Elliott, R., Rubinsztein, J. S., Sahakian, B. J., et al. (2002) The neural basis of mood-congruent processing biases in depression. *Arch Gen Psychiatry*, **59**, 597–604.

Ernst, M., Dickstein, D. P., Munson, S., et al. (2004) Reward-related processes in pediatric bipolar disorder: a pilot study. *J Affect Disord*, **82** Suppl 1, S89–S101.

Ferrier, I. N., Chowdhury, R., Thompson, J. M., et al. (2004) Neurocognitive function in unaffected first-degree relatives of patients with bipolar disorder: a preliminary report. *Bipolar Disord*, **6**, 319–22.

Flor-Henry, P. (1969) Schizophrenic-like reactions and affective psychoses associated with temporal lobe epilepsy: etiological factors. *Am J Psychiatry*, **126**, 400–4.

Frazier, J. A., Breeze, J. L., Makris, N., et al. (2005a) Cortical gray matter differences identified by structural magnetic resonance imaging in pediatric bipolar disorder. *Bipolar Disord*, **7**, 555–69.

Frazier, J. A., Chiu, S., Breeze, J. L., et al. (2005b) Structural brain magnetic resonance imaging of limbic and thalamic volumes in pediatric bipolar disorder. *Am J Psychiatry*, **162**, 1256–65.

Fried, I., Cameron, K. A., Yashar, S., et al. (2002) Inhibitory and excitatory responses of single neurons in the human medial temporal lobe during recognition of faces and objects. *Cereb Cortex*, **12**, 575–84.

Gallelli, K. A., Wagner, C. M., Karchemskiy, A., et al. (2005) *N*-Acetylaspartate levels in bipolar offspring with and at high-risk for bipolar disorder. *Bipolar Disord*, **7**, 589–97.

Galvan, A., Hare, T. A., Parra, C. E., et al. (2006) Earlier development of the accumbens relative to orbitofrontal cortex might underlie risk-taking behavior in adolescents. *J Neurosci*, **26**, 6885–92.

Geller, B., Badner, J. A., Tillman, R., et al. (2004) Linkage disequilibrium of the brain-derived neurotrophic factor Val66Met polymorphism in children with a prepubertal and early adolescent bipolar disorder phenotype. *Am J Psychiatry*, **161**, 1698–700.

Glahn, D. C., Bearden, C. E., Caetano, S., et al. (2005) Declarative memory impairment in pediatric bipolar disorder. *Bipolar Disord*, **7**, 546–54.

Goodwin, G. M., Cavanagh, J. T., Glabus, M. F., et al. (1997) Uptake of 99mTc-exametazime shown by single photon emission computed tomography before and after lithium withdrawal in bipolar patients: associations with mania. *Br J Psychiatry*, **170**, 426–30.

Gorrindo, T., Blair, R. J., Budhani, S., et al. (2005) Deficits on a probabilistic response-reversal task in patients with pediatric bipolar disorder. *Am J Psychiatry*, **162**, 1975–7.

Hamakawa, H., Kato, T., Murashita, J., et al. (1998) Quantitative proton magnetic resonance spectroscopy of the basal ganglia in patients with affective disorders. *Eur Arch Psychiatry Clin Neurosci*, **248**, 53–8.

Hwang, J., Lyoo, I. K., Dager, S. R., et al. (2006) Basal ganglia shape alterations in bipolar disorder. *Am J Psychiatry*, **163**, 276–85.

Jastrowitz, M. (1888) Beitrage zür localisation im grosshirn and uber deren praktische verwerthung. *Dtsch Med Wochenshr*, **14**, 81.

Kato, T., Hamakawa, H., Shioiri, T., et al. (1996) Choline-containing compounds detected by proton magnetic resonance spectroscopy in the basal ganglia in bipolar disorder. *J Psychiatry Neurosci*, **21**, 248–54.

Kaur, S., Sassi, R. B., Axelson, D., et al. (2005) Cingulate cortex anatomical abnormalities in children and adolescents with bipolar disorder. *Am J Psychiatry*, **162**, 1637–43.

Keri, S., Kelemen, O., Benedek, G., et al. (2001) Different trait markers for schizophrenia and bipolar disorder: a neurocognitive approach. *Psychol Med*, **31**, 915–22.

Killgore, W. D., Oki, M., and Yurgelun-Todd, D. A. (2001) Sex-specific developmental changes in amygdala responses to affective faces. *NeuroReport*, **12**, 427–33.

Kronhaus, D. M., Lawrence, N. S., Williams, A. M., et al. (2006) Stroop performance in bipolar disorder: further evidence for abnormalities in the ventral prefrontal cortex. *Bipolar Disord*, **8**, 28–39.

Kruger, S., Alda, M., Young, L. T., et al. (2006) Risk and resilience markers in bipolar disorder: brain responses to emotional challenge in bipolar patients and their healthy siblings. *Am J Psychiatry*, **163**, 257–64.

Leibenluft, E., Charney, D. S., Towbin, K. E., et al. (2003) Defining clinical phenotypes of juvenile mania. *Am J Psychiatry*, **160**, 430–7.

Leibenluft, E., Rich, B. A., Vinton, D. T., et al. (2007) Neural circuitry engaged during unsuccessful motor inhibition in pediatric bipolar disorder. *Am J Psychiatry*, **164**, 52–60.

Lembke, A. and Ketter, T. A. (2002) Impaired recognition of facial emotion in mania. *Am J Psychiatry*, **159**, 302–4.

Lopez-Larson, M. P., Del Bello, M. P., Zimmerman, M. E., et al. (2002) Regional prefrontal gray and white matter abnormalities in bipolar disorder. *Biol Psychiatry*, **52**, 93–100.

Lyoo, I. K., Kim, M. J., Stoll, A. L., et al. (2004) Frontal lobe gray matter density decreases in bipolar I disorder. *Biol Psychiatry*, **55**, 648–51.

Machado, C. J. and Bachevalier, J. (2003) Non-human primate models of childhood psychopathology: the promise and the limitations. *J Child Psychol Psychiatry*, **44**, 64–87.

MacMaster, F. P., Carrey, N., Sparkes, S., et al. (2003) Proton spectroscopy in medication-free pediatric attention-deficit/hyperactivity disorder. *Biol Psychiatry*, **53**, 184–7.

Malhi, G. S., Lagopoulos, J., Sachdev, P. S., et al. (2005) An emotional Stroop functional MRI study of euthymic bipolar disorder. *Bipolar Disord*, **7** Suppl 5, 58–69.

Manji, H. K., Moore, G. J., and Chen, G. (2000) Clinical and preclinical evidence for the neurotrophic effects of mood stabilizers: implications for the pathophysiology and treatment of manic-depressive illness. *Biol Psychiatry*, **48**, 740–54.

McClure, E. B., Pope, K., Hoberman, A. J., et al. (2003) Facial expression recognition in adolescents with mood and anxiety disorders. *Am J Psychiatry*, **160**, 1172–4.

McClure, E. B., Treland, J. E., Snow, J., et al. (2005a) Memory and learning in pediatric bipolar disorder. *J Am Acad Child Adolesc Psychiatry*, **44**, 461–9.

McClure, E. B., Treland, J. E., Snow, J., et al. (2005b) Deficits in social cognition and response flexibility in pediatric bipolar disorder. *Am J Psychiatry*, **162**, 1644–51.

Moore, C. M., Biederman, J., Wozniak, J., et al. (2006) Differences in brain chemistry in children and adolescents with attention deficit hyperactivity disorder with and without comorbid bipolar disorder: a proton magnetic resonance spectroscopy study. *Am J Psychiatry*, **163**, 316–18.

Moore, G. J., Bebchuk, J. M., Parrish, J. K., et al. (1999) Temporal dissociation between lithium-induced changes in frontal lobe *myo*-inositol and clinical response in manic-depressive illness. *Am J Psychiatry*, **156**, 1902–8.

Moore, G. J., Bebchuk, J. M., Wilds, I. B., et al. (2000) Lithium-induced increase in human brain grey matter. *Lancet*, **356**, 1241–2.

Morecraft, R. J., Geula, C., and Mesulam, M. M. (1992) Cytoarchitecture and neural afferents of orbitofrontal cortex in the brain of the monkey. *J Comp Neurol*, **323**, 341–58.

Murphy, F. C., Sahakian, B. J., Rubinsztein, J. S., et al. (1999) Emotional bias and inhibitory control processes in mania and depression. *Psychol Med*, **29**, 1307–21.

Noga, J. T., Vladar, K., and Torrey, E. F. (2001) A volumetric magnetic resonance imaging study of monozygotic twins discordant for bipolar disorder. *Psychiatry Res*, **106**, 25–34.

Ohara, K., Isoda, H., Suzuki, Y., et al. (1998) Proton magnetic resonance spectroscopy of the lenticular nuclei in bipolar I affective disorder. *Psychiatry Res*, **84**, 55–60.

Oppenheim, H. (1889) Zür pathologie der grosshirngeschwulste. *Arch Psychiatry*, 560.

Parikh, V., Khan, M. M., and Mahadik, S. P. (2004) Olanzapine counteracts reduction of brain-derived neurotrophic factor and TrkB receptors in rat hippocampus produced by haloperidol. *Neurosci Lett*, **356**, 135–9.

Patel, N. C., Del Bello, M. P., Cecil, K. M., et al. (2006) Lithium treatment effects on *myo*-inositol in adolescents with bipolar depression. *Biol Psychiatry*, **60**, 998–1004.

Pavuluri, M. N., O'Connor, M. M., Harral, E., et al. (2007) Affective neural circuitry during facial emotion processing in pediatric bipolar disorder. *Biol Psychiatry*, **62**, 158–67.

Pavuluri, M. N., Schenkel, L. S., Aryal, S., et al. (2006) Neurocognitive function in unmedicated manic and medicated euthymic pediatric bipolar patients. *Am J Psychiatry*, **163**, 286–93.

Rich, B. A., Bhangoo, R. K., Vinton, D. T., et al. (2005) Using affect-modulated startle to study phenotypes of pediatric bipolar disorder. *Bipolar Disord*, **7**, 536–45.

Rich, B. A., Schmajuk, M., Perez-Edgar, K. E., et al. (2007) Different psychophysiological and behavioral responses elicited by frustration in pediatric bipolar disorder and severe mood dysregulation. *Am J Psychiatry*, **164**, 309–17.

Rich, B. A., Vinton, D. T., Roberson-Nay, R., et al. (2006) Limbic hyperactivation during processing of neutral facial expressions in children with bipolar disorder. *Proc Natl Acad Sci U S A*, **103**, 8900–5.

Rubia, K., Overmeyer, S., Taylor, E., et al. (2000) Functional frontalisation with age: mapping neurodevelopmental trajectories with fMRI. *Neurosci Biobehav Rev*, **24**, 13–19.

Sanches, M., Roberts, R. L., Sassi, R. B., et al. (2005a) Developmental abnormalities in striatum in young bipolar patients: a preliminary study. *Bipolar Disord*, **7**, 153–8.

Sanches, M., Sassi, R. B., Axelson, D., et al. (2005b) Subgenual prefrontal cortex of child and adolescent bipolar patients: a morphometric magnetic resonance imaging study. *Psychiatry Res*, **138**, 43–9.

Sassi, R. B., Brambilla, P., Hatch, J. P., et al. (2004) Reduced left anterior cingulate volumes in untreated bipolar patients. *Biol Psychiatry*, **56**, 467–75.

Sassi, R. B., Stanley, J. A., Axelson, D., et al. (2005) Reduced NAA levels in the dorsolateral prefrontal cortex of young bipolar patients. *Am J Psychiatry*, **162**, 2109–15.

Sheline, Y. I. (2000) 3D MRI studies of neuroanatomic changes in unipolar major depression: the role of stress and medical comorbidity. *Biol Psychiatry*, **48**, 791–800.

Silverstone, P. H., Wu, R. H., O'Donnell, T., et al. (2002) Chronic treatment with both lithium and sodium valproate may normalize phosphoinositol cycle activity in bipolar patients. *Hum Psychopharmacol*, **17**, 321–7.

Wilke, M., Kowatch, R. A., Del Bello, M. P., et al. (2004) Voxel-based morphometry in adolescents with bipolar disorder: first results. *Psychiatry Res*, **131**, 57–69.

Winsberg, M. E., Sachs, N., Tate, D. L., et al. (2000) Decreased dorsolateral prefrontal *N*-acetyl aspartate in bipolar disorder. *Biol Psychiatry*, **47**, 475–81.

11

Anxiety and depressive disorders

Daniel S. Pine

Introduction

Through four sections, the current chapter reviews data from brain imaging experiments in pediatric anxiety and depressive disorders. The first section delineates basic clinical features of the relevant conditions, outlines the goals of imaging research, and reviews data on underlying neural circuits implicated in anxiety and depressive disorders. The final three sections review specific findings from relevant brain imaging studies.

Background

Relevant conditions

The current chapter summarizes data for three classes of pediatric mental disorder and associated phenotypes, relying on classifications from the current, fourth edition of the *Diagnostic and Statistical Manual* (DSM-IV). The chapter examines data in: (1) pediatric anxiety disorders and (2) major depressive disorder (MDD) as well as in (3) youth who face high risk for these conditions.

Anxiety disorders involve abnormalities in fear and anxiety. The term "fear" refers to the psychological state associated with direct exposure to overt danger from a stimulus capable of harming the organism, leading the organism to avoid it.

The term "anxiety" refers to the psychological state associated with exposure to cues that predict impending danger, as opposed to the presence of an overt threat. DSM-IV recognizes more than 10 anxiety syndromes, each associated with abnormal fear and anxiety. Some, such as panic disorder, occur predominantly among adults. Others, such as separation anxiety disorder, typically manifest among children and adolescents.

Each syndrome involves fear or anxiety that is "abnormal," based on its capacity to cause extreme levels of distress and interfere with function. Longitudinal research demonstrates strong relationships among anxiety manifest during early childhood, adolescence, and adulthood (Perez-Edgar and Fox, 2005; Pine et al., 1998). As a result, the current chapter reviews data in child and adolescent anxiety disorders as they relate to data in adult anxiety disorders. Moreover, longitudinal research demonstrates strong relationships among the individual anxiety disorders, as defined in DSM-IV. This raises unanswered questions on the degree to which the distinct anxiety disorders in DSM-IV represent truly distinct clinical entities, as opposed to minor variations on a core syndrome. In terms of specific pediatric anxiety disorders, the current chapter reviews data in three conditions in which there are some imaging data from children: generalized anxiety disorder (GAD), social phobia (SOPH), and post-traumatic stress disorder (PTSD). Chapter 12

Neuroimaging in Developmental Clinical Neuroscience, eds. Judith M. Rumsey and Monique Ernst. Published by Cambridge University Press. © Cambridge University Press 2009.

reviews data on obsessive-compulsive disorder (OCD), which is considered an anxiety disorder but appears to differ from other anxiety disorders.

Depressive disorders involve patterns of sadness, irritability, or reduced capacity to experience pleasure, occurring in tandem with associated disruptions in cognition and homeostatic regulation. The current chapter focuses on MDD. Chapter 10 reviews data on bipolar disorder. Major depressive disorder exhibits strong relationships with anxiety disorders, based on longitudinal and family-history data (Pine et al., 1998; Weissman et al., 2005, 2006). These data raise questions on the degree to which anxiety and MDD represent distinct conditions. In general, pediatric anxiety can be distinguished from pediatric MDD (Costello et al., 2002); brain-imaging studies do implicate common neural circuitry in MDD and anxiety disorders, though with some evidence of distinct perturbations.

A final group of juveniles has been defined in brain-imaging studies based on their high risk for anxiety disorders or MDD. In terms of environmental risks, a growing body of research examines associations between various traumatic experiences and perturbations in neural circuitry. This work considers associations both with traumatic exposure, independent of associated symptoms in the child, as well as with PTSD, the best-described clinical disorder emerging following trauma. In terms of family-genetic risks, two approaches have been used to identify at-risk juveniles. One approach identifies children based on their response to evocative events. Depending on the pattern of behavior and age of study, these children have been termed "high reactive" or "behaviorally inhibited" (Fox et al., 2005; Kagan, 1994; Perez-Edgar and Fox, 2005). These temperament types show strong evidence of genetic influence and strong associations with MDD and anxiety disorders later in life (Goldsmith and Lemery, 2000). A second approach identifies children based on histories in their parents, either of anxiety disorders or MDD, which confer high risk in the child for anxiety or MDD. Similarly, studies in adults link polymorphisms in specific genes implicated in anxiety and MDD to

perturbations in brain structure or function (Pezawas et al., 2005). While brain-imaging studies are only beginning to apply these approaches in children, considerable work documents comparable clinical associations with specific genes in children as found in adults (Fox et al., 2005b).

Goal of brain-imaging research

One vital and clinically relevant goal of many brain-imaging studies is to generate useful insights into pathophysiology. Such insights might lead to novel diagnostic tests that predict clinical outcome, identify underlying family-genetic risk, or delineate response to therapeutics. However, at least for research specifically in pediatric anxiety and MDD, despite claims to the contrary, brain-imaging techniques currently provide no clinically useful insights. In fact, it is likely to be many years before such insights emerge.

Given these limitations, the utility of imaging research lies in its ability to constrain theory based on understandings of brain–behavior associations. Neuroscience research with rodents and non-human primates delineates functional aspects of distinct neural circuits. This work provides an unparalleled precision in mechanistic understandings of behavior, as it is shaped by neural process. Particularly marked advances have emerged in research on aspects of emotionally modulated behaviors. Theories attempting to explain the origins of pediatric anxiety and MDD might benefit from these mechanistic understandings. Those theories that ascribe between-person differences to functions in well-understood neural circuits would be expected to produce stronger predictions in terms of long-term outcomes and treatment response.

Mechanistic understandings might inform research on pediatric anxiety and MDD if parallels in brain–behavior relationships could be demonstrated among humans, rodents, and non-human primates. Brain-imaging research provides the best opportunity for this. Research on dementia illustrates the advantages of mutually reinforcing dialogue between basic science and brain-imaging

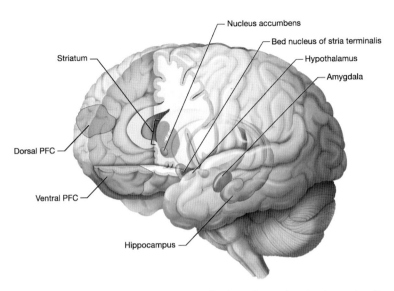

Striatum

Dorsal PFC

Ventral PFC

Hippocampus

Nucleus accumbens

Bed nucleus of stria terminalis

Hypothalamus

Amygdala

Figure 11.1. Brain regions implicated in pediatric anxiety and major depressive disorders.

approaches. Thus, work in basic science implicated specific neural circuits in declarative memory formation (Squire and Kandel, 1999), and brain-imaging studies extended these findings by demonstrating perturbations in these systems among patients (Bookheimer et al., 2000; Small et al., 2006). Work in pediatric anxiety and depressive disorders will benefit from a similar approach examining functional aspects of neural circuits implicated in emotional modulation of behavior in rodents and primates.

Neural circuits in emotionally modulated behavior

Studies in rodents and non-human primates examine two aspects of emotionally modulated behaviors relevant to imaging research in pediatric anxiety and MDD. This work comprises studies of the brain response to threats, stimuli that an organism will expend effort to avoid, and studies of the brain response to decision-making tasks where an organism is attempting to maximize rewards, stimuli that an organism will approach. Figure 11.1 illustrates the circuitry associated with these

responses. Of importance for work on pediatric anxiety and MDD, considerable work in animal models demonstrates strong developmental influences on these circuits (Gross and Hen, 2004). Specifically, both genetic and environmental influences produce longer-lasting effects on these circuits in immature, relative to mature, organisms.

Brain response to threats

Much work delineating response in rodents and non-human primates to threats relies on the fear-conditioning paradigm. In this paradigm, neutral conditioned stimuli (CS+), such as tones or lights, are paired with aversive unconditioned stimuli, such as electric shocks (UCS). Following these pairings, the CS+ comes to elicit threat-relevant responses appropriate to the UCS. The neural circuit engaged by this process centers on the amygdala, a medial temporal lobe (MTL) collection of nuclei. While debate continues on the precise role played by the amygdala in fear conditioning, most theories acknowledge amygdala involvement in diverse learning-related processes beyond fear conditioning, including those involving either punishments or rewards (Blair et al., 2005; Davis, 1998; LeDoux, 1998). The amygdala appears to play a particularly important role in modulating attention during these learning processes, allowing an organism to learn to associate specific CS stimuli with specific motivationally salient stimuli (i.e., "stimulus-reinforcement" learning).

Work on the formation of stimulus-reinforcement associations in fear conditioning has stimulated research on related processes. For example, considerable work examines brain regions that become engaged when modulating stimulus-reinforcement associations. Thus, the prefrontal cortex (PFC) is required to represent time-related contextual aspects of these associations, as occurs during extinction (Bouton, 2002). In particular, ventral and medial expanses of the PFC show rich anatomical connections with the amygdala (Pine, 2003); these regions may become engaged in the service of regulating amygdala responses to threats. Similarly, the hippocampus is required to represent spatial contextual aspects of stimulus-reinforcement associations, though this structure may mediate other behaviors relevant to anxiety and depressive disorders (Santarelli et al., 2003). This MTL structure possesses a rich array of steroid receptors, both for sex steroids and for glucocorticoids (Pine, 2003). Emotional processes may relate to alterations in hippocampal structure or function through the effects of emotional processes on steroids.

Finally, emerging work also delineates associations between neural architecture and response to innate threats, stimuli that organisms will avoid even in the absence of prior exposures. Thus, work in non-human primates implicates the amygdala in the response to intrinsically dangerous stimuli, such as snakes (Kalin 2004), whereas work in rodents implicates the bed nucleus of the stria terminalis in response to intrinsically dangerous contexts, such as brightly lit rooms (Davis, 1998). Similarly, other work in rodents implicates components of the striatum, a subcortical collection of nuclei consistently implicated in motor behavior and reward processing, in response to socially relevant stressors (Nestler and Carlezon, 2006).

Brain response to rewards

Considerable work delineating responses in rodents and non-human primates to rewarding stimuli relies on paradigms in which organisms are required to perform responses to obtain these rewards (i.e., "stimulus-response" learning). In these paradigms, rodents or primates typically are presented with various behavioral "choices." In some experiments, activity in one or another brain structure is monitored as organisms make these choices and receive the relevant rewards. In other experiments, the effects of lesions on these choices are documented. Based on this work, reward-related behavior appears to be mediated by a distributed neural circuit encompassing ventral aspects of the PFC, the amygdala, and the key components of the striatum, including the nucleus accumbens (Baxter and Murray, 2002). Within this circuit, the current value of a stimulus is thought to be coded by the amygdala and updated by the ventral PFC. The function of the striatum, which is modulated by dopaminergic transmission, takes place at the transition stage between stimulus value processing and motor response and contributes critically to the motivation to act. Dopaminergic function modulates responses to stimuli by signaling expectation errors, i.e., dopaminergic neuronal firing increases in response to unexpected rewards and decreases the omission of expected rewards (Pasupathy and Miller, 2005; Schultz, 2001).

Neural circuits and behavior in anxiety and depressive disorders

Brain imaging work in pediatric anxiety or MDD benefits from integration with research outside of imaging. For example, studies on overt behaviors manifest in the laboratory provide connections to theories that precisely delineate the nature of brain–behavior relationships in health and psychopathology. Thus, one productive avenue for brain imaging work is to focus on exploring neural correlates of behaviors that can be most directly compared among rodents, nonhuman primates, and humans. Such work could delineate pathology–behavior associations in humans where basic work outlines circuits implicated in threat- and reward-modulated behavior.

From this perspective, a key intermediate step toward elucidating structure–function relationships in illness involves focusing on information-processing functions of the relevant circuitry. A full discussion of this issue is beyond the scope of this chapter. However, particularly for the discussion of functional magnetic resonance imaging (fMRI), studies in pediatric illness are reviewed with an eye toward delineating relevant information-processing functions. For example, initial research implicating the MTL in dementia focused on documenting mnemonic perturbations, manifest in the laboratory. In pediatric anxiety and MDD, researchers should focus both on underlying neural circuitry and the precise patterns of associated information-processing perturbations, as manifest in the laboratory. In general, the most extensive work in this area documents abnormalities in memory in MDD and abnormalities in either threat–attention interactions or threat-appraisal processes in anxiety.

Morphometry

Structural MRI provides the best current methods for comparing morphometry of brain structures among healthy adults or children and individuals with various anxiety or depressive disorders. While considerable work examines morphometry in adult conditions, relatively few studies focus on pediatric anxiety or MDD. Moreover, despite the availability of considerable data in adults, findings for most structures actually show inconsistent associations between alterations in morphometry and disorders. This work focuses on structures implicated in neural response to threats and rewards illustrated in Figure 11.1: components of the PFC, basal ganglia or striatum, and the MTL. Given the strength of data in adults, coupled with developmental relationships among adult and pediatric conditions,

this section begins with a review of data in adults and then considers the degree to which data in youth appear similar.

Data in adults

Comprehensive reviews of findings in adult anxiety and depressive disorders can be found elsewhere (Bonelli et al., 2006; Drevets, 2003; Nemeroff et al., 2006; Rauch et al., 2003; Videbech and Ravnkilde, 2004). Results from these reviews on data in adults for each specific brain structure are briefly summarized.

Data on hippocampal morphometry probably provide the strongest evidence of volumetric abnormalities in any adult mood and anxiety disorders. Sufficient research examines associations with adult MDD to support a quantitative meta-analysis demonstrating moderate reductions in volume among MDD patients (Videbech and Ravnkilde, 2004). Similarly strong findings in traumatized adults implicate the hippocampus in PTSD (Nemeroff et al., 2006). Nevertheless, even in these areas, major questions arise concerning mechanisms that account for these observed associations. While some suggest that hippocampal changes in MDD or PTSD reflect the effects of chronic stress or repeated episodes of illness, others suggest that hippocampal changes reflect risk for MDD or PTSD (Gross and Hen, 2004; Nemeroff et al., 2006). Insufficient data exist to support strong generalizations for other adult anxiety or depressive disorders.

For other subcortical structures, findings appear inconsistent. Findings in the amygdala, the other key MTL structure, appear less consistent than those for the hippocampus. Thus, evidence of both increased and decreased volume has been reported in MDD (Hastings et al., 2004; Lange and Irle, 2004), with various factors being considered moderators of such differential findings (Sheline, 2003). Similarly, inconsistent evidence of amygdala volume perturbations emerges in research on PTSD and other adult anxiety disorders (Rauch et al., 2003). Comparably inconsistent findings for basal ganglia morphometry emerge in research on adult MDD. Prior findings note both increased and decreased volume within specific basal ganglia structures, with no clear consensus about the consistency of findings (Bonelli et al., 2006).

The PFC represents the only cortical structure consistently examined in morphometry research among adults with anxiety and depressive disorders. Two findings emerge with some degree of consistency. First, data in adult MDD implicate reduced ventral PFC volumes in the condition (Drevets 2001), particularly in ventral medial expanses of the PFC that have been implicated in various emotional processes, such as extinction of conditioned fear. Nevertheless, even here the data appear somewhat inconsistent, where findings might be moderated by various factors, such as familial risk, genetics, or gender (Hastings et al., 2004; Pezawas et al., 2005). Secondly, data in PTSD implicate volume reductions in dorsal medial PFC, encompassing the anterior cingulate, as well as in the ventral cingulate (Nemeroff et al., 2006), though here too not all findings are consistent (Corbo et al., 2005).

Data in juveniles

As of this writing, insufficient data have emerged to justify any strong conclusions on neuromorphometry in pediatric anxiety or MDD. Evidence in adults provides a backdrop against which available studies can be interpreted and future studies might be planned.

In terms of the hippocampus, data in pediatric disorders stand in contrast to those in adults. Thus, no evidence emerges of associations between hippocampal morphometry and any pediatric anxiety or depressive disorder, including PTSD, other anxiety disorders, or MDD (De Bellis, 2001; Millham et al., 2005). In contrast, some inconsistent evidence of perturbed amygdala volume emerges, with one study finding enlarged volume in pediatric anxiety (De Bellis et al., 2000), one finding no difference (MacMillan et al., 2003), and two other

studies finding reduced volume, one in anxiety and the other in MDD (Millham et al., 2005; Rosso et al., 2005). Findings also appear inconsistent for the basal ganglia and for PFC volumes, with reports of either increased or decreased volume in distinct PFC components for various anxiety and depressive disorders (Millham et al., 2005; Nolan et al., 2002).

Functional magnetic resonance imaging

Functional magnetic resonance imaging (fMRI) represents one of the most promising imaging techniques for research on pediatric anxiety and MDD. This promise arises from the fact that fMRI combines excellent temporal and spatial resolution for structures implicated in emotion regulation with outstanding safety and tolerability. Functional MRI research on both adult and pediatric mental syndromes is accumulating at an astonishing pace, to the point where fMRI will soon rival morphometry as the most frequently employed imaging modality for studies of pathophysiology. Given the pace of these advances, a comprehensive review of all fMRI research is beyond the scope of this chapter. Rather, this section will first summarize recent work in adult anxiety and MDD, focusing most closely on findings relevant to juveniles. The section will then review work on pediatric anxiety and depressive disorders.

Data in adults

Much like work on morphometry, fMRI studies of adult anxiety and MDD have systematically begun to examine functional aspects of specific neural structures. The focus in this work examines the same set of neural structures shown in Figure 11.1 and examined with morphometry: the MTL, PFC, and striatum.

Whereas findings in the hippocampus appear most compelling in morphometry, amygdala dysfunction is the most consistent finding in fMRI research on adult anxiety and MDD. As a result, this finding is reviewed most comprehensively. Studies among healthy adults attempting to extend work in rodents and non-human primates delineate a range of stimuli and circumstances capable of engaging the amygdala. This includes various forms of emotional stimuli, such as the threat of shock, emotional words, money, social stimuli, or aversive pictures (Haxby et al., 2002; Rauch et al., 2003; Wager et al., 2003).

In studies among adult anxiety or MDD, enhanced amygdala activation stands as the most consistent observation. This finding emerges with some consistency in three disorders: MDD, PTSD, and SOPH (Amir et al., 2005; Birbaumer et al., 1998; Fu et al., 2004; Kilts et al., 2006; Lorberbaum et al., 2004; Phan et al., 2006; Protopopescu et al., 2005; Rauch et al., 2003; Sheline, 2003; Siegle et al., 2006; Stein et al., 2002; Veit et al., 2002; Straube et al., 2004; Tillfors et al., 2001). Interestingly, adults with either specific phobias or OCD do *not* show enhanced amygdala activation on the same paradigms previously shown to differentiate adults with MDD, PTSD, or SOPH from healthy adults (Cannistraro et al., 2004; Wright et al., 2003). Thus, amygdala dysfunction shows some specificity across disorders.

Findings appear particularly consistent in SOPH. Table 11.1 summarizes results from seven fMRI studies and two positron emission tomography (PET) studies. Each of these nine studies used some form of social challenge to compare activations in adults with SOPH and healthy adults. Seven of the nine found enhanced amygdala responses in SOPH. Thus, the weight of the evidence links adult SOPH to enhanced amygdala response to social stimuli.

The most consistent findings document enhanced response to facial photographs, in five of six studies. However, even here, a few key questions emerge. First, the degree to which enhanced response is specific to the type of face emotion remains unclear, as one of the negative studies relied on only disgusted faces for the negative-valence photographs whereas other studies relied on angry or fearful faces. Secondly, the degree to which attention constrains between-group differences in amygdala response remains unclear. Recent studies in healthy adults demonstrate a robust effect on amygdala activation of manipulating attention. Particularly, distracting stimuli reduce, in healthy adults, the amygdala response to negative face emotions (Mitchell et al., 2006). Amygdala hypersensitivity in adult SOPH emerges across diverse paradigms that engage the amygdala under various attention-constraining conditions. Thus, differences have been found when task conditions require adults to explicitly attend to face emotions (Phan et al., 2006), as well as when task conditions divert attention away from face emotion (Stein et al., 2002). Only one study in adults with SOPH directly compared groups across attention conditions. This study reported stronger between-group differences when attention was directed away from emotional aspects of faces (Straube et al., 2004). Thirdly, the degree to which amygdala hyperactivity reflects state or trait factors remains unclear. In support of state effects, studies find strong reductions in amygdala activation following SOPH treatment, with degree of reduction correlating with degree of clinical improvement (Furmark et al., 2002). However, in support of trait effects, data in healthy subjects with differential genetic risk implicate a polymorphism in the serotonin transporter gene in both structural and functional changes in the amygdala (Pezawas et al., 2005).

Findings for other subcortical regions are less consistent than those observed in the amygdala. Very few studies examine hippocampal function in either adult anxiety or depressive disorders, and no consistent evidence elucidates the nature of perturbed hippocampal function in MDD or PTSD, despite the consistency of data in morphometry studies (Nemeroff et al., 2006; Sheline, 2003). Of note, however, four studies in Table 11.1 found greater parahippocampal activation in SOPH. Similarly, very few studies in adult anxiety or MDD examine the basal ganglia. Studies in MDD and in SOPH do demonstrate perturbed functions, albeit with different studies demonstrating either enhanced or reduced function (Epstein et al., 2006; Sareen et al., 2006; Sheline, 2003; Tremblay et al., 2005).

Table 11.1 Brain-imaging studies comparing activations in adult social phobia (SOPH) and healthy adults (CTRL) during exposure to evocative social stimuli or situations

Study	Method	Paradigm	Sample size (*n*)	Amygdala finding	Other notable findings
Birbaumer et al. (1998)	fMRI	Conditioning of neutral faces	7 SOPH 5 CTRL	SOPH > CTRL	–
Veit et al. (2002)	fMRI	Conditioning of neutral faces	4 SOPH 7 CTRL 4 Psychopaths	SOPH > CTRL and Psychopath	Greater fear reported in SOPH than psychopath Greater orbitofrontal activation in SOPH than CTRL and psychopath
Stein et al. (2002)	fMRI	Gender identification during block-design face-viewing task	15 SOPH 15 CTRL	SOPH > CTRL	Greater activation in SOPH in parahippocampal area, Brodmann's areas 9 and 47
Lorberbaum et al. (2004)	fMRI	Public speaking task	8 SOPH 6 CTRL	SOPH > CTRL	Greater anxiety in SOPH. Greater activation in CTRL in Brodmann's areas 24 and 9
Straube et al. (2004)	fMRI	Face-type and emotion identification during event-related face-viewing task	10 SOPH 10 CTRL	SOPH > CTRL	SOPH rated angry faces as more arousing Amygdala difference restricted to implicit, face-type-rating task. SOPH also exhibited greater insula and parahippocampal activation
Phan et al. (2006)	fMRI	Emotion identification during block-design face-viewing task	10 SOPH 10 CTRL	SOPH > CTRL	SOPH also exhibited greater cingulate and parahippocampal activation
Amir et al. (2005)	fMRI	Valence rating during block-design face-viewing task of disgusted faces	11 SOPH 11 CTRL	No activation reported	SOPH rated some faces more negatively SOPH exhibited greater activation in cingulate, parahippocampal area, and a range of PFC regions
Tillfors et al. (2001)	PET	Public speaking task	18 SOPH 6 CTRL	SOPH > CTRL	Greater activation in CTRL in insula and orbitofrontal cortex
Kilts et al. (2006)	PET	Script-driven imagery & mental arithmetic	12 SOPH 6 CTRL	Activity decrease in SOPH. No change in CTRL	Greater activation in CTRL in Brodmann's area 47

Notes:
Abbreviations: CTRL, control; fMRI, functional magnetic resonance imaging; PET, positron emission tomography; PFC, prefrontal cortex; SOPH, social phobia.

In terms of cortical regions, a series of studies have compared PFC engagement among healthy subjects and subjects with anxiety or MDD. When subjects are presented with mildly evocative stimuli, some findings document enhanced activation in ventral and medial expanses of the PFC in PTSD, SOPH, and MDD, particularly within the cingulate gyrus or portions of orbital and ventrolateral PFC. This includes three studies in SOPH shown in Table 11.1. However, on some paradigms involving evocative stimuli, as well as in some paradigms involving neutral or cognitively demanding tasks, patients with PTSD or MDD can also show reduced medial, dorsolateral, and orbital or ventrolateral PFC engagement (Kilts et al., 2006; Nemeroff et al., 2006; Rauch et al., 2003; Sheline, 2003; Siegle et al., 2006). Thus, while PFC dysfunction is frequently observed, both the direction and specific regions of the perturbation vary across disorders and even within a disorder across studies.

Data in juveniles

An emerging body of work uses fMRI to examine neural correlates of anxiety and MDD in juveniles. While provocative findings are beginning to emerge, there is also a pressing need for more work on developmental aspects of normal emotional processes. Considerable work demonstrates developmental changes in many vital emotional processes both in humans and other animals (Gross and Hen, 2004; Nelson et al., 2002; Steinberg et al., 2006). Functional MRI studies have only begun to demonstrate consistent engagement in healthy juveniles of neural circuits implicated in adult anxiety and depressive disorders.

As with data in adults, the most consistent findings in pediatric anxiety and MDD examine amygdala function. Eight fMRI papers examined the association between amygdala activity and pediatric anxiety disorders, MDD, or related conditions, with six of eight showing some sign of increased amygdala sensitivity in affected or at-risk juveniles.

Three fMRI papers relied on a block design. This design has the advantage of maximizing sensitivity to psychological processes engaged for sustained periods of time, which may also carry statistical power advantages in fMRI. In the first study (Thomas et al., 2001), enhanced amygdala activation was documented during the viewing of fearful, relative to neutral, faces. This enhanced activation occurred in pediatric GAD or panic disorder, relative to healthy age-matched subjects, whereas reduced amygdala activation, relative to both anxious and healthy subjects, was documented in pediatric MDD patients (Thomas et al., 2001). The second study examined subclinical variations in social anxiety symptoms among 16 adolescents, none with clinically significant psychopathology (Killgore and Yurgelun-Todd, 2005). Again, high social anxiety symptoms related to enhanced amygdala activation to fearful faces. The third study examined adults, classified based on their childhood histories of behavioral inhibition, a risk factor for anxiety disorders and MDD (Schwartz et al., 2003). This study found an enhanced amygdala response in formerly inhibited individuals in a comparison of novel neutral faces contrasted with neutral faces that had been previously viewed.

Five fMRI papers relied on an event-related design. This design has the disadvantage of examining relatively brief events separated in time, which may yield low statistical power. Nevertheless, this design has the major advantage of allowing precise experimental control for differences in performance on a cognitive or emotional task used during scanning. As noted above, it is vital for fMRI research to be conducted in tandem with behavioral studies, conducted in the laboratory. Ideally, fMRI studies will examine perturbations in the scanner that also manifest in the laboratory and everyday life. However, in this situation, only event-related fMRI (erfMRI) allows investigators to control for between-group differences in behavior that otherwise would raise problematic experimental confounds.

Figure 11.2 illustrates an example from one of five erfMRI studies. In this study, adolescents with MDD, anxiety disorders, or no psychopathology were scanned during a face-memory encoding task (Roberson-Nay et al., 2006). As hypothesized, adolescents with MDD exhibit a poor ability to

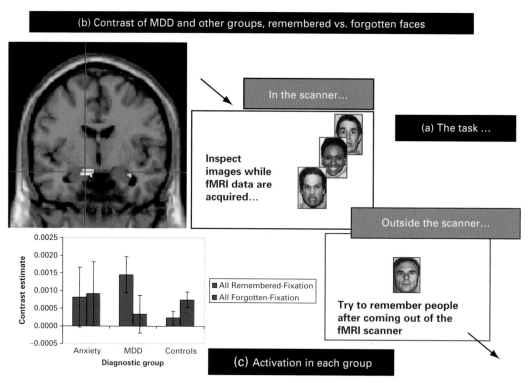

Figure 11.2. Memory encoding and emotion. The figure summarizes data from Roberson-Nay et al. (2006). A task used in this study of healthy, depressed, and anxious subjects is depicted. In this task, subjects view and are asked to rate a series of emotionally evocative facial photographs while functional magnetic resonance imaging (fMRI) data are collected. Following scanning, subjects are asked to identify faces previously seen during scanning, and brain regions engaged during successful relative to unsuccessful encoding are contrasted. The topography of group differences in engagement during successful vs. unsuccessful memory encoding is shown at the top left, illustrating a difference in the left amygdala. Differences in the amygdala emerge due to a particularly large difference in depressed adolescents during successful relative to unsuccessful encoding (bottom left).

remember evocative faces, and this deficit may actually explain prior findings noting reduced amygdala activation in pediatric MDD using a block design (Thomas et al., 2001): if analyses had selectively considered forgotten faces in Figure 11.2, MDD would have been characterized as showing reduced amygdala responses. However, with the erfMRI design, brain activation across groups can be contrasted while directly comparing correctly remembered with forgotten faces. As shown in Figure 11.2, by controlling for the performance confound associated with poor task performance in MDD, this contrast reveals the hypothesized enhanced amygdala activation in adolescent MDD. Thus, the

degree to which pediatric MDD or anxiety disorders exhibits enhanced or reduced activation on a task may depend on the precise aspects of the task examined.

Among the four remaining erfMRI studies, two documented amygdala hyperactivation (Forbes et al., 2006; McClure et al., 2007), whereas two others did not (Guyer et al., 2006; Monk et al., 2006). One study demonstrated an enhanced amygdala response to monetary reward in adolescent MDD, though only under select task conditions (Forbes et al., 2006). Thus, this study, much like the study depicted in Figure 11.2, also demonstrates that amygdala hypersensitivity in adolescent MDD is sensitive to the context of the imaging

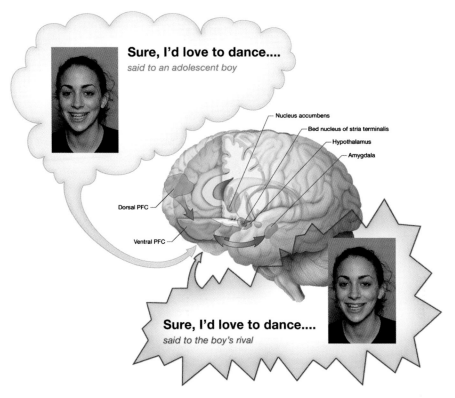

Figure 11.3. The influence of context. Context can influence the interpretation and associated neural response to stimuli that involve highly similar visual and auditory input. The same statement uttered in two distinct social contexts can be experienced as emotionally uplifting or devastating. Individual differences in such contextual effects on psychological and associated neural processes are thought to relate to individual differences in risk for, and expressions of, mood and anxiety disorders.

experiment: the finding emerges for some specific psychological contexts but not for others, emphasizing the need for future studies to utilize erfMRI that can experimentally control context. The second study directly showed that the enhanced amygdala response to evocative faces in pediatric anxiety is modulated by the attention conditions of the task (McClure et al., 2007). Specifically, this study experimentally manipulated the attention context. The study then found that adolescents with GAD, relative to healthy adolescents, exhibited enhanced amygdala activation when viewing negative-valence faces and attending to their own internal and subjective feelings of fear. This between-group difference did not emerge, however, when adolescents viewed these same faces and attended to physical features of the faces or passively viewed the faces.

Data on contextual influences raise questions on the nature of abnormalities associated with emotional disorders. These questions appear all the more prescient given data in rodents suggesting that contextual effects exert more lasting influences on MTL function in immature relative to mature organisms (Gross and Hen, 2004). Emotions experienced in adolescence also are particularly sensitive to contextual influences, reflecting the particular circumstances of situations and their meanings for individual adolescents, who experience life through the filter of specific psychological and physiological vulnerabilities (Csikszentmihalyi, 1984).

To illustrate this point, Figure 11.3 depicts one example whereby contextual influences might shape both emotional reactions and neural circuitry responses. In this example, an adolescent boy might experience upsetting emotions when witnessing the smile on the face of a potential girlfriend, encountered in a gymnasium. Such negative emotions might emerge if the girl's smile was directed toward a rival peer who had just invited her to dance. Moreover, these emotions might be particularly intense if the boy typically avoids talking to girls, lest he experience extreme anxiety. Widely different emotions might emerge in this same boy, in this same gym, faced with the very same smile, but under different circumstances. He might experience tremendous joy after mustering the courage to ask the girl to dance and encountering her smile, now directed toward him as opposed to his rival. So, context is fundamental for the emotional impact of stimuli. By using neuroimaging paradigms that manipulate viewing context, it should be possible to show that differential effects relate to both brain-emotion circuitry and adolescent psychopathology. As shown in Figure 11.3, these differential effects are likely to be reflected in neural activity encompassing a specific neural circuit, involving the amygdalae and wide expanses of the PFC.

While amygdala findings emerge as the most consistent observation in fMRI studies of pediatric anxiety and MDD, other between-group differences emerge, though with less regularity. Thus, two studies found signs of enhanced PFC activation in anxiety during face-viewing tasks, either in ventrolateral or medial PFC (McClure et al., 2007; Monk et al., 2006), whereas a third study found signs of both reduced and enhanced ventral PFC activation in MDD during distinct components of a reward task (Forbes et al., 2006). This reward study also found reduced striatal activation, in contrast to data in behavioral inhibition, which is associated with enhanced striatal activation (Guyer et al., 2006).

Other imaging modalities

While the current chapter focuses most extensively on morphometry and fMRI, other imaging modalities provide some advantages over these techniques. Results from studies using two main techniques will be reviewed briefly. These techniques comprise magnetic resonance spectroscopy (MRS) and electrophysiology.

Magnetic resonance spectroscopy

Magnetic resonance spectroscopy allows investigators to derive quantitative indices of select neurochemicals that exhibit unique magnetic resonance peaks. In general, relative to morphometry, fewer studies use MRS in either adult or pediatric anxiety and MDD. As a result, this section only reviews data from the few relevant pediatric studies.

For MRS studies, the most consistent evidence of neurochemical dysfunction emerges within the PFC, particularly medial portions of this brain region, encompassing the cingulate gyrus. Thus, prior work in pediatric MDD documents a reduced glutamine/glutamate peak, indicative of reduced glutamatergic activity, within the cingulate gyrus, as well as conflicting evidence of alterations in choline within various PFC regions (Caetano et al., 2005; Farchione et al., 2002; Rosenberg et al., 2005; Steingard et al., 2000). While such findings might be interpreted as indicative of cholinergic abnormalities, any firm conclusions are limited by the inconsistencies across studies, in terms of both methodology and findings. The only other MRS study in pediatric anxiety or depressive disorders reported evidence in pediatric PTSD of reduced cingulate N-acetylaspartate (NAA), a marker of neuronal viability (De Bellis et al., 2000). These findings are consistent with other evidence of cingulate abnormalities in adult PTSD.

Electrophysiology

Various indicators of brain function can be derived by monitoring the electroencephalogram (EEG). Two specific indicators have been used with some consistency in work on anxiety and MDD: quantitative EEG (qEEG) and event-related potentials

(ERPs). These indices have two significant advantages. First, the relative affordability of these techniques facilitates research in large samples and across many laboratories. Secondly, the excellent temporal resolution of EEG affords the opportunity to examine neural processes at the sub-second level. The current chapter briefly reviews a select group of papers in anxiety and depressive disorders relying on these techniques.

Quantitative EEG allows investigators to derive a quantitative index of fluctuations in the various frequency bands of the EEG. The main focus in work with qEEG on anxiety and MDD has concerned aspects of lateralization in power within the alpha band of the qEEG power spectrum. Considerable work using this technique among adults demonstrates enhanced activity in right frontal regions both in MDD and anxiety disorders (Davidson et al., 2002). This frontal asymmetry has been hypothesized to reflect broad dispositional tendencies, with right laterality being associated with withdrawal-related emotions. In terms of posterior asymmetries, MDD and some adult anxiety disorders exhibit distinct profiles, with MDD involving asymmetry favoring the left hemisphere and some anxiety disorders involving asymmetry favoring the right hemisphere (Bruder et al., 2005). Event-related potentials generate comparable evidence of altered posterior asymmetry in adult MDD and anxiety disorders (Bruder et al., 2002).

Electrophysiological studies in pediatric anxiety and MDD generate broadly consistent results. Thus, children with behavioral inhibition exhibit perturbed frontal asymmetry and ERPs (Fox et al., 2005a). An abnormal ERP response to angry faces also has been noted in shy children, particularly when they inherit the short allele of the serotonin transporter protein (Battaglia et al., 2005). Moreover, both adolescents with ongoing MDD as well as adolescents at high risk for MDD exhibit perturbed posterior asymmetry, though the evidence of abnormal frontal asymmetry appears less consistent than in studies of adults (Bruder et al., 2005; Kentgen et al., 2000; Shankman et al., 2005).

Conclusion

This chapter reviews findings in pediatric anxiety and MDD. Brain-imaging studies extend a wealth of basic science research on emotion to implicate similar neural circuits in pediatric anxiety and MDD. Such findings should encourage a mutually reinforcing dialogue between basic scientists examining neural correlates of emotional development and clinicians studying pediatric anxiety and MDD.

REFERENCES

Amir, N., Klumpp, H., Elias, J., Bedwell, J. S., Yanasak, N., Miller L. S. (2005) Increased activation of the anterior cingulate cortex during processing of disgust faces in individuals with social phobia. *Biol Psychiatry*, **57**(9), 975–81.

Battaglia, M., Ogliari, A., Zanoni, A., et al. (2005) Influence of the serotonin transporter promoter gene and shyness on children's cerebral responses to facial expressions. *Arch Gen Psychiatry*, **62**(1), 85–94.

Baxter, M. G., Murray, E. A. (2002) The amygdala and reward. *Nat Rev Neurosci*, **3**(7), 563–73.

Birbaumer, N., Grodd, W., Diedrich, O., et al. (1998) fMRI reveals amygdala activation to human faces in social phobics. *NeuroReport*, **9**(6), 1223–6.

Blair, J., Mithcell, D., and Blair, K. (2005) *The Psychopath: Emotion and the Brain*. Oxford: Blackwell Publishing.

Bonelli, R. M., Kapfhammer, H. P., Pillay, S. S., and Yurgelun-Todd, D. A. (2006) Basal ganglia volumetric studies in affective disorder: what did we learn in the last 15 years? *J Neural Transm*, **113**(2), 255–68.

Bookheimer, S. Y., Strojwas, M. H., Cohen, M. S., et al. (2000) Patterns of brain activation in people at risk for Alzheimer's disease. *N Engl J Med*, **343**(7), 450–6.

Bouton, M. E. (2002) Context, ambiguity, and unlearning: sources of relapse after behavioral extinction. *Biol Psychiatry*, **52**(10), 976–86.

Bruder, G. E., Kayser, J., Tenke, C. E., et al. (2002) Cognitive ERPs in depressive and anxiety disorders during tonal and phonetic oddball tasks. *Clin Electroencephalogr*, **33**(3), 119–24.

Bruder, G. E., Tenke, C. E., Warner, V., et al. (2005) Electroencephalographic measures of regional hemispheric activity in offspring at risk for depressive disorders. *Biol Psychiatry*, **57**(4), 328–35.

Caetano, S. C., Fonseca, M., Olvera, R. L., et al. (2005) Proton spectroscopy study of the left dorsolateral prefrontal cortex in pediatric depressed patients. *Neurosci Lett*, **384**(3), 321–6.

Cannistraro, P. A., Wright, C. I., Wedig, M. M., et al. (2004) Amygdala responses to human faces in obsessive-compulsive disorder. *Biol Psychiatry*, **56**(12), 916–20.

Corbo, V., Clement, M. H., Armony, J. L., Pruessner, J. C., and Brunet, A. (2005) Size versus shape differences: contrasting voxel-based and volumetric analyses of the anterior cingulate cortex in individuals with acute posttraumatic stress disorder. *Biol Psychiatry*, **58**(2), 119–24.

Costello, E., Pine, D. S., Hammen, C., et al. (2002). Development and natural history of mood disorders. *Biol Psychiatry*, **52**(6), 529–42.

Csikszentmihalyi, M. L. R. (1984) *Being Adolescent: Conflict and Growth in the Teenage Years*. New York, NY: Basic Books.

Davidson, R. J., Lewis, D. A., Alloy, L. B., et al. (2002) Neural and behavioral substrates of mood and mood regulation. *Biol Psychiatry*, **52**(6), 478–502.

Davis, M. (1998) Are different parts of the extended amygdala involved in fear versus anxiety?" *Biol Psychiatry*, **44**(12), 1239–47.

De Bellis, M. D. (2001) Developmental traumatology: the psychobiological development of maltreated children and its implications for research, treatment, and policy. *Dev Psychopathol*, **13**(3), 539–64.

De Bellis, M. D., Casey, B. J., Dahl, R. E., et al. (2000) A pilot study of amygdala volumes in pediatric generalized anxiety disorder. *Biol Psychiatry*, **48**(1), 51–7.

De Bellis, M. D., Keshavan, M. S., Spencer, S., and Hall, J. (2000) *N*-Acetylaspartate concentration in the anterior cingulate of maltreated children and adolescents with PTSD. *Am J Psychiatry*, **157**(7), 1175–7.

Drevets, W. C. (2001) Neuroimaging and neuropathological studies of depression: implications for the cognitive-emotional features of mood disorders. *Curr Opin Neurobiol*, **11**(2), 240–9.

Drevets, W. C. (2003) Neuroimaging abnormalities in the amygdala in mood disorders. *Ann N Y Acad Sci*, **985**, 420–44.

Epstein, J., Pan, H., Kocsis, J. H., et al. (2006) Lack of ventral striatal response to positive stimuli in depressed versus normal subjects. *Am J Psychiatry*, **163**(10), 1784–90.

Farchione, T. R., Moore, G. J., and Rosenberg, D. R. (2002) Proton magnetic resonance spectroscopic imaging in pediatric major depression. *Biol Psychiatry*, **52**(2), 86–92.

Forbes, E. E., Christopher May, J., Siegle, G. J., et al. (2006) Reward-related decision-making in pediatric major depressive disorder: an fMRI study. *J Child Psychol Psychiatry*, **47**(10), 1031–40.

Fox, N. A., Henderson, H. A., Marshall, P. J., Nichols, K. E., and Ghera, M. M. (2005a) Behavioral inhibition: linking biology and behavior within a developmental framework. *Annu Rev Psychol*, **56**, 235–62.

Fox, N. A., Nichols, K., Henderson, H., et al. (2005b) Evidence for a gene-environment interaction in predicting behavioral inhibition in middle childhood. *Psychol Sci*, **16**(12), 921–6.

Fu, C. H., Williams, S. C., Cleare, A. J., et al. (2004) Attenuation of the neural response to sad faces in major depression by antidepressant treatment: a prospective, event-related functional magnetic resonance imaging study. *Arch Gen Psychiatry*, **61**(9), 877–89.

Furmark, T., Tillfors, M., Marteinsdottir, I., et al. (2002) Common changes in cerebral blood flow in patients with social phobia treated with citalopram or cognitive-behavioral therapy. *Arch Gen Psychiatry*, **59**(5), 425–33.

Goldsmith, H. H. and Lemery K. S. (2000) Linking temperamental fearfulness and anxiety symptoms: a behavior-genetic perspective. *Biol Psychiatry*, **48**(12), 1199–209.

Gross, C. and Hen, R. (2004) The developmental origins of anxiety. *Nat Rev Neurosci*, **5**(7), 545–52.

Guyer, A. E., Nelson, E. E., Perez-Edgar, K., et al. (2006) Striatal functional alteration in adolescents characterized by early childhood behavioral inhibition. *J Neurosci*, **26**(24), 6399–405.

Hastings, R. S., Parsey, R. V., Oquendo, M. A., Arango, V., and Mann, J. J. (2004) Volumetric analysis of the prefrontal cortex, amygdala, and hippocampus in major depression. *Neuropsychopharmacology*, **29**(5), 952–9.

Haxby, J. V., Hoffman, E. A., and Gobbini, M. I. (2002) Human neural systems for face recognition and social communication. *Biol Psychiatry*, **51**(1), 59–67.

Kagan, J. (1994) *Galen's Prophecy*. New York, NY: Basic Books.

Kalin, N. H. (2004) Studying non-human primates: a gateway to understanding anxiety disorders. *Psychopharmacol Bull*, **38**(1), 8–13.

Kentgen, L. M., Tenke, C. E., Pine, D. S., Fong, R., Klein, R. G., and Bruder, G. E. (2000) Electroencephalographic asymmetries in adolescents with major depression:

influence of comorbidity with anxiety disorders. *J Abnorm Psychol*, **109**(4), 797–802.

Killgore, W. D. and Yurgelun-Todd, D. A. (2005) Social anxiety predicts amygdala activation in adolescents viewing fearful faces. *NeuroReport*, **16**(15), 1671–5.

Kilts, C. D., Kelsey, J. E., Knight, B., et al. (2006) The neural correlates of social anxiety disorder and response to pharmacotherapy. *Neuropsychopharmacology*, **31**(10), 2243–53.

Lange, C., Irle, E. (2004) Enlarged amygdala volume and reduced hippocampal volume in young women with major depression. *Psychol Med*, **34**(6), 1059–64.

LeDoux, J. (1998) Fear and the brain: where have we been, and where are we going? *Biol Psychiatry*, **44**(12), 1229–38.

Lorberbaum, J. P., Kose, S., Johnson, M. R., et al. (2004) Neural correlates of speech anticipatory anxiety in generalized social phobia. *NeuroReport*, **15**(18), 2701–5.

MacMillan, S., Szeszko, P. R., Moore, G. J., et al. (2003) Increased amygdala: hippocampal volume ratios associated with severity of anxiety in pediatric major depression. *J Child Adolesc Psychopharmacol*, **13**(1), 65–73.

McClure, E. B., Monk, C. S., Nelson, E. E., et al. (2007) Abnormal attention modulation of fear circuit function in pediatric generalized anxiety disorder. *Arch Gen Psychiatry*, **64**(1), 97–106.

Millham, M. P., Nugent, A. C., Drevets, W. C. et al. (2005) Selective reduction in amygdala volume in pediatric generalized anxiety disorder: a voxel-based morphometry investigation. *Biol Psychiatry*, **57**, 961–6.

Mitchell, D. G., Nakic, M., Fridberg, D., Kamel, N., Pine, D. S., and Blair, R. J. (2006) The impact of processing load on emotion. *NeuroImage*, **34**(3), 1299–309.

Monk, C. S., Nelson, E. E., McClure, E. B., et al. (2006) Ventrolateral prefrontal cortex activation and attentional bias in response to angry faces in adolescents with generalized anxiety disorder. *Am J Psychiatry*, **163**, 1091–7.

Nelson, C. A., Bloom, F. E., Cameron, J. L., Amaral, D., Dahl, R. E., and Pine, D. (2002) An integrative, multidisciplinary approach to the study of brain-behavior relations in the context of typical and atypical development. *Dev Psychopathol*, **14**(3), 499–520.

Nemeroff, C. B., Bremner, J. D., Foa, E. B., Mayberg, H. S., North, C. S., and Stein, M. B. (2006) Posttraumatic stress disorder: a state-of-the-science review. *J Psychiatr Res*, **40**(1), 1–21.

Nestler, E. J., Carlezon, W. A. Jr. (2006) The mesolimbic dopamine reward circuit in depression. *Biol Psychiatry*, **59**(12), 1151–9.

Nolan, C. L., Moore, G. J., Madden, R., et al. (2002) Prefrontal cortical volume in childhood-onset major depression: preliminary findings. *Arch Gen Psychiatry*, **59**(2), 173–9.

Pasupathy, A., Miller, E. K. (2005) Different time courses of learning-related activity in the prefrontal cortex and striatum. *Nature*, **433**(7028), 873–6.

Perez-Edgar, K., Fox, N. A. (2005) Temperament and anxiety disorders. *Child Adolesc Psychiatr Clin N Am*, **14**(4), 681–706, viii.

Pezawas, L., Meyer-Lindenberg, A., Drabant, E. M., et al. (2005) 5-HTTLPR polymorphism impacts human cingulate-amygdala interactions: a genetic susceptibility mechanism for depression. *Nat Neurosci*, **8**(6), 828–34.

Phan, K. L., Fitzgerald, D. A., Nathan, P. J., and Tancer, M. E. (2006) Association between amygdala hyperactivity to harsh faces and severity of social anxiety in generalized social phobia. *Biol Psychiatry*, **59**(5), 424–9.

Pine, D. S. (2003) Developmental psychobiology and response to threats: relevance to trauma in children and adolescents. *Biol Psychiatry*, **53**(9), 796–808.

Pine, D. S., Cohen, P., Gurley, D., Brook, J., and Ma, Y. (1998) The risk for early-adulthood anxiety and depressive disorders in adolescents with anxiety and depressive disorders. *Arch Gen Psychiatry*, **55**(1), 56–64.

Protopopescu, X., Pan, H., Tuescher, O., et al. (2005) Differential time courses and specificity of amygdala activity in posttraumatic stress disorder subjects and normal control subjects. *Biol Psychiatry*, **57**(5), 464–73.

Rauch, S. L., Shin, L. M., and Wright, C. I. (2003) Neuroimaging studies of amygdala function in anxiety disorders. *Ann N Y Acad Sci*, **985**, 389–410.

Roberson-Nay, R., McClure, E. B., Monk, C. S., et al. (2006) Increased amygdala activity during successful memory encoding in adolescent major depressive disorder: an FMRI study. *Biol Psychiatry*, **60**(9), 966–73.

Rosenberg, D. R., Macmaster, F. P., Mirza, Y., et al. (2005) Reduced anterior cingulate glutamate in pediatric major depression: a magnetic resonance spectroscopy study. *Biol Psychiatry*, **58**(9), 700–4.

Rosso, I. M., Cintron, C. M., Steingard, R. J., et al. (2005) Amygdala and hippocampus volumes in pediatric major depression. *Biol Psychiatry*, **57**(1), 21–6.

Santarelli, L., Saxe, M., Grosse, C., et al. (2003) Requirement of hippocampal neurogenesis for the behavioral effects of antidepressants. *Science*, **301** (5634), 805–9.

Sareen, J., Campbell, D. W., Leslie, W. D., et al. (2006) Striatal function in generalized social phobia: a functional magnetic resonance imaging study. *Biol Psychiatry*, **61**(3), 396–404.

Schultz, W. (2001) Reward signaling by dopamine neurons. *Neuroscientist*, **7**(4), 293–302.

Schwartz, C. E., Wright, C. I., Shin, L. M., Kagan, J., and Rauch, S. L. (2003) Inhibited and uninhibited infants "grown up": adult amygdalar response to novelty. *Science*, **300**(5627), 1952–3.

Shankman, S. A., Tenke, C. E., Bruder, G. E., Durbin, C. E., Hayden, E. P., and Klein, D. N. (2005) Low positive emotionality in young children: association with EEG asymmetry. *Dev Psychopathol*, **17**(1), 85–98.

Sheline, Y. I. (2003) Neuroimaging studies of mood disorder effects on the brain. *Biol Psychiatry*, **54**(3), 338–52.

Siegle, G. J., Thompson, W., Carter, C. S., Steinhauser, S. R., and Thase, M. A. (2006) Increased amygdala and decreased dorsolateral prefrontal BOLD responses in unipolar depression: related and independent features. *Biol Psychiatry*, **61**(2), 198–209.

Small, G. W., Kepe, V., Ercoli, L. M., et al. (2006) PET of brain amyloid and tau in mild cognitive impairment. *N Engl J Med*, **355**(25), 2652–63.

Squire, L. R. K., Kandel, E. R. (1999) *Memory: From Mind to Molecules*. New York, NY: Scientific American Press.

Stein, M. B., Goldin, P. R., Sareen, J., Zorrilla, L. T., and Brown, G. G. (2002) Increased amygdala activation to angry and contemptuous faces in generalized social phobia. *Arch Gen Psychiatry*, **59**(11), 1027–34.

Steinberg, L., Dahl, R., Keating, D., et al. (2006) The study of developmental psychopathology in adolescence: integrating affective neuroscience with the study of context. In: Cicchetti, D., Cohen, D. J., and Hoboken, N. J. (eds.) *Developmental Psychopathology*. Volume 2. *Developmental Neuroscience*, 2nd edn. New York: John Wiley and Sons.

Steingard, R. J., Yurgelun-Todd, D. A., Hennen, J., et al. (2000) Increased orbitofrontal cortex levels of choline in depressed adolescents as detected by in vivo proton magnetic resonance spectroscopy. *Biol Psychiatry*, **48**(11), 1053–61.

Straube, T., Kolassa, I. T., Glauer, M., Mentzel, H. J., and Miltner, W. H. (2004) Effect of task conditions on brain responses to threatening faces in social phobics: an event-related functional magnetic resonance imaging study. *Biol Psychiatry*, **56**(12), 921–30.

Thomas, K. M., Drevets, W. C., Dahl, R. E., et al. (2001) Amygdala response to fearful faces in anxious and depressed children. *Arch Gen Psychiatry*, **58**(11), 1057–63.

Tillfors, M., Furmark, T., Marteinsdottir, I., et al. (2001) Cerebral blood flow in subjects with social phobia during stressful speaking tasks: a PET study. *Am J Psychiatry*, **158**(8), 1220–6.

Tremblay, L. K., Naranjo, C. A., Graham, S. J., et al. (2005) Functional neuroanatomical substrates of altered reward processing in major depressive disorder revealed by a dopaminergic probe. *Arch Gen Psychiatry*, **62**(11), 1228–36.

Veit, R., Flor, H., Erb, M., et al. (2002) Brain circuits involved in emotional learning in antisocial behavior and social phobia in humans. *Neurosci Lett*, **328**(3), 233–6.

Videbech, P. and Ravnkilde, B. (2004) Hippocampal volume and depression: a meta-analysis of MRI studies. *Am J Psychiatry*, **161**(11), 1957–66.

Wager, T. D., Phan, K. L., Liberzon, I., and Taylor, S. F. (2003) Valence, gender, and lateralization of functional brain anatomy in emotion: a meta-analysis of findings from neuroimaging. *NeuroImage*, **19**(3), 513–31.

Weissman, M. M., Wickramaratne, P., Nomura, Y., Warner, V., Pilowsky, D., and Verdeli, H. (2006) Offspring of depressed parents: 20 years later. *Am J Psychiatry*, **163**(6), 1001–8.

Weissman, M. M., Wickramaratne, P., Nomura, Y., et al. (2005) Families at high and low risk for depression: a 3-generation study. *Arch Gen Psychiatry*, **62**(1), 29–36.

Wright, C. I., Martis, B., McMullin, K., Shin, L. M., and Rauch, S. L. (2003) Amygdala and insular responses to emotionally valenced human faces in small animal specific phobia. *Biol Psychiatry*, **54**(10), 1067–76.

12

Disturbances of fronto-striatal circuits in Tourette syndrome and obsessive-compulsive disorder

Rachel Marsh, Daniel A. Gorman, Jason Royal and Bradley S. Peterson

Introduction

The symptoms of Tourette syndrome (TS) and pediatric-onset obsessive-compulsive disorder (OCD) lie on a spectrum of involuntary, semi-involuntary, and compulsive behaviors. Tourette syndrome is characterized by motor and vocal tics that wax and wane in severity (American Psychiatric Association, 1994). Obsessive-compulsive disorder is defined by the presence of recurrent, distressing, and intrusive thoughts, ideas or images (obsessions) together with their repetitive behavioral counterparts (compulsions) (American Psychiatric Association, 1994). The symptoms of OCD are frequently observed in patients with TS, and the presence of tics in childhood and adolescence often predicts the presence of OCD symptoms in late adolescence and adulthood (Peterson et al., 2001a). Understanding the underlying neurobiology of TS and OCD requires knowledge of their developmental trajectories, the ways in which their phenotypes are similar or dissimilar, and the ways in which symptoms differ between children and adults.

Phenomenology, natural history, and epidemiology

Tics are fragments of behaviors that often occur in response to bodily or environmental cues. They are usually experienced as an unendurable urge that is relieved only by the execution of a tic (Leckman, 2002; Leckman and Riddle, 2000; Peterson and Klein, 1997). The tics of TS vary in complexity, ranging from brief, meaningless, and abrupt movements and sounds (simple tics) to those that are longer, more involved, and seemingly more goal-directed (complex tics) (Leckman, 2003). These urges and a person's preoccupation with them are reminiscent of the obsessional urges to act that typically precede compulsive behaviors.

The modal age of onset of tics is six years. Tics affect 10%–20% of children at some time in their lives, with a ratio of boys to girls of approximately 3:1 or 4:1 (Costello et al., 1996; Peterson et al., 2001a). Tics typically begin at a low frequency and with minimal forcefulness, so that parents often attribute the behaviors to their child's "habit." In most children, tics disappear in weeks to months, but approximately 1%–2% of children have tics that persist for a year or more and that are thus defined as "chronic" (Jankovic, 2001; Nomoto and Machiyama, 1990). By definition, only children with both chronic motor and vocal tics are said to have TS.

Tic severity tends to fluctuate unpredictably over minutes, hours, days, and weeks. For tics that become chronic, the typical course involves a gradual increase in number, frequency, and forcefulness in early childhood, peaking in severity at age 10 or

Neuroimaging in Developmental Clinical Neuroscience, eds. Judith M. Rumsey and Monique Ernst. Published by Cambridge University Press. © Cambridge University Press 2009.

11, and then declining gradually during adolescence. By age 18 years, tics are substantially reduced in roughly 90% of patients with TS, and more than 40% are symptom-free (Bloch et al., 2005; Burd et al., 2001; Leckman et al., 1998). For many individuals with TS, obsessive thoughts and compulsive rituals emerge during late childhood or early adolescence, several years after the onset of motor tics (Leckman et al., 1994).

Family and twin studies indicate that TS and OCD are genetically related (Pauls et al., 1986; Peterson and Klein, 1997). Obsessive-compulsive disorder affects 1%–3% of the population worldwide, and the age of onset appears to be bimodal, with one mode of onset at 10–12 years of age and the other in early adulthood (Geller et al., 1998). About half of all OCD patients have onset of the disorder by age 15 (Karno et al., 1988), and, compared with the adult-onset form, child-onset OCD is more likely to occur in the context of a personal or family history of a tic disorder (known as "tic-related" OCD) (Pauls et al., 1995; Geller et al., 1998). When tics and obsessive-compulsive symptoms are present together, their severities correlate positively, suggesting an underlying common modulator of severity over the short-term (Lin et al., 2002). In contrast to tics, however, child-onset OCD symptoms tend to persist into late adolescence and adulthood, when they are usually more functionally debilitating than are tics alone (Bloch et al., 2006; Swedo et al., 1989a).

Factor analytic studies have identified clusters of OCD symptoms: (1) checking compulsions and aggressive, sexual, religious, and somatic obsessions; (2) symmetry and ordering; (3) cleanliness and washing; and (4) hoarding (Baer, 1994; Leckman et al., 1997). Some evidence suggests that tic-related OCD is more likely than OCD without tics to include aggressive, religious, and sexual obsessions as well as checking, counting, ordering, touching, and hoarding compulsions (Leckman et al., 1994). These symptoms may therefore represent a subtype of OCD that is characterized by earlier onset and that shares a common etiology with tic disorders. Other evidence, however, suggests that youth with

OCD have similar symptom presentations, whether or not they have a chronic tic disorder (Storch et al., 2008).

The presence of tics increases an individual's risk of having OCD, and the presence of OCD increases the risk of having a tic disorder. The risk of OCD in individuals with TS is estimated at 40%–60%, and the risk of TS or another tic disorder in those with OCD is 7%–26% (Cohen and Leckman, 1994; Geller et al., 1998). Both tic disorders and OCD often coexist with attention deficit/hyperactivity disorder (ADHD) in children, increasing the likelihood of both disruptive behaviors and learning problems (Scahill et al., 2005). In a minority of individuals, these behaviors may derive from an autoimmune disorder in which infection with group A β-hemolytic streptococcus (GABHS) produces GABHS antibodies that cross-react with neural antigens in the basal ganglia, a phenomenon known as "**p**ediatric **a**utoimmune **n**europsychiatric **d**isorders **a**ssociated with **s**treptoccocal infection" ("PANDAS") (Singer, 2005; Swedo et al., 1998).

Neural circuitry

Neuroimaging studies increasingly suggest that the neural bases of TS and OCD reside in anatomical and functional disturbances of cortico-striato-thalamo-cortical (CSTC) circuits, neural pathways that mediate, among other functions, behavioral inhibition and impulse control (Miller and Cohen, 2001; Pasupathy and Miller, 2005). The CSTC circuits, composed of loops between cortical and subcortical brain regions, have at least four components—those initiating from and projecting back to sensorimotor cortex, orbitofrontal cortex (OFC), limbic and anterior cingulate cortices, and association cortices (see Table 12.1). These cortical regions map excitatory input onto specific regions of the striatum (caudate and putamen), which in turn project inhibitory input onto either the internal or external segment of the pallidum. From the internal pallidum, information flows to the thalamus via a single inhibitory projection in the direct pathway. From

Table 12.1 Cortico-striato-thalamo-cortical (CTSC) pathways

CSTC component	Sensorimotor pathways	Orbitofrontal pathways	Association pathways	Limbic system pathways
Cortical afferents	Somatosensory	Orbitofrontal	DLPFC	AC
	Primary motor	STG	Posterior parietal	Hippocampus
	SMA	ITG	Arcuate premotor	Entorhinal cortex
		AC		STG, ITG
Striatum	Dorsolateral putamen	Ventral caudate	Dorsolateral caudate	Ventral caudate
	Dorsolateral caudate	Ventral putamen		Ventral putamen
				NAcc
GP/SNr	Ventrolateral GPi	Dorsomedial GPi	Dorsomedial GPi	Rostrolateral GPi
	Caudolateral SNr	Rostromedial SNr	Rostrolateral SNr	Ventral pallidum
				Rostrodorsal SNr
Thalamic nuclei	Ventrolateral	Medial dorsal	Ventral anterior	Medial dorsal
	Centromedial			
	Intralaminar			
Cortical projections	Supplementary motor	Orbitofrontal	DLPFC	AC

Notes:

AC, anterior cingulate; DLPFC, dorsolateral prefrontal cortex; GPi, globus pallidus (internal segment); ITG, inferior temporal gyrus; NAcc, nucleus accumbens; SMA, supplementary motor area; SNr, substantia nigra pars reticulata; STG, superior temporal gyrus.

the external pallidum, it flows via an inhibitory projection to the subthalamic nucleus and then via an excitatory projection to the thalamus in the indirect pathway. From the thalamus, both the direct and indirect pathways send excitatory projections back to the cortex (Figure 12.1).

Careful consideration of circuit diagrams for these connections (Figure 12.1) reveals that, in normal functioning, cortical activity excites the striatum, which in the direct pathway will in turn excite the thalamus via two successive inhibitory projections, and the thalamic excitation will in turn excite the cortex. In the indirect pathway, in contrast, striatal excitation will inhibit the thalamus via the successive chain of inhibition, excitation, and inhibition. Inhibition of the thalamus in the indirect pathway will then reduce activity in the cortex. Thus, during normal functioning of the striatum, the initiation of movement is thought to require activation of striatal neurons in the direct pathway.

Activation of striatal neurons in the indirect pathway, in contrast, is thought to act as a "brake" on motor patterns that are generated in the cortex and brainstem (Albin and Mink, 2006). Removal of this inhibitory influence of the indirect pathway releases the brake on motor pattern activity, thereby allowing execution of the intended motor behavior (Albin and Mink, 2006). Dopaminergic projections from the midbrain to the basal ganglia excite the inhibitory fibers of the direct pathway via D1 receptors, and thus enhance the overall excitatory influences of the direct pathway on the cortex, whereas those projections via D2 receptors inhibit the inhibitory fibers of the striatum that project to the external pallidum in the indirect pathway, thereby reducing the braking functions of that pathway (Figure 12.1) (Graybiel, 1990).

This dopaminergic transmission is thought to be abnormal in individuals with TS. Positron emission tomography (PET) and single photon emission

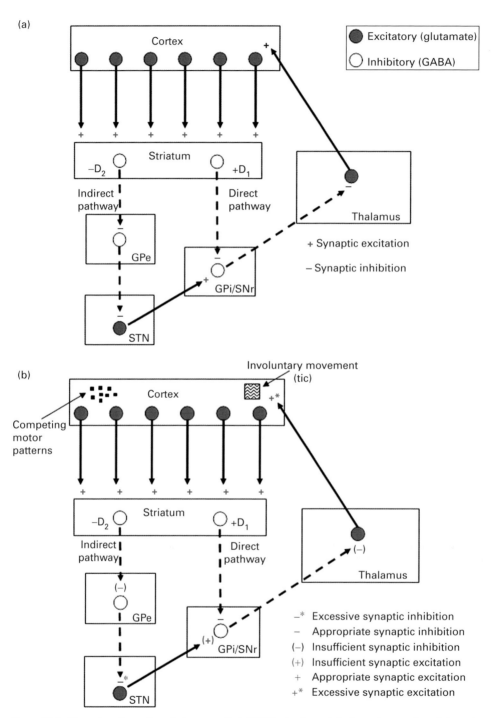

Figure 12.1. Normal cortico-striato-thalamo-cortical (CTSC) circuits and the hyperinnervation of the striatum by dopaminergic neurons in Tourette syndrome (TS). (a) Normal CTSC circuits: striatal neurons containing GABA

computed tomography (SPECT) studies suggest, for example, that the dopamine D2 receptors are excessively sensitive in TS (Wolf et al., 1996; Wong et al., 1997). In addition, dopamine D2 receptor antagonists decrease the severity of tic symptoms (Singer, 2005). Together these observations suggest that excess dopaminergic activity in the striatum may excessively stimulate the direct pathway and inhibit the indirect pathway, and each of these effects would tend to increase movement propensity in persons with TS (Albin and Mink, 2006; Leckman et al., 2006; Peterson and Thomas, 2000) (Figure 12.1). Similar effects may also contribute to the genesis of more complex movements and thoughts in this population, perhaps accounting in part for the phenotypic similarities between complex tics and compulsions and for the covariation in the severity of these symptoms in persons who have both tics and obsessive-compulsive symptoms. This possibility is consistent with evidence that dopaminergic transmission is abnormal in persons with OCD (Denys et al., 2004) and with evidence that serotonin, the neurotransmitter most strongly implicated in the pathophysiology of OCD (Thomsen, 2000), likely modulates dopaminergic function within the striatum (Kapur and Remington, 1996).

Anatomical disturbances in Tourette syndrome and obsessive-compulsive disorder

Tourette syndrome

Selected anatomical findings from studies of TS are summarized in Tables 12.2 and 12.3. These findings include decreased caudate volumes in both children and adults with TS, abnormalities that are thought to represent trait disturbances and the likely site of origin of tic behaviors in persons with TS (Peterson et al., 2003). The observation of a decreased number and proportion of caudate neurons that were positive for the calcium-binding protein parvalbumin in the tissue of TS subjects suggests a cellular explanation for the smaller caudate volumes in persons with TS (Kalanithi et al., 2005). Larger dorsal prefrontal cortices (DPFC) and smaller sizes of the corpus callosum have also been reported in children, but not in adults, with TS relative to age-matched controls. The larger dorsal prefrontal volumes and smaller callosal size seen in children accompany less severe tic symptoms (Peterson et al., 2001b; Plessen et al., 2004). Compared to age-matched controls, patients with TS also have larger volumes of the hippocampus and amygdala (Peterson et al., 2007).

Caption for Figure 12.1. (*cont.*)

(gamma-aminobutyric acid) are inhibitory, whereas projections from the thalamus to the cortex are glutamatergic and excitatory. The ***direct outflow pathway*** transmits inhibitory signals from the striatum onto the internal segment of globus pallidus and substantia nigra pars reticulata (GPi/SNr), onto to the thalamus before projecting back to the cortex. The ***indirect pathway*** transmits inhibitory signals from the striatum to the globus pallidus pars externa (GPe) with a second inhibitory projection to the subthalamic nucleus (STN). STN neurons transmit excitatory signals to the GPi/SNr before sending inhibitory signals to the thalamus and then back to the cortex (Saint-Cyr, 2003; Wichmann and DeLong, 1996). (b) Excessive dopaminergic sensitivity in persons with TS: In the direct pathway, dopaminergic hyperinnervation stimulates D_1 receptors (located on GABAergic postsynaptic neurons) that project from the GPi/SNr to the thalamus, blocking inhibitory interneurons within the thalamus, thereby enhancing glutamatergic excitation of the cortex and increasing motor output. Striatal neurons within the indirect pathway express D_3 dopamine receptors which decrease the effect of cortical input to striatal neurons (Surmeier et al., 1993). In this pathway, the excitatory output from the STN to the GPi/SNr likely enhances rather than blocks inhibitory projections of these basal ganglia output neurons, thereby reducing thalamocortical excitation. Dopaminergic sensitivity may produce, through both the direct and the indirect pathways in the basal ganglia, multiple effects that result in disinhibition of thalamo-cortical excitatory projections and thus disinhibition of behavior in TS-related conditions. Excitatory (glutamatergic) projections are indicated by solid lines, and inhibitory (GABAergic) projections by dotted lines. GPe, external segment of the globus pallidus; STN, subthalamic nucleus; GPi/SNr, internal segment of the globus pallidus or the substantia nigra pars reticulata.

Table 12.2 Selected findings from anatomical studies of Tourette syndrome (TS)

| Investigator | Sample size | | Age group | Findings |
	TS	NC		
Peterson et al. (2001b)	155	131	Children and adults	Larger dorsal prefrontal and parieto-occipital volumes in TS children
Peterson et al. (2003)	154	130	Children and adults	Smaller volume of the caudate nucleus across all ages in TS
Kim and Peterson (2003)	97	64	Children	Smaller cavum septum pellucidum in TS children and adults
Plessen et al. (2004)	158	121	Children and adults	Smaller overall corpus callosum size in TS children
Peterson et al. (2007)	154	128	Children and adults	Larger overall volumes of hippocampus and amygdala in TS children

Notes:
Smaller caudate volumes and smaller cavum septum pellucidum in both children and adults with TS may predispose an individual to develop TS. Larger prefrontal, parieto-occipital, hippocampus and amygdala volumes, as well as smaller corpus callosum size, likely reflect compensatory, neuroplastic responses in TS children. NC, normal, age-matched control.

Table 12.3 Comparison of findings in children/adolescents and adults with Tourette syndrome (TS)

	Children/adolescents*	Adults*	References
Anatomical MRI			
Cavum septum pellucidum	TS < controls	TS = controls	Kim and Peterson (2003)
Frontal cortex	TS > controls	TS < controls	Peterson et al. (2001b)
Parietal cortex	TS > controls	TS < controls	Peterson et al. (2001b)
Corpus callosum	TS < controls	TS > controls	Plessen et al. (2004)
Caudate nucleus	TS < controls	TS < controls	Peterson et al. (2003)
Lenticular nucleus	TS = controls	TS < controls	Peterson et al. (2003)
Hippocampus	TS > controls	TS < controls	Peterson et al. (2007)
Amygdala	TS > controls	TS < controls	Peterson et al. (2007)
Functional MRI (on self-regulatory control tasks)			
Default mode deactivation	TS > controls	TS < controls	Marsh et al. (2007)
Fronto-striatal activation	TS > controls	TS > controls	
Performance	TS = controls	TS = controls	

Notes:
Anatomical and functional findings often diverge between children and adults with TS, likely because of reduced compensatory responses in adults.
*Anatomical findings: >, structure larger in TS subjects; <, smaller in TS; =, structure is the same size in TS and controls. Functional findings: >, greater activation/deactivation in TS; <, less activation/deactivation in TS; =, activation/ deactivation does not differ between TS and controls.

Detailed analyses of the surfaces of these regions suggest that the increased volumes derive primarily from the head and medial surface of the hippocampus over the length of the dentate gyrus, and from the dorsal and ventral surfaces of the amygdala, over its basolateral and central nuclei. Volumes of these subregions declined with age in the TS group but not in the controls, such that the subregions were significantly larger in the TS children, but significantly smaller in TS adults than in controls. In both children and adults with TS, volumes of these subregions correlated inversely with the severity of tic and OCD symptoms, suggesting that enlargement of these subregions may have a compensatory and neuromodulatory influence over symptoms.

Likewise, we suspect that larger DPFC volumes in children with TS represent a compensatory or adaptive process that serves to attenuate tics. The continuous need to suppress tics is known to activate prefrontal regions (Peterson et al., 1998), thereby stimulating activity-dependent plasticity in these regions that provides enhanced neural resources for the control of tic symptoms. Likewise, exaggerated axonal pruning during development may contribute to the smaller size of the corpus callosum detected in children with TS, which would serve to limit the transmission of excitatory activity across this interhemispheric commissure. Reduced excitatory input onto inhibitory interneurons in prefrontal cortices would in turn enhance excitatory activity within prefrontal cortices, thus helping children with TS to exert control over their tics (Plessen et al., 2004), presumably by enhancing activity in the indirect "braking" pathway of CSTC circuits. In contrast, the significantly smaller prefrontal volumes reported in adults with TS may represent an inability to generate this activity-dependent plastic response (Peterson et al., 2001b). Thus, anatomical findings in the frontal cortex and hippocampus diverge in children and adults with TS (Table 12.3), likely because of deficient compensatory responses in the subpopulation of patients who experience TS into adulthood. In addition, an anatomical structure in the brain that develops during fetal life, the cavum septum pellucidum, is smaller in both children and adults with TS than in healthy controls, suggesting an embryological origin to the brain disturbances that may produce TS (Kim and Peterson, 2003).

Lastly, a recent anatomical imaging study that has identified abnormal thinning of the sensorimotor, primary motor, and premotor cortices in children with TS (with the degree of thinning being proportional to the severity of tic symptoms) (Sowell et al., 2008), suggesting that tics may arise from the sensory and motor cortices. This possibility is supported by the finding that activity in paralimbic regions (including the anterior cingulate and insular cortex), supplementary motor area, and parietal operculum correlates with the experience of the "premonitory urges" that precede tics (Bohlhalter et al., 2006). In addition, these urges are similar to the urges to move that have been described by patients who have undergone electrical stimulation of the sensorimotor cortex (Lim et al., 1994). Thus, tics likely arise from abnormal structure and function of the sensorimotor portions of the CSTC loop.

Obsessive-compulsive disorder

Structural studies of the frontal cortex and basal ganglia in adults with OCD have yielded mixed findings. Some studies have reported no difference in caudate volume between patients and controls (Aylward et al., 1996; Kellner et al., 1991), while others have reported smaller caudate nuclei in persons with OCD (Luxenberg et al., 1988; Robinson et al., 1995). Findings of smaller volumes of the putamen (Rosenberg et al., 1997c) and globus pallidus (Szeszko et al., 2004a) in children with OCD could be attributable to the inclusion of children with tic-related OCD (children having either a personal or family history of tics). These basal ganglia structures were also found to be smaller in children and adults with TS who also had OCD compared with control subjects, but not in individuals with TS who did not have OCD (Peterson et al., 2003). Nevertheless, the common findings of striatal abnormalities in children with TS and OCD may point to a common neural substrate across these disorders.

The largest anatomical MRI study of adults with OCD to date revealed reduced gray matter in the medial frontal gyrus, OFC, and insulo-opercular region, all of which are portions of the limbic system, as well as increased gray matter in the ventral striatum, another limbic structure intimately connected with the OFC (Pujol et al., 2004). The ventral striatum is involved in reward expectancy, a process mediated by the dopaminergic system (Schultz, 2002). A role for this region in the pathophysiology of OCD is consistent with the disturbances in reward perception reported by patients – feeling, for example, that the door might be unlocked, but without sufficient relief of that feeling from sensory confirmation that the door is locked, which should be rewarding (Aouizerate et al., 2004; Remijnse et al., 2006).

Inconsistent findings from anatomical studies of the frontal cortex in adults with OCD include some reports of reductions (Atmaca et al., 2007; Kang et al., 2004; Luxenberg et al., 1988; Szeszko et al., 1999) and other reports of no differences (Riffkin et al., 2005) in OFC volumes in persons with OCD compared with healthy adults. Volumes of the anterior cingulate cortex (ACC) are normal (Kang et al., 2004) or nearly normal (Atmaca et al., 2007; Pujol et al., 2004). In contrast, studies of children with OCD suggest the presence of increased ACC volumes (Rosenberg and Hanna, 2000; Rosenberg and Keshavan, 1998; Szeszko et al., 2004a). Findings of increased thalamic volumes appear consistent across recent studies of treatment-naive children (Gilbert et al., 2000) and adults (Atmaca et al., 2007) with OCD. Others have reported volumetric abnormalities in ventral, but not dorsolateral, prefrontal cortex in children and adolescents with OCD who were tic-free (Rosenberg and Keshavan, 1998). Unpublished data from our laboratory, however, have revealed smaller volumes of dorsal prefrontal cortices in tic-free children with OCD that correlate inversely with symptom severity. In contrast, children with chronic tic disorders but no OCD had larger volumes of DPFC, consistent with findings in children with TS (Peterson et al., 2001b). These findings suggest that pediatric-onset OCD may differ from chronic tic disorders in its neuroanatomical basis. Whereas chronic tic disorders likely arise from dysfunction in the striatum, child-onset OCD may arise primarily from dysfunction within prefrontal cortices.

Taken together, these findings suggest in general that the neurobiological underpinnings of OCD differ in children (who by definition have childhood-onset illness) and adults (the majority of whom have adult-onset illness), consistent with the differing heritability and symptom profiles in child- and adult-onset illness (Geller, 2006; Leckman et al., 2003). Adults seem to have anatomical disturbances in OFC-related limbic circuits, whereas children seem to have disturbances in the basal ganglia, thalamus, and cingulate portions of CSTC circuits. The differences in findings between children and adults with OCD are summarized in Tables 12.4 and 12.5.

Some compulsions may be difficult to differentiate behaviorally from complex tics. Compulsions, however, seem to be a consequence of abnormal activity in the OFC portions of the CSTC loop, rather than the sensorimotor portion that seems to produce tics. As discussed above, previous imaging studies have reported, for example, increased metabolism and blood flow to the OFC (Alptekin et al., 2001; Baxter et al., 1987; Nordahl et al., 1989; Rubin et al., 1995; Swedo et al., 1989b), and deep brain stimulation of the ventral striatum modulates activity of the OFC pathway (Rauch et al., 2006), in persons with OCD. Thus the identification of the direct and indirect pathways within CSTC circuits and the recognition that these circuits have differentially segregated cortical inputs and outputs (Table 12.1) may provide a neuroanatomical basis for the differential production of simple tics, complex tics, or compulsions in persons with TS or OCD (Mink, 2001).

Functional disturbances in Tourette syndrome and obsessive-compulsive disorder

Tourette syndrome

A developmental functional magnetic resonance imaging (fMRI) study of TS children and adults

Table 12.4 Findings from selected anatomical studies of children and adolescents with obsessive-compulsive disorder (OCD)

| Investigator | Sample size | | Age group | Findings |
	OCD	NC		
Rosenberg et al. (1997c)	19	19	Children and adolescents	Smaller putamen volumes in OCD children correlated inversely with obsessive symptom severity
Rosenberg and Keshavan (1998)	21	21	Children and adolescents	Increased AC volumes associated with reduced striatal volumes. No difference in DLPFC volumes
MacMaster et al. (1999)	21	21	Children and adolescents	Increased corpus callosal signal intensity in the genu region, suggestive of increased VPFC size and myelin sheath thickness in OCD children
Giedd et al. (2000)	34	82	Children	Larger basal ganglia volumes in streptococcal-infected children with OCD
Gilbert et al. (2000)	21	21	Children and adolescents	Increased thalamic volumes in younger, treatment-naive OCD children
Szeszko et al. (2004b)	23	27	Children	More AC gyrus gray matter and smaller globus pallidus volumes in OCD children

Notes:

Anatomical findings suggest the presence of fronto-striatal abnormalities in children with OCD.

Abbreviations: AC, anterior cingulate; DLPFC, dorsolateral prefrontal cortex; NC, normal, age-matched control; VPFC, ventral prefrontal cortex.

(8–52 years of age) found that children with TS and healthy controls performed similarly on the Stroop, a task that requires resisting the interference of a prepotent response (word reading) when executing a less automatic response (naming the color in which letters are written) (Marsh et al., 2007). Behavioral performance improved with increasing age in patients with TS, just as it did in healthy controls, likely reflecting the maturation of neural systems that subserve inhibitory control. The imaging findings, in contrast, showed that the adults with TS rely on exaggerated activation of fronto-striatal regions in order to achieve this normal performance on the task (Marsh et al., 2007), consistent with the previously reported deficits in neural plasticity and inhibitory reserve in adults with TS (Figure 12.2).

Another fMRI study reported that the willful suppression of tics in adults with TS was associated with increased activity of the frontal cortex and caudate nucleus, and with decreased activity of the putamen, pallidum, and thalamus (Peterson et al., 1998). Increased activity in the frontal cortex was associated with increased activity in the caudate which, in turn, was associated with greater decreases in activity of the putamen, globus pallidus, and thalamus. The magnitude of the activations (caudate) or deactivations (putamen, pallidum, and thalamus) during tic suppression correlated with the severity of tic symptoms in the month preceding the scan. Taken together, these findings suggest that a greater ability of the frontal cortex to suppress activity in the basal ganglia helps to lessen tic severity outside of the scanner. Consistent with these fMRI findings during tic suppression, a recent EEG study reported elevated functional coherence between sensorimotor, prefrontal, and mesial frontal cortices in adults with TS relative to controls during the voluntary suppression of tics, suggesting that these regions work together in the service of suppressing tics (Serrien et al., 2005). In addition, individuals with TS had greater EEG coherence in this same mesio-frontal network than in healthy controls during

Table 12.5 Comparison of recent findings from children versus adults with obsessive-compulsive disorder (OCD)

	Children	Adults
Anatomical MRI		
DLPFC	OCD < controls (Peterson et al., unpublished)	–
OFC	–	OCD < controls (Szeszko et al., 1999)
MPFC	–	OCD < controls (Pujol et al., 2004)
ACC	OCD > controls (Szeszko et al., 2004a)	OCD = controls (Kang et al., 2004)
Caudate nucleus	OCD = controls (Szeszko et al., 2004a; Peterson et al., unpublished)	variable
Putamen	OCD = controls (Szeszko et al., 2004a)	OCD > controls (Pujol et al., 2004)
Globus pallidus	OCD < controls (Szeszko et al., 2004a)	–
Nucleus accumbens	–	OCD > controls
Thalamus	OCD > controls (Gilbert et al., 2000)	OCD > controls (Atmaca et al., 2007)
Functional MRI in adults (during symptom provocation, Mataix-Cols et al., 2004)		
Washing/contamination	Ventromedial prefrontal activation, OCD > controls	
Checking and symmetry/ordering	Putamen, thalamus, dorsal PFC activation, OCD > controls	
Hoarding	Left precentral gyrus, right OFC activation, OCD > controls	

Notes:
Anatomical findings in children and adults with OCD likely diverge due to critical periods of abnormal maturation of ventral prefronto-striato-thalamic circuitry.
*Anatomical findings: >, structure larger in OCD subjects; <, smaller in OCD; =, structure is the same size in OCD and controls. Functional findings: >, greater activation in OCD; <, less activation in OCD; =, activation does not differ between TS and controls.

response inhibition on a go/no-go task, implicating sensorimotor–prefrontal connections in inhibitory control functions that extend beyond tic suppression (Serrien et al., 2005). Together these findings suggest that the frontal cortex plays an important role in suppressing or modulating the severity of tic and inhibitory behaviors in persons with TS.

Obsessive-compulsive disorder

Resting-state PET or SPECT studies in adults with OCD have reported hyperactivity of the lateral OFC, caudate nucleus, anterior cingulate cortex, and thalamus (Baxter, 1992; Rubin et al., 1992; Swedo et al., 1989b). Increased activity within these regions during the experimental provocation of symptoms detected with fMRI (Breiter et al., 1996; McGuire et al., 1994; Rauch et al., 1994; Saxena

et al., 1998) has suggested a causal relationship of fronto-striatal hyperactivity with OCD symptoms (Evans et al., 2004). Based on similar findings of exaggerated activity in adults with TS either during the suppression of tics or during tasks that require inhibitory control, the excess activity in fronto-striatal circuits in adults with OCD could represent attempts to modulate or control the symptoms of OCD within an inefficient inhibitory system. Consistent with this interpretation are prior neuropsychological findings of impaired inhibitory control in individuals with OCD (Evans et al., 2004), and reports that the excess activity in fronto-striatal circuits in adults with OCD disappears following successful behavioral or pharmacological therapies (Rauch et al., 2002; Saxena et al., 2001).

A preliminary PET study of adults with OCD reported that during a continuous performance

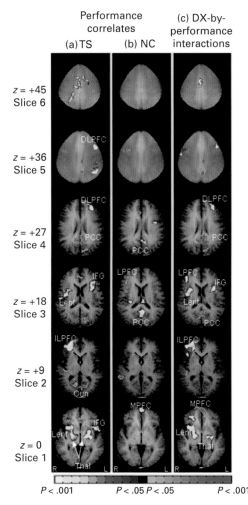

Performance correlates   (c) DX-by-performance interactions
(a) TS   (b) NC

z = +45
Slice 6

z = +36
Slice 5

z = +27
Slice 4

z = +18
Slice 3

z = +9
Slice 2

z = 0
Slice 1

P < .001   P < .05 P < .05   P < .001

Figure 12.2. Functional magnetic resonance imaging (fMRI) voxel-wise correlations of Stroop-related activations with Stroop interference scores in patients with TS and control subjects. Performance correlates in (a) individuals with Tourette syndrome (TS, $n = 66$, age range = 8.4–52.6 years) and (b) normal controls (NC, $n = 70$, age range = 7.1–57.2 years). Positive correlations with performance are in yellow and inverse correlations are in purple/blue. In TS (a), greater activation of fronto-striatal regions (Brodmann's areas 9/46, 45/46; lenticular nucleus; and thalamus) is accompanied by poorer performance (higher interference scores). Diagnosis-by-performance interactions (c) were detected in fronto-striatal and posterior cingulate regions. The TS subjects likely needed to engage fronto-striatal regions more to

task, tic-related OCD symptoms (e.g., checking compulsions and religious, aggressive, and sexual obsessions) correlated with regional cerebral blood flow (rCBF) in the striatum, whereas non-tic-related symptoms (e.g., washing and cleaning) correlated with rCBF in DLPFC (Rauch et al., 1998). These findings suggest that the dysfunction in pediatric OCD, in which tic-related symptoms are most common, may be based primarily in striatal portions of CSTC circuits, whereas dysfunction in adult-onset OCD may be based more in frontal portions of those circuits. Providing some evidence against this conclusion, however, are the findings of a resting-state SPECT study of rCBF in early- and late-onset OCD patients and healthy controls, which revealed decreased blood flow in the thalamus, ACC, and bilateral inferior PFC only in the early-onset patients (Busatto et al., 2001). Similarly, a study of a small sample of children and adolescents with OCD reported inverse correlations between age of onset and frontal rCBF, implicating frontal cortex in earlier-onset OCD (Castillo et al., 2005).

Recently, an fMRI study using a variety of symptom provocation paradigms in adults suggested that individual symptom domains of OCD (e.g., washing, checking, and hoarding) are mediated by distinct components of the fronto-striatal circuits, each of which has been implicated previously in cognitive and emotional processing (Mataix-Cols et al., 2004). After viewing washing-related pictures, for example, patients with OCD, relative to healthy controls, showed greater activation in bilateral ventromedial prefrontal regions, consistent with prior fMRI findings from a small group of patients with predominantly washing symptoms who showed greater activation than controls in ventrolateral

Caption for Figure 12.2. (*cont.*)
maintain task performance at a level that was similar to performance in controls. Cun, cuneate; DLPFC, dorsolateral prefrontal cortex; IFG, inferior frontal gyrus; ILPFC, inferolateral prefrontal cortex; Lent, lenticular nucleus; LPFC, lateral prefrontal cortex; MPFC, mesial prefrontal cortex; PCC, posterior cingulate cortex; Thal, Thalamus.

prefrontal cortex when viewing disgust-inducing pictures (Shapira et al., 2003). In contrast, patients showed greater activation in the putamen, globus pallidus, thalamus, and dorsal prefrontal cortex after viewing checking-related pictures and greater activation of the left precentral gyrus and right OFC after viewing hoarding-related pictures. Furthermore, these neural responses were associated positively with severity scores on the symptom dimension in question (washing, checking, or hoarding) (Mataix-Cols et al., 2004).

These findings suggest that although the symptoms of checking and symmetry/ordering are mediated by the same fronto-striatal regions as have been implicated in TS, the contamination/washing symptoms of OCD are likely mediated by ventromedial prefrontal cortices and associated limbic regions (Table 12.5). This conclusion is consistent with findings of amygdala activation during the presentation of contamination-related pictures in adult patients who had primarily washing symptoms (van den Heuvel et al., 2004). Limbic (bilateral amygdala) activity observed when adults with OCD attend to disease-specific emotional information is consistent with the clinical observation that compulsive behaviors typically fail to relieve obsessions (i.e., compulsive symptoms are insufficiently rewarding), and that individuals with OCD have difficulty changing their rigid and repetitive behaviors. A recent fMRI study of reversal learning in adults with OCD, in fact, reported functional disturbances in a motivational (OFC-ventral striatum) circuit that is involved in reward processing (Remijnse et al., 2006). To our knowledge, no studies have yet characterized the role of this OFC-ventral striatal circuit and anatomically connected limbic areas in childhood-onset OCD.

Stimulus-response learning in Tourette syndrome and obsessive-compulsive disorder

When we do something repeatedly, as when learning to drive a car or ride a bicycle, our brains organize behavior into action sequences (Graybiel, 2005) that tend to become "habitual" motor routines that can be complex, prolonged, and automatic. Animal and human lesion studies suggest that structures in the striatum (the caudate and putamen) support the learning of these motor skills, procedures, and stimulus-response (S-R) associations or "habits" (Jog et al., 1999; Packard and Knowlton, 2002). In contrast, mesial temporal lobe structures (the hippocampus and amygdala) support the formation of declarative and emotional memories (Eichenbaum, 2000).

We and others have hypothesized that tics might be a manifestation of S-R learning gone awry. Observations that prompted formulation of this hypothesis include: the phenomenological similarity of tics with habits, the documented anatomical and functional abnormalities of the striatum in persons with TS, and the role of the striatum in S-R learning. Probabilistic classification learning (PCL) is a form of S-R learning in humans that circumvents the use of declarative memory by probabilistically associating cues with specific outcomes. One version of a PCL task is a weather prediction game that requires the gradual learning of associations between certain stimuli and a specific response. Subjects try to predict rain or sunshine based on the presentation of a varying combination of a set of cards on a computer screen. Each card is independently and probabilistically related to the outcomes (rain or shine), each of which occurs equally often. A response is considered correct on a particular trial only if the selected outcome is more strongly associated with the cue combination that appears on that trial. Because of the probabilistic nature of the task, subjects usually believe that they are simply guessing at the outcome. Declarative memory of a previous trial is not as useful in improving performance as is information gleaned across many trials. Compared to healthy controls, both children and adults with TS were impaired at S-R learning on the weather prediction task (Marsh et al., 2004). The more impaired their S-R learning on the task, the more severe were their tic symptoms. In contrast, the same individuals were unimpaired in their acquisition of perceptual-motor skills (pursuit rotor or mirror tracing) that are thought to be based in either the lenticular nucleus or cerebellum (Marsh et al., 2005). Others have observed similar deficits in S-R learning, despite evidence of intact explicit memory, in persons with TS (Keri et al., 2002).

Together, these findings suggest that individuals with TS are impaired in forming desired S-R associations. Perhaps tics represent inappropriate S-R associations between the somatosensory urges that precede tics and the tics themselves (sensory stimulus and motor response). Alternatively, core disturbances in the striatum of persons with TS (e.g., excessive dopaminergic activity or anatomical and cellular disturbances (Kalanithi et al., 2005)) may impair the chunking of behavioral fragments into the complex action sequences that a given S-R association comprises (Graybiel, 2005). Electrophysiological recordings in animals have shown that the learning of action sequences is associated with gradual changes in the task-related firing patterns of neuronal ensembles in the striatum (Jog et al., 1999). The striatum develops a neuronal representation of the new firing pattern after the learned action sequence is consolidated. This change in firing pattern likely represents the "chunking together" of behavioral fragments of a larger complex action sequence into a single, coherently executed behavior (Jog et al., 1999). In addition, intact dopaminergic innervation is important for these "chunking functions" of the striatum (Matsumoto et al., 1999). Thus tics may be the product of anatomical or neurochemical disturbances within the striatum that predispose an individual to impaired S-R learning and to the expression of fragmented motor and vocal behaviors. Disturbances in fronto-striatal projections may then release from inhibitory control this predisposition to behavioral fragmentation and then manifest as motor or vocal tics.

In contrast to these findings in persons with TS, adults with OCD do not demonstrate impaired performance on tasks that are mediated by the dorsal striatum (Rauch et al., 2006). Numerous neuropsychological studies do suggest, however, that children and adults with OCD perform poorly on the Stroop, go/no-go, and occulomotor suppression tasks, and that the degree of impairment in inhibitory control correlates positively with the severity of OCD symptoms (Bannon et al., 2002; Rankins et al., 2006; Rosenberg et al., 1997a, 1997b; Van der Linden et al., 2005). We therefore conclude that OCD symptoms, at least in the adult form of the disorder, may be the consequence of impaired inhibitory control over the intrusive thoughts that precede the compulsive behaviors or rituals that originate in non-striatal portions of CSTC circuits (Chamberlain et al., 2005). To our knowledge, no published studies have yet employed tasks that probe functioning of striatal-based learning in childhood-onset OCD. In addition, the age of onset is frequently unclear in studies of adults with OCD. Thus, the extant data do not allow us to draw conclusions regarding the role of striatal-based learning systems in the neurodevelopmental trajectory of pediatric-onset OCD.

Conclusions

In typically developing children, the maturation of fronto-striatal systems contributes to the ability to inhibit unwanted behaviors. In TS and OCD, however, disturbances in fronto-striatal systems may impair inhibitory behavioral control over the generation of impulses to move or to perform the complex, inappropriate behaviors that we know as tics and compulsions. Whereas individuals with OCD are impaired at tasks that require the engagement of behavioral control based primarily within prefrontal cortex portions of CSTC circuits, children and adults with TS perform normally on these tasks, likely because they engage compensatory responses within these prefrontal regions.

Tics may represent inappropriate "habits" or fragmented motor and vocal behaviors arising from anatomical or neurochemical disturbances within the striatum. Neuroimaging findings in fact suggest that sensorimotor, motor, and premotor cortex and the dorsal striatum, which together make up the motor portions of CSTC circuits, are primarily involved in the pathogenesis of TS. In contrast, compulsions may represent behavioral responses aimed at relieving anxiety. This relief from anxiety may be felt as rewarding, however transient, consistent with evidence suggesting that the pathophysiology of OCD involves preferentially ventral frontal cortical and limbic structures, including OFC and ventral striatum that mediate reward processing.

Few studies have investigated the integrity of fronto-striatal systems in pediatric-onset OCD. The common neural and genetic bases of TS and childhood-onset OCD, however, along with the phenomenological similarities between tics and compulsions, suggest that these neuropsychiatric disorders are phenotypic variants of similar or overlapping disease processes. Future studies should clarify the roles of fronto-striatal subsystems within CSTC circuits in the development of TS and OCD. One approach would be to map the developmental trajectories of regional brain activity in individuals with TS and OCD, compared with healthy controls, during performance of tasks that require both behavioral inhibitory control and reward processing. In addition, research that probes striatal functions in TS and OCD may further substantiate the hypothesis that tics and compulsions represent fragmented motor plans caused by striatal dysfunction and S-R learning gone awry. Finally, translational imaging paradigms based on animal models of TS, OCD, S-R learning, and reward processing will likely improve our knowledge of the neurobiology of TS and OCD, leading to new, rationally based therapies.

ACKNOWLEDGMENTS

This work was supported in part by NIMH grants MH01232, MH59139, MH068318, K02-74677 (B. S.P), and K01-MH077652 (R. M.), the Suzanne Crosby Murphy Endowment at Columbia University College of Physicians and Surgeons, and the Thomas D. Klingenstein and Nancy D. Perlman Family Fund.

REFERENCES

Albin, R. L. and Mink, J. W. (2006) Recent advances in Tourette syndrome research. *Trends Neurosci*, **29**, 175–82.

Alptekin, K., Degirmenci, B., Kivircik, B., et al. (2001) Tc-99m HMPAO brain perfusion SPECT in drug-free obsessive-compulsive patients without depression. *Psychiatry Res*, **107**, 51–6.

American Psychiatric Association (1994) *Diagnostic and Statistical Manual of Mental Disorders*, 4th edn. Washington, DC: American Psychiatric Association.

Aouizerate, B., Guehl, D., Cuny, E., et al. (2004) Pathophysiology of obsessive-compulsive disorder: a necessary link between phenomenology, neuropsychology, imagery and physiology. *Prog Neurobiol*, **72**, 195–221.

Atmaca, M., Yildirim, H., Ozdemir, H., et al. (2007) Volumetric MRI study of key brain regions implicated in obsessive-compulsive disorder. *Prog Neuropsychopharmacol Biol Psychiatry*, **31**, 46–52.

Aylward, E. H., Harris, G. J., Hoehn-Saric, R., et al. (1996) Normal caudate nucleus in obsessive-compulsive disorder assessed by quantitative neuroimaging. *Arch Gen Psychiatry*, **53**, 577–84.

Baer, L. (1994) Factor analysis of symptom subtypes of obsessive compulsive disorder and their relation to personality and tic disorders. *J Clin Psychiatry*, **55**, 18–23.

Bannon, S., Gonsalvez, C. J., Croft, R. J., et al. (2002) Response inhibition deficits in obsessive-compulsive disorder. *Psychiatry Res*, **110**, 165–74.

Baxter, L. R., Jr. (1992) Neuroimaging studies of obsessive compulsive disorder. *Psychiatr Clin North Am*, **15**, 871–84.

Baxter, L. R. J., Phelps, J. M., Mazziotta, J. C., et al. (1987) Local cerebral glucose metabolic rates in obsessive-compulsive disorder: a comparison with rates in unipolar depression and normal controls. *Arch Gen Psychiatry*, **44**, 211–18.

Bloch, M. H., Leckman, J. F., Zhu, H., et al. (2005) Caudate volumes in childhood predict symptom severity in adults with Tourette syndrome. *Neurology*, **65**, 1253–8.

Bloch, M. H., Peterson, B. S., Scahill, L., et al. (2006) Adulthood outcome of tic and obsessive-compulsive symptom severity in children with Tourette syndrome. *Arch Pediatr Adolesc Med*, **160**, 65–9.

Bohlhalter, S., Goldfine, A., Matteson, S., et al. (2006) Neural correlates of tic generation in Tourette syndrome: an event-related functional MRI study. *Brain*, **129**, 2029–37.

Breiter, H. C., Rauch, S. L., Kwong, K. K., et al. (1996) Functional magnetic resonance imaging of symptom provocation in obsessive-compulsive disorder. *Arch Gen Psychiatry*, **53**, 595–606.

Burd, L., Kerbeshian, P. J., Barth, A., et al. (2001) Long-term follow-up of an epidemiologically defined cohort of patients with Tourette syndrome. *J Child Neurol*, **16**, 431–7.

Busatto, G. F., Buchpiguel, C. A., Zamignani, D. R., et al. (2001) Regional cerebral blood flow abnormalities in early-onset obsessive-compulsive disorder: an exploratory SPECT study. *J Am Acad Child Adolesc Psychiatry*, **40**, 347–54.

Castillo, A. R., Buchpiguel, C. A., de Araujo, L. A., et al. (2005) Brain SPECT imaging in children and adolescents with obsessive-compulsive disorder. *J Neural Transm*, **112**, 1115–29.

Chamberlain, S. R., Blackwell, A. D., Fineberg, N. A., et al. (2005) The neuropsychology of obsessive compulsive disorder: the importance of failures in cognitive and behavioural inhibition as candidate endophenotypic markers. *Neurosci Biobehav Rev*, **29**, 399–419.

Cohen, D. J. and Leckman, J. F. (1994) Developmental psychopathology and neurobiology of Tourette's syndrome [review]. *J Am Acad Child Adolesc Psychiatry*, **33**, 2–15.

Costello, E. J., Angold, A., Burns, B. J., et al. (1996) The Great Smoky Mountains study of youth. Goals, design, methods, and the prevalence of DSM-III-R disorders. *Arch Gen Psychiatry*, **53**, 1129–36.

Denys, D., Zohar, J., and Westenberg, H. G. (2004) The role of dopamine in obsessive-compulsive disorder: preclinical and clinical evidence. *J Clin Psychiatry*, **65** [Suppl 14], 11–17.

Eichenbaum, H. (2000) A cortical-hippocampal system for declarative memory. *Nat Rev Neurosci*, **1**, 41–50.

Evans, D. W., Lewis, M. D., and Iobst, E. (2004) The role of the orbitofrontal cortex in normally developing compulsive-like behaviors and obsessive-compulsive disorder. *Brain Cogn*, **55**, 220–34.

Geller, D., Biederman, J., Jones, J., et al. (1998) Is juvenile obsessive-compulsive disorder a developmental subtype of the disorder: a review of the pediatric literature. *J Am Acad Child Adolesc Psychiatry*, **37**, 420–7.

Geller, D. A. (2006) Obsessive-compulsive and spectrum disorders in children and adolescents. *Psychiatr Clin North Am*, **29**, 353–70.

Giedd, J. N., Rapoport, J. L., Garvey, M. A., et al. (2000) MRI assessment of children with obsessive-compulsive disorder or tics associated with streptococcal infection. *Am J Psychiatry*, **157**, 281–3.

Gilbert, A. R., Moore, G. J., Keshavan, M. S., et al. (2000) Decrease in thalamic volumes of pediatric patients with obsessive-compulsive disorder who are taking paroxetine. *Arch Gen Psychiatry*, **57**, 449–56.

Graybiel, A. M. (1990) Neurotransmitters and neuromodulators in the basal ganglia. *Trends Neurosci*, **13**, 244–54.

Graybiel, A. M. (2005) The basal ganglia: learning new tricks and loving it. *Curr Opin Neurobiol*, **15**, 638–44.

Jankovic, J. (2001) Differential diagnosis and etiology of tics. *Adv Neurol*, **85**, 15–29.

Jog, M. S., Kubota, Y., Connolly, C. I., et al. (1999) Building neural representations of habits. *Science*, **286**, 1745–9.

Kalanithi, P. S., Zheng, W., Kataoka, Y., et al. (2005) Altered parvalbumin-positive neuron distribution in basal ganglia of individuals with Tourette syndrome. *Proc Natl Acad Sci U S A*, **102**, 13307–12.

Kang, D. H., Kim, J. J., Choi, J. S., et al. (2004) Volumetric investigation of the frontal-subcortical circuitry in patients with obsessive-compulsive disorder. *J Neuropsychiatry Clin Neurosci*, **16**, 342–9.

Kapur, S. and Remington, G. (1996) Serotonin-dopamine interaction and its relevance to schizophrenia. *Am J Psychiatry*, **153**, 466–76.

Karno, M., Golding, J. M., Sorenson, S. B., et al. (1988) The epidemiology of obsessive-compulsive disorder in five US communities. *Arch Gen Psychiatry*, **45**, 1094–9.

Kellner, C. H., Jolley, R. R., Holgate, R. C., et al. (1991) Brain MRI in obsessive-compulsive disorder. *Psychiatry Res*, **36**, 45–9.

Keri, S., Szlobodnyik, C., Benedek, G., et al. (2002) Probabilistic classification learning in Tourette syndrome. *Neuropsychologia*, **40**, 1356–62.

Kim, K. J. and Peterson, B. S. (2003) Cavum septi pellucidi in Tourette syndrome. *Biol Psychiatry*, **54**, 76–85.

Leckman, J. and Riddle, M. (2000) Tourette's syndrome: when habit-forming systems form habits of their own? *Neuron*, **28**, 349–54.

Leckman, J. F. (2002) Tourette's syndrome. *Lancet*, **360**, 1577–86.

Leckman, J. F. (2003) Phenomenology of tics and natural history of tic disorders. *Brain Dev*, **25** [Suppl 1], S24–8.

Leckman, J. F., Grice, D. E., Boardman, J., et al. (1997) Symptoms of obsessive-compulsive disorder. *Am J Psychiatry*, **154**, 911–17.

Leckman, J. F., Pauls, D. L., Zhang, H., et al. (2003) Obsessive-compulsive symptom dimensions in affected sibling pairs diagnosed with Gilles de la Tourette syndrome. *Am J Med Genet*, **116**, 60–8.

Leckman, J. F., Vaccarino, F. M., Kalanithi, P. S., et al. (2006) Annotation: Tourette syndrome: a relentless drumbeat – driven by misguided brain oscillations. *J Child Psychol Psychiatry*, **47**, 537–50.

Leckman, J. F., Walker, D. E., Goodman, W. K., et al. (1994) "Just right" perceptions associated with compulsive behavior in Tourette's syndrome. *Am J Pschiatry*, **151**, 675–80.

Leckman, J. F., Zhang, H., Vitale, A., et al. (1998) Course of tic severity in Tourette's syndrome: the first two decades. *Pediatrics*, **102**, 14–19.

Lim, S. H., Dinner, D. S., Pillay, P. K., et al. (1994) Functional anatomy of the human supplementary sensorimotor area: results of extraoperative electrical stimulation. *Electroencephalogn Clin Neurophysiol*, **91**, 179–93.

Lin, H., Yeh, C. B., Peterson, B. S., et al. (2002) Assessment of symptom exacerbations in a longitudinal study of children with Tourette's syndrome or obsessive-compulsive disorder. *J Am Acad Child Adolesc Psychiatry*, **41**, 1070–7.

Luxenberg, J. S., Swedo, S. E., Flament, M. F., et al. (1988) Neuroanatomical abnormalities in obsessive-compulsive disorder detected with quantitative X-ray computed tomography. *Am J Pschiatry*, **145**, 1089–93.

MacMaster, F. P., Dick, E. L., Keshavan, M. S., et al. (1999) Corpus callosal signal intensity in treatment naive pediatric obsessive compulsive disorder. *Prog Neuropsychopharmacol Biol Psychiatry*, **23**, 601–12.

Marsh, R., Alexander, G. M., Packard, M. G., et al. (2004) Habit learning in Tourette syndrome: a translational neuroscience approach to a developmental psychopathology. *Arch Gen Psychiatry*, **61**, 1259–68.

Marsh, R., Alexander, G. M., Packard, M. G., et al. (2005) Perceptual-motor skill learning in Gilles de la Tourette syndrome. Evidence for multiple procedural learning and memory systems. *Neuropsychologia*, **43**, 1456–65.

Marsh, R., Zhu, H., Wang, Z., et al. (2007) A developmental fMRI study of self-regulatory control in Tourette's syndrome. *Am J Psychiatry*, **164**, 955–66.

Mataix-Cols, D., Wooderson, S., Lawrence, N., et al. (2004) Distinct neural correlates of washing, checking, and hoarding symptom dimensions in obsessive-compulsive disorder. *Arch Gen Psychiatry*, **61**, 564–76.

Matsumoto, N., Hanakawa, T., Maki, S., et al. (1999) Role of [corrected] nigrostriatal dopamine system in learning to perform sequential motor tasks in a predictive manner. *J Neurophysiol*, **82**, 978–98.

McGuire, P. K., Bench, C. J., Frith, C. D., et al. (1994) Functional anatomy of obsessive-compulsive phenomena. *Br J Psychiatry*, **164**, 459–68.

Miller, E. K. and Cohen, J. D. (2001) An integrative theory of prefrontal cortex function. *Annu Rev Neurosci*, **24**, 167–202.

Mink, J. W. (2001) Basal ganglia dysfunction in Tourette's syndrome: a new hypothesis. *Pediatric Neurol*, **25**, 190–8.

Nomoto, F., Machiyama, Y. (1990) An epidemiological study of tics. *Jpn J Psychiatry Neurol*, **44**, 649–55.

Nordahl, T. E., Benkelfat, C., Semple, W. E., et al. (1989) Cerebral glucose metabolic rates in obsessive compulsive disorder. *Neuropsychopharmacology*, **2**, 23–8.

Packard, M. G. and Knowlton, B. J. (2002) Learning and memory functions of the basal ganglia. *Annu Rev Neurosci*, **25**, 563–93.

Pasupathy, A. and Miller, E. K. (2005) Different time courses of learning-related activity in the prefrontal cortex and striatum. *Nature*, **433**, 873–6.

Pauls, D. L., Alsobrook, J. P., 2nd, Goodman, W., et al. (1995) A family study of obsessive-compulsive disorder. *Am J Psychiatry*, **152**, 76–84.

Pauls, D. L., Towbin, K. E., Leckman, J. F., et al. (1986) Gilles de la Tourette's syndrome and obsessive-compulsive disorder. Evidence supporting a genetic relationship. *Arch Gen Psychiatry*, **43**, 1180–2.

Peterson, B. and Klein, J. (1997) In: Peterson, B. S. (ed.) *Child Psychiatry Clinics of North America: Neuroimaging*. Vol. 6 (April). Philadelphia, PA: W. B. Saunders, pp. 343–64.

Peterson, B. S. and Thomas, P. (2000) Tourette's syndrome: what are we really imaging? In: Ernst, M., Rumsey, J. (eds.) *Functional Neuroimaging in Child Psychiatry*. Cambridge: Cambridge University Press, pp. 242–65.

Peterson, B. S., Choi, H. A., Hao, X., et al. (2007) Morphologic features of the amygdala and hippocampus in children and adults with Tourette syndrome. *Arch Gen Psychiatry*, **64**, 1281–91.

Peterson, B. S., Pine, D. S., Cohen, P., et al. (2001a) A prospective, longitudinal study of tic, obsessive-compulsive, and attention deficit-hyperactivity disorders in an epidemiological sample. *J Am Acad Child Adolesc Psychiatry*, **40**, 685–95.

Peterson, B. S., Skudlarski, P., Anderson, A. W., et al. (1998) A functional magnetic resonance imaging study of tic suppression in Tourette syndrome. *Arch Gen Psychiatry*, **55**, 326–33.

Peterson, B. S., Staib, L., Scahill, L., et al. (2001b) Regional brain and ventricular volumes in Tourette syndrome. *Arch Gen Psychiatry*, **58**, 427–40.

Peterson, B. S., Thomas, P., Kane, M. J., et al. (2003) Basal ganglia volumes in patients with Gilles de la Tourette syndrome. *Arch Gen Psychiatry*, **60**, 415–24.

Plessen, K. J., Wentzel-Larsen, T., Hugdahl, K., et al. (2004) Altered interhemispheric connectivity in individuals with Tourette's disorder. *Am J Psychiatry*, **161**, 2028–37.

Pujol, J., Soriano-Mas, C., Alonso, P., et al. (2004) Mapping structural brain alterations in obsessive-compulsive disorder. *Arch Gen Psychiatry*, **61**, 720–30.

Rankins, D., Bradshaw, J. L., and Georgiou-Karistianis, N. (2006) The semantic Simon effect in Tourette's syndrome and obsessive-compulsive disorder. *Brain Cogn*, **61**(3), 225–34.

Rauch, S. L., Dougherty, D. D., Malone, D., et al. (2006) A functional neuroimaging investigation of deep brain stimulation in patients with obsessive-compulsive disorder. *J Neurosurg*, **104**, 558–65.

Rauch, S. L., Dougherty, D. D., Shin, L. M., et al. (1998) Neural correlates of factor-analyzed OCD symptom dimensions: a PET study. *CNS Spectrums*, **3**, 37–43.

Rauch, S. L., Jenike, M. A., Alpert, N. M., et al. (1994) Regional cerebral blood flow measured during symptom provocation in obsessive-compulsive disorder using oxygen 15-labeled carbon dioxide and positron emission tomography. *Arch Gen Psychiatry*, **51**, 62–70.

Rauch, S. L., Shin, L. M., Dougherty, D. D., et al. (2002) Predictors of fluvoxamine response in contamination-related obsessive compulsive disorder: a PET symptom provocation study. *Neuropsychopharmacology*, **27**, 782–91.

Remijnse, P. L., Nielen, M. M., van Balkom, A. J., et al. (2006) Reduced orbitofrontal-striatal activity on a reversal learning task in obsessive-compulsive disorder. *Arch Gen Psychiatry*, **63**, 1225–36.

Riffkin, J., Yucel, M., Maruff, P., et al. (2005) A manual and automated MRI study of anterior cingulate and orbito-frontal cortices, and caudate nucleus in obsessive-compulsive disorder: comparison with healthy controls and patients with schizophrenia. *Psychiatry Res*, **138**, 99–113.

Robinson, D., Wu, H., Munne, R. A., et al. (1995) Reduced caudate nucleus volume in obsessive-compulsive disorder. *Arch Gen Psychiatry*, **52**, 393–8.

Rosenberg, D. R. and Hanna, G. L. (2000) Genetic and imaging strategies in obsessive-compulsive disorder: potential implications for treatment development. *Biol Psychiatry*, **48**, 1210–22.

Rosenberg, D. R. and Keshavan, M. S. (1998) Toward a neurodevelopmental model of of obsessive-compulsive disorder. *Biol Psychiatry*, **43**, 623–40.

Rosenberg, D. R., Averbach, D. H., O'Hearn, K. M., et al. (1997a) Oculomotor response inhibition abnormalities in pediatric obsessive-compulsive disorder. *Arch Gen Psychiatry*, **54**, 831–8.

Rosenberg, D. R., Dick, E. L., O'Hearn, K. M., et al. (1997b) Response-inhibition deficits in obsessive-compulsive disorder: an indicator of dysfunction in frontostriatal circuits. *J Psychiatry Neurosci*, **22**, 29–38.

Rosenberg, D. R., Keshavan, M. S., O'Hearn, K. M., et al. (1997c) Frontostriatal measurement in treatment-naive children with obsessive-compulsive disorder. *Arch Gen Psychiatry*, **54**, 824–30.

Rubin, R. T., Ananth, J., Villanueva-Meyer, J., et al. (1995) Regional [133]Xenon cerebral blood flow and cerebral ^{99m}Tc-HMPAO uptake in patients with obsessive-compulsive disorder before and during treatment. *Biol Psychiatry*, **38**, 429–37.

Rubin, R. T., Villanueva-Meyer, J., Ananth, J., et al. (1992) Regional xenon 133 cerebral blood flow and cerebral technetium 99m HMPAO uptake in unmedicated patients with obsessive-compulsive disorder and matched normal control subjects. Determination by high-resolution single-photon emission computed tomography [see comments]. *Arch Gen Psychiatry*, **49**, 695–702.

Saint-Cyr, J. A. (2003) Frontal-striatal circuit functions: context, sequence, and consequence. *J Int Neuropsychol Soc*, **9**, 103–27.

Saxena, S., Bota, R. G., and Brody, A. L. (2001) Brain-behavior relationships in obsessive-compulsive disorder. *Semin Clin Neuropsychiatry*, **6**, 82–101.

Saxena, S., Brody, A. L., Schwartz, J. M., and Baxter, L. R. (1998) Neuroimaging and frontal-subcortical circuitry in obsessive-compulsive disorder. *Br J Psychiatry* Suppl, **35**, 26–37.

Scahill, L., Sukhodolsky, D. G., Williams, S. K., et al. (2005) Public health significance of tic disorders in children and adolescents. *Adv Neurol*, **96**, 240–8.

Schultz, W. (2002) Getting formal with dopamine and reward. *Neuron*, **36**, 241–63.

Serrien, D. J., Orth, M., Evans, A. H., et al. (2005) Motor inhibition in patients with Gilles de la Tourette syndrome: functional activation patterns as revealed by EEG coherence. *Brain*, **128**, 116–25.

Shapira, N. A., Liu, Y., He, A. G., et al. (2003) Brain activation by disgust-inducing pictures in obsessive-compulsive disorder. *Biol Psychiatry*, **54**, 751–6.

Singer, H. S. (2005) Tourette's syndrome: from behaviour to biology. *Lancet Neurol*, **4**, 149–59.

Sowell, E., Kan, E., Yoshii, J., et al. (2008) Thinning of sensorimotor cortices in children with Tourette syndrome. *Nat Neurosci*, **11**, 637–9.

Storch, E. A., Stigge-Kaufman, D., Marien, W. E., et al. (2008) Obsessive-compulsive disorder in youth with and without a chronic tic disorder. *Depress Anxiety*, **25**(9), 761–7.

Surmeier, D. J., Reiner, A., Levine, M. S., et al. (1993) Are neostriatal dopamine receptors co-localized? *Trends Neurosci*, **16**, 299–305.

Swedo, S. E., Leonard, H. L., Garvey, M., et al. (1998) Pediatric autoimmune neuropsychiatric disorders associated with streptococcal infections: clinical description of the first 50 cases. *Am J Psychiatry*, **155**, 264–71.

Swedo, S. E., Rapoport, J. L., Leonard, H., et al. (1989a) Obsessive-compulsive disorder in children and adolescents. Clinical phenomenology of 70 consecutive cases. *Arch Gen Psychiatry*, **46**, 335–41.

Swedo, S. E., Schapiro, M. B., Grady, C. L., et al. (1989b) Cerebral glucose metabolism in childhood-onset obsessive-compulsive disorder. *Arch Gen Psychiatry*, **46**, 518–23.

Szeszko, P. R., MacMillan, S., McMeniman, M., et al. (2004a) Brain structural abnormalities in psychotropic drug-naive pediatric patients with obsessive-compulsive disorder. *Am J Psychiatry*, **161**, 1049–56.

Szeszko, P. R., MacMillan, S., McMeniman, M., et al. (2004b) Amygdala volume reductions in pediatric patients with obsessive-compulsive disorder treated with paroxetine: preliminary findings. *Neuropsychopharmacology*, **29**, 826–32.

Szeszko, P. R., Robinson, D., Alvir, J. M. J., et al. (1999) Orbital frontal and amygdala volume reductions in obsessive-compulsive disorder. *Arch Gen Psychiatry*, **56**, 913–19.

Thomsen, P. H. (2000) Obsessive-compulsive disorder: pharmacological treatment. *Eur Child Adolesc Psychiatry*, **9**, 176–84.

van den Heuvel, O. A., Veltman, D. J., Groenewegen, H. J., et al. (2004) Amygdala activity in obsessive-compulsive disorder with contamination fear: a study with oxygen-15 water positron emission tomography. *Psychiatry Res*, **132**, 225–37.

Van der Linden, M., Ceschi, G., Zermatten, A., et al. (2005) Investigation of response inhibition in obsessive-compulsive disorder using the Hayling task. *J Int Neuropsychol Soc*, **11**, 776–83.

Wichmann, T., DeLong, M. R. (1996) Functional and pathophysiological models of the basal ganglia. *Curr Opin Neurobiol*, **6**, 751–8.

Wolf, S. S., Jones, D. W., Knable, M. B., et al. (1996) Tourette syndrome: prediction of phenotypic variation in monozygotic twins by caudate nucleus D2 receptor binding. *Science*, **273**, 1225–7.

Wong, D. F., Singer, H. S., Brandt, J., et al. (1997) D2-like dopamine receptor density in Tourette syndrome measured by PET. *J Nucl Med*, **38**, 1243–7.

13

From genes to brain to behavior: the case of fragile X syndrome

Susan M. Rivera and Allan L. Reiss

Introduction

In this chapter, we show how neuroimaging can help us understand complex relationships among genetic, brain, and behavioral factors. To this aim, we will use as a model a single-gene disorder that is very well understood: fragile X syndrome. Because of the wealth of information that exists on the molecular, neuroanatomical, and behavioral aspects of this disorder, great strides have been made in understanding the complex interplay among these scientific levels of description, as well as the resulting phenotypes. This, in turn, has begun to guide treatment of the disorder in ways that are far more specific than was previously possible. While the focus throughout the chapter will be on this single-gene disorder and its phenotypic variants, we hope to use this as a "methodological roadmap" and model for understanding other disorders influenced by genetic factors.

Fragile X syndrome

Fragile X syndrome (FXS) is the most common inherited cause of mental retardation. It is caused by a trinucleotide repeat expansion $(CGG)_n$ in the $5'$ untranslated region of the fragile X mental retardation 1 gene (*FMR1*) located at Xq27.3. The "full mutation," present in individuals having more than 200 CGG repeats, involves methylation, which stops the synthesis of the *FMR1* protein (FMRP) (Fu et al., 1991; Pieretti et al., 1991; Snow et al., 1993; Verkerk et al., 1991; Yu et al., 1991). Fragile X syndrome is therefore caused by an absence or deficit of FMRP (Tassone et al., 1999). The physical features of FXS include macroorchidism (large testes), a long, narrow face and prominent ears, and mild cardiac, neuroendocrine, and connective tissue problems. However, these physical characteristics can be highly variable in cognitively affected individuals and may not be present at all in young males and females with the full mutation. Males with the full mutation typically exhibit moderate to severe mental retardation, while females as a group show less significant and more variable impairment as a result of the second, normally functioning X chromosome. The cognitive profile of FXS includes deficits in visuospatial processing and working memory, visual-motor coordination, and arithmetic skills (Baumgardner et al., 1995; Freund and Reiss, 1991; Mazzocco et al., 2006). In addition to cognitive impairment, individuals with FXS demonstrate a behavioral phenotype characterized by hyperarousal, social anxiety and withdrawal, social deficits with peers, abnormalities in communication, unusual responses to sensory stimuli, stereotypic behavior, gaze aversion, inattention, impulsivity, and hyperactivity (Bregman et al., 1988; Cohen et al., 1988, 1989, 1991; Hagerman et al., 1991; Hessl et al., 2001; Reiss and Freund, 1992;

Neuroimaging in Developmental Clinical Neuroscience, eds. Judith M. Rumsey and Monique Ernst. Published by Cambridge University Press. © Cambridge University Press 2009.

Sudhalter et al., 1990). The severity of the fragile X phenotype depends mainly on the degree of abnormal methylation of the *FMR1* gene and, in females, the degree of skewing of normal X chromosome inactivation (Martinez et al., 2005).

The association of autism with FXS has been somewhat controversial though most investigators find an increased prevalence and severity of autistic behaviors in individuals with FXS compared to IQ-matched persons with idiopathic developmental disability. For example, 25%–40% of individuals with the full mutation meet criteria for autistic disorder (Bailey et al., 1998b; Kaufmann et al., 2004; Philofsky et al., 2004; Rogers et al., 2001); however, a range of autistic symptoms is present in many individuals with FXS who do not meet full diagnostic criteria for autistic disorder.

Since the identification of the gene responsible for FXS in 1991 (Pieretti et al., 1991), an explosion of research has emerged investigating the relationships between molecular variables, behavior, and the brain in FXS. In understanding these relationships, it is important to highlight the complex interplay between various molecular variables. The repeat size and methylation-dependent expression of both messenger RNA (mRNA) and FMRP protein are known to directly influence outcomes, including cognitive function. This is true in both the full mutation (defined by more than 200 CGG repeats) and the "premutation," which is defined by ~50–200 CGG repeats (Allen et al., 2005; Kaufmann et al., 1999; Koukoui and Chaudhuri, 2007). Furthermore, the *FMR1* gene gives rise to at least two distinct molecular pathogenic mechanisms (protein deficiency vs. RNA toxicity) and attendant neurochemical processes, depending on the size of the CGG repeat and the sex of the affected individual. It is therefore more useful to think of a spectrum of involvement beginning with individuals with the full fragile X mutation, where FMRP is generally low or absent, with a gene dose–response curve in females as a consequence of variable X chromosomal activation (fraction of normal X allele active). Next on this continuum are individuals who are mosaic for mutations in the *FMR1* gene (whereby some cells

express FMRP and others do not). Depending on the amount of FMRP being expressed, these individuals can exhibit varying severity of the characteristic FXS full-mutation phenotype. Carriers of the premutation (with ~50–200 CGG repeats and absence of aberrant methylation) typically express normal levels of FMRP, but those in the upper portion of the premutation range appear to be at risk for exhibiting lower levels of FMRP and higher than normal *FMR1* mRNA (Tassone et al., 2000a).

In association with the molecular phenomenon of excess *FMR1* mRNA, one could add to this continuum a recently defined late-onset progressive neurologic disorder that has been reported in some older men with the fragile X premutation (Berry-Kravis et al., 2003; Hagerman et al., 2001, 2004; Hall et al., 2005; Jacquemont et al., 2003, 2004). This syndrome has been termed fragile X-associated tremor/ataxia syndrome (FXTAS). Symptoms of FXTAS include intention tremor, gait ataxia, neuropathy, parkinsonian features, cognitive decline, and dementia. The pathogenesis of FXTAS is thought to result from overexpression and toxicity of *FMR1* mRNA (Jacquemont et al., 2007) (see Figure 13.1). As a result of these characteristics, fragile X provides a unique model for developing a "molecules to mind" explanation of a neurogenetic disorder that can then be used to generate hypotheses about the genetic basis of disorders with less clear molecular mechanisms.

FMR1 protein and brain development

Fragile X mental retardation protein is found in both the dendrites and synapses of neurons (Devys et al., 1993; Feng et al., 1997) where it is predominantly associated with actively translating ribosomes during protein synthesis (Khandjian et al., 1996). During normal development, FMRP is produced at synapses in response to synaptic activation, and it has been found to be increased in the brain undergoing active synaptogenesis in response to motor learning or enriched environments (Irwin et al., 2005). In individuals with FXS, reductions or absence of FMRP cause developmental changes at the neuronal

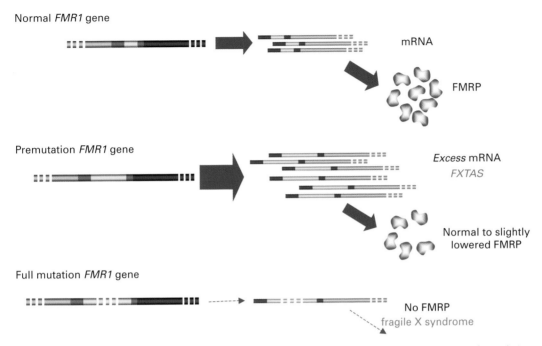

Normal *FMR1* gene

mRNA

FMRP

Premutation *FMR1* gene

Excess mRNA
FXTAS

Normal to slightly
lowered FMRP

Full mutation *FMR1* gene

No FMRP
fragile X syndrome

Figure 13.1. The relationship between fragile X mental retardation (*FMR1*) gene activity, fragile X mental retardation protein (FMRP) production, and molecular pathogenic mechanisms in fragile X syndrome. In a normal gene (<55 CGG repeats), mRNA leads to the production of normal amounts of *FMR1* protein (FMRP). In the FX premutation (55–200 CGG repeats), an excess level of messenger RNA (mRNA) is produced, resulting in normal to slightly lowered FMRP and potentially leading to the adult neurological disorder of fragile X-associated tremor/ataxia syndrome (FXTAS). In individuals with the full mutation (>200 CGG repeats), the absence of mRNA and FMRP leads to the developmental disorder fragile X syndrome.

level, predominantly impairments in spine maturation and a failure of normal synaptic pruning. Indeed, evidence of the deleterious effects of suboptimal levels of FMRP on the structure and function of both dendrites and synapses exists from studying human postmortem tissue (Hinton et al., 1991; Irwin et al., 2000; Rudelli et al., 1985) and from observing cortical neurons of an *FMR1*-knockout mouse (Braun and Segal, 2000; Oostra and Hoogeveen, 1997; Pieretti et al., 1991). Consistent with the abnormal neuron phenotypes found in both fragile X patients and FMRP-deficient mice, several FMRP mRNA targets that encode proteins involved in axon guidance or synaptic functions have been identified using microarrays (Brown et al., 2001). These neurodevelopmental processes lead to both structural and

functional irregularities that can be visualized using brain imaging methodologies.

In what follows we will review findings that demonstrate the "genes to brain to behavior" approach by combining molecular with either behavioral or brain imaging research in the full fragile X mutation, premutation, and FXTAS phenotypes that exist as part of the spectrum mentioned above.

From genes to behavior

Findings from studies of the full mutation

A large number of published studies have shown that FMRP depletion is significantly related to global cognitive deficits and behavioral problems, both in males and females with the full mutation (Bailey et al., 1998a; Dyer-Friedman et al., 2002; Kaufmann et al., 1999; Tassone et al., 1999). The results from a longitudinal study of young males with the fragile X full mutation showed that FMRP level is significantly related to the level of cognitive-behavioral development assessed by the Battelle Developmental Inventory (Bailey et al., 2001). Furthermore, in research comparing fully methylated versus partially methylated (mosaic) males, those who were fully methylated were found to be more likely to show a decrease in IQ over time (Merenstein et al., 1996). Even more specifically, Wright-Talamante et al. (1996) reported that there was no significant IQ decline in young males with less than 50% methylation of the full mutation, suggesting that a small to moderate amount of FMRP production partially protects against significant IQ decline. Loesch et al. (2004) also demonstrated a strong relationship between FMRP depletion and overall cognitive deficit, as well as specific deficits in processing speed, short-term memory, and the ability to control attention, especially in the context of regulating goal-directed behavior, in subjects with the fragile X full mutation. With respect to behavior, a common and significant problem observed in many males with FXS is the tendency to demonstrate autonomic hyperarousal in the face of environmental stressors, particularly in social contexts. Hyperarousal in FXS is manifest as overt symptoms of anxiety, turning away of the face and body from others, stereotypic motor and language characteristics, and attempts to escape from the stressful conditions. As might be surmised from this description, such behaviors can be a detriment to the establishment of developmentally appropriate peer relationships. Motoric restlessness and impulsive behavior are also quite common in males with FXS, particularly during the preschool and early school-age years.

In females with the full mutation, strong evidence has been demonstrated for the relationships between specific cognitive scores and the activation ratio – the ratio of affected/unaffected activated X chromosomes, which is highly correlated with

FMRP (Abrams et al., 1994; Reiss et al., 1995b). For example, in a study of molecular and phenotypic correlations in females with fragile X, it was found that the X inactivation ratio was strongly and positively correlated with a composite measure of executive function (Sobesky et al., 1996), suggesting that these essential cognitive skills are especially sensitive to levels of FMRP. Like males with FXS, females with the full mutation are also at risk for behavioral difficulties, though manifestations of hyperarousal and hyperactivity may be less severe.

Findings from studies of the premutation

As described in the section above, the association of cognitive and behavioral dysfunction with the molecular finding of reduced FMRP has been clearly established. In contrast, there is far less certainty about molecular or brain mechanisms that may put individuals with the fragile X premutation at higher risk for cognitive and behavioral dysfunction. Further complicating this area of investigation is an increasing awareness that the premutation should not be considered a homogeneous molecular diagnostic category. In particular, the concept of a "continuum" of effects may apply to individuals with the premutation as well as to the entire spectrum of effects associated with FMR1 mutations. Finally, environmental influences may be particularly relevant for individuals with the premutation. The great majority of mothers of children with the full mutation carry the premutation. Thus, caregiver stress and burden related to having one or more children with serious developmental disability come into play when considering the assessment of psychological outcomes in this group.

Not surprisingly, findings pertaining to cognitive impairment and molecular variables in individuals with the premutation have been inconsistent. Many studies have shown no differences in neuropsychological or behavioral profiles between premutation carriers and non-carriers (Franke et al., 1998; Johnston et al., 2001; Kaufmann et al., 1999; Myers et al., 2001; Reiss et al., 1993). Other studies of both men and women suggest that some individuals with the premutation may demonstrate subtle, yet detectable, neurocognitive problems (Cornish et al., 2005; Loesch et al., 2003a, 2003b; Moore et al., 2004a). However, the aforementioned studies failed to demonstrate a correlation between severity of cognitive impairment and CGG repeat length.

Perhaps the most persuasive evidence of cognitive involvement in females with the fragile X premutation comes from a recent study by Allen et al. (2005). This study utilized a large sample size (66 males and 217 females) and used *FMR1* repeat size as a continuous variable, rather than using a dichotomous designation of premutation versus full mutation. Results indicated a small, yet significant negative effect from increasing CGG repeat on verbal IQ, explaining approximately 4% of the variance in this measure.

Fragile X-associated tremor/ataxia syndrome findings

A number of studies have now been conducted on the progressive neurologic syndrome associated with the fragile X premutation known as FXTAS, in which associations between molecular factors and behavior have been demonstrated as well. Many of the features of FXTAS, both neuropathologic and radiologic, have been shown to be correlated with CGG repeat length in males. For example, in male premutation carriers with FXTAS, increasing numbers of intranuclear inclusions in neuronal and astrocytic cells have been observed with increasing CGG repeat length (Greco et al., 2006). The fact that elevated *FMR1* mRNA has been found in peripheral blood leukocytes of carriers (Tassone et al., 2000b, 2000c) coupled with findings of the presence of *FMR1* mRNA within the nuclear inclusions in FXTAS brains (Tassone et al., 2004), supports an RNA toxic gain-of-function model for FXTAS pathogenesis (for a review, see: Hagerman and Hagerman 2004). This RNA toxic gain-of-function mechanism, in which the degree of clinical involvement increases with increasing CGG repeat length, also predicts that those patients with larger repeat sizes will show an earlier onset of clinical involvement. This hypothesis has been supported by work from Tassone et al. (2007), who observed highly significant correlations between the ages of onset of both tremor and ataxia symptoms and the size of the CGG repeat.

What we have summarized above represents information that has been gained from comparing molecular variables with measures of behavior and symptomatology. Linking these variables has been invaluable in furthering our understanding of the phenotypic consequences of these genetic anomalies. Brain imaging technology has further allowed researchers to make these linkages even more specific by relating molecular variables to variations in brain morphology. In the next section we will summarize these findings, again delineating what has been discovered across the full spectrum of fragile X involvement.

From genes to brain

Findings from studies of the full mutation

There has been a great deal of work employing structural imaging techniques in individuals with the full mutation. A number of structural abnormalities have been observed in this group, including hypoplasia of the cerebellar vermis, increased size of the fourth and lateral ventricles (Eliez et al., 2001; Franke et al., 1998; Johnston et al., 2001; Kaufmann et al., 1999; Mostofsky et al., 1998; Reiss et al., 1988, 1991, 1993, 1995a), larger caudate nuclei, and significantly increased thalamic volume in girls (Eliez et al., 2001; Reiss et al., 1995a). In addition, white matter connectivity has been assessed in FXS using diffusion tensor imaging (DTI). Compared with controls, subjects with FXS demonstrate evidence of aberrant white matter structure (reduced fractional anisotropy), mostly in fronto-striatal and parietal sensorimotor tracts (Barnea-Goraly et al., 2003). This finding suggests that low levels of FMRP may contribute to morphological changes in white matter tracts, possibly due to an influence on neuronal growth and targeting as a result of reduced or absent FMRP.

A recent volumetric neuroimaging study examined children (age 2–7) with the fragile X full mutation, mosaicism, and control groups of children with developmental delay or Down syndrome (Kates et al., 2002). This study reported relative reductions in temporal lobe gray matter, along with relative enlargement of parietal white matter volume, the latter of which was seen only in individuals with FXS and not in control groups with either developmental language delay or Down syndrome. Interestingly, the parietal white matter enlargement was seen only in participants with the full mutation and not in a group with mosaicism. This is a strong indicator that the reduction or absence of FMRP in the full mutation group was responsible for this enlargement, possibly corresponding to maturational and synaptic pruning failures.

Findings from studies of the premutation

Similar to studies examining potential cognitive effects, there are some data suggesting that the fragile X premutation (independent of FXTAS – see next section) may be associated with variations in brain morphology. For example, a brain magnetic resonance imaging (MRI) study (Moore et al., 2004b) examining gray matter density in 20 male premutation carriers and 20 age- and IQ-matched controls found significantly reduced gray matter density in several regions, including the cerebellum, amygdalo-hippocampus complex, and thalamus in the premutation group. Within this group, increased age, increased CGG repeat size, and decreases in the percentage of blood lymphocytes expressing FMRP were associated with decreased gray matter density in the amygdalo-hippocampus complex. Though a significant association between *FMR1* mRNA and brain morphology was not observed in this study, the fact that CGG repeat size and FMRP were correlated with brain structure supports a putative gene–brain–behavior mechanism of clinical involvement in male premutation carriers. More research spanning these domains is needed to establish whether the mechanism of involvement is analogous to FXS and involves

reduced FMRP (for example in premutation carriers with high repeat number), is analogous to FXTAS and involves toxic elevation of *FMR1* mRNA, or whether there are multiple genetic and environmental influences on brain function and behavior in this group.

Fragile X-associated tremor/ataxia syndrome findings

Another controlled study of adult male premutation carriers, in this case with and without FXTAS, involved a molecular analysis of *FMR1* expression, quantitative neuroimaging, and cognitive testing (Cohen et al., 2006). The study reported significant whole-brain, cerebrum, and cerebellar volume loss, as well as increases in whole-brain white matter hyperintensity volume associated with FXTAS. These changes correlated with CGG repeat number and became more severe with age. Associations were also observed between CGG repeat length and cognitive ability in the premutation carriers, including the sample without FXTAS, suggesting that molecular abnormalities may contribute to cognitive decline prior to manifestation of obvious structural abnormalities.

With these studies we have come a long way in understanding, not only the phenotypic consequences of this single-gene disorder, but how these factors relate to abnormalities in specific brain regions. What follows is a review of studies that bring this relationship "full circle" to understanding how gene alterations lead to specific brain abnormalities, which in turn result in behaviors and symptoms related to the phenotypes expressed.

From genes to brain to behavior

Findings from studies of the full mutation

There are a number of examples in the literature, across the fragile X phenotypes that we have been discussing, that have demonstrated direct links between genetic factors, localized brain function,

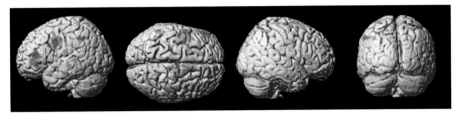

Figure 13.2. An example of the relationship between genes, brain and behavior. Brain areas (prefrontal and parietal) which show, for participants with fragile X syndrome, a significant correlation between FMRP and brain activation for 3-operand arithmetic equations. Adapted from Rivera et al., 2002, *Human Brain Mapping,* **16** (4), 206–218.

and the ensuing cognitive impairments associated with dysfunction occurring in those brain regions. Several functional MRI (fMRI) studies have now demonstrated a "dose–response" effect of FMRP on brain activation. One such study examined the neural substrate of visuospatial working memory in females with FXS using standard 1-back and 2-back tasks (Kwon et al., 2002). Behaviorally, subjects with the full mutation performed significantly worse on the more difficult, 2-back task than did age-matched controls. In terms of brain activation, comparison subjects showed a significant increase in the inferior frontal gyrus, middle frontal gyrus, superior parietal lobule, and supramarginal gyrus on the 2-back compared to the 1-back task, while subjects with FXS showed no change in activation between the two. Furthermore, molecular measures correlated with brain activation on this task since significant correlations were found during the 2-back task, between FMRP expression and activation in the right inferior and bilateral middle frontal gyri and the bilateral supramarginal gyri.

In an fMRI study of mental arithmetic in females with the full mutation, Rivera et al. (2002) found that, in response to increasing arithmetic complexity (i.e., going from 2-operand to 3-operand addition and subtraction problems), participants with FXS did not recruit the prefrontal-parietal-cerebellar network known to be involved in arithmetic processing in unaffected participants. With respect to molecular measures, this investigation showed that as levels of FMRP increased in individuals with FXS, so did task-related activation in areas that are involved in arithmetic processing in typically developing subjects, providing evidence of a direct relationship between decreased FMRP expression and impairments in mental arithmetic performance in persons with FXS (see Figure 13.2).

Menon et al. (2004) used fMRI with a response inhibition task (go/no-go) in 10- to 22-year-old females with the full mutation and age- and gender-matched typically developing controls. Although behavioral performance on the go/no-go task was equivalent in the two groups, females with FXS showed abnormal activation patterns in several cortical and subcortical regions, with significantly reduced activation in the supplementary motor area, anterior cingulate and midcingulate cortex, basal ganglia, and hippocampus. The investigators also found neural responses in the right ventrolateral prefrontal cortex (PFC) and bilateral striatum that correlated with the level of *FMR1* gene expression. In addition to task-related activation impairments, reduced levels of "deactivation" were observed in the ventromedial PFC, and, furthermore, these reductions were correlated with the level of FMRP. As a whole, these results provide direct evidence that decreased FMRP expression underlies impairments in cognitive performance in persons with the full mutation.

Findings from studies of the premutation

Potential gene–brain–behavior relationships are also beginning to emerge in studies of those with the fragile X premutation. Recently, Hessl and colleagues reported findings from an fMRI study of amygdala function in 12 adult men with the premutation (who did not exhibit clinical evidence of FXTAS) who were compared to a group of 13 premutation-negative men who were matched on age and IQ (Hessl et al., 2007). When viewing fearful facial expressions compared to viewing scrambled faces (fear-control contrast), the premutation group showed less overall activation as well as significantly different patterns of activation compared to controls. The control group showed strong activation in the superior temporal sulcus (STS) bilaterally, left and right lateral orbitofrontal gyrus, bilateral insula, and amygdala. These areas, usually associated with social cognition or emotion processing, were not activated in the premutation group. Follow-up region-of-interest (ROI) analyses confirmed that premutation carriers failed to activate the amygdala, whereas the control group showed robust bilateral amygdala activation. In the premutation group, neither CGG repeat length nor *FMR1* mRNA was significantly associated with amygdala activation; however, we did find that these measures were negatively associated with left insular activation during this task. Though the male premutation participants in this study (average age of 43 years) did not demonstrate overt symptoms of FXTAS, the findings of aberrant brain activation in this group might reflect presymptomatic brain changes associated with elevated mRNA instead of, or in addition to, specific pathogenic effects on the brain associated with the premutation.

Fragile X-associated tremor/ataxia syndrome findings

We have also recently completed fMRI studies in males with FXTAS using tasks involving the cerebellum, as well as prefrontal and parietal cortices (Rivera et al., unpublished data). The results of this study show a dissociation between cerebellar activity for a motor timing task and a cognitive, mental arithmetic task. Relative to controls, premutation carriers exhibit *hyperactivation* of the cerebellum (particularly more inferior/posterior and contralateral regions) while performing a motor timing task, and *hypoactivation* of the cerebellum during simple mental arithmetic. This dissociation suggests cerebellar dysfunction that is more than just a diminished capacity for functional activation and more specifically points to potential neuropathogenic mechanisms for this dysfunction.

Therapeutic advances in fragile X syndrome

Advances in our understanding of the development of FXS have led to a number of targeted therapeutic treatments in FXS. One example of these advances is the metabotropic glutamate receptor 5 (mGluR) theory of FXS, which posits that exaggerated signaling in mGluR pathways may underlie many of the cognitive, behavioral, and neurological symptoms of FXS (Bear et al., 2004). In the absence of FMRP, excessive mGluR-mediated dendritic translation is predicted to lead to excessive internalization of α-amino-3-hydroxy-5-methylisoxazole-4-proprionic acid (AMPA) receptors, excessive synaptic weakening, and the structurally immature-appearing elongated dendritic processes, which have been documented in both the *FMR1*-knockout mouse and in post-mortem brain tissue of humans with FXS. These insights into the defects in synaptic integrity and plasticity in FXS have led to the proposal of several pharmacotherapeutic targets in FXS to attempt to normalize synaptic connectivity, including AMPA receptor activation. One such compound is an AMPA receptor-positive modulator (ampakine). Clinical trials of ampakine, which can enhance synaptic strength and may partially correct the synaptic transmission defect in FXS, are now ongoing (Berry-Kravis et al., 2006). The hope is that this treatment can lead to improvement in cognitive and behavioral functioning in individuals with FXS.

Likewise, fMRI studies have also begun to guide therapeutic treatments for FXS. For example, imaging studies have indicated that the basal forebrain and hippocampus show significantly reduced activation during a memory encoding task (Greicius et al., 2004). These brain areas are ones in which the neurochemical acetylcholine is found in high concentrations and in which the highest *FMR1* transcription is found (Abitbol et al., 1993). Such findings have led to clinical trials of the medication donepezil (Kessler et al., under review), which has been shown to enhance acetylcholine function in the brain, to determine whether the compound will have a beneficial effect on behavior or cognition in individuals with FXS.

Conclusions

In this chapter, we have used the model of FXS, a single-gene disorder that has a range of phenotypic variants, to demonstrate a gene-to-brain-to-behavior approach in understanding neuropathological development. Because knowledge of the specific molecular basis and the neurobiology of fragile X has grown tremendously, it also represents an important genetic model for other neurodevelopmental disorders. Symptomatic commonalities among FXS and other pervasive developmental disorders such as autism and Rett syndrome may reflect an overlap in underlying neural circuits and pathways and hence shared pathophysiological mechanisms. Therefore, the possibility exists that new therapeutics developed to treat FXS also may have efficacy in treating individuals with these other disorders. Autism, for example, occurs in approximately 30% of children with FXS with an additional 20% meeting the criterion for pervasive developmental disorder-not otherwise specified (PDD-NOS) (Hatton et al., 2006; Kaufmann et al., 2004; Rogers et al., 2001). The remaining 50% of children with FXS who do not meet criteria for autism spectrum disorders often exhibit autistic symptoms including poor eye contact, unusual hand mannerisms such as hand flapping, and tactile defensiveness. Because 2%–6% of individuals with autism will have the fragile X mutation (Persico and Bourgeron, 2006; Reddy, 2005; Wassink et al., 2001), FXS is the most common known single-gene disorder associated with autism at this time. Targeted treatments for the neuropathology and neurobiological abnormalities of FXS may

thus turn out to be helpful in treating autism spectrum disorders. Therein lies the promise of a truly successful roadmap for the "molecules to mind" brand of translational research, in which converging research in molecular, behavioral and neuroscience disciplines, across multiple model systems, will lead us in the direction of new therapeutics for complex human diseases.

REFERENCES

Abitbol, M., Menini, C., Delezoide, A. L., Rhyner, T., Vekemans, M., and Mallet, J. (1993) Nucleus basalis magnocellularis and hippocampus are the major sites of FMR-1 expression in the human fetal brain. *Nat Genet*, **4**(2), 147–53.

Abrams, M. T., Reiss, A. L., Freund, L. S., Baumgardner, T. L., Chase, G. A., and Denckla, M. B. (1994) Molecular-neurobehavioral associations in females with the fragile X full mutation. *Am J Med Genet*, **51**(4), 317–27.

Allen, E. G., Sherman, S., Abramowitz, A., et al. (2005) Examination of the effect of the polymorphic CGG repeat in the FMR1 gene on cognitive performance. *Behav Genet*, **35**(4), 435–45.

Bailey, D. B., Jr., Hatton, D. D., and Skinner, M. (1998a) Early developmental trajectories of males with fragile X syndrome. *Am J Ment Retard*, **103**(1), 29–39.

Bailey, D. B., Jr., Hatton, D. D., Tassone, F., Skinner, M., and Taylor, A. K. (2001) Variability in FMRP and early development in males with fragile X syndrome. *Am J Men Retard*, **106**(1), 16–27.

Bailey, D. B., Jr., Mesibov, G. B., Hatton, D. D., Clark, R. D., Roberts, J. E., and Mayhew, L. (1998b) Autistic behavior in young boys with fragile X syndrome. *J Autism Dev Disord*, **28**(6), 499–508.

Barnea-Goraly, N., Eliez, S., Hedeus, M., et al. (2003) White matter tract alterations in fragile X syndrome: preliminary evidence from diffusion tensor imaging. *Am J Med Genet B Neuropsychiatr Genet*, **118**(1), 81–8.

Baumgardner, T., Reiss, A., Freund, L. S., and Abrams, M. T. (1995) Specification of the neurobehavioral phenotype in males with fragile X syndrome. *Pediatrics*, **95**(5), 744–752.

Bear, M. F., Huber, K. M., and Warren, S. T. (2004) The mGluR theory of fragile X mental retardation. *Trends Neurosci*, **27**(7), 370–7.

Berry-Kravis, E., Krause, S. E., Block, S. S., et al. (2006) Effect of CX516, an AMPA-modulating compound, on cognition and behavior in fragile X syndrome: a controlled trial. *J Child Adolesc Psychopharmacol*, **16**(5), 525–40.

Berry-Kravis, E., Lewin, F., Wuu, J., et al. (2003) Tremor and ataxia in fragile X premutation carriers: blinded videotape study. *Ann Neurol*, **53**(5), 616–23.

Braun, K. and Segal, M. (2000) FMRP involvement in formation of synapses among cultured hippocampal neurons. *Cereb Cortex*, **10**(10), 1045–52.

Bregman, J. D., Leckman, J. F., and Ort, S. I. (1988) Fragile X syndrome: genetic predisposition to psychopathology. *J Autism Dev Disord*, **18**(3), 343–54.

Brown, V., Jin, P., Ceman, S., et al. (2001) Microarray identification of FMRP-associated brain mRNAs and altered mRNA translational profiles in fragile X syndrome. *Cell*, **107**(4), 477–87.

Cohen, I. L., Brown, W. T., Jenkins, E. C., et al. (1989) Fragile X syndrome in females with autism. *Am J Med Genet*, **34**(2), 302–3.

Cohen, I. L., Fisch, G. S., Sudhalter, V., et al. (1988) Social gaze, social avoidance, and repetitive behavior in fragile X males: a controlled study. *Am J Ment Retard*, **92**(5), 436–46.

Cohen, I. L., Sudhalter, V., Pfadt, A., Jenkins, E. C., Brown, W. T., and Vietze, P. M. (1991) Why are autism and the fragile-X syndrome associated? Conceptual and methodological issues. *Am J Hum Genet*, **48**(2), 195–202.

Cohen, S., Masyn, K., Adams, J., et al. (2006) Molecular and imaging correlates of the fragile X-associated tremor/ataxia syndrome. *Neurology*, **67**(8), 1426–31.

Cornish, K., Kogan, C., Turk, J., et al. (2005) The emerging fragile X premutation phenotype: evidence from the domain of social cognition. *Brain Cogn*, **57**(1), 53–60.

Devys, D., Lutz, Y., Rouyer, N., Bellocq, J. P., and Mandel, J. L. (1993) The FMR-1 protein is cytoplasmic, most abundant in neurons and appears normal in carriers of a fragile X premutation. *Nat Genet*, **4**(4), 335–40.

Dyer-Friedman, J., Glaser, B., Hessel, D., et al. (2002) Genetic and environmental influences on the cognitive outcomes of children with fragile X syndrome. *J Am Acad Child Adolesc Psychiatry*, **41**(3), 237–44.

Eliez, S., Blasey, C. M., Freund, L. S., Hastie, T., and Reiss, A. L. (2001) Brain anatomy, gender and IQ in children and adolescents with fragile X syndrome. *Brain*, **124**(Pt 8), 1610–18.

Feng, Y., Gutekunst, C. A., Eberhart, D. E., Yi, H., Warren, S. T., and Hersch, S. M. (1997) Fragile X mental retardation protein: nucleocytoplasmic shuttling and association with somatodendritic ribosomes. *J Neurosci*, **17**(5), 1539–47.

Franke, P., Leboyer, M., Gänsicke, M., et al. (1998) Genotype-phenotype relationship in female carriers of the premutation and full mutation of FMR-1. *Psychiatry Res*, **80**(2), 113–27.

Freund, L. S. and Reiss, A. L. (1991) Cognitive profiles associated with the fra(X) syndrome in males and females. *Am J Med Genet*, **38**(4), 542–7.

Fu, Y. H., Kuhl, D. P., Pizzuti, A., et al. (1991) Variation of the CGG repeat at the fragile X site results in genetic instability: resolution of the Sherman paradox. *Cell*, **67**(6), 1047–58.

Greco, C. M., Berman, R. F., Martin, R. M., et al. (2006) Neuropathology of fragile X-associated tremor/ataxia syndrome (FXTAS). *Brain*, **129**(Pt 1), 243–55.

Greicius, M. D., Boyett-Anderson, J. M., Menon, V., and Reiss, A. L. (2004) Reduced basal forebrain and hippocampal activation during memory encoding in girls with fragile X syndrome. *NeuroReport*, **15**(10), 1579–83.

Hagerman, P. J. and Hagerman, R. J. (2004) The fragile-X premutation: a maturing perspective. *Am J Hum Genet*, **74**(5), 805–16.

Hagerman, R. J., Amiri, K., and Cronister, A. (1991) Fragile X checklist. *Am J Med Genet*, **38**(2–3), 283–7.

Hagerman, R. J., Leavitt, B. R., Farzin, F., et al. (2004) Fragile-X-associated tremor/ataxia syndrome (FXTAS) in females with the FMR1 premutation. *Am J Hum Genet*, **74**(5), 1051–6.

Hagerman, R. J., Leehey, M., Heinrichs, W., et al. (2001) Intention tremor, parkinsonism, and generalized brain atrophy in male carriers of fragile X. *Neurology*, **57**(1), 127–30.

Hall, D. A., Berry-Kravis, E., Jacquemonts, S., et al. (2005) Initial diagnoses given to persons with the fragile X associated tremor/ataxia syndrome (FXTAS). *Neurology*, **65**(2), 299–301.

Hatton, D. D., Sideris, J., Skinner, M., et al. (2006) Autistic behavior in children with fragile X syndrome: prevalence, stability, and the impact of FMRP. *Am J Med Genet A*, **140**(17), 1804–13.

Hessl, D., Dyer-Friedman, J., Glaser, B., et al. (2001) The influence of environmental and genetic factors on behavior problems and autistic symptoms in boys and girls with fragile X syndrome. *Pediatrics*, **108**(5), E88.

Hessl, D., Rivera, S., Koldewyn, K., et al. (2007) Amygdala dysfunction in men with the fragile X premutation. *Brain*, **130**(Pt 2), 404–16.

Hinton, V. J., Brown, W. T., Wisniewski, K., and Rudelli, R. D. (1991) Analysis of neocortex in three males with the fragile X syndrome. *Am J Med Genet*, **41**(3), 289–94.

Irwin, S. A., Christmon, C. A., Grossman, A. W., et al. (2005) Fragile X mental retardation protein levels increase following complex environment exposure in rat brain regions undergoing active synaptogenesis. *Neurobiol Learn Mem*, **83**(3), 180–7.

Irwin, S. A., Galvez, R., and Greenough, W. T. (2000) Dendritic spine structural anomalies in fragile-X mental retardation syndrome. *Cereb Cortex*, **10**(10), 1038–44.

Jacquemont, S., Hagerman, R. J., Leehey, M. A., et al. (2003) Fragile X premutation tremor/ataxia syndrome: molecular, clinical, and neuroimaging correlates. *Am J Hum Genet*, **72**(4), 869–78.

Jacquemont, S., Hagerman, R. J., Leehey, M. A., et al. (2004) Penetrance of the fragile X-associated tremor/ataxia syndrome in a premutation carrier population. *JAMA*, **291**(4), 460–9.

Jacquemont, S., Hagerman, R. J., Hageman, P. J., and Leehey, M. A. (2007) Fragile-X syndrome and fragile X-associated tremor/ataxia syndrome: two faces of FMR1. *Lancet Neurol*, **6**(1), 45–55.

Johnston, C., Eliez, S., Dyer-Friedman, J., et al. (2001) Neurobehavioral phenotype in carriers of the fragile X premutation. *Am J Med Genet*, **103**(4), 314–19.

Kates, W. R., Folley, B. S., Lanham, D. C., Capone, G. T., and Kaufmann, W. E. (2002) Cerebral growth in Fragile X syndrome: review and comparison with Down syndrome. *Microsc Res Tech*, **57**(3), 159–67.

Kaufmann, W. E., Abrams, M. T., Chen, W., and Reiss, A. L. (1999) Genotype, molecular phenotype, and cognitive phenotype: correlations in fragile X syndrome. *Am J Med Genet*, **83**(4), 286–95.

Kaufmann, W. E., Cortell, R., Kau, A. S. et al. (2004) Autism spectrum disorder in fragile X syndrome: communication, social interaction, and specific behaviors. *Am J Med Genet*, **129A**(3), 225–34.

Khandjian, E. W., Corbin, F., Woerly, S. and Rousseau, F. (1996) The fragile X mental retardation protein is associated with ribosomes. *Nat Genet*, **12**(1), 91–3.

Koukoui, S. D. and Chaudhuri, A. (2007) Neuroanatomical, molecular genetic, and behavioral correlates of fragile X syndrome. *Brain Res Rev*, **53**(1), 27–38.

Kwon, H., Reiss, A. L., and Menon, V. (2002) Neural basis of protracted developmental changes in visuo-spatial working memory. *Proc Natl Acad Sci U S A*, **99**(20), 13336–41.

Loesch, D. Z., Bui, Q. M., Grigsby, J., et al. (2003a) Effect of the fragile X status categories and the fragile X mental retardation protein levels on executive functioning in males and females with fragile X. *Neuropsychology*, **17**(4), 646–57.

Loesch, D. Z., Huggins, R. M., Bui, Q. M., et al. (2003b) Effect of fragile X status categories and FMRP deficits on cognitive profiles estimated by robust pedigree analysis. *Am J Med Genet A*, **122**(1), 13–23.

Loesch, D. Z., Huggins, R. M., and Hagerman, R. J. (2004) Phenotypic variation and FMRP levels in fragile X. *Ment Retard Dev Disabil Res Rev*, **10**(1), 31–41.

Martinez, R., Bonilla-Henao, V., Jiménez, A., et al. (2005) Skewed X inactivation of the normal allele in fully mutated female carriers determines the levels of FMRP in blood and the fragile X phenotype. *Mol Diagn*, **9**(3), 157–62.

Mazzocco, M. M., Singh Bhatia, N., and Lesniak-Karpiak, K. (2006) Visuospatial skills and their association with math performance in girls with fragile X or Turner syndrome. *Child Neuropsychol*, **12**(2), 87–110.

Menon, V., Leroux, J., White, C. D., and Reiss, A. L. (2004) Frontostriatal deficits in fragile X syndrome: relation to FMR1 gene expression. *Proc Natl Acad Sci U S A*, **101**(10), 3615–20.

Merenstein, S. A., Sobesky, W. E., Taylor, A. K., Riddle, J. E., Tran, H. X., and Hagerman, R. J. (1996) Molecular-clinical correlations in males with an expanded FMR1 mutation. *Am J Med Genet*, **64**(2), 388–94.

Moore, C. J., Daly, E. M., Schmitz, N., et al. (2004a) A neuropsychological investigation of male premutation carriers of fragile X syndrome. *Neuropsychologia*, **42**(14), 1934–47.

Moore, C. J., Daly, E. M., Tassone, F., et al. (2004b) The effect of pre-mutation of X chromosome CGG trinucleotide repeats on brain anatomy. *Brain*, **127**(Pt 12), 2672–81.

Mostofsky, S. H., Mazzocco, M. M., Aakalu, G., Warsofsky, I. S., Denckla, M. B., and Reiss, A. L. (1998) Decreased cerebellar posterior vermis size in fragile X syndrome: correlation with neurocognitive performance. *Neurology*, **50**(1), 121–30.

Myers, G. F., Mazzocco, M. M., Maddalena, A., and Reiss, A. L. (2001) No widespread psychological effect of the fragile X premutation in childhood: evidence from a preliminary controlled study. *J Dev Behav Pediatr*, **22**(6), 353–9.

Oostra, B. A. and Hoogeveen, A. T. (1997) Animal model for fragile X syndrome. *Ann Med*, **29**(6), 563–7.

Persico, A. M. and Bourgeron, T. (2006) Searching for ways out of the autism maze: genetic, epigenetic and environmental clues. *Trends Neurosci*, **29**(7), 349–58.

Philofsky, A., Hepburn, S. L., Hayes, A., Hagerman, R., and Rogers, S. J. (2004) Linguistic and cognitive functioning and autism symptoms in young children with fragile X syndrome. *Am J Ment Retard*, **109**(3), 208–18.

Pieretti, M., Zhang, F. P., Fu, Y. H. et al. (1991) Absence of expression of the FMR-1 gene in fragile X syndrome. *Cell*, **66**(4), 817–22.

Reddy, K. S. (2005) Cytogenetic abnormalities and fragile-X syndrome in Autism Spectrum Disorder. *BMC Med Genet*, **6**, 3.

Reiss, A. L., and Freund L. (1992) Behavioral phenotype of fragile X syndrome: DSM-III-R autistic behavior in male children. *Am J Med Genet*, **43**(1–2), 35–46.

Reiss, A. L., Abrams, M. T., Greenlaw, R., Freund, L., and Denckla, M. B. (1995a) Neurodevelopmental effects of the FMR-1 full mutation in humans. *Nat Med*, **1**(2), 159–67.

Reiss, A. L., Freund, L., Abrams, M. T., Boehm, C., and Kazazian, H. (1993) Neurobehavioral effects of the fragile X premutation in adult women: a controlled study. *Am J Hum Genet*, **52**(5), 884–94.

Reiss, A. L., Freund, L. S., Baumgardner, T. L., Abrams, M. T., and Denckla, M. B. (1995b) Contribution of the fMR1 gene mutation to human intellectual dysfunction. *Nat Genet*, **11**(3), 331–4.

Reiss, A. L., Freund, L., Tseng, J. E., and Joshi, P. K. (1991) Neuroanatomy in fragile X females: the posterior fossa. *Am J Hum Genet*, **49**(2), 279–88.

Reiss, A. L., Patel, S., Kumar, A. J., and Freund, L. (1988) Preliminary communication: neuroanatomical variations of the posterior fossa in men with the fragile X (Martin-Bell) syndrome. *Am J Med Genet*, **31**(2), 407–14.

Rivera, S. M., Menon, V., White, C. D., Glaser, B., and Reiss, A. L. (2002) Functional brain activation during arithmetic processing in females with fragile X Syndrome is related to FMR1 protein expression. *Hum Brain Mapp*, **16**(4), 206–18.

Rogers, S. J., Wehner, E. A., and Hagerman, R. (2001) The behavioral phenotype in fragile X: Symptoms of autism in very young children with fragile X syndrome, idiopathic autism, and other developmental disorders. *J Dev Behav Pediatri*, **22**(6), 409–417.

Rudelli, R. D., Brown, W. T., Wisniewski, K., et al. (1985) Adult fragile X syndrome. Clinico-neuropathologic findings. *Acta Neuropathol*, **67**(3–4), 289–95.

Snow, K., Doud, L. K., Hagerman, R., Pergolizzi, R. G., Erster, S. H., and Thibodeau, S. N. (1993) Analysis of a CGG sequence at the FMR-1 locus in fragile X families and in the general population. *Am J Hum Genet*, **53**(6), 1217–28.

Sobesky, W. E., Taylor, A. K., Pennington, B. F., et al. (1996) Molecular/clinical correlations in females with fragile X. *Am J Med Genet*, **64**(2), 340–5.

Sudhalter, V., Cohen, I. L., Silverman, W., and Wolf-Schein, E. G. (1990) Conversational analyses of males with fragile X, Down syndrome, and autism: comparison of the emergence of deviant language. *Am J Ment Retard*, **94**(4), 431–41.

Tassone, F., Adams, J., Berry-Kravis, E. M., et al. (2007) CGG repeat length correlates with age of onset of motor signs of the fragile X-associated tremor/ataxia syndrome (FXTAS). *Am J Med Genet B Neuropsychiatr Genet*, **144**(4), 566–9.

Tassone, F., Hagerman, R. J., Chamberlain, W. D., and Hagerman, P. J. (2000c) Transcription of the FMR1 gene in individuals with fragile X syndrome. *Am J Med Genet*, **97**(3), 195–203.

Tassone, F., Hagerman, R. J., Iklé, D. N., et al. (1999) FMRP expression as a potential prognostic indicator in fragile X syndrome. *Am J Med Genet*, **84**(3), 250–61.

Tassone, F., Hagerman, R. J., Loesch, D. Z., Lachiewicz, A., Taylor, A. K., and Hagerman, P. J. (2000a) Fragile X males with unmethylated, full mutation trinucleotide repeat expansions have elevated levels of FMR1 messenger RNA. *Am J Med Genet*, **94**(3), 232–6.

Tassone, F., Hagerman, R. J., Taylor, A. K., Gane, L. W., Godfrey, T. E., and Hagerman, P. J. (2000b) Elevated levels of FMR1 mRNA in carrier males: a new mechanism of involvement in the fragile-X syndrome. *Am J Hum Genet*, **66**(1), 6–15.

Tassone, F., Iwashashi, C., and Hagerman, P. J. (2004) FMRI RNA within the intranuclear inclusions of fragile X-associated tremor/ataxia syndrome (FXTAS). *RNA Biol*, **1**(2), 103–5.

Verkerk, A. J., Pieretti, M., Sutcliffe, J. S., et al. (1991) Identification of a gene (FMR-1) containing a CGG repeat coincident with a breakpoint cluster region exhibiting length variation in fragile X syndrome. *Cell*, **65**(5), 905–14.

Wassink, T. H., Piven, J., and Patil, S. R. (2001) Chromosomal abnormalities in a clinic sample of individuals with autistic disorder. *Psychiatr Genet*, **11**(2), 57–63.

Wright-Talamante, C., Cheema, A., Riddle, J. E., Luckey, D. W., Taylor, A. K., and Hagerman, R. J. (1996) A controlled study of longitudinal IQ changes in females and males with fragile X syndrome. *Am J Med Genet*, **64**(2), 350–5.

Yu, S., Pritchard, M., Kremer, E., et al. (1991) Fragile X genotype characterized by an unstable region of DNA. *Science*, **252**(5010), 1179–81.

Alcohol exposure and the developing human brain

James M. Bjork

Introduction

This review will deal with two critical aspects of the effect of alcohol on the developing brain. The first deals with in utero exposure of the fetus to alcohol and the resulting effects. Covered are: (1) the behavioral and cognitive characteristics of subjects with prenatal alcohol exposure and (2) neuroimaging findings of prenatal alcohol effects. The second portion of this chapter considers adolescent alcohol use disorder (AUD), including: (1) adolescent AUD diagnosis and pathophysiology, (2) cognitive deficits in AUD, and (3) brain structural and functional abnormalities in adolescent AUD.

Although many adolescents experiment with marijuana and other non-alcohol drugs, this review focuses on alcohol effects for three reasons: (1) the incidence of disorders of alcohol use far exceeds rates of any other abused substance, especially in adolescence; (2) neuroimaging findings on human developmental drug exposure have addressed alcohol effects more than the effects of any other drugs; and (3) the near-universal rates of heavy alcohol use among abusers of other drugs preclude confident attribution of altered brain development to non-alcohol drugs.

Prenatal alcohol exposure

Behavioral and cognitive consequences of prenatal alcohol exposure

Alcohol exposure in utero can have devastating developmental effects in humans (Day and Richardson, 1991), most notably microcephaly (Archibald et al., 2001), general mental retardation (Rasmussen et al., 2006), and specific executive-cognitive deficits (Rasmussen, 2005). Syndromes of prenatal alcohol exposure have been termed "fetal alcohol spectrum disorders" (FASD) and are dichotomized into: (1) "fetal alcohol syndrome" (FAS), which is primarily characterized by reduced head circumference and craniofacial anomalies, and the milder "fetal alcohol effect" (FAE), characterized by specific cognitive and behavioral dysfunction, but no craniofacial abnormalities. Rodent experiments have indicated that growth of the whole brain (Maier et al., 1997) and development of the cerebellum (Maier et al., 1999) and hippocampus (Livy et al., 2003) (structures critical for learning) are especially sensitive to third-trimester human-equivalent prenatal alcohol exposure.

There is no known threshold that can be specified for a safe dose of alcohol during pregnancy, as even

Neuroimaging in Developmental Clinical Neuroscience, eds. Judith M. Rumsey and Monique Ernst. Published by Cambridge University Press. © Cambridge University Press 2009.

low doses may have effects on brain development (Savage et al., 2002). However, not all children exposed in utero to alcohol show discernible effects, raising the issue of additional factors (e.g., genetic, maternal alcohol metabolism, etc.) that increase the risk of adverse effects (May and Gossage, 2001). The unreliability of self-report and other methods for determining exposures and the uncertain criteria for effects that are more subtle than FAS, e.g., cognitive or neurodevelopmental problems in the absence of facial anomalies, represent methodological limitations. There is, however, general agreement that heavy exposure on any occasion is associated with higher risk than more limited drinking (e.g., single drink) on a number of occasions (Maier and West, 2001).

In contrast to the literature on in utero alcohol exposure, after controlling for psychosocial variables, prenatal exposure to cocaine (Arendt et al., 2004; Frank et al., 2001; Singer et al., 2004) and marijuana (Fried and Smith, 2001) have only been shown to cause deficits in attentional and visuomotor function, but not a reduction in global IQ. Recently, Smith and colleagues reported that young adults (18–22 years) with prenatal marijuana exposure did not show appreciable deficits in performance on working memory (Smith et al., 2006) or response inhibition (Smith et al., 2004) tasks compared to controls. Rather, the exposed subjects showed a reorganization of task-elicited brain activation in service of normal performance.

Morphological consequences of prenatal alcohol exposure

Riley, Sowell, and colleagues have documented numerous anomalies in the brain structure of children known to have been exposed to heavy alcohol use in utero (Table 14.1; reviewed in Riley et al. (2004)). Critically, these include morphological differences in the ventral aspects of the frontal lobe that have been shown to play a role in behavior control (Bechara et al., 2000). First, subjects with FAS (and to a lesser extent FASD) have reduced volumes of the cranial vault relative to controls (Riley et al., 2004). Intracranial volume (ICV), defined by

the inner surface of the skull, is determined by whole-brain growth during development and is generally fixed by age 15 (Courchesne et al., 2000). Since cross-sectional comparisons have matched cases and controls only on age and gender, it is possible that abnormal brain development in children with FAS resulted at least in part from poor nutrition (Ivanovic et al., 2000) and other environmental factors characteristic of substance-abusing parents. Alcohol administration to pregnant rats under controlled laboratory conditions, however, has repeatedly been shown to cause brain abnormalities in offspring, e.g., Livy et al. (2003).

Altered corpus callosum (CC) development is also considered to be a hallmark of FAS (Jones and Smith, 1973). Thinning of the anterior and posterior CC has been found in FAS (Riley et al., 1995), with partial or complete absence in severe cases. FAS/FASD has also been characterized by an anteroventral shift of the location of the posterior CC (Sowell et al., 2001a), with the magnitude of this displacement being correlated with verbal learning task performance. Diffusion tensor imaging (DTI) studies, which assess the coherence (fractional anisotropy: FA) of white matter (axon) fiber bundles, have also shown disrupted white matter cytoarchitecture in portions of the CC in children (Wozniak et al., 2006) and young adults (Ma et al., 2005) with FAS/FASD. Children with FAS/FASD performing a finger localization task showed selectively increased rates of errors in trials that required inter-hemispheric communication (Roebuck et al., 2002). In addition, errors in performance in those trials were related to CC volume reduction.

The anterior cerebellar vermis and the caudate nucleus are also disproportionately reduced in size in subjects with FAS/FASD (Archibald et al. 2001, Mattson et al., 1992, 1996;). Given the role of the cerebellum in motor control and learning and the caudate in reward cue processing (Bjork et al., 2004; Knutson et al., 2001), their dysmorphology may play a role in reduced cognitive and motor functioning and increased rates of AUD in individuals with FAS (Baer et al., 1998).

Finally, adolescent subjects with FAS and FASD show disproportionately reduced parietal lobe

Table 14.1 Summary of neuroimaging findings with prenatal alcohol exposure

Study	Subjects	Methodology	Compared to controls, prenatal alcohol-exposed subjects showed
Riley et al. (1995)	8- to 18-year-olds with FAS ($n = 11$) or FASD ($n = 2$); age-matched controls ($n = 12$)	CC outlining of midsagittal structural MRI images	Agenesis of the CC in two FAS subjects, proportionally reduced volumes of anterior and posterior CC
Mattson et al. (1996)	8- to 19-year-olds with FAS ($n = 6$); 8- to 18-year-old controls ($n = 7$)	Pixel-based volume estimates from structural MRI images	Reduced cerebral vault, diencephalon, and basal ganglia, proportionately reduced basal ganglia and caudate
Archibald et al. (2001)	Adolescents (mean age 11.4) with craniofacial FAS ($n = 14$), adolescents (mean age 14.8) with non-craniofacial FASD ($n = 12$); age-matched controls ($n = 41$)	Segmentation of structural MRI images into gray/white matter and cerebrospinal fluid	In FAS (but not FASD), reduced total cerebral and cerebellar vault, gray and white matter volumes; proportionately reduced parietal gray and white matter, caudate, and whole cerebral white matter
Sowell et al. (2001a)	8- to 22-year-olds with heavy prenatal alcohol exposure ($n = 21$); 8- to 25-year-old controls ($n = 21$)	Cortical surface mapping and CC outlining of structural MRI images	Total intracranial volume (including cerebellum) reduced by 12%; CC volume reduced; posterior CC localized more anteroventrally
Sowell et al. (2001b)	8- to 22-year-olds with heavy prenatal alcohol exposure ($n = 21$); 8- to 25-year-old controls ($n = 21$)	Voxel-based morphometric analysis of structural MRI images	Increased gray matter density but reduced white matter density in superior temporal and inferior parietal lobe
Sowell et al. (2002a)	8- to 22-year-olds with heavy prenatal alcohol exposure ($n = 21$); 8- to 25-year-old controls ($n = 21$)	Cortical surface mapping and gray matter density of structural MRI images	Narrowing of the brain at the inferior parietal lobe, in conjunction with increased gray matter density; reduced cortical surface (area) in inferior frontal lobe
Sowell et al. (2002b)	8- to 22-year-olds with heavy prenatal alcohol exposure ($n = 21$); 8- to 25-year-old controls ($n = 83$)	Cortical surface mapping of structural MRI images	Reduced R-L asymmetry at the temporo-parietal junction
Ma et al. (2005)	18- to 25-year-olds with FAS ($n = 9$), 18- to 22-year-old controls ($n = 7$)	DTI of the CC	Reduced diffusion fractional anisotropy and increased diffusion coefficient in anterior and posterior CC
Wozniak et al. (2006)	10- to 13-year-olds with non-craniofacial FASD ($n = 14$); age-matched controls ($n = 13$)	DTI of the CC	Increased mean diffusivity of the CC isthmus (posterior)

Notes:
CC, corpus callosum, DTI, diffusion tensor imaging; FAS, fetal alcohol syndrome; FASD, fetal alcohol spectrum disorder; MRI, magnetic resonance imaging.

volume after controlling for whole brain size (Archibald et al., 2001), with proportionally greater loss of white matter in the cerebrum relative to gray matter. In another report, adolescents with FAS and FASD showed relatively increased gray matter at the peri-Sylvian junction of temporal and parietal cortex (Sowell et al., 2001b). Analyses of the cortical surface revealed significant size and shape abnormalities which were maximal in the inferior parietal cortex in FAS/FASD subjects (i.e., a narrowing of the brain in this region) and a reduction of cortical surface in left orbitofrontal lobes (Sowell et al., 2002a). Finally, whereas the brains of control adolescents showed a displacement in the posterior direction of the left superior temporal and inferior parietal cortices relative to the right, this normative asymmetry was reduced in adolescents with FAS/FASD (Sowell et al., 2002b). Considering that normal adolescent brain development is characterized by a "pruning" of superfluous synapses within gray matter (increased communication selectivity) together with expanded white matter (increased signal propagation), it is possible that the FAS brain is suboptimally developed for efficient interneuronal communication.

Adolescent alcohol use disorder

Diagnosis and pathophysiology

Substance use disorders (SUD), as defined in the *Diagnostic and Statistical Manual of Mental Disorders*, 4th edition (DSM-IV, American Psychiatric Association, 1994), comprise: (1) substance abuse – heavy use that disrupts normal life functioning and fulfillment of obligations and (2) substance dependence – compulsive use characterized by physiological tolerance to the substance, with collateral withdrawal symptoms. Substance use disorders have been considered neuro*developmental* disorders (Chambers et al., 2003) because substance abusers often begin exposing their brains to the abused substance during adolescence (Hingson et al., 2006), when the brain is still developing (cf.

Chapter 1 by Lu and Sowell), and because adolescent substance experimentation may result from reduced impulse control characteristic of adolescence (Spear, 2000). For example, adolescents show reduced recruitment of error-monitoring circuitry when reward pursuit has a possibility of being penalized (see Figure 14.1).

Alcohol is the drug of choice among American adolescents (www.monitoringthefuture.org; www.nida.nih.gov/DrugPages/MTF.html). For example, 88.9% of recent alcohol initiates were under 21. Among adolescents aged 12–17 years, 28.2% of respondents reported recent (30 day) alcohol use, with 18.8% of respondents reporting binge use (≥ 5 drinks per occasion). In contrast, the next highest substances in incidence of use were tobacco (13.1%), marijuana (6.8%), and non-medical prescription drug use (3.3%) (National Survey on Drug Use and Health; www.oas.samhsa.gov/nsduh.htm). Accordingly, AUD makes up the vast majority of SUD diagnoses (Stinson et al., 2005). Adolescent alcohol use is so prevalent that it can be considered normative in that over 80% of high school seniors report having used alcohol at least once (reviewed in Deas and Brown, 2006).

Most adolescent alcohol experimenters do not develop AUD, and a substantial portion of adolescents who develop AUD eventually remit to subclinical drinking (Clark, 2004). Underage drinking is nevertheless a significant public health issue. In a recent epidemiological survey, a lifetime incidence of alcohol dependence was reported by 47% of interviewees who started drinking before age 14, but by only 9% of interviewees who began drinking at or after age 21 (Hingson et al., 2006). Adolescent-onset alcohol and marijuana use also progress more rapidly to dependence than does adult-onset use (Clark et al., 1998). Critically, adolescent alcohol use also increases motor vehicle accidents, risky sexual behavior, and aggressive behaviors (Clark, 2004; Miller et al., 2006).

Adolescents who develop AUD are also more likely to abuse other drugs in the future. Seminal longitudinal surveys of American high-schoolers have demonstrated that the initial consumption

Developmental differences in brain activation by risky rewards

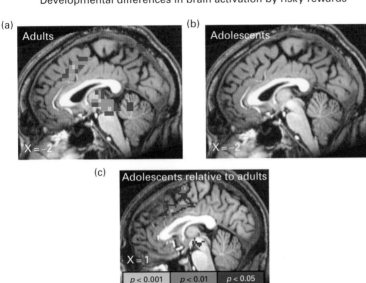

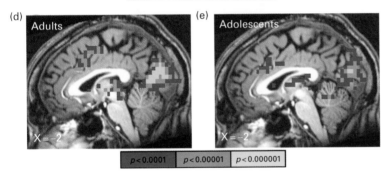

Figure 14.1. Blunted brain response to risky rewards in adolescents. Adolescents showed blunted brain activation by risky rewards in a monetary game of "chicken" (Bjork et al., 2007), where subjects forfeited trial winnings if they tried to accrue money past a secret, varying, time limit. Error-monitoring circuitry in posterior mesiofrontal cortex (PMC) was recruited in adults (a), but not adolescents (b), when subjects accrued monetary reward at risk of reward forfeiture (contrasted with activation by accrual of guaranteed money). The direct comparison of this contrast (c) illustrated impaired risk-elicited PMC activation in adolescents (blue overlay). Severe potential penalties (forfeiture of that trial's reward plus loss of some previous winnings) were necessary to elicit comparable PMC activation between adults (d) and adolescents (e). Reprinted from Bjork et al. (2007) *J Neurosci* with permission. Copyright 2007 by the Society for Neuroscience.

of cigarettes and alcohol is associated with an increased likelihood of subsequent consumption of soft (marijuana), and then hard (cocaine) illicit drugs at follow-up, even after controlling for psychosocial variables predictive of substance use generally (Yamaguchi and Kandel, 1984). This observation fostered the "gateway" model of the pathophysiological development of SUD (Kandel, 1975; Kandel and Faust, 1975). A *causative* relationship in the progression of drug use is supported by findings that adolescent rats pretreated with nicotine later showed greater responding for cocaine reward compared to animals pretreated with saline (McQuown et al., 2006), as well as sensitization to the hyperactivity elicited by cocaine (Collins and Izenwasser, 2004; Kelley and Rowan, 2004) and amphetamine (Collins et al., 2004). In humans, adolescent marijuana use among discordant twin pairs predicts subsequent use of other illicit drugs in young adulthood (Lynskey et al., 2003). Alternatively, the "common risk" model (Tarter et al., 2006; Vanyukov et al., 2003) posits that progressively harder drug use results from neurobiologically based trait impulsivity (Kreek et al., 2005) and dysfunctional environmental factors interacting with increasing illicit drug availability from adolescence to adulthood.

Diagnosis of AUD in adolescents is controversial in that the subjective symptom criteria for alcohol abuse and dependence diagnoses do not equally apply for adolescents and adults. For example, the criteria for alcohol abuse include persistent heavy drinking despite social impairment and legal troubles and/or drinking when driving or operating heavy machinery. In adolescence, heavy drinking can serve as an adaptive social lubricant and is not likely to interfere with occupation-related hazards. Moreover, children and adolescents have shown blunted adverse effects of intoxication (Spear, 2002) and withdrawal symptoms (Martin et al., 1995), which is critical since withdrawal-elicited drinking is a cardinal symptom of alcohol dependence. Another cardinal symptom for alcohol dependence is consuming more drinks than intended. This is problematic in that adolescents may lack specific plans to limit alcohol use and deliberately drink to get drunk (Chung and Martin, 2005). Accordingly, patient care providers also underestimate incidence of DSM-IV-defined SUD symptoms in their adolescent patients relative to detailed patient self-report (e.g. Wilson et al., 2004). Adolescents with alcohol-related problems who do not meet criteria for AUD have been characterized as "diagnostic orphans" (Pollock and Martin, 1999), whose incidence is similar to that of formally diagnosed adolescents.

Attributing brain structural and functional abnormalities to AUD is also problematic. Adolescents with AUD show a high incidence of comorbid and premorbid mood and behavior disorders (Couwenbergh et al., 2006; Kandel et al., 1999; Wilens et al., 2004). A critical issue in the neuroimaging study of SUD therefore is attributing individual differences directly to substance exposure (or specific vulnerability to SUD) as opposed to morphological and functional differences that are more closely associated with behavioral and affective symptomatology that lead to substance use. For example, depressed children, who are at increased risk of adolescent alcohol use (Wu et al., 2006), show disproportionately reduced bilateral amygdalar (Rosso et al., 2005) and hippocampal (MacMaster and Kusumakar, 2004) volumes similar to those reported in AUD. Moreover, subclinical (Killgore and Yurgelun-Todd, 2006) and clinical (McClure et al., 2007) anxiety symptomatology has correlated with enhanced limbic responsiveness to fearful faces. This may present a confound in that subjects with AUD show impaired classification of facial expressions that does not remit with abstinence (Foisy et al., 2007).

The incidence of comorbid disruptive behavior disorders (DBD) in adolescents with AUD has been over 80% in some surveys (reviewed in Couwenbergh et al., 2006). Large-scale studies of twins have identified a core heritable impulse control dysfunction (Fu et al., 2002; Slutske et al., 1998), which can account for SUD comorbidity with behavior disorders (Kreek et al., 2005) and for the use of multiple drugs (Kendler et al., 2003; Vanyukov et al., 2003). In non-drinking adolescents, inferior parietal recruitment by response inhibition in a go/no-go task correlated positively with expectancies of negative effects of alcohol use, but negatively with expectancies of positive effects (Anderson et al., 2005). These findings suggest that differences in brain structure or activation between adolescents with and without SUD may relate more closely to premorbid (often heritable, Todd and Botteron, 2002) behavioral traits. In summary, adolescents who meet formal criteria for an AUD are likely to be especially severe drinkers, who likely show a host of comorbid mood and behavior symptomatology.

Cognitive deficits in adolescent alcohol use disorder

Neuroimaging surveys of adolescents with AUD have been stimulated by findings of decrements in cognitive performance in AUD (Bates et al., 2002), which can remit with abstinence. Adolescents with SUD have generally shown (primarily non-verbal) deficits similar to those seen in adults with SUD, including lower full-scale IQ than controls (Moss et al., 1994). Tapert and Brown (1999) performed multiple neurocognitive assessments of clinically treated adolescents with AUD and at least one comorbid SUD, but no premorbid psychiatric disorder other than conduct disorder. Adolescents who had either relapsed or accelerated their substance use at the 4-year follow-up assessment did not differ from abstainers on neurocognitive performance at the time of their first (baseline) assessment, but showed significant decrements in attentional performance at follow-up. Moreover, withdrawal symptomatology within 3 months prior to follow-up was inversely proportional to visuo-spatial functioning (e.g., block design and visual reproduction) after controlling for trait baseline variables.

That recent withdrawal episodes would be specifically linked to impaired cognitive performance at follow-up is in accord with extensive findings of dose-dependent behavioral and neurochemical sequelae (notably glutamatergic hyperexcitability and neurodegeneration) of alcohol withdrawal in rats (Fadda and Rossetti, 1998), which produces behavioral sensitization with repeated withdrawal episodes (Maier and Pohorecky, 1989). Critically, alcohol administered in intermittent large bolus doses (binges) to the developing rodent brain caused markedly greater cellular and morphological damage than a larger total quantity administered continuously at a lower concentration (Bonthius and West, 1990; Maier and West, 2001). Alcohol withdrawal effects are especially relevant for adolescents, because they typically binge-drink. Although adolescents, aged 12–17, on average report half the drinking days per month (4.6) as

do adults 26 years and older (9.06), they consume significantly more drink-equivalents per drinking occasion (4.65) than do adults (2.65) (SAMHSA National Survey on Drug Use and Health, 2003; www.samhsa.gov).

Structural and functional brain abnormalities in adolescent alcohol use disorder

Cross-sectional (Bjork et al., 2003) and longitudinal (Pfefferbaum et al., 1998) magnetic resonance imaging (MRI) studies of adults have shown that chronic alcohol intoxication accelerates gray matter atrophy beyond normative atrophy with aging, where anterior superior temporal lobe atrophy is disproportionately severe (Pfefferbaum et al., 1998). Consequently, in studies demonstrating reduced brain volumes in abusers of non-alcohol drugs (e.g., Aasly et al., 1993), it is difficult to eliminate the confounding incidence of alcohol use, which is nearly universal among 18- to 35-year-old cocaine users (Grant and Harford, 1990).

As has been reported in adults (Beresford et al., 2006), adolescents with alcohol dependence had smaller hippocampal volumes after controlling for intracranial or whole-brain volume (De Bellis et al., 2000; Nagel et al., 2005). Medina et al. (2007) reported no differences in proportional (relative to whole-brain volume) hippocampal size between adolescent controls and adolescents who used both alcohol and marijuana, and both groups showed similar left-greater-than-right (L > R) asymmetry of hippocampal volumes. Adolescents with alcohol-only use, however, showed R > L hippocampus asymmetry and had smaller left hippocampal volumes, and these differences directly corresponded with the number of AUD symptoms. In addition, as with adults (Kril and Halliday, 1999), adolescents with AUD showed relatively reduced prefrontal cortex volumes (De Bellis et al., 2005), which may produce, in part, the executive cognitive deficits of adolescents with AUD. In the De Bellis et al. studies, hippocampal (De Bellis et al., 2000) and prefrontal cortical (De Bellis et al., 2005) gray matter volumes were also inversely related to the severity of heavy drinking. In light of findings of decreased hippocampal pyramidal neurons and synapses in mice chronically treated with alcohol (reviewed in Fadda and Rossetti, 1998), alcohol-induced alterations of hippocampal development may be a neurodegenerative underpinning of memory deficits characteristic of adolescents with AUD (Brown et al., 2000).

The causality in these symptom-volume correlations, however, remains unclear. In the Nagel et al. (2005) study, the AUD group had no psychiatric comorbidity other than conduct disorder and showed no decrement in neurocognitive performance. Among these "cleaner" AUD subjects, there was also no correlation between alcohol use severity and hippocampal volume. Those authors raised the possibility that adolescents with the greatest cumulative alcohol exposure simply had the smallest (premorbid) volumes prior to drinking. Adult alcohol-dependent subjects, however, showed dynamic recovery of gray matter (Pfefferbaum et al., 1998) and motor and cognitive functioning (Sullivan et al., 2000a) following remission in a longitudinal study, in which working memory recovery correlated with morphological recovery (Sullivan et al., 2000b). Future longitudinal studies showing similar morphological recovery with abstinence in adolescence would not only suggest that chronic alcohol administration in adolescence *causes* volume reductions, but would also strongly motivate initiatives to curtail adolescent drinking.

Functional neuroimaging studies of brain activation provide complementary evidence that brain circuitry also reorganizes in response to chronic alcohol exposure. Adults with alcohol dependence have shown enhanced limbic responsiveness to substance-related visual stimuli, as well as altered regional patterns of working-memory-task- (WMT-) elicited brain activation relative to controls, despite similar task performance in visual (Pfefferbaum et al., 2001) and verbal (Desmond et al., 2003) WMT. A WMT typically require subjects to either compare the stimulus they are viewing to a stimulus presented shortly before or to respond to features of previously presented stimuli irrespective of the

currently presented stimulus. These are challenging probes of brain function because they require stimulus processing, memory encoding and retention, and decision-making about responding. In a series of studies (reviewed in Table 14.2), Susan Tapert and colleagues have recently extended adult findings to adolescent and young adult populations.

First, Tapert and colleagues reported that young women with alcohol dependence showed impaired performance and blunted, right-lateralized parietal, middle frontal, right postcentral, and left superior frontal cortex recruitment in a contrast between blocks of a spatial WMT task and a simpler vigilance task (Tapert et al., 2001). In contrast, the WMT administered to adolescents with and without AUD showed an opposite effect, in that adolescents with AUD showed greater activation during WMT in bilateral superior parietal cortices (Tapert et al., 2004b), despite similar task performance. Subjects with AUD also showed a diminished response in the left precentral gyrus and cerebellar declive. Activation differences were greater in AUD subjects who reported more general withdrawal symptoms and alcohol consumption. Critically, these AUD subjects were rigorously selected, with a substantial number of applicants excluded for comorbid drug use (21%) or psychiatric symptoms (39%) (Tapert, personal communication), and they did not differ from controls on a host of neurocognitive and behavioral measures.

A subsequent study examined gender differences in brain activation alterations with AUD (Caldwell et al., 2005). Again, AUD subjects showed increased bilateral activation in the inferior parietal lobe despite similar spatial WMT task accuracy, with additional increased activation in superior frontal, right middle frontal, fusiform, and middle temporal gyri. Subjects with AUD had decreased activation in the inferior frontal gyrus, left precentral gyrus, and cerebellum. Activation differences were gender dependent, with significant gender × diagnosis interactions in several cortical regions indicating a more severe deviation in female subjects with AUD. This is in accord with greater alcohol-mediated brain atrophy in adult female alcoholics compared to males (Hommer et al., 2001).

Table 14.2 Summary of neuroimaging findings with self-administered alcohol exposure

Study	Subjects	Methodology	Compared to controls, alcohol-abusing subjects showed
De Bellis et al. (2000)	Adolescents and young adults age 13–21 with AD ($n = 12$) and controls ($n = 24$)	Semiautomated segmentation of structural MRI images	Reduced absolute and proportional right and left hippocampus volume
Tapert et al. (2003)	14- to 17-year-olds with AUD ($n = 15$); demographically matched social drinking controls ($n = 15$)	Functional MRI responses elicited by beverage advertisements for preferred alcohol beverage and ads for non-alcohol beverages	Increased activation by preferred alcohol images in (primarily left) frontal cortex, bilateral amygdala and cuneus
Tapert et al. (2004a)	Women age 18–24 with AD ($n = 8$), age-matched female social drinking controls ($n = 9$)	Functional MRI responses elicited by alcohol-associated words and neutral words	Increased activation by alcohol-associated words in bilateral insular cortex, subcallosal anterior cingulate cortex, and left prefrontal cortex
Tapert et al. (2004b)	14- to 17-year-olds with AUD ($n = 15$); demographically matched non-drinking controls ($n = 19$)	Functional MRI responses elicited by spatial WMT and control vigilance task	Increased activation by WMT in bilateral superior parietal lobe, but reduced activation in cerebellum and occipital and L precentral gyri
De Bellis et al. (2005)	14- to 20-year-olds with adolescent-onset AUD ($n = 14$); 13- to 21-year-old controls ($n = 28$)	Semiautomated segmentation of structural MRI images	Reduced total frontal cortex and frontal white matter volumes, but increased frontal cerebrospinal fluid volumes; reduced cerebellar volumes in males
Caldwell et al. (2005)	14- to 17-year-olds with AUD ($n = 18$); demographically matched non-drinking controls ($n = 21$)	Functional MRI responses elicited by spatial WMT and control vigilance task	Increased activation by WMT in temporal, inferior parietal, and middle and superior frontal gyri, but reduced activation in cerebellum, inferior frontal, and precentral gyri; more pronounced alterations in female subjects
Schweinsburg et al. (2005)	15- to 17-year-olds with AUD ($n = 15$); comorbid AUD and MAUD ($n = 15$); demographically matched non-drinking controls ($n = 19$)	Functional MRI responses elicited by spatial WMT and control vigilance task	MAUD subjects showed less activation in left temporal and inferior frontal lobes, but more activation in right superior and middle frontal gyri; MAUD subjects showed greater attention-elicited *de*activation of mesofrontal cortex. Additional differences between MAUD and AUD-only subjects

Table 14.2 (*cont.*)

Study	Subjects	Methodology	Compared to controls, alcohol-abusing subjects showed
Nagel et al. (2005)	15- to 17-year-olds with AUD (*n* = 14); demographically matched non-drinking controls (*n* = 17)	Semiautomated segmentation of structural MRI images	Reduced volume of left hippocampus relative to total ICV
Medina et al. (2007)	15- to 17-year-old social drinkers (Alc-only; *n* = 16); 15- to 19-year-old users of both alcohol and marijuana (MJ + Alc; *n* = 26); 15- to 18-year-old non-using controls (*n* = 21)	Semiautomated calculation of ICV and manual calculation of hippocampus volume from structural MRI	Reduced proportional left hippocampal volume; L > R volume asymmetry in controls and MJ + Alc users but not in Alc-only subjects

Notes:
AD, alcohol dependence; AUD, alcohol use disorder; ICV, intracranial volume; MAUD, marijuana use disorder; MRI, magnetic resonance imaging; WMT, working memory task.

To examine functional brain alterations caused by adolescent marijuana use, while controlling for high rates of alcohol use in marijuana users, Schweinsburg et al. (2005) compared cerebral activation by the WMT in adolescents with comorbid alcohol and marijuana use disorders to activation in the AUD adolescents previously studied by Tapert et al. (2004b). These adolescents did not differ in alcohol drinking parameters or task performance and showed generally similar psychometric characteristics as the previous AUD-only subjects. The comorbid subjects showed reduced inferior frontal and temporal recruitment, but increased mesofrontal recruitment during a WMT relative to AUD-only subjects. Collectively, these studies suggest that by adolescence, chronic substance use triggers a reorganization of brain circuitry in pursuit of adequate cognitive performance.

This group also investigated whether adolescents and young adults with AUD showed increased alcohol cue-elicited activation in reward-related limbic structures shown to be responsive to drug cues in adults. In young women with AUD, alcohol-related words (contrasted with neutral words) elicited increased left-lateralized activation of subgenual and anterior cingulate cortex and increased bilateral activation of uncus, insula, and pre-cuneus (Tapert et al., 2004a). Conversely, AUD subjects had reduced alcohol-word-elicited activation in bilateral frontal and temporal lobes, and subgenual cortex activation correlated with increased alcohol craving after the task.

Compared to controls, adolescents with AUD showed greater amygdalar and midline and left-lateralized activation in frontal, parietal, and temporal gyri when shown photographs of alcohol beverages (contrasted with responses to photographs of non-alcohol beverages) (Tapert et al., 2003). Conversely, AUD subjects showed reduced activation by alcohol images only in right middle and inferior frontal gyri. Alterations in posterior cingulate activation were most pronounced in AUD adolescents with the heaviest drinking. Critically, these increased activations in AUD subjects occurred in limbic regions commonly recruited by motivationally salient stimuli across a variety of incentives (Koob, 2006) and strongly suggest that alcohol images acquired increased motivational significance in the subjects with AUD.

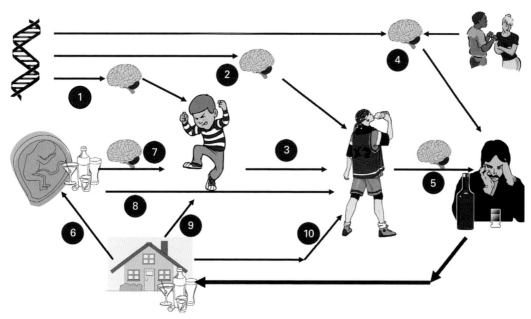

Figure 14.2. Etiological pathways of alcohol dependence. Neurobiological (brain icon) and environmental factors interact to cause alcohol dependence (AD). (1) Heritable neurobiological traits governing gross morphological and synaptic brain development increase risk of disruptive behavior disorders in childhood. (2) Heritable neurobiological traits conferring disproportionately positive versus negative consequences of alcohol intoxication promote binge-drinking in adolescence. (3) Disruptive behavior disorders in childhood increase odds of adolescent alcohol use, such as by increased novelty seeking and exposure to deviant peers. (4) Individuals with overactive stress-response neurocircuitry use anxiolytic effects of alcohol intoxication to cope with environmental stressors. (5) Alcohol self-administration during adolescence alters neurocircuitry governing the stress response, executive cognitive functioning, and reward-association learning to increase the odds of lifetime development of AD. In households with alcohol-dependent parents (6), alcohol-dependent pregnant women expose developing fetus to alcohol, (7) fetal alcohol exposure causes altered brain developmental trajectory to promote cognitive and behavioral impairment in childhood, (8) fetal alcohol exposure causes neurobiological adaptations that increase motivational salience of alcohol to promote binge-drinking in adolescence, (9) incompetent/abusive parenting by intoxicated parents increases risk of disruptive behavior disorders in childhood, and (10) increased alcohol availability, coupled with behavioral modeling and other domestic environmental stressors, increases adolescent binge-drinking.

Conclusion

Pediatric neuroimaging findings illustrate how both prenatal exposure and self-administration of alcohol alter the development of fronto-cortical and subcortical brain regions that govern behavioral control (Crews et al., 2007) and represent a mechanism by which hereditary biological factors and environment can interact to contribute to AUD (Figure 14.2).

In particular: (1) prenatal exposure of the developing brain to alcohol results in persistent dysmorphology, (2) whole-brain and structure-specific reductions found in adults with alcoholism are already found in adolescents with AUD, (3) altered recruitment of neural circuitry in service of successful cognitive performance in AUD occurs by adolescence, and (4) increased drug cue-elicited limbic activation in AUD is also already present by adolescence. These findings provide vivid correlational data in support of policy initiatives to interdict drinking during pregnancy as well as drinking by adolescents.

REFERENCES

Aasly, J., Storsaeter, O., Nilsen, G., Smevik, O., and Rinck P. (1993) Minor structural brain changes in young drug abusers. A magnetic resonance study. *Acta Neurol Scand*, **87**, 210–14.

American Psychiatric Association (1994) *Diagnostic and Statistical Manual of Mental Disorders*, 4th edn. Arlington, VA: American Psychiatric Association.

Anderson, K. G., Schweinsburg, A., Paulus, M. P., Brown, S. A., and Tapert S. (2005) Examining personality and alcohol expectancies using functional magnetic resonance imaging (fMRI) with adolescents. *J Stud Alcohol*, **66**, 323–31.

Archibald, S. L., Fennema-Notestine, C., Gamst, A., Riley, E. P., Mattson, S. N., and Jernigan, T. L. (2001) Brain dysmorphology in individuals with severe prenatal alcohol exposure. *Dev Med Child Neurol*, **43**, 148–54.

Arendt, R. E., Short, E. J., Singer, L. T., et al. (2004) Children prenatally exposed to cocaine: developmental outcomes and environmental risks at seven years of age. *J Dev Behav Pediatr*, **25**, 83–90.

Baer, J. S., Barr, H. M., Bookstein, F. L., Sampson, P. D., and Streissguth, A. P. (1998) Prenatal alcohol exposure and family history of alcoholism in the etiology of adolescent alcohol problems. *J Stud Alcohol*, **59**, 533–43.

Bates, M. E., Bowden, S. C., and Barry, D. (2002) Neurocognitive impairment associated with alcohol use disorders: implications for treatment. *Exp Clin Psychopharmacol*, **10**, 193–212.

Bechara, A., Damasio, H., and Damasio, A. R. (2000) Emotion, decision making and the orbitofrontal cortex. *Cereb Cortex*, **10**, 295–307.

Beresford, T. P., Arciniegas, D. B., Alfers, J., et al. (2006) Hippocampus volume loss due to chronic heavy drinking. *Alcohol Clin Exp Res*, **30**, 1866–70.

Bjork, J. M., Grant, S. J., and Hommer, D. W. (2003) Cross-sectional volumetric analysis of brain atrophy in alcohol dependence: effects of drinking history and comorbid substance use disorder. *Am J Psychiatry*, **160**, 2038–45.

Bjork, J. M., Knutson, B., Fong, G. W., Caggiano, D. M., Bennett, S. M., and Hommer, D. W. (2004) Incentive-elicited brain activation in adolescents: similarities and differences from young adults. *J Neurosci*, **24**, 1793–802.

Bjork, J. M., Smith, A. R., Danube, C. L., and Hommer, D. W. (2007) Developmental differences in posterior mesofrontal cortex recruitment by risky rewards. *J Neurosci*, **27**, 4839–49.

Bonthius, D. J. and West, J. R. (1990) Alcohol-induced neuronal loss in developing rats: increased brain damage with binge exposure. *Alcohol Clin Exp Res*, **14**, 107–18.

Brown, S. A., Tapert, S. F., Granholm, E., and Delis, D. C. (2000) Neurocognitive functioning of adolescents: effects of protracted alcohol use. *Alcohol Clin Exp Res*, **24**, 164–71.

Caldwell, L. C., Schweinsburg, A. D., Nagel, B. J., Barlett, V. C., Brown, S. A., and Tapert, S. F. (2005) Gender and adolescent alcohol use disorders on BOLD (blood oxygen level dependent) response to spatial working memory. *Alcohol Alcohol*, **40**, 194–200.

Chambers, R. A., Taylor, J. R., and Potenza, M. N. (2003) Developmental neurocircuitry of motivation in adolescence: a critical period of addiction vulnerability. *Am J Psychiatry*, **160**, 1041–52.

Chung, T. and Martin, C. S. (2005) What were they thinking? Adolescents' interpretations of DSM-IV alcohol dependence symptom queries and implications for diagnostic validity. *Drug Alcohol Depend*, **80**, 191–200.

Clark, D. B. (2004) The natural history of adolescent alcohol use disorders. *Addiction*, **99** [Suppl 2], 5–22.

Clark, D. B., Kirisci, L., and Tarter, R. E. (1998) Adolescent versus adult onset and the development of substance use disorders in males. *Drug Alcohol Depend*, **49**, 115–21.

Collins, S. L. and Izenwasser, S. (2004) Chronic nicotine differentially alters cocaine-induced locomotor activity in adolescent vs. adult male and female rats. *Neuropharmacology*, **46**, 349–62.

Collins, S. L., Montano, R., and Izenwasser, S. (2004) Nicotine treatment produces persistent increases in amphetamine-stimulated locomotor activity in periadolescent male but not female or adult male rats. *Brain Res Dev Brain Res*, **153**, 175–87.

Courchesne, E., Chisum, H. J., Townsend, J., et al. (2000) Normal brain development and aging: quantitative analysis at in vivo MR imaging in healthy volunteers. *Radiology*, **216**, 672–82.

Couwenbergh, C., van den Brink, W., Zwart, K., Vreugdenhil, C., van Wijngaarden-Cremers, P., and van der Gaag, R. J. (2006) Comorbid psychopathology in adolescents and young adults treated for substance use disorders: a review. *Eur Child Adolesc Psychiatry*, **15**, 319–28.

Crews, F., He, J., and Hodge, C. (2007) Adolescent cortical development: a critical period of vulnerability for addiction. *Pharmacol Biochem Behav*, **86**(2), 189–99.

Day, N. L. and Richardson, G. A. (1991) Prenatal alcohol exposure: a continuum of effects. *Semin Perinatol*, **15**, 271–9.

De Bellis, M. D., Clark, D. B., Beers, S. R., et al. (2000) Hippocampal volume in adolescent-onset alcohol use disorders. *Am J Psychiatry*, **157**, 737–44.

De Bellis, M. D., Narasimhan, A., Thatcher, D. L., Keshavan, M. S., Soloff, P., and Clark, D. B. (2005) Prefrontal cortex, thalamus, and cerebellar volumes in adolescents and young adults with adolescent-onset alcohol use disorders and comorbid mental disorders. *Alcohol Clin Exp Res*, **29**, 1590–600.

Deas, D. and Brown, E. S. (2006) Adolescent substance abuse and psychiatric comorbidities. *J Clin Psychiatry*, **67**, e02.

Desmond, J. E., Chen, S. H., DeRosa, E., Pryor, M. R., Pfefferbaum, A., and Sullivan, E. V. (2003) Increased frontocerebellar activation in alcoholics during verbal working memory: an fMRI study. *NeuroImage*, **19**, 1510–20.

Fadda, F. and Rossetti, Z. L. (1998) Chronic ethanol consumption: from neuroadaptation to neurodegeneration. *Prog Neurobiol*, **56**, 385–431.

Foisy, M. L., Kornreich, C., Fobe, A., et al. (2007) Impaired emotional facial expression recognition in alcohol dependence: do these deficits persist with midterm abstinence? *Alcohol Clin Exp Res*, **31**, 404–10.

Frank, D. A., Augustyn, M., Knight, W. G., Pell, T., and Zuckerman, B. (2001) Growth, development, and behavior in early childhood following prenatal cocaine exposure: a systematic review. *J Am Med Assoc*, **285**, 1613–25.

Fried, P. A. and Smith, A. M. (2001) A literature review of the consequences of prenatal marijuana exposure. An emerging theme of a deficiency in aspects of executive function. *Neurotoxicol Teratol*, **23**, 1–11.

Fu, Q., Heath, A. C., Bucholz, K. K., et al. (2002) Shared genetic risk of major depression, alcohol dependence, and marijuana dependence: contribution of antisocial personality disorder in men. *Arch Gen Psychiatry*, **59**, 1125–32.

Grant, B. F. and Harford, T. C. (1990) Concurrent and simultaneous use of alcohol with cocaine: results of national survey. *Drug Alcohol Depend*, **25**, 97–104.

Hingson, R. W., Heeren, T., and Winter, M. R. (2006) Age at drinking onset and alcohol dependence: age at onset,

duration, and severity. *Arch Pediatr Adolesc Med*, **160**, 739–46.

Hommer, D., Momenan, R., Kaiser, E., and Rawlings, R. (2001) Evidence for a gender-related effect of alcoholism on brain volumes. *Am J Psychiatry*, **158**, 198–204.

Ivanovic, D. M., Leiva, B. P., Perez, H. T., et al. (2000) Long-term effects of severe undernutrition during the first year of life on brain development and learning in Chilean high-school graduates. *Nutrition*, **16**, 1056–63.

Jones, K. L. and Smith, D. W. (1973) Recognition of the fetal alcohol syndrome in early infancy. *Lancet*, **2**, 999–1001.

Kandel, D. (1975) Stages in adolescent involvement in drug use. *Science*, **190**, 912–14.

Kandel, D. and Faust, R. (1975) Sequence and stages in patterns of adolescent drug use. *Arch Gen Psychiatry*, **32**, 923–32.

Kandel, D. B., Johnson, J. G., Bird, H. R., et al. (1999) Psychiatric comorbidity among adolescents with substance use disorders: findings from the MECA Study. *J Am Acad Child Adolesc Psychiatry*, **38**, 693–9.

Kelley, B. M. and Rowan, J. D. (2004) Long-term, low-level adolescent nicotine exposure produces dose-dependent changes in cocaine sensitivity and reward in adult mice. *Int J Dev Neurosci*, **22**, 339–48.

Kendler, K. S., Jacobson, K. C., Prescott, C. A., and Neale, M. C. (2003) Specificity of genetic and environmental risk factors for use and abuse/dependence of cannabis, cocaine, hallucinogens, sedatives, stimulants, and opiates in male twins. *Am J Psychiatry*, **160**, 687–95.

Killgore, W. D. and Yurgelun-Todd, D. A. (2006) Ventromedial prefrontal activity correlates with depressed mood in adolescent children. *NeuroReport*, **17**, 167–71.

Knutson, B., Adams, C. M., Fong, G. W., and Hommer, D. (2001) Anticipation of increasing monetary reward selectively recruits nucleus accumbens. *J Neurosci*, **21**, RC159.

Koob, G. F. (2006) The neurobiology of addiction: a neuroadaptational view relevant for diagnosis. *Addiction*, **101** [Suppl 1], 23–30.

Kreek, M. J., Nielsen, D. A., Butelman, E. R., and LaForge, K. S. (2005) Genetic influences on impulsivity, risk taking, stress responsivity and vulnerability to drug abuse and addiction. *Nat Neurosci*, **8**, 1450–7.

Kril, J. J. and Halliday, G. M. (1999) Brain shrinkage in alcoholics: a decade on and what have we learned? *Prog Neurobiol*, **58**, 381–7.

Livy, D. J., Miller, E. K., Maier, S. E., and West, J. R. (2003) Fetal alcohol exposure and temporal vulnerability:

effects of binge-like alcohol exposure on the developing rat hippocampus. *Neurotoxicol Teratol*, **25**, 447–58.

Lynskey, M. T., Heath, A. C., Bucholz, K. K., et al. (2003) Escalation of drug use in early-onset cannabis users vs co-twin controls. *J Am Med Assoc*, **289**, 427–33.

Ma, X., Coles, C. D., Lynch, M. E., et al. (2005) Evaluation of corpus callosum anisotropy in young adults with fetal alcohol syndrome according to diffusion tensor imaging. *Alcohol Clin Exp Res*, **29**, 1214–22.

MacMaster, F. P. and Kusumakar, V. (2004) Hippocampal volume in early onset depression. *BMC Med*, **2**, 2.

Maier, D. M. and Pohorecky, L. A. (1989) The effect of repeated withdrawal episodes on subsequent withdrawal severity in ethanol-treated rats. *Drug Alcohol Depend*, **23**, 103–10.

Maier, S. E. and West, J. R. (2001) Drinking patterns and alcohol-related birth defects. *Alcohol Res Health*, **25**, 168–74.

Maier, S. E., Chen, W. J., Miller, J. A., and West, J. R. (1997) Fetal alcohol exposure and temporal vulnerability regional differences in alcohol-induced microencephaly as a function of the timing of binge-like alcohol exposure during rat brain development. *Alcohol Clin Exp Res*, **21**, 1418–28.

Maier, S. E., Miller, J. A., Blackwell, J. M., and West, J. R. (1999) Fetal alcohol exposure and temporal vulnerability: regional differences in cell loss as a function of the timing of binge-like alcohol exposure during brain development. *Alcohol Clin Exp Res*, **23**, 726–34.

Martin, C. S., Kaczynski, N. A., Maisto, S. A., Bukstein, O. M., and Moss, H. B. (1995) Patterns of DSM-IV alcohol abuse and dependence symptoms in adolescent drinkers. *J Stud Alcohol*, **56**, 672–80.

Mattson, S. N., Riley, E. P., Jernigan, T. L., et al. (1992) Fetal alcohol syndrome: a case report of neuropsychological, MRI and EEG assessment of two children. *Alcohol Clin Exp Res*, **16**, 1001–3.

Mattson, S. N., Riley, E. P., Sowell, E. R., Jernigan, T. L., Sobel, D. F., and Jones, K. L. (1996) A decrease in the size of the basal ganglia in children with fetal alcohol syndrome. *Alcohol Clin Exp Res*, **20**, 1088–93.

May, P. A. and Gossage, J. P. (2001) Estimating the prevalence of fetal alcohol syndrome. *Alcohol Res Health*, **25**(3), 159–67.

McClure, E. B., Monk, C. S., Nelson, E. E., et al. (2007) Abnormal attention modulation of fear circuit function in pediatric generalized anxiety disorder. *Arch Gen Psychiatry*, **64**, 97–106.

McQuown, S. C., Belluzzi, J. D., and Leslie, F. M. (2006) Low dose nicotine treatment during early adolescence increases subsequent cocaine reward. *Neurotoxicol Teratol*, **29**(1), 66–73.

Medina, K. L., Schweinsburg, A. D., Cohen-Zion, M., Nagel, B. J., and Tapert, S. F. (2007) Effects of alcohol and combined marijuana and alcohol use during adolescence on hippocampal volume and asymmetry. *Neurotoxicol Teratol*, **29**, 141–52.

Miller, T. R., Levy, D. T., Spicer, R. S., and Taylor, D. M. (2006) Societal costs of underage drinking. *J Stud Alcohol*, **67**, 519–28.

Moss, H. B., Kirisci, L., Gordon, H. W., and Tarter, R. E. (1994) A neuropsychologic profile of adolescent alcoholics. *Alcohol Clin Exp Res*, **18**, 159–63.

Nagel, B. J., Schweinsburg, A. D., Phan, V., and Tapert, S. F. (2005) Reduced hippocampal volume among adolescents with alcohol use disorders without psychiatric comorbidity. *Psychiatry Res*, **139**, 181–90.

Pfefferbaum, A., Desmond, J. E., Galloway, C., Menon, V., Glover, G. H., and Sullivan, E. V. (2001) Reorganization of frontal systems used by alcoholics for spatial working memory: an fMRI study. *NeuroImage*, **14**, 7–20.

Pfefferbaum, A., Sullivan, E. V., Rosenbloom, M. J., Mathalon, D. H., and Lim, K. O. (1998) A controlled study of cortical gray matter and ventricular changes in alcoholic men over a 5-year interval. *Arch Gen Psychiatry*, **55**, 905–12.

Pollock, N. K. and Martin, C. S. (1999) Diagnostic orphans: adolescents with alcohol symptom who do not qualify for DSM-IV abuse or dependence diagnoses. *Am J Psychiatry*, **156**, 897–901.

Rasmussen, C. (2005) Executive functioning and working memory in fetal alcohol spectrum disorder. *Alcohol Clin Exp Res*, **29**, 1359–67.

Rasmussen, C., Horne, K., and Witol, A. (2006) Neurobehavioral functioning in children with fetal alcohol spectrum disorder. *Child Neuropsychol*, **12**, 453–68.

Riley, E. P., Mattson, S. N., Sowell, E. R., Jernigan, T. L., Sobel, D. F., and Jones, K. L. (1995) Abnormalities of the corpus callosum in children prenatally exposed to alcohol. *Alcohol Clin Exp Res*, **19**, 1198–202.

Riley, E. P., McGee, C. L., and Sowell, E. R. (2004) Teratogenic effects of alcohol: a decade of brain imaging. *Am J Med Genet C Semin Med Genet*, **127**, 35–41.

Roebuck, T. M., Mattson, S. N., and Riley, E. P. (2002) Interhemispheric transfer in children with heavy prenatal alcohol exposure. *Alcohol Clin Exp Res*, **26**, 1863–71.

Rosso, I. M., Cintron, C. M., Steingard, R. J., Renshaw, P. F., Young, A. D., and Yurgelun-Todd, D. A. (2005) Amygdala and hippocampus volumes in pediatric major depression. *Biol Psychiatry*, **57**, 21–6.

Savage, D. D., Becher, M., de la Torre, A. J., et al. (2002) Dose-dependent effects of prenatal ethanol exposure on synaptic plasticity and learning in mature offspring. *Alcohol Clin Exp Res*, **26**(11), 1752–8.

Schweinsburg, A. D., Schweinsburg, B. C., Cheung, E. H., Brown, G. G., Brown, S. A., and Tapert, S. F. (2005) fMRI response to spatial working memory in adolescents with comorbid marijuana and alcohol use disorders. *Drug Alcohol Depend*, **79**, 201–10.

Singer, L. T., Minnes, S., Short, E., et al. (2004) Cognitive outcomes of preschool children with prenatal cocaine exposure. *J Am Med Assoc*, **291**, 2448–56.

Slutske, W. S., Heath, A. C., Dinwiddie, S. H., et al. (1998) Common genetic risk factors for conduct disorder and alcohol dependence. *J Abnorm Psychol*, **107**, 363–74.

Smith, A. M., Fried, P. A., Hogan, M. J., and Cameron, I. (2004) Effects of prenatal marijuana on response inhibition: an fMRI study of young adults. *Neurotoxicol Teratol*, **26**, 533–42.

Smith, A. M., Fried, P. A., Hogan, M. J., and Cameron, I. (2006) Effects of prenatal marijuana on visuospatial working memory: an fMRI study in young adults. *Neurotoxicol Teratol*, **28**, 286–95.

Sowell, E. R., Mattson, S. N., Thompson, P. M., Jernigan, T. L., Riley, E. P., and Toga, A. W. (2001a) Mapping callosal morphology and cognitive correlates: effects of heavy prenatal alcohol exposure. *Neurology*, **57**, 235–44.

Sowell, E. R., Thompson, P. M., Mattson, S. N., et al. (2001b) Voxel-based morphometric analyses of the brain in children and adolescents prenatally exposed to alcohol. *NeuroReport*, **12**, 515–23.

Sowell, E. R., Thompson, P. M., Mattson, S. N., et al. (2002a) Regional brain shape abnormalities persist into adolescence after heavy prenatal alcohol exposure. *Cereb Cortex*, **12**, 856–65.

Sowell, E. R., Thompson, P. M., Peterson, B. S., et al. (2002b) Mapping cortical gray matter asymmetry patterns in adolescents with heavy prenatal alcohol exposure. *NeuroImage*, **17**, 1807–19.

Spear, L. P. (2000) The adolescent brain and age-related behavioral manifestations. *Neurosci Biobehav Rev*, **24**, 417–63.

Spear, L. P. (2002) The adolescent brain and the college drinker: biological basis of propensity to use and misuse alcohol. *J Stud Alcohol Suppl*, **14**, 71–81.

Stinson, F. S., Grant, B. F., Dawson, D. A., Ruan, W. J., Huang, B., and Saha, T. (2005) Comorbidity between DSM-IV alcohol and specific drug use disorders in the United States: results from the National Epidemiologic Survey on Alcohol and Related Conditions. *Drug Alcohol Depend*, **80**, 105–16.

Sullivan, E. V., Rosenbloom, M. J., and Pfefferbaum, A. (2000a) Pattern of motor and cognitive deficits in detoxified alcoholic men. *Alcohol Clin Exp Res*, **24**, 611–21.

Sullivan, E. V., Rosenbloom, M. J., Lim, K. O., and Pfefferbaum, A. (2000b) Longitudinal changes in cognition, gait, and balance in abstinent and relapsed alcoholic men: relationships to changes in brain structure. *Neuropsychology*, **14**, 178–88.

Tapert, S. F. and Brown, S. A. (1999) Neuropsychological correlates of adolescent substance abuse: four-year outcomes. *J Int Neuropsychol Soc*, **5**, 481–93.

Tapert, S. F., Brown, G. G., Baratta, M. V., and Brown, S. A. (2004a) fMRI BOLD response to alcohol stimuli in alcohol dependent young women. *Addict Behav*, **29**, 33–50.

Tapert, S. F., Brown, G. G., Kindermann, S. S., Cheung, E. H., Frank, L. R., and Brown, S. A. (2001) fMRI measurement of brain dysfunction in alcohol-dependent young women. *Alcohol Clin Exp Res*, **25**, 236–45.

Tapert, S. F., Cheung, E. H., Brown, G. G., et al. (2003) Neural response to alcohol stimuli in adolescents with alcohol use disorder. *Arch Gen Psychiatry*, **60**, 727–35.

Tapert, S. F., Schweinsburg, A. D., Barlett, V. C., et al. (2004b) Blood oxygen level dependent response and spatial working memory in adolescents with alcohol use disorders. *Alcohol Clin Exp Res*, **28**, 1577–86.

Tarter, R. E., Vanyukov, M., Kirisci, L., Reynolds, M., and Clark, D. B. (2006) Predictors of marijuana use in adolescents before and after licit drug use: examination of the gateway hypothesis. *Am J Psychiatry*, **163**, 2134–40.

Todd, R. D. and Botteron, K. N. (2002) Etiology and genetics of early-onset mood disorders. *Child Adolesc Psychiatr Clin N Am*, **11**, 499–518.

Vanyukov, M. M., Tarter, R. E., Kirisci, L., Kirillova, G. P., Maher, B. S., and Clark, D. B. (2003) Liability to substance use disorders: 1. Common mechanisms and manifestations. *Neurosci Biobehav Rev*, **27**, 507–15.

Wilens, T. E., Biederman, J., Kwon, A., et al. (2004) Risk of substance use disorders in adolescents with bipolar disorder. *J Am Acad Child Adolesc Psychiatry*, **43**, 1380–86.

Wilson, C. R., Sherritt, L., Gates, E., and Knight, J. R. (2004) Are clinical impressions of adolescent substance use accurate? *Pediatrics*, **114**, e536–540.

Wozniak, J. R., Mueller, B. A., Chang, P. N., Muetzel, R. L., Caros, L., and Lim, K. O. (2006) Diffusion tensor imaging in children with fetal alcohol spectrum disorders. *Alcohol Clin Exp Res*, **30**, 1799–806.

Wu, P., Bird, H. R., Liu, X., et al. (2006) Childhood depressive symptoms and early onset of alcohol use. *Pediatrics*, **118**, 1907–15.

Yamaguchi, K. and Kandel, D. B. (1984) Patterns of drug use from adolescence to young adulthood: II. Sequences of progression. *Am J Public Health*, **74**, 668–72.

15

Neuroimaging as a tool for unlocking developmental pathophysiology in anorexia and bulimia nervosa

Guido K. W. Frank and Walter H. Kaye

Introduction

The eating disorders (EDs) anorexia (AN) and bulimia nervosa (BN) are severe psychiatric disorders that most commonly have their onset during adolescence (American Psychiatric Association, 2000; Sullivan, 1995). The teenage years form a critical period of significant changes in both biological and psychosocial development. Neurobiological and genetic factors have only recently been recognized as contributing to the development of AN and BN, in addition to well-known psychological and environmental factors (Bulik, 2005; Frank and Kaye, 2005). This new understanding of the etiology of EDs lays the foundation for a developmental neuroscience perspective in ED research. In this chapter, we describe ED phenotypes and the current knowledge of their neurobiology. Several limitations should be noted. Because women are predominantly affected, research in males has been sparse, and thus this review focuses on the female population. Secondly, although AN and BN usually begin in adolescence, most available neurobiological research has studied women older than 18 years, and developmental aspects have been largely ignored.

Phenotypes of anorexia and bulimia nervosa

Anorexia nervosa is associated with emaciation (body mass index, body weight in kg/height in m²,

≤ 17.5), intense fear of gaining weight, feeling fat despite being underweight, and amenorrhea for 3 months or more (see Table 15.1). About 95% of affected individuals are female. Age of onset is most often adolescence. However, early-onset types, between ages 8 and 12 years, have been reported. Regardless of age of onset, the clinical presentation of AN is quite consistent (Atkins and Silber, 1993). A restricting type has been distinguished from a binge-eating/purging type (American Psychiatric Association, 2000).

Anorexia nervosa has the highest mortality rate of the psychiatric disorders (Sullivan, 1995). In a large cohort study ($n = 103$), at the end of a 12-year follow-up, 7.7% of AN subjects had died. Of the surviving, using a global measure of outcome, 39.6% had a poor outcome, 25.3% an intermediate outcome, and 27.5% a good outcome. Diagnostically, 52.4% showed no major DSM eating disorder, 19.0% remained anorexic, 9.5% had BN, and 19% were classified as eating disorder-not otherwise specified (ED-NOS) (Fichter et al., 2006). Predictors of poor outcome were sexual problems, impulsivity, long duration of inpatient treatment, and long disorder duration. Specialty ED inpatient care may reduce mortality (Lindblad et al., 2006).

Bulimia nervosa is characterized by recurrent binge eating and loss of control over the amount eaten, followed by behaviors to counteract weight gain, such as self-induced vomiting, use of laxatives, or over-exercising. About 90% of BN individuals are

Neuroimaging in Developmental Clinical Neuroscience, eds. Judith M. Rumsey and Monique Ernst. Published by Cambridge University Press. © Cambridge University Press 2009.

Table 15.1 Diagnostic features of anorexia and bulimia nervosa (American Psychiatric Association, 2000)

A. DSM-IV criteria for anorexia nervosa (AN)	B. DSM-IV criteria for bulimia nervosa (BN)
A. *Body weight below 85% of that expected*	A. *Recurrent episodes of binge eating:*
B. *Intense fear of gaining weight* or becoming fat, even though underweight	(1) Within any 2-hour period eating an amount of food definitely larger than most people would eat during similar time period under similar circumstances
C. *Disturbance in the way in which one's body weight or shape is experienced,* undue influence of body weight or shape on self-evaluation, or denial of the seriousness of the current low body weight	(2) A lack of control over eating during the episode
D. In postmenarchal females, *absence of at least three consecutive menstrual cycles*	B. *Recurrent inappropriate compensatory behavior* in order to prevent weight gain (self-induced vomiting; misuse of laxatives, diuretics, enemas or other medications; fasting or excessive exercise)
	C. *Both binge-eating and inappropriate compensatory behaviors occur, at least twice a week for 3 months*
	D. *Self-evaluation is unduly influenced by body shape and weight*
	E. *The disturbance does not occur exclusively during episodes of anorexia nervosa*
Subtypes	**Subtypes**
Restricting type: During current AN episode, person has not regularly engaged in binge eating or purging behavior (i.e., self-induced vomiting or the misuse of laxatives, diuretics, or enemas)	Purging type: During the current episode of BN, the person has regularly engaged in self-induced vomiting or the misuse of laxatives, diuretics or enemas
Binge-eating/purging type: During the current episode of AN the person has regularly engaged in binge-eating or purging behavior	Non-purging type: During the current episode of BN, the person has used inappropriate compensatory behaviors, such as fasting or excessive exercise, but has not regularly engaged in self-induced vomiting or the misuse of laxatives, diuretics or enemas

females, usually of normal body weight, who suffer from fears of gaining weight and preoccupations with food and body weight. After a 12-year follow-up (Fichter and Quadflieg, 2004), 70% of BN subjects had recovered, 10% had BN, 2% had died, and 13% had an ED-NOS.

Adolescents may have a better prognosis than adults (Fisher, 2003), but juvenile occurrence of an ED is often followed by adult ED (Kotler et al., 2001). Although the prognosis is generally unclear (Steinhausen et al., 2000), comorbid psychiatric disorders predict negative ED outcomes (Fichter and Quadflieg, 2004). Eating disorders are often associated with mood and anxiety disorders. Major depressive disorder co-occurs in AN (up to 40%) and, even more so, in BN (up to 80%) (Godart et al., 2007).

A large study from our group (Kaye et al., 2004) reported obsessive-compulsive disorder (OCD) in 41% and social phobia in 20% of cases with a history of AN or BN. Importantly, this study found that anxiety symptoms predated the ED onset and thus might have contributed to risk for EDs.

Risk factors

Neurobiological studies implicate serotonin, dopamine, gonadal hormones, genetic variations, and trait behaviors in the pathology of EDs. Similarly, congenital insults or disturbances of maturational processes could be risk factors.

Serotonin

Brain serotonin (5-hydroxytryptamine: 5HT) is involved in mood, anxiety, feeding and sleep (Naughton et al., 2000). Early studies used cerebrospinal fluid (CSF) 5HT levels to approximate 5HT brain levels (Stanley et al., 1985). Symptomatic AN subjects showed reductions of the CSF 5HT metabolite 5-hydroxyindole acetic acid (5HIAA) compared to healthy controls (Kaye et al., 1984). Symptomatic BN patients as a group were found to have normal CSF 5HIAA levels (Kaye and Weltzin, 1991), but the more severely affected presented lower CSF 5HIAA levels (Jimerson et al., 1992). In contrast, those who had recovered from AN and from BN showed elevated CSF 5HIAA (Kaye et al., 1991). This suggests that abnormally high 5HT brain levels could be a trait marker and ED behavior might be a means to reduce 5HT transmission.

Dopamine

Brain dopamine (DA) plays a key role in the brain's reward systems (Schultz, 2002). The DA pathways arise from the midbrain and project to three main regions: striatum, tubero-infundibular area, and prefrontal cortex. Our group found that recovered restricting-type AN had reduced concentrations of CSF homovanillic acid (HVA), the major metabolite of DA, compared to controls (Kaye et al., 1999). Whether individuals with restricting AN have intrinsically lower DA remains uncertain and in need of replication. However, if replicated, this might hold profound implications for the mechanism of reduced food reward in AN (Schultz, 2002).

Hormonal and neuropeptide influences

Gonadal hormones are low during the symptomatic state of EDs, and multiple neuropeptides are also affected. Corticotropin releasing hormone (CRH), opioids, neuropeptide Y (NPY) and peptide YY (PYY), vasopressin and oxytocin, cholecystokinin (CCK), and leptin have been reported to be altered, usually being low, but commonly remit with

recovery (Bailer and Kaye, 2003). The role of these abnormalities as causal or secondary remains to be clarified (Barbarich et al., 2003).

Genotype

Anorexia and bulimia nervosa share common genetic vulnerabilities, based on familial cross-transmission. The prevalence of AN and BN is increased 7- to 12-fold in relatives of ED probands compared to controls (Strober et al., 2000), with 60%–80% of the variance explained by genetic factors (Bulik et al., 2000). These heritability estimates are similar to those found in schizophrenia and bipolar disorder, suggesting that AN and BN may be highly genetically influenced. Various studies have linked specific chromosomes or genes with EDs. Two studies have implicated polymorphisms, that is genetic variations, in the serotonin 1D (HTR1D) and the opioid delta (OPRD1) receptor in AN (Bergen et al., 2003; Brown et al., 2006). However, the specific genetic sequence involved remains to be identified. A polymorphism in the promoter region of the 5HT2A-1438G/A receptor has been repeatedly implicated in AN and BN, and most of the evidence now indicates that the 5HT2A-1438A/A genotype is more common in restricting-type AN compared to controls (Nacmias et al., 1999). Reward dependence, a measure of one's need to please others and be rewarded by approval (Cloninger, 1987), has also been associated with the 5HT2A-1438A/A promoter genotype. Adolescent AN patients homozygous for the 5HT2A-1438A allele had lower reward dependence than subjects who had the G/G genotype (Rybakowski et al., 2006). In contrast, adults did not show such an association (Kusumi et al., 2002). Various other associations of the 5HT2A-1438 genotype have been reported. The 5HT2A-1438A/A promoter genotype has been associated with greater weight concerns and ED symptom severity (Ricca et al., 2004), as well as with lower energy and with total, monounsaturated, and saturated fat intakes (Herbeth et al., 2005). In summary, the function of the 5HT2A-1438A/A promoter polymorphism is still unclear, but evidence suggests that this receptor is related to AN subtypes and trait behavior regulation. Other studies (Bergen et al., 2005; Frieling et al., 2006; Miyasaka et al., 2006; Ribases et al., 2005) have suggested a role in EDs for catecholamine-*O*-methyl-transferase (COMT, which breaks down neuro-transmitters such as dopamine, epinephrine and norepinephrine), the 5HT transporter (reuptake of synaptic 5HT), the brain-derived neurotrophic factor (neuronal growth), ghrelin (appetite-stimulating gut hormone), DA-D2 receptor (DA release) poly-morphisms, and others. However, replication studies have produced inconsistent results. In summary, the 5HT2A-1438G/A, the serotonin HTR1D receptor, and the opioid OPRD1 receptor gene variants seem to be the best candidates for potential genetic contributors to ED pathophysiology.

Trait behaviors

Behavioral abnormalities that remain after pro-longed recovery may have existed premorbidly and could represent risk factors for ED develop-ment. After recovery, AN and BN women show persisting obsessional behaviors, inflexible think-ing, restraint in emotional expression, and a high degree of self- and impulse-control problems (Anderluh et al., 2003; Holliday et al., 2005). Symp-tomatic and recovered AN and BN patients have elevated harm avoidance (Klump et al., 2004), a construct reflecting anxious, avoiding behavior, theorized to be related to 5HT activity (Cloninger, 1987). Impaired set shifting as a marker for cogni-tive inflexibility has been hypothesized to be an ED endophenotype (Holliday et al., 2005).

The frequent treatment resistance of patients with AN or BN may reflect such cognitive inflexibil-ity and behavioral rigidity, together with being fear-ful of change even when fully aware of the severe health consequences of their disorder. In summary, heightened anxiety and cognitive inflexibility may be traits related to the development and perpetu-ation of eating disorders. These symptoms persist, but are often reduced in intensity following recovery.

Congenital risks

Prenatal complications and insults have been implicated as risk factors for EDs, although not understood mechanistically. A large prospective study reported that a number of preterm delivered babies, who presented with cephalohematomas and low birth weight, developed AN years later (Cnattingius et al., 1999). The authors suggested that structural brain damage could be considered a vulnerability factor for AN. Another study found maternal anemia, diabetes mellitus, preeclampsia, placental infarction, and neonatal cardiac problems to be risk factors for AN, whereas placental infarction, neonatal hyporeactivity, early eating difficulties, and low birth weight were risk factors for BN (Favaro et al., 2006). Smoking during pregnancy has been associated with BN in offspring by age 30, but inter-estingly not with AN (Montgomery et al., 2005). These hypothesized congenital risk factors must be viewed with caution since they represent findings from only a few studies. For instance, whether low birth weight is more related to AN (Cnattingius et al., 1999) than to BN (Favaro et al., 2006) is unclear. Favaro (Favaro et al., 2006) used a prospective design and adjusted for socioeconomic status, maternal age, marital status, parity, and multiple births, while Cnattingius (Cnattingius et al., 1999) used hospital records. Environmental factors may greatly influence study outcome and should be further investigated.

Childhood developmental risk factors

Childhood and adolescence are transition periods during which critical structural, neurochemical, hormonal, and functional cerebral changes take place. Some of the prominent changes observed in human brain development include initial volumet-ric increases in gray matter, followed by decreases around puberty (Gogtay et al., 2004). Research in rhesus monkeys suggests that an initial overproduc-tion of central dopamine (DA-D1, DA-D2), serotonin (5HT1 and 5HT2), and adrenergic (α1, α2 and β) receptors (Lidow and Rakic, 1992) in the first few months of life is followed by a gradual decrease during childhood up to puberty. As mentioned earlier, these neurotransmitter systems affect motivation to eat and satiety (Houpt, 2000). Opioid transmission is also of interest since it mediates the hedonic experience of food (Kelley, 2004). However, developmental neuroscience has not yet addressed opioid neurotransmission in ED research in humans.

Basic habits, including eating behavior, are learned from caregivers and peers during childhood and adolescence (Mikkila et al., 2005). Individual eating habits may modulate biological feedback mechanisms, such as DA neuronal priming or conditioning for certain foods in a certain context. Dieting is a significant behavioral risk factor for ED onset (Patton et al., 1999), and weight concerns were found to be predictive of ED presence in a 4-year prospective study of high school girls (Killen et al., 1996). Brain imaging might be used in adolescents to assess the extent to which "normal" dieting alters brain function and potentially contributes to ED development.

Sexual maturation also has been proposed to influence ED etiology. Early puberty and sexual experience were found to be associated with BN in girls (Kaltiala-Heino et al., 2001). During puberty, stores of subcutaneous fat tissue increase in adolescent girls; this, in turn, stimulates leptin release to the hypothalamus, and affects various neuropeptides.

Animal models that manipulate eating behavior have been used to study the role of neuropeptides in eating regulation. Neuropeptide Y neuronal activity has been found to be increased, and the cocaine- and amphetamine-regulated transcript protein (CART) has been found to be decreased in rodent models for obesity (King, 2005). Other neuropeptides of interest associated with feeding regulation include the melanocortin stimulating hormone, CCK, and the agouti-related protein (King, 2005). Disturbance at any point in a series of brain events during development could impact feeding regulation. Such developmental processes have not been studied in AN or BN.

Neurotransmitter receptor changes during development may also predispose to EDs. For example, food restriction in rodents has been shown to increase DA-D2 receptor binding (Carr et al., 2003). A study in humans from our group reported increased DA-D2/D3 receptor availability in recovered AN subjects (Frank et al., 2005). In addition, DA abnormalities during developmentally critical periods or when symptomatic with AN could persist and lead to deficits in cognitive flexibility (Floresco et al., 2006; Von Huben et al., 2006). Neurocognitive deficits that include spatial memory impairments (Fowler et al., 2006), as well as cognitive inflexibility (impaired set shifting), have been described in women with AN (Holliday et al., 2005).

Brain imaging studies of eating disorders

Structural and resting-state metabolic neuroimaging

Several structural brain imaging studies using MRI found reduced gray and white matter in symptomatic AN and BN, but recent data indicate that these abnormalities are state dependent since they seem to remit with recovery (Katzman et al., 1996; Wagner et al., 2006). Other abnormalities, which normalized with weight restoration, have been found in symptomatic AN using magnetic resonance spectroscopy (MRS) and positron emission tomography (PET). Using MRS to examine prefrontal cortex, Castro-Fornieles et al. (2007) reported a lower ratio of N-acetylaspartate (required for brain myelin synthesis and a precursor of neuronal peptides) to choline (needed for cell membrane integrity and neurotransmitters), a lower ratio of glutamate (excitatory neurotransmitter) to glutamine (derived from glutamate, constituent of proteins) and lower myo-inositol (cell membrane constituent and secondary messenger). Using PET, ill ED subjects frequently show resting state frontal, parietal, and temporal brain hypoperfusion (Frank et al., 2004). These MRS and PET findings during the ill state that reflect neuronal membrane damage and reduced hemodynamic brain responsiveness seem to be secondary to malnutrition, rather than etiological (Frank et al., 2004, 2007).

Functional neuroimaging

Most functional task activation and neurotransmitter-receptor studies in EDs have been conducted with adults. Table 15.2 summarizes these studies which are described in the next section.

Anorexia nervosa

Task-activation studies

Functional imaging studies of AN have used pictures of or actual food that triggered anxiety and resulted in the activation of frontal, temporal (including the amygdala), and cingulate cortices (Ellison et al., 1998; Gordon et al., 2001; Naruo et al., 2000; Nozoe et al., 1993, 1995) (see Table 15.2). The amygdala, in particular, has been implicated in fear conditioning (LeDoux, 2003), which may play a role in AN. In a cross-sectional comparison of symptomatic and recovered AN (Uher et al., 2003), pictures of food activated medial prefrontal and anterior cingulate regions in both recovered and symptomatic AN, but lateral prefrontal regions were activated only in recovered AN. Higher activity in the anterior cingulate and medial prefrontal cortex in both symptomatic and recovered AN, compared to healthy control women, was hypothesized to be a trait marker for AN.

Body image distortion is a key symptom of AN. Functional MRI (fMRI) studies (Seeger et al., 2002; Uher et al., 2005; Wagner et al., 2003) have examined the neural responses of AN subjects and controls to body-image-related tasks. Seeger et al.'s (2002) study used a paradigm wherein subjects saw on a monitor distorted images (increased size) of their thighs, breasts or abdominal circumference. Increased amygdala activation was seen in AN subjects ($n = 3$) compared to controls (Seeger et al., 2002). A follow-up study in a larger sample using a similar design reported increased prefrontal and inferior parietal cortex activation in the AN subjects compared to controls, but the previous amygdala finding could not be replicated (Wagner et al., 2003). This latter study indicated a hyper-responsiveness of brain areas belonging to the frontal visual attention system [Brodmann's area

(BA) 9] and parietal visuospatial processing area [inferior parietal lobule (BA 40), including the anterior portion of the intraparietal sulcus]. Another study found reduced parietal cortex activation in AN after visual presentation of line drawings of body shapes (thin, normal, overweight) (Uher et al., 2005). These conflicting body image findings are difficult to interpret, particularly in view of the different tasks being used.

Taken together, these findings suggest that neural activity in anterior cingulate, mesial temporal (including the amygdala), and parietal regions frequently differs between AN and healthy control women. Further study is needed to determine whether these abnormalities are specific to AN and if they are particularly related to body image distortion, disturbed mood, or heightened anxiety.

Receptor imaging studies

As mentioned above, the serotonin neurotransmitter system has been implicated in EDs. In particular the 5HT1A and 5HT2A receptors are believed to be involved in the modulation of mood, feeding, impulse control, sleep, and anxiety. Using PET and the radioligand [^{11}C]WAY, 5HT1A receptor binding has been found to be elevated across most brain regions in a mixed group of symptomatic restricting and binge-eating/purging-type AN subjects compared to healthy controls (Bailer et al., 2007), as well as in binge-eating/purging-type AN after recovery (Bailer et al., 2005). In contrast, recovered restricting-type AN show normal brain 5HT1A binding (Bailer et al., 2005).

For the 5HT2A receptor type, one group, using single photon emission computed tomography (SPECT) and the radioligand ^{123}I-5-I-R91150, found reduced binding in the frontal, parietal and occipital cortices of symptomatic AN (Audenaert et al., 2003). A follow-up of that study (Goethals et al., 2007) separated AN by subtype and suggested that the binge-eating/purging-type AN group ($n = 7$) has reduced parietal 5HT2A binding compared to the restricting-type AN subjects ($n = 9$), but this study did not include a control group. A study that controlled for brain volume loss found

Table 15.2 Functional brain imaging studies in anorexia and bulimia nervosa

Year	Author	Method	Activation	Age Mean (SD)	ILL	REC	n	Frontal cortex	Temporal cortex/ L	R	amygdala	Cingulate cortex L	R	Parietal cortex L	R	Occipital cortex	Other ROIs L	R	ROIs
AN "Neuroreceptor" studies																			
2003	Audenaert	SPECT 5HT2A		23(3)	AN*	AN	15		↓	nl	nl				nl		↓	↓	AVS ↑
2002	Frank	PET 5HT2A		25(6)		AN	16	nl				↓		nl			↓	↓	Raphe ↑
2004	Bailer	PET 5HT2A		25(3)		AN-B/P	10										↓		
2005	Bailer	PET 5HT1A		23(5)		AN-R	13	nl		nl	nl	nl		nl		nl			
2005	Frank	PET DAD2/D3		29(7)		AN-B/P	12	↑		↑	↑	↑		↑	↑				
2007	Bailer	PET 5HT1A		24(5)	AN*	AN	10	↑											
2007	Bailer	PET 5HT2A		25(5)	AN*		15	nl		nl	nl	nl		↑	nl	nl			
	Bailer			25(5)	AN*		15	nl											
AN "Task activation" studies																			
1995	Nozoe	SPECT	Eating food	24(8)	AN*		8	nl		nl		nl	↑	↑	↑	↑	↑	↑	
2000	Naruo	SPECT	Food images	22(8)	AN-R		7	nl							nl				
2000	Naruo	SPECT	Food images	26(3)	AN-B/P		7		nl	↑				nl	↑				
2001	Gordon	PET rCBF	Food images	20(3)	AN*		8	nl		nl	nl	nl		nl		nl			
1998	Ellison	fMRI	Food images	17(0.5)	AN*		6	nl		nl		↑	↑	↑		nl			
2002	Seeger	fMRI	Body image	15(1)	AN-R		3	nl				nl	nl			nl			
2003	Wagner	fMRI	Body image	26(3)/27(5)	AN-R		15	↑		↑	nl	↑		↑	↑				
2003	Uher	fMRI	Food images	27(5)	AN-R	AN-R		↑			↑								
2005	Uher	fMRI	Food and emotional images	27(12)	AN*		16	↑ OFC ↓ lat PF (Food)				↑ (Food)							
2005	Uher	fMRI	Body image	25(10)	AN*		13	↓						↓	↓	↓	↓		
2006	Santel	fMRI	Food images	16(2)	AN*		13												
BN "Neuroreceptor" studies																			
2001	Tauscher	SPECT 5HTT		30(4)	BN	BN	10												
2001	Kaye	PET 5HT2A		26(5)	BN		9	↓		nl		nl		nl	nl	nl			
2004	Tiihonen	PET 5HT1A		22(5)	BN		8	↑		↑		↑		↑					
2004	Goethals	SPECT 5HT2A			BN		10	nl		nl		nl		nl	nl	nl			
BN "Task activation" studies																			
1995	Nozoe	SPECT	Eating food	21(3)	BN	BN	5	nl		nl		↑	nl	nl		nl		↓	
2004	Frank	fMRI	Sweet taste	28(7)	BN	BN	10												
2005	Uher	fMRI	Food and emotional images	30(9)	BN		10	↑ OFC ↓ lat PF (Food)				↑ (Food)		↑	↑	→			
2005	Uher	fMRI	Body image	30(9)	BN		9	↓						↓	↓	↓			

Notes:

nl, normal; ↓, decreased compared to controls; ↑, increased compared to controls; AN-R, anorexia nervosa, restricting type; AN-B/P, anorexia nervosa, binging-purging type; AN*, diagnostic subgroup not specified; BN, bulimia nervosa; REC, recovered. PET, positron emission tomography; SPECT, single photon emission computed tomography; 5HT, serotonin; DA, dopamine; fMRI, functional magnetic resonance imaging; AVS, anteroventral striatum; OFC, orbitofrontal cortex; lat PF, lateral prefrontal cortex; mean age expressed in years ± SD.

normal 5HT2A receptor (radioligand [^{18}F]altanserin) availability in symptomatic restricting and binge-eating/purging-type AN (Bailer et al., 2007). After recovery, both restricting and binge-eating/purging-type AN had reduced subgenual cingulate 5HT2A binding (Bailer et al., 2004; Frank et al., 2002). The restricting-type AN group also presented with significantly reduced mesial temporal (amygdala and hippocampus) 5HT2A binding. In summary, after recovery, 5HT1A receptor binding seems to differentiate AN subtypes, whereas 5HT2A receptor binding is reduced in both restricting and binge-eating/purging-type AN in various brain regions. Since these disturbances occur after recovery, they may reflect either trait disturbances or scars from the illness.

Harm avoidance (the heritable tendency to respond intensely to aversive stimuli and inhibit behavior in order to avoid punishment) (Cloninger, 1986), a behavioral correlate of anxiety, has been found to be correlated positively with mesial temporal cortex 5HT2A binding in recovered binge-eating/purging-type AN and with mesial temporal cortex 5HT1A binding in recovered restricting-type AN. Such correlations may help explain how 5HT receptor activity, together with other neurotransmitter systems, can modulate complex behaviors such as harm avoidance. Specifically, high harm avoidance in AN could be a key trait related to abnormal 5HT1A and 5HT2A receptor expression. The associated pattern of heightened anxiety and lack of cognitive flexibility that seem to be part of the AN phenotype could affect treatment response.

Recently, we have found increased DA-D2/D3 receptor binding ([^{11}C]raclopride, PET) in the antero-ventral striatum of a group of recovered restricting and binge-eating/purging-type AN (Frank et al., 2005). Figure 15.1 shows preliminary results from 14 control women and 16 recovered restricting-type AN suggesting ventral striatal DA-D2/D3 receptor increase, extending toward the amygdala. This receptor increase could be consistent with reduced CSF DA metabolites found in the past (Kaye et al., 1999) and may further suggest DA abnormalities in AN. Interestingly, reduced DA-D2/D3 receptor binding has been found in obese subjects (Wang et al.,

2004), and the DA-D2/D3 receptor binding rate could thus be a biological correlate of the tendency to restrict food or overeat.

Bulimia nervosa

Task-activation studies

Using SPECT, Nozoe found that, compared to controls, BN subjects had greater right inferior frontal and left temporal blood flow before a meal, but similar activity after a meal (Nozoe et al., 1995). The assessed brain regions were large, and finer regional specificity was not reported. Bulimia nervosa subjects showed increased hedonic ratings for sweet stimuli compared to controls (Drewnowski et al., 1987) and therefore may have altered centrally processed taste perception. An fMRI study of 10 recovered bulimic subjects from our group (Frank et al., 2006), using a glucose challenge, found reduced anterior cingulate activity compared to six healthy control women. The anterior cingulate is involved in the anticipation of reward (Richmond et al., 2003). Whether recovered bulimic subjects have less of a cognitive reward-expectation-related brain response to sweet stimuli has yet to be tested with more specific reward-related tasks. One study investigated body image perception in BN compared to controls and found results similar to those reported in AN subjects, i.e., reduced occipital/parietal activation in response to line drawing pictures of body shapes (thin, normal, overweight) (Uher et al., 2005). This would implicate a common neural network involving parietal cortex in EDs.

Receptor imaging studies

Kaye et al. (2001) found reduced orbitofrontal 5HT2A receptor binding in recovered BN using PET and the radioligand [^{18}F]altanserin. A number of studies have implicated the orbitofrontal cortex in inhibitory processes (Robbins, 2005) and in the representation of food-related affective values (Kringelbach et al., 2003). Thus, orbitofrontal alterations may contribute to behavioral disturbances associated with BN, such as impulsivity and altered emotional processing (Steiger et al., 2001). No specific

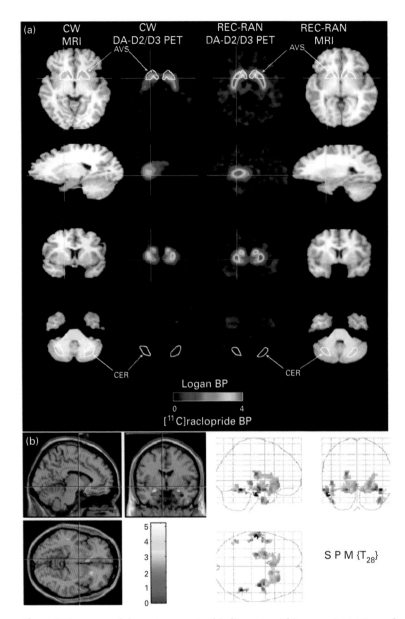

Figure 15.1. Increased dopamine receptor binding potential in anorexia. (a) Exemplary higher dopamine (DA) D2/D3 receptor binding potential as assessed with [^{11}C]raclopride and positron emission tomography (PET) in a recovered restricting-type anorexia nervosa (AN-R) woman compared to a healthy age-matched control woman (CW). AVS, Anteroventral striatum; CER, cerebellum (reference region). (b) Group comparison of 14 CW and 16 restricting-type AN using statistical parametric mapping (SPM) indicating increased DA-D2/D3 receptor binding in the ventral striatum, extending into the amygdala, in AN-R.

tasks, however, have been applied in conjunction with 5HT receptor binding that would assess such a behavior–neuroreceptor relationship. Women with BN also failed to show the negative correlations of age and 5HT2A binding found in normal controls (Kaye et al., 2001). This lack of correlation may reflect a scarring effect from the illness. Symptomatic BN patients were found to show reduced 5HT transporter binding in the thalamus and hypothalamus (Tauscher et al., 2001), but increased 5HT1A receptor binding (Tiihonen et al., 2004), most prominently in the medial prefrontal cortex, posterior cingulate, and angular gyrus of the parietal cortex.

The dynamics between 5HT transporter expression and synaptic 5HT are not well understood. Two explanatory hypotheses can be entertained: either 5HT transporter up-regulation (negative feedback) in response to low 5HT (Meyer et al., 2004), or adaptive, dynamic 5HT transporter reduction in order to adjust to hypothesized low 5HT (Parsey et al., 2006). Reduced 5HT2A binding in recovered BN subjects may represent a higher level of endogenous 5HT in the synaptic cleft or a down-regulation of the receptor. With the same schema, increased 5HT1A receptor binding during the symptomatic state may reflect a reduced 5HT synaptic level or up-regulation of the receptor (Jimerson et al., 1992; Kaye et al., 1998). Of interest, selective 5HT reuptake inhibitors (SSRIs) are effective in the treatment of BN (Steffen et al., 2006), but symptomatic BN requires higher doses of such medications compared to, for instance, patients being treated for depression (Fluoxetine Bulimia Nervosa Collaborative Study Group, 1992). This relative resistance to SSRI treatment may be related to an up-regulation of the 5HT1A autoreceptors, which inhibit 5HT release (Hensler, 2003).

Conclusions

Brain imaging methods provide unique opportunities for studying brain function in vivo in relation to the psychopathology of EDs. The typical age of onset during adolescence and skewed gender distribution suggest a strong role for developmental factors in the etiology of EDs. We are entering an exciting time when neuroimaging can be combined with other biological approaches, particularly genetics and cognitive neuroscience, to unravel pathophysiological mechanisms of disease and identify new treatment targets.

REFERENCES

American Psychiatric Association (2000) *Diagnostic and Statistical Manual of Mental Disorders: DSM-IV-TR*, 4th edn. Arlington, VA: American Psychiatric Association.

Anderluh, M. B., Tchanturia, K., Rabe-Hesketh, S., et al. (2003) Childhood obsessive-compulsive personality traits in adult women with eating disorders: defining a broader eating disorder phenotype. *Am J Psychiatry*, **160**(2), 242–7.

Atkins, D. M. and Silber, T. J. (1993) Clinical spectrum of anorexia nervosa in children. *J Dev Behav Pediatr*, **14**(4), 211–16.

Audenaert, K., Van Laere, K., Dumont, F., et al. (2003) Decreased 5-HT2a receptor binding in patients with anorexia nervosa. *J Nucl Med*, **44**(2), 163–9.

Bailer, U. F. and Kaye, W. H. (2003) A review of neuropeptide and neuroendocrine dysregulation in anorexia and bulimia nervosa. *Curr Drug Targets CNS Neurol Disord*, **2**(1), 53–9.

Bailer, U. F., Price, J. C., Meltzer, C. C., et al. (2004) Altered 5-HT2A receptor activity after recovery from bulimia-type anorexia nervosa: relationships to harm avoidance and drive for thinness. *Neuropsychopharmacology*, **29**(6), 1143–55.

Bailer, U. F., Frank, G. K., Henry, S. E., et al. (2005) Altered brain serotonin 5-HT1A receptor binding after recovery from anorexia nervosa measured by positron emission tomography and [carbonyl[11]C]WAY-100635. *Arch Gen Psychiatry*, **62**(9), 1032–41.

Bailer, U. F., Frank, G. K., Henry, S. E., et al. (2007) Exaggerated 5-HT$_{1A}$ but normal 5-HT$_{2A}$ receptor activity in individuals ill with anorexia nervosa. *Biol Psychiatry*, **61**(9), 1090–9.

Barbarich, N., Kaye, W., and Jimerson, D. (2003) Neurotransmitter and imaging studies in anorexia nervosa: new targets for treatment. *Curr Drug Targets CNS Neurol Disord*, **2**(1): 61–73.

Bergen, A., van den Bree, M. B. M., Yeager, M., et al. (2003) Candidate genes for anorexia nervosa in the 1p33–36 linkage region: serotonin 1D and delta opioid receptors display significant association to anorexia nervosa. *Mol Psychiatry*, **8**(4), 397–406.

Bergen, A. W., Yeager, M., Welch, R. A., et al. (2005) Association of multiple DRD2 polymorphisms with anorexia nervosa. *Neuropsychopharmacology*, **30**(9), 1703–10.

Brown, K. M., Bujac, S. R., Mann, E. T., et al. (2006) Further evidence of association of OPRD1 and HTR1D polymorphisms with susceptibility to anorexia nervosa. *Biol Psychiatry*, **61**(3), 367–73.

Bulik, C. M. (2005) Exploring the gene-environment nexus in eating disorders. *J Psychiatry Neurosci*, **30**(5), 335–9.

Bulik, C. M., Sullivan, P. F., Wade, T. D., et al. (2000) Twin studies of eating disorders: a review. *Int J Eat Disord*, **27**(1), 1–20.

Carr, K. D., Tsimberg, Y., Berman, Y., et al. (2003) Evidence of increased dopamine receptor signaling in food-restricted rats. *Neuroscience*, **119**(4), 1157–67.

Castro-Fornieles, J., Bargalló, N., Lázaro, L., et al. (2007) Adolescent anorexia nervosa: cross-sectional and follow-up gray matter disturbances detected with proton magnetic resonance spectroscopy. *J Psychiatry Res*, **41**(11), 952–8.

Cloninger, C. R. (1986) A unified biosocial theory of personality and its role in the development of anxiety states. *Psychiatr Dev*, **4**(3), 167–226.

Cloninger, C. (1987) A systematic method for clinical description and classification of personality variants. A proposal. *Arch Gen Psychiatry*, **44**(6), 573–88.

Cnattingius, S., Hultman, C. M., Dahl, M., et al. (1999) Very preterm birth, birth trauma, and the risk of anorexia nervosa among girls. *Arch Gen Psychiatry*, **56**(7), 634–8.

Drewnowski, A., Bellisle, F., Aimez, P., et al. (1987) Taste and bulimia. *Physiol Behav*, **41**, 621–6.

Ellison, Z., Foong, J., Howard, R., et al. (1998) Functional anatomy of calorie fear in anorexia nervosa. *Lancet*, **352**(9135), 1192.

Favaro, A., Tenconi, E., and Santonastaso, P. (2006) Perinatal factors and the risk of developing anorexia nervosa and bulimia nervosa. *Arch Gen Psychiatry*, **63**(1), 82–8.

Fichter, M. M. and Quadflieg, N. (2004) Twelve-year course and outcome of bulimia nervosa. *Psychol Med*, **34**(8), 1395–406.

Fichter, M. M., Quadflieg, N., and Hedlund, S. (2006) Twelve-year course and outcome predictors of anorexia nervosa. *Int J Eat Disord*, **39**(2), 87–100.

Fisher, M. (2003) The course and outcome of eating disorders in adults and in adolescents: a review. *Adolesc Med*, **14**(1), 149–58.

Floresco, S. B., Magyar, O., Ghods-Sharifi, S., et al. (2006) Multiple dopamine receptor subtypes in the medial prefrontal cortex of the rat regulate set-shifting. *Neuropsychopharmacology*, **31**(2), 297–309.

Fluoxetine Bulimia Nervosa Collaborative Study Group (1992) Fluoxetine in the treatment of bulimia nervosa. A multicenter, placebo-controlled, double-blind trial. *Arch Gen Psychiatry*, **49**(2), 139.

Fowler, L., Blackwell, A., Jaffa, A., et al. (2006) Profile of neurocognitive impairments associated with female in-patients with anorexia nervosa. *Psychol Med*, **36**(4), 517–27.

Frank, G. K. and Kaye, W. H. (2005) Positron emission tomography studies in eating disorders: multireceptor brain imaging, correlates with behavior and implications for pharmacotherapy. *Nucl Med Biol*, **32**(7), 755–61.

Frank, G. K., Bailer, U., Henry, S., et al. (2004) Neuroimaging studies in eating disorders. *CNS Spectr*, **9**(7), 539–48.

Frank, G. K., Bailer, U. F., Henry, S. E., et al. (2005) Increased dopamine D2/D3 receptor binding after recovery from anorexia nervosa measured by positron emission tomography and [^{11}C]raclopride. *Biol Psychiatry*, **58**(11), 908–12.

Frank, G. K. Bailer, U. F., Meltzer, C. C., et al. (2007) Regional cerebral blood flow after recovery from anoxexia or bulimia nervosa. *Int J Eat Disord*, **40**(6), 488–92.

Frank, G. K., Kaye, W. H., Meltzer C. C., et al. (2002) Reduced 5-HT2A receptor binding after recovery from anorexia nervosa. *Biol Psychiatry*, **52**(9), 896–906.

Frank, G. K., Wagner, A., Achenbach, S., et al. (2006) Altered brain activity in women recovered from bulimic-type eating disorders after a glucose challenge: a pilot study. *Int J Eat Disord*, **39**(1), 76–9.

Frieling, H., Romer, K. D., Wilhelm, J., et al. (2006) Association of catecholamine-*O*-methyltransferase and 5-HTTLPR genotype with eating disorder-related behavior and attitudes in females with eating disorders. *Psychiatr Genet*, **16**(5), 205–8.

Godart, N. T., Perdereau, F., Rein, Z., et al. (2007) Comorbidity studies of eating disorders and mood disorders. Critical review of the literature. *J Affect Disord*, **97**(1–3), 37–49.

Goethals, I., Vervaet, M., Audenaert, K., et al. (2004) Comparison of cortical 5-HT2A receptor binding in bulimia nervosa patients and healthy volunteers. *Am J Psychiatry*, **161**(10), 1916–18.

Goethals, I., Vervaet, M., Audenaert, K., et al. (2007) Differences of cortical 5-HT(2A) receptor binding index with SPECT in subtypes of anorexia nervosa: relationship with personality traits? *J Psychiatr Res*, **41**(5), 455–8.

Gogtay, N., Giedd, J. N., Lusk, L., et al. (2004) Dynamic mapping of human cortical development during childhood through early adulthood. *Proc Natl Acad Sci U S A*, **101**(21), 8174–9.

Gordon, C. M., Dougherty, D. D., Fischman, A. J., et al. (2001) Neural substrates of anorexia nervosa: a behavioral challenge study with positron emission tomography. *J Pediatr*, **139**(1), 51–7.

Hensler, J. (2003) Regulation of 5-HT$_{1A}$ receptor function in brain following agonist or antidepressant administration. *Life Sci*, **75**(15), 1665–82.

Herbeth, B., Aubry, E., Fumeron, F., et al. (2005) Polymorphism of the 5-HT2A receptor gene and food intakes in children and adolescents: the Stanislas Family Study. *Am J Clin Nutr*, **82**(2), 467–70.

Holliday, J., Tchanturia, K., Landau, S., et al. (2005) Is impaired set-shifting an endophenotype of anorexia nervosa? *Am J Psychiatry*, **162**(12), 2269–75.

Houpt, T. A. (2000) Molecular neurobiology of ingestive behavior. *Nutrition*, **16**(10), 827–36.

Jimerson, D., Lesem, M., Kaye, W., et al. (1992) Low serotonin and dopamine metabolite concentrations in cerebrospinal fluid from bulimic patients with frequent binge epidsodes. *Arch Gen Psychiatry*, **49**, 132–8.

Kaltiala-Heino, R., Rimpela, M., Rissanen, A., et al. (2001) Early puberty and early sexual activity are associated with bulimic-type eating pathology in middle adolescence. *J Adolesc Health*, **28**(4), 346–52.

Katzman, D. K., Lambe, E. K., Mikulis, D. J., et al. (1996) Cerebral gray matter and white matter volume deficits in adolescent girls with anorexia nervosa. *J Pediatr*, **129**, 794–803.

Kaye, W. H. and Weltzin, T. E. (1991) Serotonin activity in anorexia and bulimia nervosa: relationship to the modulation of feeding and mood. *J Clin Psychiatry*, Suppl, **52**, 41–8.

Kaye, W. H., Bulik, C. M., Thornton, L., et al. (2004) Comorbidity of anxiety disorders with anorexia and bulimia nervosa. *Am J Psychiatry*, **161**(12), 2215–21.

Kaye, W. H., Ebert, M. H., Raleigh, M., et al. (1984) Abnormalities in CNS monoamine metabolism in anorexia nervosa. *Arch Gen Psychiatry*, **41**(4), 350–5.

Kaye, W. H., Frank, G. K., and McConaha, C. (1999) Altered dopamine activity after recovery from restricting-type anorexia nervosa. *Neuropsychopharmacology*, **21**(4), 503–6.

Kaye, W. H., Frank, G. K., Meltzer, C. C., et al. (2001) Altered serotonin 2A receptor activity in women who have recovered from bulimia nervosa. *Am J Psychiatry*, **158**(7), 1152–5.

Kaye, W. H., Greeno, C. G., Moss, H., et al. (1998) Alterations in serotonin activity and psychiatric symptomatology after recovery from bulimia nervosa. *Arch Gen Psychiatry*, **55**(10), 927–35.

Kaye, W. H., Gwirtsman, H. E., George, D. T., et al. (1991) Altered serotonin activity in anorexia nervosa after long-term weight restoration. Does elevated cerebrospinal fluid 5-hydroxyindoleacetic acid level correlate with rigid and obsessive behavior? Isoproterenol infusion test in anorexia nervosa: assessment of pre- and post-beta-noradrenergic receptor activity. *Arch Gen Psychiatry*, **48**(6), 556–62.

Kelley, A. E. (2004) Ventral striatal control of appetitive motivation: role in ingestive behavior and reward-related learning. *Neurosci Biobehav Rev*, **27**(8), 765–76.

Killen, J. D., Taylor, C. B., Hayward, C., et al. (1996) Weight concerns influence the development of eating disorders: a 4-year prospective study. *J Consult Clin Psychol*, **64**(5), 936–40.

King, P. J. (2005) The hypothalamus and obesity. *Curr Drug Targets*, **6**(2), 225–40.

Klump, K. L., Strober, M., Bulik, C. M., et al. (2004) Personality characteristics of women before and after recovery from an eating disorder. *Psychol Med*, **34**(8), 1407–18.

Kotler, L. A., Cohen, P., Davies, M., et al. (2001) Longitudinal relationships between childhood, adolescent, and adult eating disorders. *J Am Acad Child Adolesc Psychiatry*, **40**(12), 1434–40.

Kringelbach, M. L., O'Doherty, J., Rolls, E. T., et al. (2003) Activation of the human orbitofrontal cortex to a liquid food stimulus is correlated with its subjective pleasantness. *Cereb Cortex*, **13**(10), 1064–71.

Kusumi, I., Suzuki, K., Sasaki, Y., et al. (2002) Serotonin 5-HT(2A) receptor gene polymorphism, 5-HT(2A) receptor function and personality traits in healthy subjects: a negative study. *J Affect Disord*, **68**(2–3), 235–41.

LeDoux, J. (2003) The emotional brain, fear, and the amygdala. *Cell Mol Neurobiol*, **23**(4–5), 727–38.

Lidow, M. S. and Rakic, P. (1992) Scheduling of monoaminergic neurotransmitter receptor expression in the primate neocortex during postnatal development. *Cereb Cortex*, **2**(5), 401–16.

Lindblad, F., Lindberg, L., and Hjern, A. (2006) Improved survival in adolescent patients with anorexia nervosa: a comparison of two Swedish national cohorts of female inpatients. *Am J Psychiatry*, **163**(8), 1433–5.

Meyer, J. H., Houle, S., Sagrati, S., et al. (2004) Brain serotonin transporter binding potential measured with carbon 11-labeled DASB positron emission tomography: effects of major depressive episodes and severity of dysfunctional attitudes. *Arch Gen Psychiatry*, **61**(12), 1271–9.

Mikkila, V., Rasanen, L., Raitakari, O. T., et al. (2005) Consistent dietary patterns identified from childhood to adulthood: the cardiovascular risk in Young Finns Study. *Br J Nutr*, **93**(6), 923–31.

Miyasaka, K., Hosoya, H., Sekime, A., et al. (2006) Association of ghrelin receptor gene polymorphism with bulimia nervosa in a Japanese population. *J Neural Transm*, **113**(9), 1279–85.

Montgomery, S. M., Ehlin, A., and Ekbom, A. (2005) Smoking during pregnancy and bulimia nervosa in offspring. *J Perinat Med*, **33**(3), 206–11.

Nacmias, B., Ricca, V., Tedde, A., et al. (1999) 5HT2A receptor gene polymorphisms in anorexia nervosa and bulimia nervosa. *Neurosci Lett*, **277**(2), 134–6.

Naruo, T., Nakabeppu, Y., Sagiyama, K., et al. (2000) Characteristic regional cerebral blood flow patterns in anorexia nervosa patients with binge/purge behavior. *Am J Psychiatry*, **157**(9), 1520–2.

Naughton, M., Mulrooney, J. B., and Leonard, B. E. (2000) A review of the role of serotonin receptors in psychiatric disorders. *Hum Psychopharmacol*, **15**(6), 397–415.

Nozoe, S., Naruo, T., Nakabeppu, Y., et al. (1993) Changes in regional cerebral blood flow in patients with anorexia nervosa detected through single photon emission tomography imaging. *Biol Psychiatry*, **34**(8), 578–80.

Nozoe, S., Naruo, T., Yonekura, R., et al. (1995) Comparison of regional cerebral blood flow in patients with eating disorders. *Brain Res Bull*, **36**(3), 251–5.

Parsey, R. V., Hastings, R. S., Oquendo, M. A., et al. (2006) Lower serotonin transporter binding potential in the human brain during major depressive episodes. *Am J Psychiatry*, **163**(1), 52–8.

Patton, G. C., Selzer, R., Coffey, C., et al. (1999) Onset of adolescent eating disorders: population based cohort study over 3 years. *Br Med J*, **318**(7186), 765–8.

Ribases, M., Gratacos, M., Fernandez-Aranda, F., et al. (2005) Association of BDNF with restricting anorexia nervosa and minimum body mass index: a family-based association study of eight European populations. *Eur J Hum Genet*, **13**(4), 428–34.

Ricca, V., Nacmias, B., Boldrini, M., et al. (2004) Psychopathological traits and 5-HT2A receptor promoter polymorphism (-1438 G/A) in patients suffering from anorexia nervosa and bulimia nervosa. *Neurosci Lett*, **365**(2), 92–6.

Richmond, B., Liu, Z., and Shidara, M. (2003) Neuroscience. Predicting future rewards. *Science*, **301**, 179–80.

Robbins, T. W. (2005) Chemistry of the mind: neurochemical modulation of prefrontal cortical function. *J Comp Neurol*, **493**(1), 140–6.

Rybakowski, F., Slopien, A., Dmitrzak-Weglarz, M., et al. (2006) The 5-HT2A-1438 A/G and 5-HTTLPR polymorphisms and personality dimensions in adolescent anorexia nervosa: association study. *Neuropsychobiology*, **53**(1), 33–9.

Santel S., Baving L., Krauel K., Münte T. F., and Rotte M. (2006) Hunger and satiety in anorexia nervosa: fMRI during cognitive processing of food pictures. *Brain Res*, **1114**(1), 138–48.

Schultz, W. (2002) Getting formal with dopamine and reward. *Neuron* **36**(2), 241–63.

Seeger, G., Braus, D. F., Ruf, M., et al. (2002) Body image distortion reveals amygdala activation in patients with anorexia nervosa – a functional magnetic resonance imaging study. *Neurosci Lett*, **326**, 25–8.

Stanley, M., Traskman-Bendz, L., and Dorovini-Zis, K. (1985) Correlations between aminergic metabolites simultaneously obtained from human CSF and brain. *Life Sci*, **37**(14), 1279–86.

Steffen, K. J., Roerig, J. L., Mitchell, J. E., et al. (2006) Emerging drugs for eating disorder treatment. *Expert Opin Emerg Drugs*, **11**(2), 315–36.

Steiger, H., Young, S., Kin, N., et al. (2001) Implications of impulsive and affective symptoms for serotonin function in bulimia nervosa. *Psychol Med*, **31**(1), 85–95.

Steinhausen, H. C., Seidel, R., and Winkler Metzke, C. (2000) Evaluation of treatment and intermediate and long-term outcome of adolescent eating disorders. *Psychol Med*, **30**(5), 1089–98.

Strober, M., Freeman, R., Lampert, C., et al. (2000) Controlled family study of anorexia nervosa and bulimia nervosa: evidence of shared liability and transmission of partial syndromes. *Am J Psychiatry*, **157**(3), 393–401.

Sullivan, P. F. (1995) Mortality in anorexia nervosa. *Am J Psychiatry*, **152**(7), 1073–4.

Tauscher, J., Pirker, W., Willeit, M., et al. (2001) [^{123}I]beta-CIT and single photon emission computed tomography reveal reduced brain serotonin transporter availability in bulimia nervosa. *Biol Psychiatry*, **49**(4), 326–32.

Tiihonen, J., Keski-Rahkonen, A., Lopponen, M., et al. (2004) Brain serotonin 1A receptor binding in bulimia nervosa. *Biol Psychiatry*, **55**, 871.

Uher, R., Brammer, M., Murphy, T., et al. (2003) Recovery and chronicity in anorexia nervosa: brain activity associated with differential outcomes. *Biol Psychiatry*, **54**, 934–42.

Uher, R., Murphy, T., Friederich, H. C., et al. (2005) Functional neuroanatomy of body shape perception in healthy and eating-disordered women. *Biol Psychiatry*, **58**(12), 990–7.

Von Huben, S. N., Davis, S. A., Lay, C. C., et al. (2006) Differential contributions of dopaminergic D1- and D2-like receptors to cognitive function in rhesus monkeys. *Psychopharmacology (Berl)*, **188**(4), 586–96.

Wagner, A., Greer, P., Bailer, U., et al. (2006) Normal brain tissue volumes after long-term recovery in anorexia and bulimia nervosa. *Biol Psychiatry*, **59**(3), 291–3.

Wagner, A., Ruf, M., Braus, D. F., et al. (2003) Neuronal activity changes and body image distortion in anorexia nervosa. *NeuroReport*, **14**(17), 2193–7.

Wang, G., Volkow, N., Thanos, P., et al. (2004) Similarity between obesity and drug addiction as assessed by neurofunctional imaging: a concept review. *J Addict Dis*, **23**, 39–53.

Ethical issues

Introduction to Section 3

Ethical issues and the legal and regulatory context within which research on human subjects proceeds are the topic of this section. Hermes outlines the special considerations and regulations that apply to research with children and discusses key regulatory laws and agencies, standards of care, theoretical ethics, and paradigm case studies (Chapter 16). While most ethical considerations that apply to pediatric neuroimaging research apply to other forms of research on human subjects with children as well, illustrations of ethical issues that arise in neuroimaging research will aid the reader in applying these general principles to the neuroimaging research context.

16

Legal and ethical considerations in pediatric neuroimaging research

Clinton D. Hermes

Introduction

Perhaps understandably, many clinicians first think of malpractice lawsuits when they encounter the term "legal considerations" while reading about their field. Although this chapter will touch on basic malpractice or negligence, it will focus primarily on the broader regulatory issues pertaining to research involving children in the United States. Pediatric neuroimaging research examples will be used, but most of the ethical and legal issues presented by pediatric neuroimaging research are inherent in many other types of research involving children. Moreover, our current regulatory regime is largely rooted in the philosophical debates of the National Commission for the Protection of Human Subjects of Biomedical and Behavioral Research, which authored the foundational document of modern research ethics in the United States – the Belmont Report (National Commission for the Protection of Human Subjects of Biomedical and Behavioral Research, 1979). Thus, the regulations themselves embody somewhat of an ethical consensus on the "major" issues. Here, law and ethics are intertwined.

This chapter will highlight ethical dilemmas in pediatric neuroimaging research about which there is ongoing debate. This chapter will *not* provide ethical or legal recommendations for resolving these dilemmas, except in stating others' views in that regard. This is because, as is the case in many areas of ethics and law, there are few absolutes; many conclusions depend entirely on one's ethical framework and on the facts of the specific case.

Lawsuits and regulations

A potential range of penalties can result from failing to follow federal and state laws that govern this field of research. How concerned should you be about which laws and why? None of the federal laws discussed in this chapter allows a research subject or a subject's parent to sue an individual or institution that violates these laws, but none forbids such lawsuits. Thus, it is possible for a person to sue based on a theory of negligence (in other words, a routine "tort" claim, or claim of wrongdoing) with reference to a federal law or regulation. The plaintiff might argue that failure to follow federal research law and regulation is evidence of negligence. For example, to prove negligence, the plaintiff must show that (1) the defendant had a duty (i.e., responsibility) to the plaintiff, (2) the defendant breached that duty, and (3) the breach is what caused the plaintiff's harm. The plaintiff must also be able to prove the nature and extent of the harm (i.e., "damages").

There are many duties that parties owe to one another. For example, "malpractice" is a physician's breach of a duty to render medical care consistent with prevailing standards. While this notion has not

Neuroimaging in Developmental Clinical Neuroscience, eds. Judith M. Rumsey and Monique Ernst. Published by Cambridge University Press. © Cambridge University Press 2009.

been widely tested in the courts, it is plausible that a duty can be inferred from standards established by federal law. (The Grimes case, discussed below, provides an example of such an inference.) A breach of various federal regulations and ethical proclamations has been alleged as a cause of action in a number of lawsuits, especially in the last decade, in which clinical trials litigation has risen substantially. However, these lawsuits are typically settled before trial and so the claims are rarely addressed directly by judges. Nevertheless, failure to adhere to these laws could result in a successful lawsuit or settlement by anyone who is injured by the failure, even though the laws themselves may not provide for this possibility. This is especially troublesome in the context of pediatric neuroimaging research because often there are no clear legal or ethical answers to questions faced by those in the field. This, in turn, makes researchers, institutions, and Institutional Review Boards (IRBs) vulnerable to second-guessing when issues arise. Perhaps worse, it may make researchers and IRBs reactively conservative to the detriment of good science.

Federal regulatory agencies

Lawsuits are always a possibility, but so are government enforcement actions, which can involve civil or criminal penalties. For any given law or regulation at any level of government, typically an agency or authority is charged with enforcing it. For example, the Office for Human Research Protections (OHRP), within the Department of Health and Human Services (DHHS), has primary responsibility for ensuring compliance with the Code of Federal Regulations (C.F.R.), Title 45, Part 46 (discussed below) (US Department of Health and Human Services, 2005). These and other regulatory bodies are described in Table 16.1. The OHRP has more than once issued findings of non-compliance to institutions where the IRB approved pediatric studies without appropriate evidence that the studies met the relevant criteria. While OHRP does not

typically take harsh enforcement actions, it has the ability (which it has exercised in the past) to suspend an institution's entire portfolio of federally funded research until a matter of non-compliance is corrected to its satisfaction.

The Food and Drug Administration (FDA), which regularly audits IRBs and investigators for compliance, can also take enforcement action to prohibit an IRB from approving FDA-regulated research or an individual researcher from conducting an FDA-regulated study. Such enforcement actions can have disastrous consequences for the research, the researchers, and the institution. In addition, it should be noted that FDA regulates the use and marketing of radiological devices, such as magnetic resonance scanners.

The OHRP and FDA are by no means the only agencies that can have a hand in enforcing federal research regulations, but they are among the most important. Others include the Department of Justice, the DHHS Office of the Inspector General, the National Institutes of Health (NIH), and other funding agencies.

Federal regulation of human subjects research

Title 21 Parts 50 and 56, and Title 45 Part 46, Subpart D of the Code of Federal Regulations

To the surprise of many, the federal government only indirectly regulates human subjects research. In other words, there is no general federal law that requires oversight and protections in experimentation on children per se. Instead, the federal government regulates research by attaching conditions to obtaining certain privileges. For example, to conduct any pediatric research with funds from DHHS, including the NIH and the Centers for Disease Control and Prevention, an institution must usually comply with 45 C.F.R. Part 46. Researchers or institutions that receive no federal funding are not required to comply with these regulations.

Table 16.1 Agencies and applicable laws

Regulatory organizations	Abbreviations	Functions
US Department of Health and Human Services	DHHS	DHHS is the United States government's principal agency for protecting the health of all Americans and providing essential human services. FDA, OHRP, National Institutes of Health (NIH), and Centers for Disease Control (CDC), among others, are all within DHHS
Office for Human Research Protections	OHRP	OHRP supports the nation's system for protecting volunteers in research that is conducted or supported by DHHS. One part of this support is OHRP's enforcement of the Common Rule for DHHS. OHRP's enforcement powers usually extend to non-federally funded research because of institutions' FWAs (below)
Secretary's Advisory Committee on Human Research Protections	SACHRP	Within OHRP/DHHS, this committee provides expert advice and recommendations to the Secretary of DHHS on issues related to the protection of human research subjects
Food and Drug Administration	FDA	FDA regulations apply to any research involving a drug or device used in any way other than in accordance with its FDA-approved labeling. In the absence of any approved marketing applications for the drug/device, any clinical investigation involving the drug or device may be subject to FDA's jurisdiction[a]
National Commission for the Protection of Human Subjects of Biomedical and Behavioral Research	National Commission	On July 12, 1974, the National Research Act was signed into law, thereby creating the National Commission. One of the charges to the National Commission was to identify the basic ethical principles that should underlie the conduct of biomedical and behavioral research involving human subjects and to develop guidelines to assure that such research is conducted in accordance with those principles In 1978, the National Commission's report was published. It was named the "Belmont Report," for the Belmont Conference Center, where the National Commission met when drafting the report. The Belmont Report remains the cornerstone of the US ethical and regulatory framework for human subjects research
Federal laws and regulations		
Code of Federal Regulations Title 45, Part 46, or Title 21, Parts 50 and 56	45 C.F.R. § 1, or the "Common Rule"	Technically known as the "Federal Policy for the Protection of Human Subjects," this basic legal framework for protecting human subjects – requiring informed consent and IRB review – derives from the Belmont Report and is replicated identically across almost 20 different federal agencies. This widespread acceptance that has resulted in the Federal Policy being called the "Common Rule"

Table 16.1 (*cont.*)

Federal laws and regulations		
Code of Federal Regulations Title 45, Part 46, Subpart D, or Title 21, Part 50, Subpart D	45 C.F.R. § 1_, 21 C.F.R. § 5_, or "Subpart D"	The DHHS (Title 45) or FDA (Title 21) regulations containing special protections for children who are research subjects
Federalwide Assurance	FWA	With few exceptions, every institution that participates in human subjects research conducted or supported by any agency within DHHS must have an FWA on file with OHRP. The FWA is a standard OHRP form that affected institutions complete. For historical reasons, most institutions promise in their FWAs to comply with the Common Rule for all of their research (not just federally supported research). OHRP's jurisdiction is, in practice, much broader than what is stated in the Common Rule
Health Insurance Portability and Accountability Act of 1996	HIPAA	HIPAA, under which privacy regulations were put into effect in 2003, restricts the use and disclosure of identifiable health information, including protected health information (PHI) for research purposes

Note:

[a]Note that "drug" and "clinical investigation" have specific meanings under 21 C.F.R. Part 312 and the Federal Food, Drug, and Cosmetic Act that may differ from common understandings of those terms. For example, according to 21 USC. §321(g) (1), a "drug" includes any "articles (other than food) intended to affect the structure or any function of the body of man or other animals." Thus, various chemicals, biologics, and the like may be considered "drugs" for these purposes by the FDA, even if there is no intent that the article will be brought to market or used therapeutically. Moreover, a research study involving such an article need not have as its primary aim the study of the safety or efficacy of the article. For a controversial FDA warning letter espousing this broad view, see http://www.fda.gov/cder/warn/2003/02-hfd-45-0303.pdf (March 31, 2003 letter to Dr. Alkis Togias, Johns Hopkins Asthma and Allergy Center, regarding a protocol involving hexamethonium bromide). This presents challenging questions with respect to whether, for example, use of an unapproved or off-label contrast agent in a study of brain physiology (i.e., a study not intended to produce data for FDA purposes) falls under the FDA regulations described above.

In theory then, research could be conducted on children in the United States without federal oversight, as long as (1) the research is not federally funded and does not involve an unapproved drug or device, and (2) the data are not to be submitted to the FDA. However, in practice, the federal regulations extend further. For an institution to obtain federal funding for *any* human subjects research, it must have on file with OHRP a currently active Federalwide Assurance (FWA) that binds the institution to conduct *all* of its human subjects research in accordance with a stated set of ethical principles. While OHRP does not keep statistics on how many institutions choose which principles to follow for non-federally funded research, it is widely accepted that almost all institutions choose 45 C.F.R. Part 46. So, in this way, 45 C.F.R. Part 46 extends to all human subjects research conducted at most institutions that receive federal research grants.

Compliance with 45 C.F.R. Part 46 does not exempt an institution from additional regulations for any particular study. If FDA regulations apply (e.g., because a "next generation" unapproved imaging device is involved), the institution must comply with those regulations as well. The requirements in 21 C.F.R. Parts 50 and 56 (the FDA regulations) are similar to those in 45 C.F.R. Part 46, but not identically. For example, under 45 C.F.R. Part 46, the standard requirement for informed consent (or parental permission and child assent) for research subjects can be waived by an IRB under certain circumstances (e.g., for some minimal risk studies in which it is impractical to obtain consent). But FDA regulations do not generally permit IRB waivers of informed consent because they pertain almost exclusively to clinical trials in which such waivers would be unethical (US Department of Health and Human Services, 2006). Thus, a study that is subject to both sets of regulations is generally not eligible for a waiver of informed consent because such a waiver would violate FDA regulations. In addition, whether either or both sets of regulations apply affects which agencies have oversight authority over the study and therefore which agency guidance documents might apply.

Importantly, 45 C.F.R. Part 46 is divided into four subparts. Subpart A, widely known as the Common Rule because of its acceptance by almost 20 federal agencies, sets forth the basic standards that govern all human subjects research, from clinical trials to non-exempt sociological and behavioral studies. These standards include essential elements of informed consent, IRB composition and initial review, continuing IRB review, and reporting of certain unanticipated problems and non-compliance.

Subparts B, C, and D set forth special standards to protect pregnant women, human fetuses, and neonates; prisoners; and children, respectively. The FDA has requirements similar to those of 45 C.F.R. Part 46, Subpart D (see 21 C.F.R. Part 50, Subpart D), but it does not have regulations equivalent to those of Subparts B and C. This chapter will focus primarily on Subparts A and D (and by implication the FDA equivalents). All section references are to 45 C.F.R. Part 46.

Subpart D of 45 C.F.R. Part 46 permits research involving children only if adequate provisions are made for soliciting the assent of children and the permission[a] of their parents or guardians (discussed below) *and* the research falls into one of the following four categories:

1. *The IRB finds that the research poses no greater than minimal risk to the child (Section 404).* "Minimal risk" means that the probability and magnitude of harm or discomfort anticipated in the research are not greater in and of themselves than those ordinarily encountered in daily life or during the performance of routine or psychological examinations or tests. Research approved under Section 404 will virtually always have a favorable risk-benefit ratio if scientifically meritorious due to the minimal risk level and the knowledge to be gained.

2. *The IRB finds that the research poses more than minimal risk to the child, but presents the prospect of direct benefit to the individual subject (Section 405).* The risk must be justified by the anticipated benefit to the subjects, and the relation of the anticipated benefit to the risk must be at least as favorable as that of available alternative approaches.

3. *The IRB finds that the research poses more than minimal risk to the child and that the intervention or procedure does not hold out the prospect of direct benefit for, or likely contribution to, the well-being of the individual subject, but is likely to yield generalizable knowledge about the subject's disorder or condition (Section 406).* In this case, the amount of risk must represent only a minor increase over minimal risk. However, the experience of the intervention or procedure for the child must be reasonably commensurate with other experiences inherent in their actual or expected clinical or social situations. And, the intervention or procedure must also be likely to

---

[a] This chapter uses the terminology of the Common Rule and Subpart D. While an adult may be capable of giving "informed consent," a child is not. A child's agreement to participate in research is thus referred to as "assent." Similarly, national and international research ethics codes hold as anathema the notion that one person can give "consent" for another to participate in research. A parent's agreement that his or her child may participate in research is thus referred to as "permission."

yield generalizable knowledge about the child's disorder or condition that is of vital importance for its understanding or amelioration.

4. *If none of the above categories apply, the IRB can find that the research presents a reasonable opportunity to further the understanding, prevention, or alleviation of a serious problem affecting the health or welfare of children (Section 407).* Then, the Secretary of DHHS, after consultation with a panel of experts in pertinent disciplines (e.g., science, medicine, education, ethics, or law) and after opportunity for public review and comment (often referred to as the 407 review process), determines that either:

• the research falls into one of the other three categories, or

• the research presents a reasonable opportunity to further the understanding, prevention, or alleviation of a serious problem affecting the health or welfare of children and the research will be conducted in accordance with sound ethical principles.

Ethical considerations built into Subpart D

Correctly applying terms such as "prospect of direct benefit," "minimal risk," "minor increase over minimal risk," and even "children" is one of the most challenging aspects of determining which (if any) of the above criteria are satisfied by a particular research proposal. Each term is crucial in one or more of the above criteria, and a specific definition or application is not always clear-cut. The following example will illustrate the point.

A researcher wants to conduct a study of brain development that will require two MRIs under conscious sedation for otherwise healthy toddlers who are siblings of children with autism. The toddlers are between 2 and 3 years old, a difficult age for conducting these procedures without using sedation. The primary purpose of the MRIs is not to benefit the children themselves, but to inform early diagnosis and intervention efforts for children who may be at risk by virtue of having a sibling with autism. Contingent on grant funding, an add-on

study, in which the effectiveness of a manualized behavioral therapy regimen would be tested in the same children (with prior parental permission), is planned; however, the add-on study is not currently before the IRB for approval.

For the IRB to determine whether to approve the first study, it must determine into which category the study might fit. Thus, the IRB would first assess whether the study poses "no greater than minimal risk to children." "Minimal risk" is defined above, but it remains one of the most hotly debated terms in the Common Rule: does "ordinarily encountered in daily life" mean by children in general, or by children with similar eligibility criteria (or in the same community) as those to be enrolled in the study? What types of examinations or tests are "routine," and can this vary based on the characteristics of the children to be studied (e.g., age, gender, ethnicity, socioeconomic background)? Must the proposed research procedures be part of (or very similar to) routine childhood examinations in order to qualify as minimal risk?

To these questions, there are no legal answers or answers with which all ethicists would agree, although certain recommendations provide some guidance (Ernst, 1999). For example, the Institute of Medicine (IOM) recommends that risk be evaluated "in relation to the usual experiences of average, healthy, usual children" (Institute of Medicine of the National Academies, 2004). The Secretary's Advisory Committee for Human Research Protections (SACHRP) has charged a subcommittee with examining all controversial terms (including "minimal risk") under Subpart D – an ongoing effort – and SACHRP's predecessor created a table of common research interventions categorized by risk level (National Human Research Protections Advisory Committee, 2002). However, because there is no clear definition of "minimal risk," there is substantial variability in how IRBs apply the term, and IRBs on the more permissive end of the spectrum may pose greater liability risks to themselves and their institutions.

Returning to the above example, let us assume that the IRB finds the eligibility screening questionnaires and the MRIs are minimal risk, but that the conscious sedation is more than minimal risk. The study would therefore not qualify for approval under Section 404. Would the study be approvable under Section 405? How remote may the "prospect" of benefit be, and how "direct" must it be? How "likely" is it that the MRIs will contribute to the subjects' well-being?

In the example presented, the researcher may indicate the prospect of detecting nascent tumors or other brain abnormalities as potential benefits for the subjects. However, the purpose of the proposed research is not to test whether any particular diagnostic procedure or intervention is beneficial, and there is no reason to believe that these children are more likely than average to have nascent tumors or other clinically significant abnormalities. The prevailing view in this case would be that any benefit related to the possible detection of an abnormality is incidental, rather than an intended purpose of the research. If these children were more likely than others to have brain abnormalities with clear clinical implications that would warrant intervention, then routine magnetic resonance imaging scans to detect such abnormalities would be the standard of care or at least the topic of outcomes research in their own right.

This discussion assumes that early detection of a treatable abnormality would be an actual benefit to the subjects, even if the possibility of early detection may not be a "direct benefit" in the context of Subpart D. This may not always be the case, however, since it must be weighed against the risks of "false positives" (and the attendant financial and psychological costs) and the costs associated with treatment (what if the subject is uninsured?), among other risks. The issue of how to handle "incidental" findings is fraught with ethical complications, as discussed very ably by others (Illes et al., 2006). In brief, the author agrees with the emerging recognition that such incidental findings should be anticipated in advance of any neuroimaging research study, and that the IRB should carefully consider an investigator's proposed procedures for informing subjects of the

possibility of incidental findings and disclosing them to the subjects (or their primary care physicians) when appropriate. The author further believes that careful pre-enrollment screening and exclusion procedures should be used whenever possible to reduce the possibility of an incidental finding (e.g., inquiring about previous head trauma), and that children whose parents who would not wish to be informed of a clinically relevant incidental finding should not be used as control subjects.

Most IRBs would not qualify the detection of incidental findings in a research MRI study as a prospect of direct benefit, but this view is not legally or ethically compelled. If the IRB were to conclude differently and also found (under Section 405) that there is a sufficient benefit to counterbalance the risk posed by the study procedures, how should the IRB determine whether the risk-benefit ratio is at least as favorable as that presented by "available alternative approaches"? In the context of neuroimaging and the early detection of brain abnormalities, the IRB presumably must consider not only what technology is geographically within reach of the subjects, but also what is affordable in the non-research context, what is accepted in the medical community, and whether a family can practically obtain it.

These are not, of course, the only examples of when the regulations are – intentionally or not – vague enough to lead to differing conclusions among IRBs regarding the same research proposal and thus, by extension, whether such research is legally or ethically acceptable. The example does not contain nearly enough factual detail to answer the questions posed above, and it is the author's experience that IRB debates and determinations are extremely dependent on such facts. The example and discussion are used merely to illustrate how the law regulates this type of research: somewhat vague guidelines are provided to a group of experts and lay people – the local IRB – who apply their collective judgment and pass on whether the research meets the legal (and, by extension, the ethical) guidelines they must follow.

Component analysis

It is not only the application of specific terms and phrases within Subpart D that varies among IRBs and results in debate and sometimes confusion. The *process* of making these decisions is also not uniform. In the author's experience, IRBs often make these determinations with respect to the research proposal as a whole, an approach that is supported by the headings in Sections 404–407 ("Research not involving greater than minimal risk," "Research involving greater than minimal risk but presenting the prospect of direct benefit to the individual subjects," etc.).

However, the texts of these same sections hint at a different approach, calling for analysis of the risks and benefits of specific interventions or procedures in the research. The latter approach – called "component analysis" and espoused by the National Commission, the National Bioethics Advisory Commission, and SACHRP – holds that "each research procedure in a study must be evaluated independently in terms of potential benefits and risks to subjects" and that "different procedures in a single trial may be approved or disapproved under different Subpart D categories" (Fisher and Kornetsky, 2005a). Thus, if one procedure in a study is more than minimal risk and non-therapeutic (e.g., sedation of toddlers in the example above), and another procedure in the same study is "minimal risk" and therefore approvable under Section 404 (e.g., cognitive psychological tests administered to the children), the most protective section should govern.

Moreover, under component analysis, the benefits associated with one procedure or arm of a study cannot be used to offset the risks of another procedure or arm of the same study. For example, in a study involving both therapeutic and non-therapeutic procedures, non-therapeutic procedures that are more than a minor increase over minimal risk cannot be justified solely by the potential benefits that subjects may obtain from the therapeutic procedures. Instead, the risks of the non-therapeutic procedures must be independently justified by

being minimized and reasonable in relation to the knowledge those procedures will contribute. Some ethicists argue that this "component analysis" approach is a best practice (and one that can result in decisions different from a "total package" analysis), albeit one that is neither consistently applied nor strictly mandated by the regulations (Weijer and Miller, 2004).

Assent and permission

As stated above, all four categories of permissible research on children require "adequate" provisions for soliciting the assent of children and the permission of their parents or guardians. What is adequate in each case depends on the specifics of the research proposal and the category into which it falls.

First, not all children are capable of assenting: a 2-year-old child would not be capable of assessing even basic information about the risks and benefits of enrolling in a research study; nor would an older child who is developmentally delayed. Thus, the IRB is charged with first determining whether a proposed pediatric study population is capable of providing meaningful assent. Most IRBs require assent beginning with developmentally normal and healthy 7- or 8-year-olds, generally consistent with the National Commission's recommendations. If some or all of the children to be enrolled cannot reasonably provide assent, then the IRB may waive this requirement.

Second, sometimes a child and his or her parents do not always agree on whether the child should participate in a study. Parental permission can be waived in certain circumstances (e.g., research involving abused children) if it is subject only to the Common Rule and not to FDA regulations. If parental permission is not waived (or if the research is regulated by the FDA), a child may not enroll in a study if a parent or guardian refuses. If parental permission is obtained, a child's refusal to assent to an intervention or procedure (or to "withdraw assent," e.g., by refusing to get into position for entry into a scanner) is also *usually* to be honored. However, if the intervention or procedure holds out a prospect of direct benefit that is important to the health or well-being of the child and it is available only in the context of the research, the assent of the child is not necessary for proceeding. An IRB may also waive the assent requirement altogether in certain circumstances (e.g., in circumstances in which informed consent could be waived for an adult subject).

Finally, the categorization of research determines whether both parents must give permission for the child to participate or whether the permission of one parent suffices. Specifically, only one parent's permission is required for studies approvable under Sections 404 and 405. Both parents must give permission for studies approvable under Sections 406 and 407, unless one parent is deceased, unknown, incompetent, or not reasonably available, or only one parent has legal responsibility for the care and custody of the child.

Other potentially applicable Federal laws and regulations

While the principal federal regulations that govern pediatric research are the Common Rule and its FDA equivalent, other laws also govern certain aspects of research. Of particular note are the privacy regulations published under the Health Insurance Portability and Accountability Act of 1996 (HIPAA) (US Department of Health and Human Services, 2006b). These regulations, which took effect in 2003, impose detailed restrictions on the use and disclosure of identifiable health information held by certain so-called covered entities, which includes most healthcare providers. The regulations are not limited to the use and disclosure of "protected health information" (PHI) for research purposes, but research is not exempt from the regulations.

A substantial portion of neuroimaging research conducted in the United States is affected by the HIPAA regulations in some fashion. Magnetic resonance imaging (MRI) scanners are often located at

hospitals or physician group practices, which are covered entities under HIPAA. As a result, for a researcher to generate, analyze, and share a child's PHI (e.g., MRI images that contain identifying information – such as a medical record number or date), the researcher is required to obtain from each child's personal representative (e.g., parent or guardian) a written release that complies with all HIPAA regulations. This release, called an authorization, is not the same as the parental permission or child assent required by the Common Rule. For example, it does not typically discuss the procedures, risks, and benefits associated with the research; instead, it pertains solely to the use and disclosure of PHI. Authorizations have their own content requirements and can (but need not) be combined with the parental permission form.

Apart from the complexities that can arise out of this authorization requirement and a related HIPAA requirement that parents have a right to access certain of their children's records, HIPAA contains provisions that govern the relationship between parents (or guardians) and minor children. Except in situations of abuse or neglect, a parent or guardian has sole authority to act on behalf of an unemancipated minor with respect to the minor's PHI, to the extent that the parent or guardian has the authority, under applicable state law, to make healthcare decisions for the minor. However, the minor has sole authority to act with respect to his or her PHI regarding a particular healthcare encounter if allowed by law to consent to that encounter without additional (e.g., parental) consent. This is generally the same result as that under Subpart D, which defines children as "persons who have not attained the legal age for consent to treatments or procedures involved in the research, under the applicable law of the jurisdiction in which the research will be conducted," with one important difference. Under Subpart D, even when parental permission is required, the minor's assent is required unless the IRB has waived it. Under HIPAA, that minor would have no legally mandated input into the use and disclosure of his or her PHI for the research.

State law: the Common Rule and HIPAA revisited

The Common Rule, the FDA regulations, and HIPAA do not generally preempt state laws that provide for additional individual protections. In addition, those regulations sometimes specifically defer to state law.

State statutes

A small number of states, including New York, California, and Maryland, have laws that pertain to most or all human research carried out within the state. In addition, state law can play an important part in determining whether Subpart D applies. Specifically, Subpart D applies only to research involving "children." As mentioned above, someone is considered a child if he or she has not attained the legal age for consent to treatments or procedures involved in the research under the law of the state in which the research will be conducted. For young children, this definition presents few issues, but for adolescents, it has raised a number of questions.

Consider the following issues, presented at the August 1, 2005, SACHRP Subpart D subcommittee meeting (Fisher and Kornetsky, 2005b):

- Is a child considered an adult under Subpart D if the investigator is studying psychosocial or biological correlates of a condition or disorder for which applicable law permits him/her independently to consent to treatment?
- Is a child considered an adult only if the questions or biological tests are similar to procedures used for clinical diagnosis, monitoring, or treatment follow-up for a condition for which his/her independent consent is allowed?
- If a child is considered a "mature minor" for a specific treatment or procedure, but the research procedure is not normally conducted as part of the treatment, can parental permission be waived?

The Office for Human Research Protections has recently provided some limited guidance on this topic by stating "if research . . . involves solely treatments or procedures for which minors can give consent outside the research context (e.g., related to sexually transmitted diseases or pregnancy), such individuals would not meet the definition of children" under Subpart D (US Department of Health and Human Services, 2007). While this guidance may be helpful in certain instances, it does not address all questions. For example, in Massachusetts, a minor may "give consent to his medical or dental care at the time such care is sought if . . . he reasonably believes himself to be suffering from or to have come in contact with any disease defined as dangerous to the public health . . . (for example, HIV); provided, however, that such minor may only consent to care which relates to the diagnosis or treatment of such disease" (M.G.L. ch. 112, § 12F in the General Laws of the Commonwealth of Massachusetts). If a researcher in Massachusetts wants to include teenage HIV-positive subjects in an MRI study of dementia related to acquired immunodeficiency syndrome (AIDS), would the study fall under Subpart D? The minors enrolled can undoubtedly consent to MRIs as part of their AIDS-related care and observation, but what if the protocol-related blood draw is not part of the minors' regular treatment?

State court cases and the Grimes opinion

In examining the laws of the various states, one must look not only at statutes and regulations, but also at court decisions pertaining to pediatric research. Perhaps the best known in this area comes from the Maryland Court of Appeals (Maryland's highest court). *Grimes* v. *Kennedy Krieger Institute, Inc.* (*Grimes* v. *Kennedy Krieger Institute, Inc.*, 2001) involved children enrolled in an observational study of the effectiveness of certain lead paint abatement procedures. Some children were allegedly allowed to remain in housing units with lead paint. Researchers at the Kennedy Krieger Institute monitored blood lead levels of children living in rental housing in which various partial lead abatement procedures had been performed. The Institute was alleged to have failed to obtain parental permission for the children to participate in a research study. Because of technical reasons, the opinion in the *Grimes* case may not be legally binding, but the court's opinion is nonetheless instructive:

In our view, otherwise healthy children should not be the subjects of nontherapeutic experimentation or research that has the potential to be harmful to the child . . . What right does a parent have to knowingly expose a child not in need of therapy to health risks or otherwise knowingly place a child in danger, even if it can be argued it is for the greater good? . . . It is, simply, and we hope, succinctly put, not in the best interest of any healthy child to be intentionally put in a nontherapeutic situation where his or her health may be impaired, in order to test methods that may ultimately benefit all children. (pp. 850–855) (*Grimes* v. *Kennedy Krieger Institute, Inc.*, 2001).

Thus, the ethical debate that was played out during the meetings of the National Commission was recently revived in a legal forum, where all the countervailing ethical concerns were reconsidered. Although Section 406 and Section 407 studies are permissible under federal law, they have been the subject of judicial scrutiny and some degree of judicial skepticism in Maryland, which may be mirrored elsewhere in the future.

Standards of care

As noted previously, there is a possibility that research regulations could serve as a benchmark for a "standard of care" in litigation. Standards of care – which establish duties, which can in turn establish negligence if those duties are breached – can also be derived from other sources. For example, providing patients with enough information to make informed decisions about medical treatment is considered a standard of care in the medical profession, the violation of which can constitute malpractice. This notion has often been used in the context of clinical trial litigation. Inadequate disclosure of risks during the informed consent process can also result in a lawsuit for battery (harmful, intentional contact), in addition to negligence. While a plaintiff's consent is a defense against battery, if the informed consent process

is defective, the plaintiff can argue that his or her consent to participate was induced by fraud (or misrepresentation by omission), which makes this defense ineffective. Finally, some courts have held that an informed consent document creates a contractual relationship between the institution/researcher and the research subject (support for this can be found in the *Grimes* case, among others), and this view allows injured subjects to sue under theories of breach of contract as well.

The uncertain position of the law in most jurisdictions on these types of claims can have numerous implications with respect to the issue of incidental findings in neuroimaging research. If an informed consent discussion does not include the risks of detecting an unexpected abnormality, the limits of the scan, and the potential distress and expense that the disclosure of the incidental finding could cause, a subject who is told about an incidental finding (particularly a false positive) could sue for negligence and battery. That said, even if a child and his or her parents agree that no information will be provided about any incidental findings (but especially if this discussion does not take place or is not well documented), the researcher's failure to report an incidental finding might be construed as a breach of trust. If the child is not excluded from participation, a plaintiff could argue that the failure to follow-up on the finding constitutes such a breach, given the special relationship between researcher and subject or between doctor and patient (a notion supported by the *Grimes* case). If the parents and child elect as part of the consent process to hear about any incidental findings, and the researcher fails to disclose a finding for whatever reason, a claim may be brought for negligence and breach of contract if the abnormality is detected during a future clinical examination.

Theoretical ethics: the Belmont Report and its philosophical roots

Before discussing particular ethical dilemmas in pediatric neuroimaging research, a brief introduction to ethical frameworks is in order. The Common Rule and the FDA regulations – and indeed all of modern research ethics in the United States – are derived in large part from the National Commission's Belmont Report. The Belmont Report, in turn, is largely informed by deontological theories of ethics. Principles of justice, fairness, honesty, and autonomy represent the core of deontology. The Belmont Report also gives a passing nod to consequentialist theories of ethics, such as utilitarianism. These competing theories are described briefly below.

Deontology, literally the study of duty, can be found in philosophical theories from such diverse philosophers as Aristotle, Immanuel Kant, and John Rawls. Whereas Aristotle focused on the duties to which one must adhere to be an excellent person or professional, Kant relied on his formulation of the "Categorical Imperative" (for Kant, the only absolute duty) and the rationality and dignity of persons. Kant's Categorical Imperative requires that the moral actor (or, in the case of incompetent actors, their guardians) never treat another rational person merely as a means to an end but as the "supreme limiting condition in the use of all means" (i.e., as an end in and of him- or herself) (Kant, 1785). Thus, for Kant, the Belmont principle of "respect for persons" would be absolute, and certain types of research currently allowable under federal regulations (e.g., deception studies) would not be permitted. It is not clear from Kant's writings whether children (or adolescents) would be accorded the status of full rational beings, and thus how his Categorical Imperative would apply to Section 406 and Section 407 studies or to Section 405 studies in which the requirement for assent was waived.

Questions of justice also belong to the deontological system. In discussing arrangements for a just society, for example, John Rawls argues that if inequalities occur, they should benefit those who are worst off, because that is the result everyone would choose if they were behind a "veil of ignorance" and were unaware of the circumstances they would face in life (Rawls, 1971). This notion of justice has obvious implications for subject selection. Rawls might argue that, all other considerations being equal, those children who are least well-off socially, economically, or medically deserve special consideration when subjects are selected for clinical trials that might be of benefit to them and, conversely, they should not be selected if no direct benefit can be expected. The latter, at least, is not the outcome achieved by Section 406, which allows non-therapeutic research on children to learn about their "disorder or condition." This was one basis for the objection to the National Commission's findings raised by Commissioner Robert Turtle in his dissent (National Commission for the Protection of Human Subjects of Biomedical and Behavioral Research, 1977).

Deontology's rival ethical system is consequentialism, a popular form of which is utilitarianism. Depending on the philosopher, creating the greatest happiness or good or reducing suffering are paramount ethical concerns. Justice, autonomy, and other duties are of no concern to true utilitarians. Consequentialism is partially reflected in the Belmont Report's notion of "beneficence." At first blush, the utilitarian theory appears to provide an excellent philosophical justification for allowing many types of research on children who cannot provide true consent because society benefits from the knowledge gained. However, there are other, perhaps competing, utilitarian principles that are less "research-friendly." For example, any research that might be distressing to sentient creatures may be impermissible unless it were overcome by the amount of overall good it would offer.

Strict application of a consequentialist paradigm can be problematic on several fronts. First, it can lead to the very type of grossly unethical experiments condemned by the Nuremberg Code (*Trials of War Criminals before the Nuremberg Military Tribunals under Control Council Law No. 10*, 1949), the Declaration of Helsinki (World Medical Association, 2004), and similar international consensus documents. Second, utilitarianism seeks to balance goods against harms mathematically, but both can be difficult to predict at the outset of a study. Given the number of variables to weigh (unpredictable benefits, downstream consequences of how injured subjects may perceive researchers, etc.), consequentialist theories may be difficult to apply in practice.

Deontological theories can be inconsistent (as between competing duties) and conflict with consequentialism, so the Belmont Report (National Commission for the Protection of Human Subjects of Biomedical and Behavioral Research, 1979), which reflects portions of several moral philosophies, does not present an ethical roadmap that would result in any two people necessarily reaching the same conclusion about a particular case. Instead, it provides considerations that researchers and IRBs should take into account when reviewing a particular study. Given this situation, rather than trying strictly to adhere to any one particular ethical system, some philosophers instead rely on "paradigm cases." This is, in essence, what one must do in following the Belmont Report. Similar to case law in the legal profession, paradigm case studies look at what has been justified in previous instances and why, and brings those moral considerations to bear on the particular case in question. It is useful to present paradigm case questions for pediatric neuroimaging research, to raise ethical issues that might be encountered in this field, and to discuss how others have resolved them.

Paradigm case studies and applied ethics

Returning to the earlier example, a determination of whether the proposed study is ethical raises several questions. Do parents in general want to learn about incidental findings? Are parents (or older children) able to comprehend and assess the considerations in making this determination? What is the likelihood of "false positives"? What is the potential for psychological harm suffered by parents and children who are awaiting a clinical interpretation of their incidental finding? Will the research study absorb the costs associated with a radiological opinion, and, if not, does the ethical status of the study depend on the insurance status of the

participants? Should the IRB consider the potential benefits of the yet-to-be-approved add-on study?

One can imagine a number of other questions that ought to be addressed to develop an informed opinion about the ethics of the study. The nature of the questions raised by pediatric neuroimaging research will undoubtedly change as technology advances. For example, will we be able to predict a young child's intellectual strengths and weaknesses and future behavioral functioning?

This highlights the need for case-specific ethical analyses. Institutional Review Board members or other reviewers will bring their own biases to bear in assessing the relative importance of these questions and the knowledge to be gained from the research. Whatever ethical perspectives drive the reviewers, few would adopt absolute ethical theories that exclude the types of questions posed above.

The regulations discussed in this chapter reflect a broad consensus on the types of ethical considerations to be addressed. Notwithstanding this, the ambiguity and flexibility inherent in the regulations recognize that absolutist ethical principles are not widely accepted and that paradigm cases must guide the research community.

REFERENCES

Ernst, M. (1999) PET in child psychiatry: the risks and benefits of studying normal healthy children. *Prog Neuropsychopharmacol Biol Psychiatry*, **23**(4), 561–70.

Fisher, C. B. and Kornetsky, S. Z. (2005a) *6th Report for SACHRP Consideration Clarifying 45 C.F.R. Subpart D Definitions* (meeting presentation, 18 April 2005). US Department of Health and Human Services, SACHRP Research Involving Children Subcommittee. Viewed 14 July 2007. http://www.hhs.gov/ohrp/sachrp/mtgings/present/Children_files/frame.htm.

Fisher, C. B. and Kornetsky, S. Z. (2005b) *7th Report for SACHRP Consideration Clarifying 45 C.F.R. Subpart D Definitions* (meeting presentation, 1 August 2005). US Department of Health and Human Services, SACHRP Research Involving Children Subcommittee. Viewed 14 July 2007. http://www.hhs.gov/ohrp/sachrp/mtgings/mtg08-05/present/definitions_files/frame.htm.

Grimes v. *Kennedy Krieger Institute, Inc.*, 782 A2d 807 (Md. App. Ct., 2001).

Illes, J., Kirschen, M. P., Edwards, E., et al. (2006) Incidental findings in brain imaging research. *Science*, **311**, 783–4.

Institute of Medicine of the National Academies, Committee on Clinical Research Involving Children, Board on Health Sciences Policy (2004) In: Field, M. J., Behrman, R. E. (eds) *Ethical Conduct of Clinical Research Involving Children*. Washington, DC: National Academies Press.

Kant, I. (1785) In: Abbott, T. K. (ed.) *Fundamental Principles of the Metaphysic of Morals*. Eng. transl., Adelaide, Australia. Viewed 14 July 2007. http://etext. library.adelaide.edu.au/k/kant/immanuel/k16prm/.

National Commission for the Protection of Human Subjects of Biomedical and Behavioral Research (1977) *Report and Recommendations: Research Involving Children* (1977). DHEW Publication (OS) 77-0004, p. 146 et seq.

National Commission for the Protection of Human Subjects of Biomedical and Behavioral Research (1979) *The Belmont Report: Ethical Principles and Guidelines for the Protection of Human Subjects of Research*. Viewed 14 July 2007. http://ohsr.od.nih.gov/guidelines/ belmont.html.

National Human Research Protections Advisory Committee (NHRPAC) (2002) *Report from NHRPAC: Clarifying Specific Portion of 45 C.F.R. 46 Subpart D that Governs Children's Research*. Viewed 14 July 2007. http://www.hhs.gov/ohrp/nhrpac/documents/ nhrpac16.pdf.

Rawls, J. (1971) *A Theory of Justice*. Cambridge, MA: Belknap Press of Harvard University Press.

Trials of War Criminals before the Nuremberg Military Tribunals under Control Council Law No. 10 (1949). Vol. 2. Washington, DC: US Government Printing Office, pp. 181–2.

US Department of Health and Human Services (2005) *45 C.F.R. Part 46: Protection of Human Subjects*. Revised 23 June 2005. Viewed 14 July 2007. http://www.hhs.gov/ ohrp/humansubjects/guidance/45cfr.46.htm.

US Department of Health and Human Services (2006) *21 C.F.R.: Food and Drugs*. Revised 1 April 2006. Viewed 14 July 2007. http://www.accessdata.fda.gov/scripts/ cdrh/cfdocs/cfcfr/cfrsearch.cfm.

US Department of Health and Human Services (2007) Frequently asked questions: research with children. Answer ID 1016. Viewed 2 August 2008. www.hhs.gov/ ohrp/researchfaq.html.

US Department of Health and Human Services, Office for Civil Rights (2006b) *HIPAA Administrative Simplification: Regulation Text, 45 C.F.R. Parts 160, 162, and 164*. Unofficial version, as amended through 16 February 2006. Viewed 2 August 2008. www.hhs.gov/ ocr/AdminSimpRegText.pdf.

Weijer, C., Miller, P. B. (2004) When are research risks reasonable in relation to anticipated benefits? *Nature Med*, **10**, 570–3.

World Medical Association (2004) *The Declaration of Helsinki: Ethical Principles for Medical Research Involving Human Subjects* (2004). Viewed 14 July 2007. http://www.nihtraining.com/ohsrsite/guidelines/ helsinki.html.

Techniques and integration with other research approaches

Introduction to Section 4

This section is dedicated to the major imaging techniques used in pediatric neuroimaging and their integration with other neuroscience approaches. All of the major imaging techniques widely used in children, including healthy, typically developing children, employ magnetic resonance (MR) techniques and scanners. These include blood-oxygen-level-dependent (BOLD) functional MRI (fMRI) and magnetic resonance spectroscopy (MRS), techniques that have seen increasing use in pediatric imaging over the past decade, and diffusion tensor imaging (DTI) and arterial spin labeling perfusion MRI (ASL), techniques that are now beginning to be used either alone or in conjunction with other imaging modalities. Two prime scientific endeavors that integrate neuroimaging with other research methods are the neuroimaging of treatment effects and imaging genetics. This section will inform readers of the physiological principles underlying MR-based imaging techniques, recent advances in them, specific applications in developmental clinical neuroscience, methodological issues, and the potential and challenges for continued improvements.

Levita, Jones and Casey provide an update on BOLD fMRI methods with an emphasis on pediatric applications (Chapter 17). Silveri, Yurgelun-Todd, and Renshaw cover spectroscopy techniques and their use for studying brain chemistry and metabolism in pediatric populations (Chapter 18). Kim describes the basis of DTI, its application in developmental clinical neuroscience for studying white matter development, and its multimodal integration with fMRI (Chapter 19). In Chapter 20, Wang, Rao and Detre address recent developments in arterial spin labeling perfusion imaging, an approach that provides quantitative mapping of regional cerebral blood flow. Pliszka and Glahn review the use of neuroimaging for investigating the neural mechanisms underlying treatment responses toward the discovery of biomarkers (Chapter 21). Finally, Goldman, Buzas and Xu review the emerging field of imaging genetics, which promises to yield intermediate phenotypes useful for understanding complex behavioral variation and development (Chapter 22).

This section not only provides a useful overview of these techniques and approaches, but specifically addresses their roles in developmental clinical neuroscience. The reader will gain from this section a perspective on the unique contribution, strengths, and limitations of each approach and the power of their integration.

17

BOLD fMRI: an update with emphasis on pediatric applications

Liat Levita, Rebecca M. Jones and B. J. Casey

Introduction

It has been just over a decade since contemporary neuroimaging tools, such as functional magnetic resonance imaging (fMRI), were first applied to developmental questions. These tools provide invaluable information on how brain anatomy, function, and connectivity change during development. Applications of this methodology to pediatric populations have helped to establish developmental trajectories in typically developing individuals, providing insight on delays and differences in brain development in atypical pediatric populations. This chapter highlights the use of this technique and its challenges when studying pediatric populations and is organized into three sections. First we provide an overview of the basic principles and methodology of fMRI. Secondly, we discuss its application to pediatric populations, including the potential challenges inherent to such studies and recommendations for addressing them in experimental design and analysis. In the last section, we describe advances in the field that may be applied to significantly enhance the study of pediatric brain development.

Principles and methodology of functional magnetic resonance imaging

Principles of magnetic resonance imaging

Magnetic resonance imaging (MRI) uses strong magnetic fields to create images of biological tissues,

taking advantage of the magnetic resonance properties of hydrogen molecules present in the human body. Thus, when an individual lies with his/her head in the MRI scanner, the spin-axes of the hydrogen nuclei in the brain line up parallel to the magnetic field. At specified time points, a radio-frequency (RF) pulse is broadcast which causes the axes of the nuclei to tilt by a certain angle with respect to the main magnetic field. However, only those nuclei which "resonate" at the RF pulse frequency will be affected. As the nuclei relax, each gives out a characteristic pulse that changes over time, and it is this signal that is used to form the MR images. For example, hydrogen nuclei in fats have a different microenvironment than those in water and thus transmit different pulses. Such pulse differences in RF signals provide a means of discriminating between gray matter, white matter, and cerebral spinal fluid in structural images of the brain, and for fMRI, a means of imaging differences in regional blood oxygenation levels (Huettel et al., 2004).

Blood-oxygen-level-dependent contrast

The technique of fMRI provides an indirect measure of brain activity by measuring regional changes in oxygenated hemoglobin. The MR signal of blood differs slightly with the level of oxygenation; deoxygenated hemoglobin attenuates the MR signal. Neural activity causes an increase in demand for oxygen, and the vascular system actually overcompensates

Neuroimaging in Developmental Clinical Neuroscience, eds. Judith M. Rumsey and Monique Ernst. Published by Cambridge University Press. © Cambridge University Press 2009.

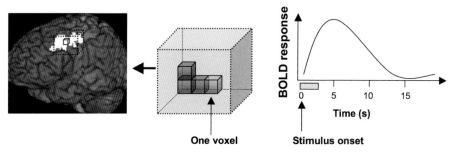

Figure 17.1. Brain magnetic resonance imaging (MRI) and blood-oxygen-level-dependent (BOLD) response. With MRI, the brain is scanned in a series of slices that can than be reconstructed into 3D space. Each slice is characterized by its thickness (1–5 mm) and is composed of units of volume, voxels, used to construct the image. The dimensions of voxels for functional MRI (fMRI) are commonly 9 mm^3. The intensity of the BOLD signal in the imaged voxel is our measure of interest. The picture on the left depicts a functional map of the brain reconstructed into 3D space; the area with statistically significant activity is indicated by white. On the right is a cartoon illustration of the hemodynamic response function (HRF) in one voxel during an fMRI experiment in which a stimulus (gray bar) is shown for 2 s, following which the HRF peaks at 5–6 s.

for this, increasing the amount of oxygenated hemoglobin relative to deoxygenated hemoglobin, which leads to an increase in MR signal. This change in MR signal produces the blood-oxygen-level-dependent (BOLD) contrast, whereby higher BOLD signal intensities arise from an increase in the concentration of oxygenated hemoglobin relative to the concentration of deoxygenated hemoglobin.

The time course of the human BOLD response to an event is called the hemodynamic response function (HRF). An illustration of a typical HRF is shown in Figure 17.1. Here, after participants view a 2-s stimulus, the BOLD response rises and peaks at approximately 5–6 s after stimulus onset, and then returns to baseline. It is the slow prolonged nature of the hemodynamic response that limits the temporal resolution of fMRI.

What is actually being measured by the BOLD contrast? By simultaneously acquiring electrophysiological and fMRI data in monkeys, Logothetis and colleagues showed that the prolonged hemodynamic response that occurs after stimulus onset reflects a local increase in neural activity as assessed by the local field potentials, but not by the spiking activity of single or multiple neurons (Logothetis et al., 2001; Bandettini and Ungerleider, 2001). Local field potentials represent regional

neuronal processing, including the activity of excitatory and inhibitory interneurons, slow waveforms such as synaptic potentials, afterpotentials of somatodendritic spikes, and voltage-gated membrane oscillations. The results of the combined physiological and fMRI experiments in non-human primates suggest that the BOLD signal mainly reflects inputs into an area and the processing of this input information by the local circuitry.

Magnetic strength

The increase in the magnetic field strength of scanners is perhaps one of the most significant advances in MRI technology. The strength of the magnetic field used for MRI is specified in units of Tesla (T), and common field strengths for human studies range from 1.5 to 3 T, although 7-T and 9-T magnets are increasingly being used for human research purposes. As an example of the magnetic strength of these scanners, a 1.5-T MR system has a magnetic field 30 000 times stronger than the pull of gravity on the Earth's surface. Higher magnetic fields are a powerful method to increase signal-to-noise ratio, BOLD contrast, and spectral resolution. However, disadvantages include potential reduction of contrast in anatomic images and enhanced

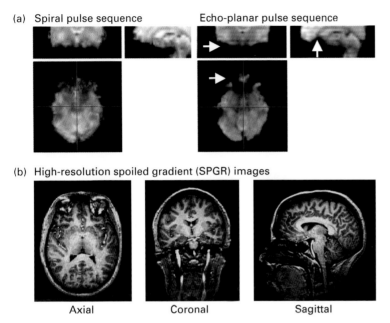

(a) Spiral pulse sequence   Echo-planar pulse sequence

(b) High-resolution spoiled gradient (SPGR) images

Axial   Coronal   Sagittal

Figure 17.2. Reduction in frontal susceptibility with spiral (left) relative to echo-planar pulse (right) sequences for 3-T scanner. (a) The commonly observed signal losses in the orbitofrontal and temporal lobes and in regions close to the brainstem (as indicated by arrows) in echo-planar pulse sequence are fully recovered using the spiral sequence (images acquired only at base of brain). (b) Cross-sectional anatomic images (axial, coronal, and sagittal planes) at the location marked by a blue cross on the axial images in (a) (courtesy of the Sackler Institute for Developmental Psychobiology at Weill Medical College of Cornell University).

susceptibility (attenuation and blurring of signal) associated with respiration, motion and air/tissue interfaces near the sinuses and ear canals (Hu and Norris, 2004). Technical advances to address these challenges, e.g., optimized scan sequences such as spiral sequences help attenuate artifacts.

Spiral imaging is a technique for fast image acquisition that uses sinusoidally changing gradients to trace a corkscrew-like trajectory through k-space. The k-space is a temporary memory of the spatial frequency information in two or three dimensions of an object being scanned. "Gradient" refers to gradient field and/or gradient coil. Inside the magnet are three gradient coils that produce the desired gradient (magnetic) fields. These fields are used to alter, collectively and sequentially, the influence of the static field on the imaged object through which selective spatial excitation

and spatial encoding is achieved. The advantage of using spiral sequences at higher magnetic field strengths is illustrated in Figure 17.2a, which shows less susceptibility in orbitofrontal regions in all three planes using one of these sequences (a spiral in/out sequence, Glover and Thomason, 2004) relative to a traditional echo-planar imaging (EPI) scan sequence in a 3-T magnet.

Experimental design

During fMRI experiments, individuals engage in behavioral tasks while a change in the BOLD signal is measured over time. Changes in neural (i.e., electrical) activity occur on a millisecond time scale compared to blood oxygenation changes, which occur over an average of 6–12 s. Consequently, there is a delay between neural activity and the

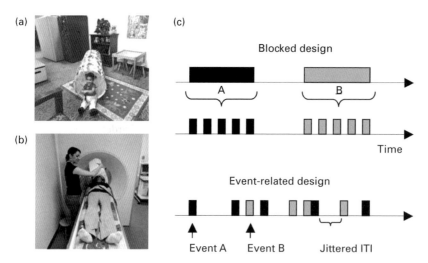

Figure 17.3. Acclimation to the scanner environment and basic task designs. (a) Some young children like to play in a toy tunnel before they try the MRI simulator. (b) The scanner simulator includes a replica of the head coil and stimulus display device just above the opening of the bore. While in the simulator, the child wears headphones and sees a video display of a movie (courtesy of the Sackler Institute for Developmental Psychobiology at Weill Medical College of Cornell University). (c) Two basic types of fMRI experimental designs: blocked design and event-related design. ITI, intertrial interval.

change in BOLD signal that limits the temporal resolution of fMRI. This temporal delay imposes specific constraints for study design and analysis. Ultimately, the experimental design will depend on balancing constraints and the behavior of interest, as well as the limitations of the MR environment.

Blocked and event-related fMRI designs

Many of the first proof-of-concept fMRI experiments were block designs with two conditions: time periods of a control task or rest condition were interleaved with periods of an activation task (e.g., Kwong et al., 1992). In a block design, each condition is typically presented continuously for a period of 10–120 s (Figure 17.3c, top panel). Block designs are used for assessing BOLD signal magnitude differences between conditions because the time length of the blocks allows the hemodynamic response to reach maximal values, while the inter-block intervals are long enough for the hemodynamic response to return to baseline during rest. Block fMRI designs are particularly appropriate

for studying neural activity that is not influenced by practice or immediate experience, i.e., retinotopy or object perception experiments (Kanwisher et al., 1996; Tootell et al., 1998). However, block designs are less useful for studies that examine the temporal dynamics and neural underpinnings of specific components of behavior. For example, the neural correlates of correct versus incorrect responses in tasks involving suboptimal performance cannot be dissociated in a block design. The development of the event-related design in fMRI experiments addresses this challenge.

Event-related designs allow different events (trials or stimuli) to be presented in arbitrary sequences during the experiment (Figure 17.3c, bottom panel). Slow event-related designs, which typically use an interstimulus interval of 10 s or more, allow the hemodynamic response to return to baseline before the next event or trial occurs. Rapid event-related designs do not allow the hemodynamic response to fully return to baseline (Dale, 1999). Consequently, during analysis, it is necessary to parse apart (deconvolve) the resulting interwoven/overlapping (convolved) hemodynamic responses. These designs require that the rapidly presented events be interspersed randomly and separated by jittered (variable duration) intertrial intervals (Figure 17.3c, bottom panel). A limitation of rapid event-related designs is that there is less statistical power with which to detect activation compared to block designs, resulting in a need for lengthy data acquisitions from subjects. This is a trade-off for obtaining more refined information about cognitive and neural processes.

Subtractive and additive logic: importance of control conditions

In fMRI, brain activity is considered relative to another condition, as the signal depends on changes in the ratio of oxygenated to deoxygenated hemoglobin. The absolute signal is thus meaningless on its own. Therefore, fMRI must rely on relative measures, and to determine the temporal sequence of cognitive and neural processes during fMRI, investigators have applied mental chronometry methodologies (Posner, 1978). Mental chronometry can be defined as the study of the temporal sequencing of information processing in the human brain. There are two traditional chronometric approaches, subtractive and additive, and both have been applied elegantly in imaging studies of cognition in children (e.g., Galvan et al., 2006; Ladouceur et al., 2006).

When applying subtractive logic (Donders, 1868; 1969), cognitive tasks are designed so that the test condition and its comparison condition are identical in all ways except in the one cognitive process of interest. For instance, when examining the neural basis of attention switching, one would need an activation and a control task that required attention, but only one of the tasks would have the component of a switch in the type of information being attended (e.g., from attention to color versus shape of the stimuli). The control condition would include encoding, attending, and making a response to stimulus, without the switch component. Just as subtraction of the reaction times for the two types of trials yields a measure of the time required by the switching process, so too would a subtraction of the MR signal during the two types of trials indicate which brain regions are involved in this process. However, one can never guarantee that the performance of a simpler task differs in only one component, as subjects may use different strategies for easier versus harder tasks.

Another approach for linking cognitive and neural processes is to use additive factors logic (Sternberg, 1969). This strategy permits designs in which the process under scrutiny can be parametrically manipulated. For example, when examining the neural processes related to memory search, one could manipulate the number of items that the subjects have to hold in memory. The classic Sternberg memory search task paradigm is an excellent example. The subject is presented with one, two or three items to remember (e.g., pictures of a cat, horse, and dog) on a display. The items disappear and then, after a short delay, an item is presented on the screen (e.g., monkey) and the subject must respond yes or no as to whether the item was in the memory set. Comparing reaction times for one item relative to two or three will provide the amount of time it takes to search memory. Likewise identifying those brain regions that show activation differences with increasing memory load will identify regions supporting memory search (e.g., Braver et al., 1997).

In contrast to the above two methodologies, some fMRI studies have used rest periods or simple fixation as a baseline for contrast, rather than a comparison condition. However, in recent years, a number of neuroimaging studies have reported that regional activity can be greater during rest or simple fixation conditions than during activation conditions (Raichle et al., 2001; Shulman et al., 1997), and may reflect significant mental processing (Stark and Squire, 2001). This activity, as measured by BOLD fMRI, is organized in multiple highly specific functional anatomical networks, which have been termed resting state, or default mode, networks. The resting state networks strongly overlap with sensory–motor, visual, auditory, attention, and language networks that are commonly modulated during active behavioral tasks (Damoiseaux et al., 2006; De Luca et al., 2006). For example, the default network involves portions of the cingulate cortex (especially posterior and ventral regions) that consistently show greater activity during resting states than during behavioral tasks. These patterns of activity are thought to reflect organized intrinsic brain activity, which may serve to maintain information for interpreting, responding to, and predicting environmental demands, in contrast to stimulus-driven activity (Dosenbach et al., 2007; Fox et al., 2005; Peter, 2005; Raichle and Snyder, 2007). The examination of this pattern of activity in pediatric populations is still largely unexplored (but see Tomarken et al., 2004; Zang et al., 2007). A sole reliance on resting state baselines with no behavioral demands or measures of what the subject is doing raises concerns that subjects may be engaging in a variable number of uncontrolled mental processes (e.g., thinking about the task they just performed, day dreaming, thinking about self, etc.) and that such designs will yield less specific information than studies using comparisons based on chronometrics as described above.

Analytic strategies

A number of approaches have been developed to test for statistical differences in patterns of brain activity as a function of the experimental manipulation. Traditionally these statistical tests have been hypothesis driven. However, more contemporary methods have begun to use data-driven approaches that make fewer a priori asumptions. For example, principal component and independent component analyses (PCA and ICA, respectively) (Carlson et al., 2003; Haynes and Rees, 2006; LaConte et al., 2005; McKeown and Sejnowski, 1998; O'Toole et al., 2005; Rajapakse et al., 2006) represent significant advances in image analysis, which can be used with free-viewing paradigms that permit unconstrained/natural conditions during scanning, e.g., watching a movie, listening to music, or playing a video game

(e.g., Bartels and Zeki, 2004; Janata et al., 2002; Mathiak and Weber, 2006).

The majority of studies, however, rely on hypothesis-driven general linear model (GLM) analyses, which are based on whole-brain voxelwise tests of differences or region-of-interest (ROI) analysis. The GLM analyses compare BOLD signal change in each voxel under different task conditions. Furthermore, to determine which regions are reliably active, many research groups also use cluster or contiguity thresholds, since the likelihood of a single voxel showing significance by chance is significantly greater than the likelihood of activation in contiguous voxels (Forman et al., 1995). With this method, no a priori hypotheses about regional differences are necessary. In contrast, ROI analyses focus on pre-specified brain regions based on anatomical or functional criteria for hypothesis testing.

Currently, software for image analysis varies widely. Some of the public domain software packages include AFNI (Analysis of Functional Neuro-Images, Cox, 1996), FSL (The FMRIB Software Library, Smith et al., 2004), FreeSurfer (Fischl and Dale, 2000) and SPM (Statistical Parametric Mapping, Friston, 2007). A more complete list with the capabilities of each package is presented in Appendix A. Notably, these platforms are continually evolving and current choice rests on individual experimental questions and needs. A recent advance includes the integration of two software analysis packages, FSL and FreeSurfer, which has provided a much needed platform to marry both cortical functional and structural imaging data. FreeSurfer is a set of automated tools for reconstruction of the brain's cortical surface from structural MRI data, and fMRI data from FSL can now be overlaid onto the reconstructed surface. Together, these tools provide for multi-subject statistics in a standard spherical surface space, thus eliminating a source of variability (e.g., inter-subject variations in cortical folding). Furthermore, some software packages (e.g., AFNI) are better than others in handling statistical interactions particularly with variables of interest that are continuous and multifactorial.

Special considerations when imaging pediatric populations

The use of fMRI with children can be challenging due to practical, methodological, and analytical issues. To address anatomical, physiological, and psychological differences between children and adults, a number of precautions and procedures are worth noting which are particularly relevant for studies of very young children and infants (Miller, 2007; Souweidane et al., 1999; Yamada et al., 2000). For a review of fMRI of the fetus and the newborn, see Seghier et al. (2006) and Cowan and Rutherford (2005).

Scanning procedure for children

Many aspects of fMRI can be difficult for children, e.g., assessment in a medical environment, large and noisy equipment, and confinement in a small space. Discomfort during scanning may affect neural activation through decreased attention to instructions, decreased task performance, and engagement of emotional and stress-related systems. A key approach for avoiding or minimizing these problems involves prior acclimation to the procedure. Familiarization and simulation of the scan allows the experimenter to train the child to remain still, to determine the child's ability to successfully complete a scan, and to reduce the child's anxiety during the actual experiment (Rosenberg et al., 1997).

Some sites use "mock scanners" to simulate the scanning procedure for participants (Figure 17.3a, b). Motion compliance training typically requires the use of an automated motion detector to track a child's head movement as he/she watches a video in the simulated environment. If the head movement exceeds a predetermined threshold, the system interrupts the video to alert the child that he or she has moved. Motion compliance training shapes the child's behavior by incrementally decreasing the movement threshold to within a range acceptable for scanning. For additional information on automated head-tracking systems

see http://www.sacklerinstitute.org/cornell/assays_and_tools/.

Even with motion compliance training, it is difficult for participants, particularly children and patients, to stay completely still while being scanned. Therefore, motion correction is a critical data-processing step. Motion correction is usually performed using an automated registration algorithm, which estimates rigid-body motion at each time point compared to a reference point, such as the first or middle image collected in a series (Bammer et al., 2007; Friston et al., 1996; Slifer et al., 1993). However, these techniques can fail to fully correct for the effects of bulk head motion that can occur while performing a task (Field et al., 2000). One way to address these issues is to incorporate head motion in the GLM analysis (Friston et al., 1994). The change in the BOLD signal over time in each voxel is modeled in terms of a set of regressors: regressors of interest (task stimuli) and regressors of no-interest such as the estimated motion parameters obtained from a motion-correction algorithm (Bullmore et al., 1999). Inclusion of motion parameters reduces the risk of false positives (detecting MR signal changes that are due to movement within a voxel). Including motion in the statistical model is important for group comparisons, e.g., different age groups or clinical populations such as those with attention deficit/hyperactivity disorder, as groups may differ in the amount of motion artifacts.

Addressing physiological differences between children and adults

Physiological differences between children and adults, such as in respiration, are a concern since they can introduce noise in imaging due to the movement of the lungs and diaphragm. Age-related vascular differences may contribute to altered BOLD fMRI signal responsiveness (Thomason et al., 2005). The variability in these physiological factors is greater in children of the same age than in adults of the same age. Consequently, since statistics generally assume equal variance between groups, it is important to consider physiological noise. Increasing the number of participants in pediatric studies can help provide a more reliable estimate of variability of physiological noise, but statistical comparisons still need to account for differences in noise between groups (Thomason et al., 2005). One potential solution is a pneumatic respiration belt, a device already available on many MRI scanners that monitors respiration without interfering with performance on most tasks. There are benefits for distinguishing neural activity due to the experimental task from activity correlated with variations in respiration (Birn et al., 2006), a particularly important distinction in studies of pediatric populations. However, the contribution of physiological noise to fMRI data has not yet been fully quantified, particularly in children, which leaves open the question of the degree of its interference with findings.

Spatial normalization

A critical step in image processing is spatial normalization, which provides a way to compare functional activation patterns across participants. The standard procedure is to warp individual brains into a standard stereotaxic space. However, structural variability may differ with age, gender, and psychopathology and thus has the potential to distort functional maps. To address some of these issues, an NIH-sponsored longitudinal multi-site structural MRI study of brain development is currently being conducted to characterize healthy brain development in typically developing children, from newborn through late adolescence/early adulthood (Almli et al., 2007). The resulting database should provide age-specific spatial-normalization templates for fMRI analysis of studies with infants and young children.

Spatial normalization procedures are particularly challenging when scanning very young children and infants (Matsuzawa et al., 2001; Schneider et al., 2004) or pediatric clinical groups with smaller regional and whole-brain volumes (Engelbrecht et al., 2002; Muller, 2007) than seen in typically

developing children 6 years and older (Muzik et al., 2000). Spatial normalization may be impossible for clinical populations with large lesions (e.g., hemispherectomy, brain trauma). Consequently, when inter-subject variability is too great, data processing is typically performed in individual subjects without spatial normalization across a group. For example, when functionally defining a region (e.g., primary visual cortex or Broca's area) for subsequent study or for surgical mapping, it is justified to use functional mapping within individual subjects, rather than stereotactic localization across subjects. A functional mapping approach entails identification of brain regions using specific behavioral assays (e.g., visual stimuli known to activate primary visual cortex or word generation known to activate language areas) for each individual subject. This approach has been used successfully in mapping visual cortices in healthy adults (Schneider et al., 1994) and for preoperative localization of language centers in children and adults with brain tumors (Stippich et al., 2007).

Differences in cognitive ability

When designing a task for pediatric research, performance differences across age groups or between typical and atypical developmental populations (Lucker et al., 1996) require special consideration. A number of approaches described below have been utilized to address these potential performance confounds.

Parametric manipulations of task difficulty

Parametric designs typically titrate task difficulty (Ward et al., 2003) or paradigm complexity by increasing task demands (Braver et al., 1997; Durston et al., 2002), e.g., by increasing response competition (Durston et al., 2002, 2003), memory load (Klingberg et al., 2002), and/or stimulus degradation (Grady et al., 1998; Morris et al., 1996). A benefit of this approach is that incremental changes in activity that correspond to the parametric manipulation provide convincing evidence that a specific region is being utilized by the process being manipulated in the study design (Casey et al., 2004).

Post-hoc matching of groups by performance

An alternative strategy for addressing differences in ability across groups is to assign participants to subgroups based on their behavioral performance post hoc (Marsh et al., 2007; Schlaggar et al., 2002). This allows one to distinguish between patterns of activity that are performance-related, age/group-related, or independent of performance and age/group. In other words, this methodology allows one to dissociate brain activity related to age from that related to behavioral performance. However, this technique would not be appropriate given insufficient overlap in behavioral performance of the two groups. Moreover, statistical power is greatly reduced due to the division of the sample and thus large sample sizes are required (Casey, 2002).

Correlating and co-varying age and performance

Another approach to performance differences is to determine the main effects of behavioral performance on brain activation, while co-varying age, in addition to correlating performance with activation. This approach dissociates brain activity related to age from that related to behavioral performance without reducing power due to the division of subjects into subgroups. Correlation analyses of this sort have been used successfully to address questions regarding age versus performance-driven differences in patterns of brain activity (e.g., Casey et al., 1997, 2002; Klingberg et al., 2002; Yurgelun-Todd and Killgore, 2006).

Longitudinal versus cross-sectional studies in pediatric imaging

One of the most systematic methods for accurately determining whether the immature brain engages in the same neural processes as the mature one is to use longitudinal designs to measure brain structure and/or function in the same individuals across different time points (e.g., Giedd et al., 1999; Lu et al., 2007; Sowell et al., 2004). Such longitudinal studies have been shown to be more sensitive to subtle developmental changes than cross-sectional comparisons (Durston et al., 2006). These types of studies are critical for defining developmental trajectories for cognitive and neural processes and for characterizing disruption in development that may lead to disabilities (Casey, 2004; Kotsoni et al., 2006).

Functional magnetic resonance imaging reproducibility: issues for cross-site and longitudinal pediatric studies

A recent study examined the consistency of the fMRI BOLD signal across different imaging conditions (Voyvodic, 2006). Participants repeatedly performed the same simple motor task under a variety of imaging conditions: spiral and standard echo-planar pulse sequences; 1.5- and 4.0-T magnetic field strengths. The results demonstrated that while the absolute amplitude of BOLD statistical activation signals varied significantly across time and scanning conditions, the relative spatial pattern of the BOLD signal was highly reproducible. However, during development, spatial activation patterns change as a consequence of brain maturation (Durston et al., 2006). Hence, an issue that arises with developmental longitudinal studies that occur over several years is whether to continually optimize scanner hardware and software or to keep it constant, to ensure that changes in activation are indeed a consequence of development and not an artifact from changes in scanning procedure and hardware.

Also fundamental for many pediatric research studies, longitudinal and otherwise, are cross-site collaborations, especially those that require large sample sizes or require access to clinical populations. Blood-oxygen-level-dependent activation patterns from different scanning centers may vary even with constant scanning parameters and consistent task performance (Liu et al., 2004; Yoo et al., 2005). Consequently, studies involving fMRI necessitate the creation and testing of comparable protocols and methodologies that allow for cross- and within-site replication, the latter being critical for longitudinal studies (Casey et al., 1998, 2007; Epstein et al., 2007; Fu et al., 2006).

Recent advances and future directions

Building on remarkable advances in neuroimaging over the last decade, functional neuroimaging is now on the cusp of offering new and substantial information about brain–behavior relations in both typical and atypical development by using an increasingly multidisciplinary approach in the study of brain development.

Great advances have been made when combining fMRI with other imaging modalities. An increasing number of studies have begun to use event-related potentials (ERPs) with fMRI to answer basic research questions in children and adolescents (e.g., Bucher et al., 2006) and to investigate cognitive and neural processes on a sub-second time scale (Linden, 2007). Further, following the recent sequencing of the human genome, imaging genetics is emerging as a new and exciting methodology that may allow us to examine how genetic variation can influence brain development and function (see Chapter 22).

Another advance moving us away from modular interpretations of findings in single brain regions toward interpretations that reflect brain connectivity and distributed neural networks is the use of diffusion tensor imaging (DTI, see Chapter 19). Diffusion tensor imaging allows identification and characterization of white matter tracts according to the direction and degree of anisotropic water diffusion within them. Quantifying the degree of anisotropy in terms of metrics such as the fractional anisotropy offers critical insight into white matter development. To date, only a handful of studies have examined the association between developmental changes and DTI-based measures (e.g., Nagy et al., 2004). Yet, the strength of combining fMRI and DTI is clear as it provides a way to enhance and constrain interpretations of pediatric

imaging data, as demonstrated in recent studies in our laboratory examining cognitive control mechanisms across development (Casey et al., 2007; Liston et al., 2006) (see Figure 17.4). Specifically, combining these approaches can provide information concerning the extent to which regional activity is dependent on the strength of projections to or from these regions within a neural circuit. This information moves the field away from the simplistic approach of focusing on a single structure of the brain in isolation, to a more plausible focus on a circuit or network as shown in recent studies examining cognitive development (e.g., Liston et al., 2006). Further, DTI-based irregularities in identified fiber tracts can constrain the interpretation of fMRI-based patterns of activity, as potentially due to structural abnormalities (e.g., atypical axonal organization or delayed myelination) as opposed to functional abnormalities (e.g., neurochemical or neurophysiological) in atypical cognitive development (Casey et al., 2007) (see Figure 17.4).

Another approach is transcranial magnetic stimulation (TMS, for review see, Rossini and Rossi, 2007). Transcranial magnetic stimulation is a safe non-invasive method that can be used to induce in the brain electrical currents that can depolarize neurons, or to modulate (prime or suppress) neural activity, and it heralds great promise in pediatric research. With TMS, brain activity can be disrupted, making it possible to determine which regions are essential to a process or task performance (Pascual-Leone et al., 2000). A number of studies have made use of TMS in children (Garvey and Gilbert, 2004; Lin and Pascual-Leone, 2002), but very few have combined it with fMRI (e.g., Maegaki et al., 2002; Staudt et al., 2004). Yet together, these methodologies enhance the strengths of each other, as TMS can be used to demonstrate causal relationships. Thus while fMRI can map the set of regions that are activated when a subject performs a task, this alone is not proof that those regions are actually critical for the task. If TMS-induced suppression of activity in a region impairs performance, this is much stronger evidence that the region is used in

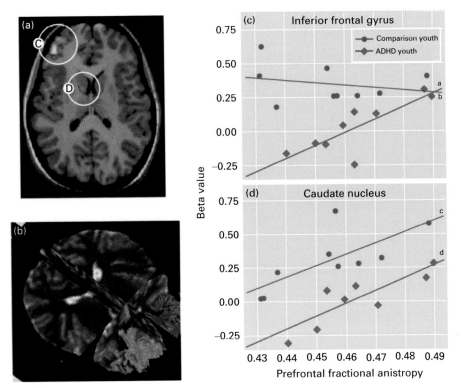

Figure 17.4. Combined use of fMRI and diffusion tensor imaging (DTI) to understand circuitry in attention deficit/hyperactivity disorder (ADHD). (a) Functional map showing regions that correlated with go/no-go task performance: the axial plane shows the two functional regions that positively correlated with go/no-go task performance, the left prefrontal cortex (C) and left striatum (D) (Epstein et al., 2007). Greater activity in these regions resulted in better performance. (b) Representative 3D illustration of fibers connecting the prefrontal and striatal regions. The striatal activation was used as a seed point with an automated tractography algorithm to delineate tracts projecting from the caudate nucleus of the striatum to the prefrontal cortex (red = left hemisphere; green = right). (c) Activity in the ventral prefrontal cortex was positively correlated with fractional anisotropy in this region in youths with ADHD, but not in controls. This suggests that fMRI-detected prefrontal activity in ADHD is closely related to myelination and/or coherence of fibers in connecting white matter tracts. Ventral prefrontal activity was indexed by beta values from regression coefficients representing an estimate of percent signal change from the mean. (d) Activity in the striatum (caudate) correlated positively with fractional anisostropy in prefrontal cortex in both ADHD and comparison youths. These results suggest that activity in fronto-striatal regions implicated in ADHD is closely related to myelination and/or coherence of the fibers connecting these regions (prefrontal cortex and striatum). Together, the results shown in (c) and (d) suggest that atypical white matter architecture or delayed myelination may contribute to ADHD. (Reprinted with permission from Casey et al., 2007. *American Journal of Psychiatry*, **164**: 11, 1729–36.)

performing the task. Future use of TMS together with fMRI holds great promise for understanding the relationship between regional brain activity and behavior across development, and possibly for therapeutic modulation of activity in specific circuits.

Conclusions

Advances in neuroimaging technology have moved the field of human development and cognitive neuroscience forward significantly. The field as a whole has begun to advance with the use of converging methods (e.g., genetics, DTI, ERPs, and fMRI) and novel experimental designs and analyses. In summary, the non-invasive nature of fMRI makes it highly attractive in pediatric research. It is a tool that has already greatly enhanced our understanding of typical and atypical brain development and will continue to offer insight into normal as well as atypical developmental trajectories. There are unique challenges and limitations with fMRI for pediatric studies, yet experimental design and analytical techniques are continually evolving to enhance the sensitivity of the methodology, to overcome many of the limitations, and to ensure the validity of study results and their interpretation in pediatric brain research.

REFERENCES

Almli, C. R., Rivkin, M. J., and McKinstry R. C. (2007) The NIH MRI study of normal brain development (objective-2): newborns, infants, toddlers, and preschoolers. *NeuroImage*, **35**, 308–25.

Bammer, R., Aksoy, M., and Liu, C. (2007) Augmented generalized sense reconstruction to correct for rigid body motion. *Magn Reson Med*, **57**, 90–102.

Bandettini, P. A. and Ungerleider, L. G. (2001) From neuron to BOLD: new connections. *Nat Neurosci*, **4**, 864–6.

Bartels, A. and Zeki, S. (2004) Functional brain mapping during free viewing of natural scenes. *Hum Brain Mapp*, **21**, 75–85.

Birn, R. M., Diamond, J. B., Smith, M. A., et al. (2006) Separating respiratory-variation-related fluctuations from neuronal-activity-related fluctuations in fMRI. *NeuroImage*, **31**, 1536–48.

Braver, T. S., Cohen, J. D., Nystrom, L. E., et al. (1997) A parametric study of prefrontal cortex involvement in human working memory. *NeuroImage*, **5**, 49–62.

Bucher, K., Dietrich, T., Marcar, V. L., et al. (2006) Maturation of luminance- and motion-defined form perception beyond adolescence: a combined ERP and fMRI study. *NeuroImage*, **31**, 1625–36.

Bullmore, E. T., Brammer, M. J., Rabe-Hesketh, S., et al. (1999) Methods for diagnosis and treatment of stimulus-correlated motion in generic brain activation studies using fMRI. *Hum Brain Mapp*, **7**, 38–48.

Carlson, T. A., Schrater, P., and He, S. (2003) Patterns of activity in the categorical representations of objects. *J Cogn Neurosci*, **15**, 704–17.

Casey, B. J. (2002) Neuroscience. Windows into the human brain. *Science*, **296**, 1408–9.

Casey, B. (2004) *Developmental Psychobiology.* Washington, DC: American Psychiatric Publishing, p. 179.

Casey, B. J., Cohen, J. D., O'Craven, K., et al. (1998) Reproducibility of fMRI results across four institutions using a spatial working memory task. *NeuroImage*, **8**, 249–61.

Casey, B. J., Davidson, M. C., Hara, Y., et al. (2004) Early development of subcortical regions involved in non-cued attention switching. *Dev Sci*, **7**, 534–42.

Casey, B., Epstein, J., Buhle, J., et al. (2007) Frontostriatal connectivity and its role in cognitive control in parent-child dyads with ADHD. *Am J Psychiatry*, **164**, 1–7.

Casey, B. J., Trainor, R. J., Orendi, J. L., et al. (1997) A developmental functional MRI study of prefrontal activation during performance of a go-no-go task. *J Cogn Neurosci*, **9**, 835–47.

Cowan, F. M. and Rutherford, M. (2005) Recent advances in imaging the fetus and newborn. *Semin Fetal Neonatal Med*, **10**, 401–2.

Cox, R. (1996) AFNI: software for analysis and visualization of functional magnetic resonance neuroimages. *Comput Biomed Res*, **29**, 162–73.

Damoiseaux, J. S., Rombouts, S. A., Barkhof, F., et al. (2006) Consistent resting-state networks across healthy subjects. *Proc Natl Acad Sci U S A*, **103**, 13848–53.

De Luca, M., Beckmann, C. F., De Stefano, N., et al. (2006) FMRI resting state networks define distinct modes of long-distance interactions in the human brain. *NeuroImage*, **29**, 1359–67.

Donders, F. (1868; 1969) On the speed of mental processes. In: Koster W. (ed.) *Attention and Performance.* (Originally published, 1868). Amsterdam: North-Holland.

Dosenbach, N. U., Fair, D. A., Miezin, F. M., et al. (2007) Distinct brain networks for adaptive and stable task control in humans. *Proc Natl Acad Sci U S A*, **104**, 11073–8.

Durston, S., Davidson, M., Thomas, K., et al. (2003) Parametric manipulation of conflict and response competition using rapid mixed-trial event-related fMRI. *NeuroImage*, **20**, 2135–41.

Durston, S., Davidson, M., Tottenham, N., et al. (2006) A shift from diffuse to focal cortical activity with development. *Dev Sci*, **9**, 1–8.

Durston, S., Thomas, K., Yang, Y., et al. (2002) A neural basis for the development of inhibitory control. *Dev Sci* 5:4, F9–F16.

Engelbrecht, V., Scherer, A., Rassek, M., et al. (2002) Diffusion-weighted MR imaging in the brain in children: findings in the normal brain and in the brain with white matter diseases. *Radiology*, 222, 410–18.

Epstein, J. N., Casey, B. J., Tonev, S. T., et al. (2007) ADHD- and medication-related brain activation effects in concordantly affected parent-child dyads with ADHD. *J Child Psychol Psychiatry*, 48, 899–913.

Field, A. S., Yen, Y. F., Burdette, J. H., et al. (2000) False cerebral activation on BOLD functional MR images: study of low-amplitude motion weakly correlated to stimulus. *AJNR Am J Neuroradiol*, 21, 1388–96.

Fischl, B. and Dale, A. M. (2000) Measuring the thickness of the human cerebral cortex from magnetic resonance images. *Proc Natl Acad Sci U S A*, 97, 11050–5.

Forman, S. D., Cohen, J. D., Fitzgerald, M., et al. (1995) Improved assessment of significant activation in functional magnetic resonance imaging (fMRI): use of a cluster-size threshold. *Magn Reson Med*, 33, 636–47.

Fox, M. D., Snyder, A. Z., Vincent, J. L., et al. (2005) The human brain is intrinsically organized into dynamic, anticorrelated functional networks. *Proc Natl Acad Sci U S A*, 102, 9673–8.

Friston, K. (ed.) (2007) *Statistical Parametric Mapping: The Analysis of Functional Brain Images*. London: Academic Press.

Friston, K., Holmes, A., Worsley, K., et al. (1994) Statistical parametric maps in functional imaging: a general linear approach. *Hum Brain Mapp*, 2, 189–210.

Friston, K. J., Williams, S., Howard, R., et al. (1996) Movement-related effects in fMRI time-series. *Magn Reson Med*, 35, 346–55.

Fu, L., Fonov, V., Pike, B., et al. (2006) Automated analysis of multi site MRI phantom data for the NIHPD project. *Med Image Comput Comput Assist Interv Int Conf Med Image Comput Comput Assist Interv*, 9, 144–51.

Galvan, A., Hare, T. A., Parra, C. E., et al. (2006) Earlier development of the accumbens relative to orbitofrontal cortex might underlie risk-taking behavior in adolescents. *J Neurosci*, 26, 6885–92.

Garvey, M. A. and Gilbert, D. L. (2004) Transcranial magnetic stimulation in children. *Eur J Paediatr Neurol*, 8, 7–19.

Giedd, J. N., Blumenthal, J., Jeffries, N. O., et al. (1999) Brain development during childhood and adolescence: a longitudinal MRI study. *Nat Neurosci*, 2, 861–3.

Glover, G. H. and Thomason, M. E. (2004) Improved combination of spiral-in/out images for BOLD fMRI. *Magn Reson Med*, 51, 863–8.

Grady, C. L., McIntosh, A. R., Bookstein, F., et al. (1998) Age-related changes in regional cerebral blood flow during working memory for faces. *NeuroImage*, 8, 409–25.

Haynes, J. D., Rees, G. (2006) Decoding mental states from brain activity in humans. *Nat Rev Neurosci*, 7, 523–34.

Hu, X. and Norris, D. G. (2004) Advances in high-field magnetic resonance imaging. *Annu Rev Biomed Eng*, 6, 157–84.

Huettel, S., Song, A., and McCarthy, G. (2004) *Functional Magnetic Resonance Imaging*. Sunderland, MA: Sinauer Associates.

Janata, P., Tillmann, B., and Bharucha, J. J. (2002) Listening to polyphonic music recruits domain-general attention and working memory circuits. *Cogn Affect Behav Neurosci*, 2, 121–40.

Kanwisher, N., Chun, M. M., McDermott, J., et al. (1996) Functional imaging of human visual recognition. *Brain Res Cogn Brain Res*, 5, 55–67.

Klingberg, T., Forssberg, H., and Westerberg, H. (2002) Increased brain activity in frontal and parietal cortex underlies the development of visuospatial working memory capacity during childhood. *J Cogn Neurosci*, 14, 1–10.

Kotsoni, E., Byrd, D., and Casey, B. J. (2006) Special considerations for functional magnetic resonance imaging of pediatric populations. *J Magn Reson Imaging*, 23, 877–86.

Kwong, K. K., Belliveau, J. W., Chesler, D. A., et al. (1992) Dynamic magnetic resonance imaging of human brain activity during primary sensory stimulation. *Proc Natl Acad Sci U S A*, 89, 5675–9.

LaConte, S., Strother, S., Cherkassky, V., et al. (2005) Support vector machines for temporal classification of block design fMRI data. *NeuroImage*, 26, 317–29.

Ladouceur, C. D., Dahl, R. E., Williamson, D. E., et al. (2006) Processing emotional facial expressions influences performance on a go/no go task in pediatric anxiety and depression. *J Child Psychol Psychiatry*, 47, 1107–15.

Lin, K. L. and Pascual-Leone, A. (2002) Transcranial magnetic stimulation and its applications in children. *Chang Gung Med J*, 25, 424–36.

Linden, D. E. (2007) What, when, where in the brain? Exploring mental chronometry with brain imaging and electrophysiology. *Rev Neurosci*, 18, 159–71.

Liston, C., Watts, R., Tottenham, N., et al. (2006) Frontostriatal microstructure modulates efficient recruitment of cognitive control. *Cereb Cortex*, **16**, 553–60.

Liu, J. Z., Zhang, L., Brown, R. W., et al. (2004) Reproducibility of fMRI at 1.5 T in a strictly controlled motor task. *Magn Reson Med*, **52**, 751–60.

Logothetis, N. K., Pauls, J., Augath, M., et al. (2001) Neurophysiological investigation of the basis of the fMRI signal. *Nature*, **412**, 150–7.

Lu, L., Leonard, C., Thompson, P., et al. (2007) Normal developmental changes in inferior frontal gray matter are associated with improvement in phonological processing: a longitudinal MRI analysis. *Cereb Cortex*, **17**, 1092–9.

Lucker, J. R., Geffner, D., and Koch, W. (1996) Perception of loudness in children with ADD and without ADD. *Child Psychiatry Hum Dev*, **26**, 181–90.

Maegaki, Y., Seki, A., Suzaki, I., et al. (2002) Congenital mirror movement: a study of functional MRI and transcranial magnetic stimulation. *Dev Med Child Neurol*, **44**, 838–43.

Marsh, R., Zhu, H., Wang, Z., et al. (2007) A developmental fMRI study of self-regulatory control in Tourette's syndrome. *Am J Psychiatry*, **164**, 955–66.

Mathiak, K. and Weber R. (2006) Toward brain correlates of natural behavior: fMRI during violent video games. *Hum Brain Mapp*, **27**, 948–56.

Matsuzawa, J., Matsui, M., Konishi, T., et al. (2001) Age-related volumetric changes of brain gray and white matter in healthy infants and children. *Cereb Cortex*, **11**, 335–42.

McKeown, M. J. and Sejnowski, T. J. (1998) Independent component analysis of fMRI data: examining the assumptions. *Hum Brain Mapp*, **6**, 368–72.

Miller, S. P. (2007) Newborn brain injury: looking back to the fetus. *Ann Neurol*, **61**, 285–7.

Morris, J. S., Frith, C. D., Perrett, D. I., et al. (1996) A differential neural response in the human amygdala to fearful and happy facial expressions. *Nature*, **383**, 812–15.

Muller, R. A. (2007) The study of autism as a distributed disorder. *Ment Retard Dev Disabil Res Rev*, **13**, 85–95.

Muzik, O., Chugani, D. C., Juhasz, C., et al. (2000) Statistical parametric mapping: assessment of application in children. *NeuroImage*, **12**, 538–49.

Nagy, Z., Westerberg, H., and Klingberg, T. (2004) Maturation of white matter is associated with the development of cognitive functions during childhood. *J Cogn Neurosci*, **16**, 1227–33.

O'Toole, A. J., Jiang, F., Abdi, H., et al. (2005) Partially distributed representations of objects and faces in ventral temporal cortex. *J Cogn Neurosci*, **17**, 580–90.

Pascual-Leone, A., Walsh, V., and Rothwell, J. (2000) Transcranial magnetic stimulation in cognitive neuroscience – virtual lesion, chronometry, and functional connectivity. *Curr Opin Neurobiol*, **10**, 232–7.

Peter, F. (2005) Spontaneous low-frequency BOLD signal fluctuations: an fMRI investigation of the resting-state default mode of brain function hypothesis. *Hum Brain Mapp*, **26**, 15–29.

Posner, M. (1978) *Chronometric Explorations of Mind*. Hillsdale, NJ: Lawrence Erlbaum Associates.

Raichle, M. E. and Snyder, A. Z. (2007) A default mode of brain function: a brief history of an evolving idea. *NeuroImage*, **37**, 1083–90; discussion 1097–9.

Raichle, M. E., MacLeod, A. M., Snyder, A. Z., et al. (2001) A default mode of brain function. *Proc Natl Acad Sci U S A*, **98**, 676–82.

Rajapakse, J. C., Tan, C. L., Zheng, X., et al. (2006) Exploratory analysis of brain connectivity with ICA. *IEEE Eng Med Biol Mag*, **25**, 102–11.

Rosenberg, D. R., Sweeney, J. A., Gillen, J. S., et al. (1997) Magnetic resonance imaging of children without sedation: preparation with simulation. *J Am Acad Child Adolesc Psychiatry*, **36**, 853–9.

Rossini, P. M. and Rossi, S. (2007) Transcranial magnetic stimulation: diagnostic, therapeutic, and research potential. *Neurology*, **68**, 484–8.

Schlaggar, B. L., Brown, T. T., Lugar, H. M., et al. (2002) Functional neuroanatomical differences between adults and school-age children in the processing of single words. *Science*, **296**, 1476–9.

Schneider, J., Il'yasov, K., Hennig, J., et al. (2004) Fast quantitative diffusion-tensor imaging of cerebral white matter from the neonatal period to adolescence. *Neuroradiology*, **46**, 258–66.

Schneider, W., Casey, B., and Noll, D. (1994) Functional MRI mapping of stimulus rate effects across visual processing stages. *Hum Brain Mapp*, **1**, 117–33.

Seghier, M. L., Lazeyras, F., and Huppi, P. S. (2006) Functional MRI of the newborn. *Semin Fetal Neonatal Med*, **11**, 479–88.

Shulman, G. L., Corbetta, M., Buckner, R. L., et al. (1997) Top-down modulation of early sensory cortex. *Cereb Cortex*, **7**, 193–206.

Slifer, K. J., Cataldo, M. F., Cataldo, M. D., et al. (1993) Behavior analysis of motion control for pediatric neuroimaging. *J Appl Behav Anal*, **26**, 469–70.

Smith, S., Jenkinson, M., Woolrich, M., et al. (2004) Advances in functional and structural MR image analysis and implementation as FSL. *NeuroImage*, **23**, 208–19.

Souweidane, M. M., Kim, K. H., McDowall, R. et al. (1999) Brain mapping in sedated infants and young children with passive-functional magnetic resonance imaging. *Pediatr Neurosurg*, **30**, 86–92.

Sowell, E. R., Thompson, P. M., Leonard, C. M., et al. (2004) Longitudinal mapping of cortical thickness and brain growth in normal children. *J Neurosci*, **24**, 8223–31.

Stark, C. E. and Squire, L. R. (2001) When zero is not zero: the problem of ambiguous baseline conditions in fMRI. *Proc Natl Acad Sci U S A*, **98**, 12760–6.

Staudt, M., Krageloh-Mann, I., Holthausen, H., et al. (2004) Searching for motor functions in dysgenic cortex: a clinical transcranial magnetic stimulation and functional magnetic resonance imaging study. *J Neurosurg*, **101**, 69–77.

Sternberg, S. (1969) *The Discovery of Processing Stages: Extensions of Donders' Method*. Amsterdam: North-Holland.

Stippich, C., Rapps, N., Dreyhaupt, J., et al. (2007) Localizing and lateralizing language in patients with brain tumors: feasibility of routine preoperative functional MR imaging in 81 consecutive patients. *Radiology*, **243**, 828–36.

Thomason, M. E., Burrows, B. E., Gabrieli, J. D., et al. (2005) Breath holding reveals differences in fMRI BOLD signal in children and adults. *NeuroImage*, **25**, 824–37.

Tomarken, A. J., Dichter, G. S., Garber, J., et al. (2004) Resting frontal brain activity: linkages to maternal depression and socio-economic status among adolescents. *Biol Psychol*, **67**, 77–102.

Tootell, R. B., Hadjikhani, N. K., Vanduffel, W., et al. (1998) Functional analysis of primary visual cortex (V1) in humans. *Proc Natl Acad Sci U S A*, **95**, 811–17.

Voyvodic, J. T. (2006) Activation mapping as a percentage of local excitation: FMRI stability within scans, between scans and across field strengths. *Magn Reson Imaging*, **24**, 1249–61.

Yamada, H., Sadato, N., Konishi, Y., et al. (2000) A milestone for normal development of the infantile brain detected by functional MRI. *Neurology*, **55**, 218–23.

Yoo, S. S., Wei, X., Dickey, C. C., et al. (2005) Long-term reproducibility analysis of fMRI using hand motor task. *Int J Neurosci*, **115**, 55–77.

Yurgelun-Todd, D. A. and Killgore, W. D. (2006) Fear-related activity in the prefrontal cortex increases with age during adolescence: a preliminary fMRI study. *Neurosci Lett*, **406**, 194–9.

Zang, Y. F., He, Y., Zhu, C. Z., et al. (2007) Altered baseline brain activity in children with ADHD revealed by resting-state functional MRI. *Brain Dev*, **29**, 83–91.

18

Magnetic resonance spectroscopy: methods and applications in developmental clinical neuroscience

Marisa M. Silveri, Deborah Yurgelun-Todd and Perry Renshaw

Introduction

In this review, we outline the basic principles of magnetic resonance spectroscopy (MRS) and describe MRS methods, the physiological significance of quantifiable brain metabolites, and limitations of the technique. We then present applications in developmental neuroscience in healthy populations and in cohorts with neuropsychiatric disorders. Finally, we highlight the contributions of MRS methods to developmental neurobiology thus far and identify promising future directions for MRS research in this field.

Principles of magnetic resonance spectroscopy

Magnetic resonance spectroscopy can be used to study any nucleus that has a magnetic moment, i.e., possesses electromagnetic properties that cause the nucleus of interest to align with a static magnetic field, e.g., hydrogen (^1H), phosphorus (^{31}P), and carbon (^{13}C). The static magnetic field reflects the strength of the magnetic resonance (MR) scanner, which typically ranges from 1.5 T to 4.0 T in human studies. For each magnetic nucleus in the brain, a given static magnetic field is associated with a particular resonance frequency, called the Larmor frequency. For example, at 1.5 T, the proton resonance (Larmor) frequency is 63.88 MHz, and at 4.0 T, 170.4 MHz (see Table 18.1).

A transient magnetic field introduced at the molecule's Larmor frequency using a radiofrequency (RF) coil, or antenna, causes lower energy spins (aligned with the static magnetic field, e.g., 3.0 T) to transition to a higher energy level (aligned against the static magnetic field). When the transient magnetic field is removed, or turned off, energy (or signal intensity) at the resonance frequency (hertz, Hz) is released from the higher energy spins as they realign with the static magnetic field, a process called relaxation or recovery (T_1 relaxation is within the longitudinal axis and T_2 is within the transverse axis). This process whereby the signals induced by free precession of nuclear spins around the static field after the RF pulse is turned off is referred to as free induction decay (FID). The FID is plotted as energy release (relaxation) as a function of time and converted via Fourier transformation to a series of peaks (or spectrum), where signal intensity can be visualized in the frequency domain (Hz).

Within a nucleus, such as hydrogen, chemically distinct groups within the molecule (e.g., for hydrogen, H_2O versus COH) possess minor differences in their resonance frequencies, due to interactions among groups within the molecule (Bovey et al., 1988). These small differences, referred to as the *chemical shift*, make it possible to differentiate distinct peaks within a molecule. Thus, as can be seen in Figure 18.1, the frequency of each individual peak within the proton spectrum frequency is

Neuroimaging in Developmental Clinical Neuroscience, eds. Judith M. Rumsey and Monique Ernst. Published by Cambridge University Press. © Cambridge University Press 2009.

Table 18.1 Relative NMR sensitivities

Nucleus	Spin quantum number[a]	NMR resonance gyromagnetic ratio (MHz/T) at		Relative sensitivity at constant field
		1.5 T	4.0 T	
^{1}H	1/2	63.87	170.40	1
^{19}F	1/2	60.08	160.40	0.83
^{7}Li	3/2	24.83	66.21	0.29
^{23}Na	3/2	16.89	45.04	0.09
^{31}P	1/2	25.88	68.80	0.06
^{13}C	1/2	16.07	42.80	0.02
^{39}K	3/2	2.99	7.97	0.0005

Note:
[a]Spin quantum number is a term used to describe the angular momentum of nuclei.

represented on a frequency scale measured in parts per million (ppm), which is a conversion from Hz that is independent of magnetic field strength. The area of the resonance intensity (area under the peak) is proportional to the concentration of molecules that contribute to the resonance.

Nuclei capable of producing a resonance signal that can be quantified using MRS (i.e., MR-visible nuclei) differ in MR sensitivity, based on their electromagnetic properties and concentrations, which must be large enough (millimolar range) to be detected. Sensitivity, as reflected in the intensity of the MR signal, can be amplified by increasing the size of the voxel (volume) sampled and/or by increasing the strength of the magnetic field to yield peaks that are quantifiable.

Increases in field strength, e.g., from 1.5 T to 3 or 4 T, over the past decade have been particularly valuable for MRS studies, as MR signals increase linearly with field strength. Therefore, the use of higher field strengths also increases sensitivity, due in part to a greater separation of metabolite peaks, or better spectral resolution, allowing for improved quantification.

Magnetic resonance spectroscopy methods

The design of an MRS experiment requires inclusion of study parameters that optimize the signal-to-noise

ratio from a clearly delineated brain region in an acquisition duration that patients can tolerate. Typical spectral acquisitions range from 10 to 50 minutes, depending on the sensitivity of the nucleus being studied and the region of interest being sampled. As with other MR techniques, patient comfort is critical for minimizing subject motion, which degrades the quality of the spectra (see also "Spectral analysis and quantification" below).

The choice of MRS method depends on the nucleus to be examined. Although the two most common methods are proton (^{1}H) and phosphorus (^{31}P) MRS, quantification of resonance intensities is also possible for carbon (^{13}C), sodium (^{23}Na), fluorine (^{19}F), and lithium (^{7}Li). Although there is continuing debate concerning the interpretation of MRS metabolite signals, an overview of common interpretations of the functional significance of quantifiable proton and phosphorus metabolites is presented in Table 18.2.

Proton (^{1}H) magnetic resonance spectroscopy

In vivo proton MRS provides a means to detect and quantify a number of important amino acids, including *N*-acetylaspartate (NAA), creatine/phosphocreatine (Cr), cytosolic choline compounds (Cho), and *myo*-inositol (mI) (Figure 18.1), as well

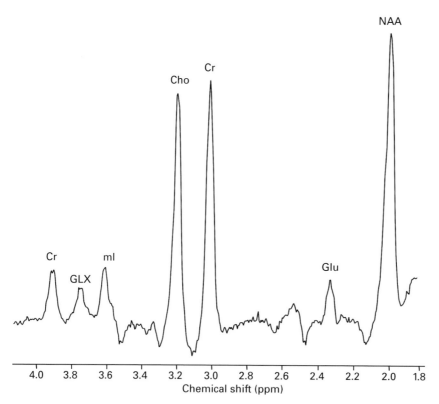

Figure 18.1. Sample ¹H spectrum. ¹H metabolites include Cr, total creatine (3.0 ppm); GLX, glutamate/glutamine; mI, *myo*-inositol (3.5 ppm); Cho, choline (3.2 ppm); Glu, glutamate; NAA, *N*-acetylaspartate (2.0 ppm).

as lactate, which is the end-product of anaerobic metabolism (glycolysis). *N*-Acetylaspartate, found primarily in neurons (Birken and Oldendorf, 1989), contributes the largest signal in the proton spectrum, at a chemical shift of 2.02 ppm. *N*-Acetylaspartate has been viewed as a neuronal marker, or an indicator of neuronal integrity, although the function of this metabolite has been widely debated (Birken and Oldendorf, 1989; Moffett et al., 2007). The singlet Cr resonances, at 3.03 ppm and 3.93 ppm, arise from the protons of Cr and phosphorylated creatine (phosphocreatine, PCr). Creatine/phosphocreatine plays a major role in energy metabolism in the brain, acting as an energy buffer by maintaining constant brain adenosine triphosphate (ATP) levels through the creatine kinase reaction and by distributing energy (via mitochondria) within the brain.

The largest Cho signal resonates at 3.22 ppm and reflects a number of choline-containing compounds such as free choline, glycerol phosphocholine and phosphocholine (GPC and PC, respectively) (Barker et al., 1994). These compounds are involved in pathways of cellular membrane synthesis and degradation. Most of the choline in brain is membrane bound as phosphatidylcholine, which is largely invisible to in vivo MRS (Miller, 1991). The most pronounced mI resonance is at 3.54 ppm. *myo*-Inositol is involved in the synthesis and turnover of phospholipid membranes, as well as in the maintenance of osmotic equilibrium (Moore et al., 1999). Magnetic resonance spectroscopy studies of multiple sclerosis and Alzheimer's disease, conditions associated with white matter pathology, have demonstrated abnormal levels of Cho and mI.

Table 18.2 Physiological significance of MRS metabolites

Physiological significance	
^{1}H MRS	
N-Acetylaspartate (NAA)	Marker of neuronal viability. Reductions indicate tissue pathology. Levels increase with brain maturation
Choline (Cho)	Involved in pathways of cellular membrane synthesis and degradation
Creatine + phosphocreatine (tCr)	Markers of cellular energetic state
myo-Inositol (mI)	Involved in phospholipid metabolism and maintenance of osmotic equilibrium
Glutamate (Glu)	Excitatory neurotransmitter and key molecule in cellular metabolism. Also a precursor for GABA synthesis
Glutamine (Gln)	Precursor for glutamate and plays a role in protein synthesis
GABA	Inhibitory neurotransmitter and plays a role in cellular metabolism
"Glx"	Combination of Glu, Gln, GABA resonances
Lactate (Lac)	By-product of anaerobic metabolism. Visible under conditions of oxygen deprivation
^{31}P MRS	
Phosphocreatine (PCr)	High-energy phosphate, contributes to the maintenance of βNTP levels
Polyphosphate regions of the spectrum nucleotriphosphates, α, γ, βNTP	Level of ATP in brain (βNTP)
Phosphomonoesters (PME) Phosphoethanolamine (PE) Phosphocholine (PC)	Building blocks of membrane phospholipids
Phosphodiesters (PDE) Glycerophosphoethanolamine (GPE) Glycerol phosphocholine (GPC)	Major catabolic products of membrane phospholipid degradation
Inorganic phosphate (Pi)	High-energy phosphate that combines with creatine to form PCr, but also is released from phosphocreatine to synthesize nucleoside triphosphates (ADP to ATP). Chemical shift of inorganic phosphate can be used to calculate intracellular pH. PCr/Pi ratio provides a measure of the energy status in brain, as it is a ratio of the most labile form of high-energy phosphate (PCr) to the ultimate breakdown product of all high-energy phosphate compounds (Pi)

Lactate, also known as lactic acid, increases markedly when the brain is deprived of oxygen even for a short period of time under conditions such as ischemia (stroke) or hypoxia, making this metabolite an important metabolic marker. Thus, although lactate is always present in the brain, the concentration tends to increase whenever anaerobic respiration increases. Lactate produces a characteristic doublet at 1.33 ppm in brain ^{1}H MR spectra, which is obscured by a resonance arising from lipids (Auer et al., 2001; Behar et al., 1994). Since lipids have shorter T_2 relaxation times than lactate, the lactate resonance can generally be detected more clearly when a long time to echo (TE) is used (Behar et al., 1994).

Additional metabolites quantifiable in the proton spectrum, albeit near the lower limit of detection, include glutamate, glutamine and gamma aminobutyric acid (GABA). Glutamine (Gln) is a precursor for glutamate (Glu), which is a major excitatory neurotransmitter found in all brain cell types, with the highest concentrations generally observed in neurons. Gamma-aminobutyric acid is the major inhibitory neurotransmitter in the mammalian brain (McCormick, 1989). Although glutamate is present in the brain at much higher concentrations than GABA, only a small fraction of brain Glu participates in neurotransmission. Because the peaks associated with these metabolites are difficult to resolve and are obscured by peaks of higher concentration metabolites in brain spectra, investigators generally report a combined "Glx" resonance intensity. However, the increasing availability of high-field MR scanners and the development of specialized editing techniques for enhancing the MR visibility of these peaks (Keltner et al., 1997; Weber et al., 1997) holds promise for separating these resonances in order to measure neurochemicals that are implicated in psychiatric disorders, such as glutamate and GABA.

Phosphorus (^{31}P) magnetic resonance spectroscopy

In vivo ^{31}P MRS provides a means of detecting high-energy phosphate metabolites and constituents of membrane synthesis, indicating the cellular bioenergetic state and the integrity and the function of cell membranes, respectively. Three phospholipid metabolites associated with high-energy intracellular metabolism are detectable: PCr, inorganic phosphate (Pi), and β-nucleoside triphosphate (NTP, primarily reflecting ATP in the brain). Under steady-state conditions, the rate of ATP synthesis equals the rate of ATP utilization, via suppression of excessive glycolysis and activation of mitochondrial oxidative phosphorylation. In the absence of additional glucose, however, ATP levels remain constant since high-energy PCr serves as a buffer for maintenance of β NTP levels and a shuttle for energy from sites of production to sites of utilization (Bessman and Geiger, 1981; Wallimann et al., 1992). Availability of PCr therefore pushes the creatine kinase reaction to generate β NTP, via conversion to

creatine and high-energy phosphate (Pi) (Wallimann et al., 1992), resulting in a drop in PCr levels while levels of ADP and Pi increase to support steady-state levels of β NTP (Gyulai et al., 1985). In this regard, the ratio of PCr relative to Pi has been shown to reflect phosphorylation potential (Nioka et al., 1990). Furthermore, the chemical shift of Pi also can be used to determine internal pH level, which plays a role in modulation of synaptic transmission and plasticity and of neuronal excitability, and can aid in the discrimination between diseased and healthy tissue.

Constituents of cell membranes detectable with ^{31}P MRS include phosphomonoesters (PME) and phosphodiesters (PDE). The brain PME resonance arises primarily from the phospholipid precursors phosphoethanolamine (PE) and phosphocholine (PC) and is derived from a total metabolite pool of approximately 3.0 mM (Pettegrew et al., 1991). These metabolites are known to be precursors of membrane phospholipid synthesis. The in vivo PDE resonance has a broad component, arising from membrane bilayers, and a narrow component, which is derived from the phospholipid catabolites GPC and glycerophosphoethanolamine (GPE). The PDE peak reflects breakdown products of membrane phospholipids. A phosphorus spectrum acquired from a healthy adult subject at 1.5 T is shown in Figure 18.2a.

Proton decoupling is a technique that can increase sensitivity for detecting ^{31}P metabolites by producing line-narrowing effects for improved resolution of the PC and PE in the *PME peak* and GPC and GPE in the *PDE peak*. A special head coil, dually tuned for both ^{1}H and ^{31}P nuclei, is required for proton decoupling. Alternatively, acquiring spectra at higher field strengths also improves ^{31}P sensitivity by increasing spectral dispersion.

Other MR-visible nuclei

While ^{1}H and ^{31}P MRS are the only techniques that have been widely applied to the study of neuropsychiatric disorders, a number of other MR visible nuclei have been used to probe brain chemistry in neurological disorders (Mason et al., 1996; Tyson

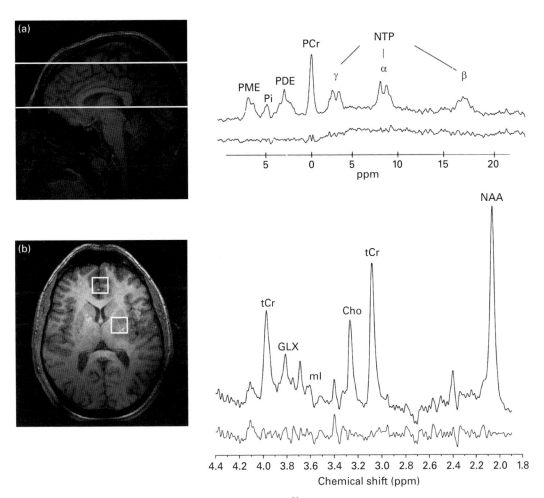

Figure 18.2. Sample ^{31}P spectrum. (a) Example of a large slab ^{31}P spectral acquisition acquired at 1.5 T from a healthy adult subject: 5 cm slab region prescribed through the orbitofrontal and occipital cortices using the central sagittal image and associated ^{31}P spectrum. (b) Example of a single-voxel ^{1}H spectral acquisition acquired at 4 T from a healthy adult subject: 2 cm × 2 cm × 2 cm (8 cm^3 or 8 ml) single voxels prescribed in the anterior cingulate cortex (ACC) and basal ganglia using an axial image and associated ^{1}H ACC spectrum. PME, phosphomonoesters; Pi, inorganic phosphate; PDE, phosphodiesters; PCr, phosphocreatine; NTP, nucleoside triphosphate; tCr, total creatine/phosphocreatine; GLX, glutamate/glutamine; mI, *myo*-inositol; Cho, choline; NAA, *N*-acetylaspartate.

et al., 1996). Sodium (^{23}Na), abundant in the form of sodium chloride and other salts, gives rise to a single resonance line. Sodium MRS has been used to probe the concentration gradient of intracellular versus extracellular sodium, which provides important information about cellular processes such as generation of neuronal impulses and regulation

of cell volume. Abnormalities or alterations in the sodium resonance have been associated with cerebral ischemia (e.g., Tyson et al., 1996).

While carbon atoms are found in almost every compound in living systems, the ordinary nuclide of carbon (^{12}C) does not possess magnetic properties, making it NMR invisible. The magnetic form of carbon (^{13}C) is present in low concentrations in all tissue. Thus, it is possible to administer ^{13}C as an MR ligand to detect compounds that have incorporated ^{13}C. ^{13}C MRS offers the opportunity to study the flux of metabolic pathways such as the tricarboxylic acid cycle (TCA), which produces useable energy generated via cerebral carbohydrate metabolism (Mason et al., 1996). Limitations of ^{13}C MRS include difficulties optimizing the signal-to-noise ratio (SNR) and spectral resolution at commonly available field strengths (1.5–3.0 T), as well as steep costs associated with ^{13}C infusion. Since ^{13}C MRS requires infusion of the magnetic form of carbon, this invasive procedure has reduced utility for pediatric research.

Fluorine (^{19}F) is one of the most sensitive MR nuclei with 83% of proton's sensitivity. The chemical shift range is more than four times that of phosphorus, and because there are no endogenous levels of fluorine in the body there is no issue with overlapping resonances. Applications of in vivo ^{19}F MRS include the ability to monitor uptake and metabolism of fluorinated anesthetics (e.g., halothane, isoflurane) and other drugs such as selective serotonin reuptake inhibitor antidepressants (SSRIs, e.g., fluoxetine, also known as Prozac). To obtain a ^{19}F spectrum, patients must be exposed to a fluorinated compound, either as an infusion during the scanning session or as part of a pharmacological treatment regimen, limiting its use in healthy children.

Lithium (^{7}Li), typically administered clinically as oral Li_2CO_3, is the treatment of choice for bipolar disorder. Thus, ^{7}Li MRS provides a technique for in vivo monitoring of lithium in human brain. To date, ^{7}Li MRS studies have been limited by low MR sensitivity, although the recent use of higher-field-strength scanners and better designed RF coils have yielded shorter acquisition times, higher sensitivity, and more precise quantification.

Anatomical localization of magnetic resonance spectroscopy

Due to improved spatial localization techniques, MRS studies have evolved from acquisitions of spectra from a large slab of tissue (Figure 18.2a), to acquisition from a predefined single voxel or multiple single voxels (Figure 18.2b), to simultaneous two- and three-dimensional acquisitions from multiple regions within a brain slice or slices (Figure 18.3). Single-voxel MRS samples a specific cube or cubes of tissue localized in a particular brain region. In practice, most single-voxel studies include from one to three voxels (with a typical minimum size of 8 cm^3), acquired sequentially in a single imaging session.

The two- and three-dimensional techniques are referred to as spectroscopic imaging (SI) or chemical shift imaging (CSI). These multi-voxel methods collect chemical shift signals (differing metabolites within a nuclear species) from a wide region of tissue selected using high resolution MRI. This array of data is later decoded into individual spectra from each voxel after the patient has been removed from the scanner. Recent studies using CSI have demonstrated considerable regional variation in the metabolites that can be measured using MRS (Wiedermann et al., 2001), thus demonstrating the value of being able to sample data from multiple regions of the brain simultaneously. However, the spatial resolution of single-voxel spectroscopy is considered superior, because the magnetic field can be optimally homogenized for the volume selected. Thus, for clinical investigations aimed at the identification of focal pathology, the single-voxel method may be most advantageous.

Due to differences in the relaxation times and sensitivity, proton and phosphorus MRS differ in their spatial localization. Stimulated echo acquisition (STEAM) and point-resolved spectroscopy (PRESS) methods are often used to collect proton MRS data (Moonen et al., 1989) from brain volumes on the order of 1–10 cm^3 (Figure 18.3a). Since the sensitivity of ^{31}P is lower than that of ^{1}H, larger tissue volumes are required to obtain an optimal signal. Spatial localization methods used to collect ^{31}P data include depth-resolved surface coil spectroscopy (DRESS) (Bottomley et al., 1984), image selected in vivo spectroscopy (ISIS) (Ordidge et al., 1986), and low resolution, two- or three-dimensional spectroscopic images (Brown et al., 1982) (Figure 18.3b).

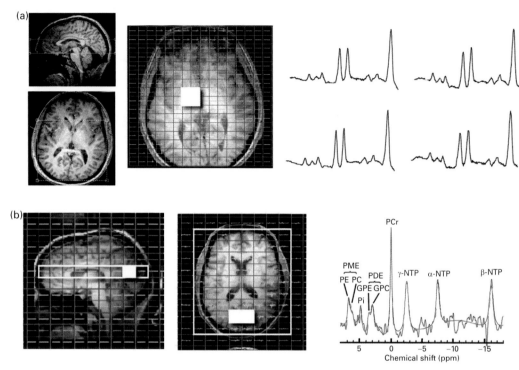

Figure 18.3. Chemical shift imaging (CSI). (a) Example of a 2D ^{1}H PRESS CSI spectral acquisition at 4 T from a healthy adult subject: multi-voxel grid [4 cm × 1.125 cm − 1.125 cm (5.1 ml) voxels] prescribed through the mesiotemporal region using sagittal and axial images and four ^{1}H spectra extracted from four voxels within the larger volume of interest. (b) Example of a 3D ^{31}P CSI spectral acquisition at 4 T from a healthy adult subject: multi-voxel grid [2.1 cm × 2.1 cm × 2.1 cm (8.8 ml) voxels] prescribed using sagittal and axial images to encompass the corpus callosum and associated ^{31}P raw spectrum (black) and fit (smooth red, using fitMAN software, Potwarka et al., 1999). Abbreviations as defined in Figure 18.2.

Spectral analysis and quantification

Quantification of brain metabolites for specific brain regions is accomplished by measurement of the area under individual peaks in the spectrum or spectra. To allow for quantification, the MRS acquisition scheme should optimize linewidth and SNR (typically by shimming, or by creating a homogeneous magnetic field in the tissue of interest). Interpretation of spectral data also requires consideration of overlapping peaks, saturation and relaxation effects, contributions of broad macromolecules, lipid signal contamination, and underlying tissue content [gray matter, white matter, and cerebrospinal fluid (CSF)] in the region of interest. Although there have been advances in strategies for addressing these concerns, they remain significant.

In addition to software developed in house at research centers, commercially available software programs for interactive and automated quantification are available. Widely accepted in the MRS field are VARiable PROjection (VARPRO), an improved recent version of VARPRO called Advanced Method for Accurate, Robust and Efficient Spectral fitting (AMARES), and Linear Combination of Model Spectra (LCModel, in vivo proton MRS only) (Provencher, 1993, 2001). These methods apply a priori knowledge of metabolite chemical shifts to calculate the area under the curve of each peak. VARPRO and AMARES use an iterative fitting procedure

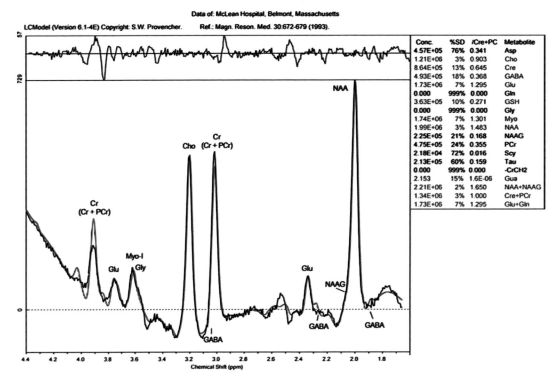

Data of: McLean Hospital, Belmont, Massachusetts

LCModel (Version 6.1-4E) Copyright: S.W. Provencher.   Ref.: Magn. Reson. Med. 30:672-679 (1993).

Conc.	%SD	/Cre+PC	Metabolite
4.57E+05	76%	0.341	Asp
1.21E+06	3%	0.903	Cho
8.64E+05	13%	0.645	Cre
4.93E+05	18%	0.368	GABA
1.73E+06	7%	1.295	Glu
0.000	999%	0.000	Gln
3.63E+05	10%	0.271	GSH
0.000	999%	0.000	Gly
1.74E+06	7%	1.301	Myo
1.99E+06	3%	1.483	NAA
2.25E+05	21%	0.168	NAAG
4.75E+05	24%	0.355	PCr
2.18E+04	72%	0.016	Scy
2.13E+05	60%	0.159	Tau
0.000	999%	0.000	-CrCH2
2.153	15%	1.6E-06	Gua
2.21E+06	2%	1.650	NAA+NAAG
1.34E+06	3%	1.000	Cre+PCr
1.73E+06	7%	1.295	Glu+Gln

Figure 18.4. ^{1}H spectrum (black), LCModel fit (red) and associated LCModel quantitative output. The printout to the right of the spectrum provides the absolute concentration (Conc.) for each metabolite, derived from a theoretical fitting routine (basis set) that incorporates the spectroscopic time to repetition (TR) and time to echo (TE) and nuclei-specific relaxation times. Because these values do not take tissue content or scanner-specific factors into account, error may be introduced. Thus, ratio data (metabolite concentration/Cr + PCr) are also calculated. An estimated standard deviation (%SD) indicates the reliability of the LCModel fit, such that %SD × 2 yields a 95% confidence interval. Values of %SD less than 20% are considered reliable; those above 50% suggest that the metabolite is unreliably detected.

based on biochemical prior knowledge of metabolite resonance frequencies. LCModel fits spectra by comparing in vivo raw spectral data with in vitro metabolite data collected under conditions identical to the in vivo MRS protocol, which are used to generate a basis set, or template, of metabolites used as a reference. As can be seen in the sample LCModel output (Figure 18.4), the raw proton spectrum acquired from a CSI voxel of a healthy adult at 4 T is represented by the black curve and the LCModel fit is represented by the red smoothed curve.

Some investigators report their findings as absolute values, while others report their data using ratios of one metabolite to another. In practice,

calculating the absolute values of metabolite concentrations requires knowledge of the T_1 and T_2 relaxation times of the molecule of interest, the time to repetition (TR) and time to echo (TE) acquisition parameters, the tissue volume of interest, and the efficiency of signal detection. As illustrated in the sample LCModel output in Figure 18.4, the first column in the table to the right of the spectrum provides the absolute concentration for each metabolite. Calculation of absolute concentrations in LCModel uses a theoretical fitting routine (basis set) that incorporates the spectroscopic imaging parameters (TR, TE) and nuclei-specific relaxation times. However, other factors specific to the volume of interest (partial volume effects or tissue content) and to the MR scanner are not incorporated in these calculations, which makes metabolite concentrations very liberal, and perhaps unreliable, estimates of true metabolite levels. Given these difficulties, it is common practice to report relative values of MRS data as metabolite ratios.

Metabolite ratios can be calculated using a number of strategies. Metabolite peaks can be quantified relative to an external standard (typically a small capsule containing known amounts of metabolites) placed within the coil, near the patient's head. This approach increases scan time, as spectra collected from the in vitro, external standard must be acquired within the same scanning session with the patient in the scanner. Alternatively, some investigators acquire reference data from a phantom following removal of the subject from the scanner. However, this method is subject to variations due to differences between the subject's head position in the coil and that of the phantom (also referred to as coil loading). The unsuppressed peak arising from water also has been used to calculate metabolite ratios, but is complicated by the need to discriminate between water in tissue versus water in CSF, which can differ by up to 30%–40%.

A more common strategy is to use a relatively stable metabolite peak as an internal reference. For instance, in proton MRS, area under the choline, NAA, or other metabolite peaks can be expressed as a ratio to the area under the creatine peak. The creatine peak represents the concentration of creatine plus phosphocreatine, which is thought to be maintained at constant levels in healthy tissue (see LCModel output, Figure 18.4, metabolites relative to Cr + PCr). It is likely that creatine levels vary between brain regions and tissue types, but also levels may differ between subject populations, complicating interpretations of metabolite ratios when creatine serves as the denominator.

An important factor to consider in interpreting metabolite ratios is the degree of estimated standard deviation (SD, Cramer Rao lower bound). For instance, located to the right of the absolute concentration values in the LCModel output, the %SD indicates the reliability of the LCModel fit (see Figure 18.4).

Limitations of magnetic resonance spectroscopy

Although studies utilizing MRS yield an abundance of information about both the structure and chemical composition of tissues, MRS technology is limited by a number of factors. Optimal SNR and a homogenous magnetic field are necessary to obtain narrow, clear resonance peaks for quantification, which may require 10–45 min of scanning, free of motion artifact. Perhaps the most significant limitation of MRS is its low sensitivity. The signal strength of a particular nucleus depends upon its inherent signal intensity and the externally applied magnetic field strength. As previously discussed, sensitivity can be improved by acquiring spectra at higher field strengths using optimized strategies for spatial localization. Choosing the denominator for determining tissue-specific metabolite ratios also is challenging, as obtaining absolute concentrations is a difficult pursuit and extra efforts are required to obtain an external reference spectrum.

In addition, because the chemical composition of gray matter differs from that of white matter, it is important to assess the tissue content of the sampled brain regions. The high resolution of anatomical MR images used to define voxels for spectral acquisition allows for tissue segmentation and subsequent calculation of relative metabolite levels in gray versus white matter (Pouwels and Frahm, 1998). However, few studies to date have reported segmented imaging data in conjunction with metabolite information (Lim et al., 1998).

Applications of magnetic resonance spectroscopy in developmental clinical neuroscience

Healthy brain development

To date, only a few studies, all cross-sectional, have been published regarding age-related differences in neurochemical metabolites. The majority of these studies focus on early to late childhood, with little data available regarding changes during adolescence (see Table 18.3).

Proton magnetic resonance spectroscopy

The limited literature reporting changes in proton metabolites during development has involved subjects less than 1 month of age up through adulthood and has used large voxels placed in areas containing predominantly white or gray matter, with only a couple of studies examining focal regions of interest. Regardless, the accumulated data thus far have begun to characterize important age-related neurochemical changes.

Across the published developmental proton MRS studies (Table 18.3), levels of NAA have typically been reported to increase with age, whether reported as a concentration or as a ratio relative to Cho, Cr or tissue water (Costa et al., 2002; Horska et al., 2002; Kadota et al., 2001; Kreis et al., 1993; Pouwels et al., 1999; van der Knaap et al., 1990). These NAA levels increase most rapidly within the first few years of life, and plateau anywhere from 2 years to 20 years depending on the region and tissue type studied. N-Acetylaspartate has been most widely accepted as a biomarker of neuronal viability, but also neuronal volume, which is not surprising given the rapid changes in brain tissue volume early in life. Production of NAA also is related to energy metabolism, which is consistent with the global increases in glucose metabolism associated with brain maturation (Cheng et al., 1997).

Findings with regard to choline are much less consistent, with some studies reporting an age-related decrease (Pouwels et al., 1999; Van Der Knaap et al., 1990) or an age-related increase (Kreis et al., 1993) in Cho. The choline resonance reflects several choline-containing compounds, each of which may have distinctly unique developmental profiles. Declines in Cho/Cr and in PME are consistent with accelerated myelination, which is a hallmark of brain development. Thus, NMR-visible choline residues become incorporated into invisible macromolecules that are associated with myelin production. Accordingly, the widely documented increases in the ratio of NAA/Cho may reflect both increases in NAA as well as decreases in Cho. Simultaneous changes in NAA and Cho are quite informative, as MRS studies of developmental delay in children have reported reduced NAA and elevated Cho levels, thought to reflect delayed myelination (Fayed and Modrego, 2005).

Reports of developmental changes in metabolites other than NAA and Cho have been quite sparse, although there is some evidence that creatine levels do not change after the second year of life (Kreis et al., 1993; Pouwels et al., 1999), but that levels are regionally specific (Pouwels et al., 1999). Developmental increases in mI also have been reported to increase with age until reaching a plateau by age 2 and are thought to reflect changes in cellular osmoregulation and signaling pathways (Kreis et al., 1993). Only one MRS study has reported sex differences in brain metabolites, with steeper age-related increases in white matter NAA/Cho being observed in males compared with females (Kadota et al., 2001).

Phosphorus magnetic resonance spectroscopy

Relatively few ^{31}P MRS studies have been conducted in pediatric populations and these have sampled large voxels, thus providing little regionally specific information. Akin to proton findings, studies of young children using ^{31}P MRS have documented rapid metabolite changes within the first 2 years of life, with high-energy metabolites and constituents of membrane synthesis and breakdown reaching adult-like levels as early as 3 years of age.

In general, a decline in the ratio of PME/PDE has been observed with time, until ~2 years of age, in both the cerebrum and the cerebellum (Boesch et al., 1989; Hanaoka et al., 1998; van der Knaap et al., 1990). Changes in the ratio of PME/PDE reflect changes in membrane synthesis and degradation. However, decreases in PME may be due in part to the incorporation of membrane precursors into membrane macromolecules which are largely

Table 18.3 MR spectroscopic findings in healthy children and adolescents

Investigators	Sample	MRS	Region of interest	Metabolites/findings
Van Der Knaap et al. (1990)	1 mo–16 yrs Mean = 5.9 yrs (n = 41) Healthy No sedation	^{1}H SV, 1.5 T 7 × 3 × 3 cm	Paraventricular region, predominantly WM	– NAA/Cho and NAA/Cr increases with age – Cho/Cr decreases with age – Most rapid changes from 1 to 3 years
Kreis et al. (1993)	35 wks–17.8 yrs (n = 109 scans) Healthy (18%) Recovered infants (21%) Cerebral pathology (61%) Chloral hydrate	^{1}H SV, 1.5 T 3–8 cm^3 young 8–16 cm^3 older	Occipital cortex (GM) Parieto-occipital (WM)	– Cr, mI increases until 2 yrs – NAA and Cho increases until 7 yrs
Pouwels et al. (1999)	0–18 yrs (n = 97) Healthy (8%) Neuropediatric (92%) 18–39 yrs adults (n = 72) Healthy <6 yrs chloral hydrate	^{1}H SV, 2.0 T 8–18 ml 8–18 ml 4–5 ml 4–6 ml 4–6 ml	Parietal GM Parieto-occipital WM Cerebellum, vermis Thalamus Basal ganglia	– NAA increases in GM, cerebellum, thalamus with age up to 2 yrs, then stabilizes – NAA constant in WM, basal ganglia – tCr no change after 1 yr, highest level in cerebellum, followed by basal ganglia, thalamus, parietal GM, parieto-occipital WM – Cho declines in WM after 5 yrs
Choi et al. (2000)	3–14 yrs Mean = 9 yrs (n = 30) Healthy Young, chloral hydrate	^{1}H SV, 1.5 T 1.8 × 2 × 2 cm	Allocortex (hippocampus, parahippocampal) Isocortex (medial, frontal and parietal)	– NAA/Cr lower in allocortex than isocortex – Cho/Cr and mI/Cr higher in allocortex than isocortex – No age effects
Costa et al. (2002)	3–18 yrs (n = 37) Healthy No sedation	^{1}H SV, 1.5 T 8 cm^3	Parieto-occipital WM Cerebellar hemisphere	– NAA/Cr, Cho/Cr lower in cerebellum than parieto-occipital WM – NAA/H$_2$O, Cr/H$_2$O and Cho/H$_2$O higher in the cerebellum than parieto-occipital WM – NAA/H$_2$O increased with age in the cerebellum and parieto-occipital WM – Cho/H$_2$O increased with age in the cerebellum

Table 18.3 (*cont.*)

Investigators	Sample	MRS	Region of interest	Metabolites/findings
Horska et al. (2002)	3–19 yrs Mean = 12.3 yrs ($n = 15$) Healthy $n = 2$ nembutal	^{1}H CSI, 1.5 T 8 cm^3	Frontal, parietal (WM) Basal ganglia (GM) Thalamus (GM)	– NAA/Cho in GM peaks at 10 yrs – NAA/Cho in WM increases with age

Kadota et al. (2001)	4–88 yrs
	Mean = 45.6 yrs
	($n = 90$)
	Healthy
	No sedation
		^{1}H CSI, 1.5 T
		1.125 cm^3
			Superior to corpus
			callosum, 6 voxels:
			bilateral WM bilateral
			mesial GM
				– NAA/Cho in WM rapidly increases until 10–20 yrs
				– NAA/Cho in WM declines starting in late thirties
				– Laterality in males for NAA/Cho in WM
				– Steeper increase in WM NAA/Cho in males
				   than in females
				– NAA/Cho in GM gradually declines linearly
				   with age

Boesch et al. (1989)	33 wks–6 yrs ($n = 40$, 48 exams) Healthy (25%) Cerebral pathology (75%) Chloral hydrate	^{31}P SV, 2.35 T	Bilateral fronto-temporal regions	– PME/PDE declines until 1.3 yrs – PCr/βNTP increases until 1.3 yrs
van der Knaap et al. (1990)	1 mo–16 yrs Mean = 5.9 yrs ($n = 41$) Healthy No sedation	^{31}P SV, 1.5 T	Paraventricular region, predominantly WM	– PME/βNTP decreases with age – PDE/βNTP, PCr/βNTP, PCr/Pi increases with age
Hanaoka et al. (1998)	4 mo–13 yrs ($n = 37$) Healthy Sedation not specified	^{31}P SV, 2.0 T 60–90 cm^3 40–60 cm^3	Bilateral fronto-parietal cerebrum Bilateral cerebellar hemispheres	– PME/PDE decreases until 2 yrs in cerebellum – PME/PDE decreases with age in cerebrum – PCr/γNTP increases 1–2 yrs, peaks at 8 yrs
Moss and Talagala (1997)	Mean = 13.4 yrs ($n = 29$) Health status and sedation not specified	^{31}P CSI, 1.5 T 3.5 cm^3	Mesial frontal lobe Mesial occipital lobe	– Males, βNTP higher in frontal than occipital – Females, βNTP higher in occipital than frontal – Males, higher PDE in occipital than frontal

Notes:

Abbreviations (other than those defined in Table 18.2):

^{1}H, proton; ^{31}P, phosphorus; SV, single voxel; CSI, chemical shift imaging; WM, white matter; GM, gray matter; ml, milliliter (e.g., 2 cm × 3 cm × 2 cm = 12 ml);
yrs, years; mo, months; wks, weeks.

NMR invisible. In contrast, increases in the ratio of PCr to both βNTP and γNTP have been documented to increase most rapidly within the first few years of life, reaching a plateau around age 8 (Boesch et al., 1989; Hanaoka et al., 1998; van der Knaap et al., 1990). Phosphocreatine serves as a high-energy buffer for the critical maintenance of cellular energetic state. Thus increases in the ratio of PCr to NTP likely reflect an increase in the PCr resource pool, while NTP levels remain constant. From the limited data on sex differences in ^{31}P metabolites during childhood and adolescence, males do not appear to differ from females in terms of overall age-related changes in metabolite levels. However, sex differences in the regional distributions of metabolites have been reported (Moss and Talagala, 1997). Higher βNTP has been observed in frontal versus occipital regions in males, whereas females exhibit higher βNTP in the occipital region relative to frontal regions. Furthermore, males demonstrate higher PDE levels in the occipital region than the frontal region. Sex differences in ^{31}P metabolites may therefore reflect the well documented sex differences in structural and functional brain changes during childhood and adolescence, as well as the differences between males and females in cognitive functioning.

Taken together, these studies suggest that the most rapid proton metabolite changes are observed early in development, reaching somewhat stable values by age 3. It is important to note that the majority of the subject pools from these studies consisted of newborns and children up to 7 years of age, with a sparse distribution of sampling from adolescent-aged subjects, making it difficult to interpret the timecourse of age-related changes in proton-containing metabolites. Furthermore, the available study findings include not only healthy children, but also those with neurological conditions, and at least half of the studies report that sedative agents were used when scanning infants and children (Boesch et al., 1989; Choi et al., 2000; Horska et al., 2002; Kreis et al., 1993; Pouwels et al., 1999), which may influence metabolite quantification.

Pediatric psychiatric disorders

As compared to MRS investigations of healthy brain development, MRS has been more widely used to examine psychopathologies in children and adolescents, including schizophrenia, mood disorders, obsessive-compulsive disorder, autism, and ADHD. These studies have provided information regarding neurochemical abnormalities useful for understanding pathophysiology, mechanisms of treatment response (see, for example, Chapters 8, 9 and 21), and risk factors for the early manifestation of psychiatric illness. While a comprehensive review of this work is beyond the scope of this chapter, the following examples illustrate the varied uses of MRS in studies of developmental neuropsychiatric disorders. It is notable, however, that in addition to traditional investigations of the proton metabolites NAA, Cho and Cr, more recently there has been growing interest in characterizing glutamate, glutamine, and GABA metabolites in childhood and adolescent psychopathology.

Proton magnetic resonance spectroscopy

Using proton MRS to examine 16 children with schizophrenia spectrum disorders (8–12 years) and a group of 12 healthy age- and sex-matched subjects, Brooks (Brooks et al., 1998) reported decreased left frontal NAA/Cr levels in children with schizophrenia spectrum disorders compared to the control group. Similarly, Chang et al. (2003) reported reduced NAA/Cr in the right dorsolateral prefrontal cortex of adolescent bipolar patients relative to age-matched healthy comparison subjects (Chang et al., 2003). Elevated mI levels in the anterior cingulate cortex (ACC) also have been reported in adolescent bipolar patients compared to controls (Davanzo et al., 2001, 2003). These findings (reduced NAA) suggest altered neuronal density or viability in these severe disorders of childhood and adolescence. Furthermore, the altered mI levels in the frontal lobe of adolescent bipolar patients are consistent with findings in adults and suggest altered phospholipid metabolism via disruption of intracellular message transduction (Moore et al., 2000).

Proton MRS has been used to examine glutamatergic functioning in adolescent bipolar disorder. In a single-voxel proton MRS study, 10 children with bipolar disorder were found to have significantly higher Glx in both the left and right frontal lobe and the basal ganglia compared with 10 healthy non-age-matched adolescent controls, with no evidence for significant bipolar-related alterations in NAA or Cho in the regions examined (Castillo et al., 2000). Single-voxel spectroscopy also has been used to isolate the Glx peak in the ACC of 10 unmedicated adolescents with type I bipolar disorder (Moore et al., 2007). Lower Glx/Cr levels were found in the ACC of unmedicated adolescents when compared with eight adolescents undergoing risperidone treatment. In that study, lower scores on the Young Mania Rating Scale (YMRS) were associated with higher Glx/Cr levels. Although the sample was limited, adolescents who were scanned following extended risperidone treatment exhibited increased Glx/Cr levels relative to their baseline.

In a study of 22 (7 unmedicated) adolescents with bipolar disorder, unmedicated adolescent patients exhibited significantly lower glutamine levels in the anterior cingulate region of the frontal cortex relative to medicated patients and healthy controls (Moore et al., 2007). However, glutamate levels did not differ between groups.

Phosphorus magnetic resonance spectroscopy

Two studies, to date, have utilized ^{31}P MRS to identify potential neurochemical risk factors for the development and manifestation of early-onset schizophrenia in offspring who have at least one schizophrenic first-degree relative (family history positive, FH+). In a study by Rzanny et al. (2003), 18 asymptomatic adolescent children or siblings of patients hospitalized for schizophrenic disorder were compared with 18 age- and sex-matched healthy controls (family history negative, FH−). Single-voxel ^{31}P MRS was used to quantify phosphorus-containing metabolites in the dorso-lateral prefrontal cortex and revealed significantly higher levels of PDE in FH+ subjects relative to FH− subjects. Keshavan and colleagues (2003) examined 16 adolescents FH+ for schizophrenia, 11 of whom had evidence of Axis I psychopathology, and 37 FH− controls. ^{31}P chemical shift imaging was used to examine bilateral frontal regions. Lower levels of PME were observed in the high-risk group, with this group difference being driven by symptomatic adolescents (FH+ with psychiatric symptoms), also consistent with studies documenting reduced PME in adult schizophrenic patients (Fukuzako, 2001).

Lithium and carbon magnetic resonance spectroscopy

Thus far, there has been an extremely limited number of MRS applications of nuclei other than proton and phosphorus to study pediatric psychiatric populations. Moore and colleagues (2002) used ^{7}Li MRS to measure in vivo brain lithium levels relative to serum lithium levels and found a lower brain:serum ratio in children and adolescents with bipolar disorder relative to adult patients, suggesting that higher dosing may be necessary at younger ages to maintain therapeutic brain lithium levels. Strauss and colleagues (2002) utilized ^{19}F MRS to establish a significant relationship between fluvoxamine or fluoxetine dosage and brain levels of fluorine in children with autism or other pervasive developmental disorders. In that study, it was demonstrated that pediatric brain levels of fluvoxamine and fluoxetine did not differ significantly from adult levels when corrected for drug dose per unit of body mass. ^{13}C MRS has also been used to characterize glutamate levels in pediatric Canavan disease (one of the leukodystrophies) (Bluml, 1999).

Conclusions and future directions

There have been several major advances in the field of MRS since the first in vivo spectrum was collected from human brain. Most notably, improved sensitivity of detection of in vivo metabolites has resulted from the utilization of higher field strength scanners (4.0 T and higher), which can enhance spectral resolution. In terms of metabolite quantification, accepted, standard software packages for the quantification of MR spectra have become widely available. There remains a significant debate regarding the optimal reference standard for determining metabolite ratios. It is likely that the resolution of this technological issue will be a major focus of research for the next several years. Because the role of metabolites in brain function continues to be debated, particularly in psychiatric disorders, improving the ability to more precisely measure and interpret in vivo metabolites is critical.

The above advances have made MRS technology more amenable to the study of pediatric populations, offering promise for understanding typical brain neurochemical development and atypical developmental trajectories associated with psychiatric disorders. Given the paucity of extant developmental MRS data, a better characterization of neurochemical development is needed to identify potential risk factors and diagnose pediatric psychiatric illness. The continued evolution of MRS is expected to contribute to our understanding of pathophysiology and mechanisms of treatment response and ultimately to contribute to advances in treatment development. Furthermore, pediatric studies are more likely than adult studies to reflect core etiologic processes, since they are free of confounds associated with chronic illness and pharmacological treatment history.

Significant future advances in understanding pediatric neurochemical development will depend on a number of factors: (1) increased uniformity in protocols across studies and sites (to enable multi-site studies); (2) the inclusion of large sample sizes to achieve statistical power to detect significant group differences (minimum of 20 subjects per group in a cross-sectional design); and (3) collection of longitudinal datasets. It also will be important to consider the influence of sex differences and brain laterality on brain metabolites. The measurement of additional nuclei (e.g., ^{13}C, ^{19}F)

and the integration of metabolite data with data from other MR imaging modalities, such as brain tissue volumes (MRI), indices of white matter microstructure integrity (diffusion tensor imaging) and neuronal activation (fMRI), will provide a more comprehensive understanding of neurochemical brain development. Taken together, MRS studies of pediatric populations hold promise for advancing developmental neuroscience by identifying the biochemical signatures associated with healthy brain development, as well as the neurobiological correlates of developmental neuropsychiatric diseases.

REFERENCES

Auer, D. P., Gossl, C., Schirmer, T., et al. (2001) Improved analysis of ^{1}H-MR spectra in the presence of mobile lipids. *Magn Reson Med*, **46**, 615–18.

Barker, P. B., Breiter, S. N., Soher, B. J., et al. (1994) Quantitative proton spectroscopy of canine brain: in vivo and in vitro correlations. *Magn Reson Med*, **32**, 157–63.

Behar, K. L., Rothman, D. L., Spencer, D. D., et al. (1994) Analysis of macromolecule resonances in ^{1}H NMR spectra of human brain. *Magn Reson Med*, **32**, 294–302.

Bessman, S. P. and Geiger, P. J. (1981) Transport of energy in muscle: the phosphorylcreatine shuttle. *Science*, **211**, 448–52.

Birken, D. L. and Oldendorf, W. H. (1989) *N*-Acetyl-L-aspartic acid: a literature review of a compound prominent in ^{1}H-NMR spectroscopic studies of brain. *Neurosci Biobehav Rev*, **13**, 23–31.

Bluml, S. (1999) In vivo quantitation of cerebral metabolite concentrations using natural abundance ^{13}C MRS at 1.5 T. *J Magn Reson*, **136**, 219–25.

Boesch, C., Gruetter, R., Martin, E., et al. (1989) Variations in the in vivo P-31 MR spectra of the developing human brain during postnatal life. Work in progress. *Radiology*, **172**, 197–9.

Bottomley, P. A., Foster, T. B., and Darrow, R. D. (1984) Depth-resolved surface coil spectroscopy (DRESS) for in vivo ^{1}H, ^{31}P, and ^{13}C NMR. *J Magn Res*, **59**, 338–43.

Bovey, F. A., Jelinski, L., and Mirau, P. A. (1988) *Nuclear Magnetic Resonance Spectroscopy*, San Diego, CA: Academic Press.

Brooks, W. M., Hodde-Vargas, J., Vargas, L. A., et al. (1998) Frontal lobe of children with schizophrenia spectrum disorders: a proton magnetic resonance spectroscopic study. *Biol Psychiatry*, **43**, 263–9.

Brown, T. R., Kincaid, B. M., and Ugurbil, K. (1982) NMR chemical shift imaging in three dimensions. *Proc Natl Acad Sci U S A*, **79**, 3523–6.

Castillo, M., Kwock, L., Courvoisie, H., et al. (2000) Proton MR spectroscopy in children with bipolar affective disorder: preliminary observations. *AJNR Am J Neuroradiol*, **21**, 832–8.

Chang, K., Adleman, N., Dienes, K., et al. (2003) Decreased *N*-acetylaspartate in children with familial bipolar disorder. *Biol Psychiatry*, **53**, 1059–65.

Cheng, M. A., Theard, M. A., and Tempelhoff, R. (1997) Intravenous agents and intraoperative neuroprotection. Beyond barbiturates. *Crit Care Clin*, **13**, 185–99.

Choi, C. G., Ko, T. S., Lee, H. K., et al. (2000) Localized proton MR spectroscopy of the allocortex and isocortex in healthy children. *AJNR Am J Neuroradiol*, **21**, 1354–8.

Costa, M. O., Lacerda, M. T., Garcia Otaduy, M. C., et al. (2002) Proton magnetic resonance spectroscopy: normal findings in the cerebellar hemisphere in childhood. *Pediatr Radiol*, **32**, 787–92.

Davanzo, P., Thomas, M. A., Yue, K., et al. (2001) Decreased anterior cingulate myo-inositol/creatine spectroscopy resonance with lithium treatment in children with bipolar disorder. *Neuropsychopharmacology*, **24**, 359–69.

Davanzo, P., Yue, K., Thomas, M. A., et al. (2003) Proton magnetic resonance spectroscopy of bipolar disorder versus intermittent explosive disorder in children and adolescents. *Am J Psychiatry*, **160**, 1442–52.

Fayed, N. and Modrego, P. J. (2005) Comparative study of cerebral white matter in autism and attention-deficit/hyperactivity disorder by means of magnetic resonance spectroscopy. *Acad Radiol*, **12**, 566–9.

Fukuzako, H. (2001) Neurochemical investigation of the schizophrenic brain by in vivo phosphorus magnetic resonance spectroscopy. *World J Biol Psychiatry*, **2**, 70–82.

Gyulai, L., Roth, Z., Leigh, J. S., Jr., et al. (1985) Bioenergetic studies of mitochondrial oxidative phosphorylation using ^{31}phosphorus NMR. *J Biol Chem*, **260**, 3947–54.

Hanaoka, S., Takashima, S., and Morooka, K. (1998) Study of the maturation of the child's brain using ^{31}P-MRS. *Pediatr Neurol*, **18**, 305–10.

Horska, A., Kaufmann, W. E., Brant, L. J., et al. (2002) In vivo quantitative proton MRSI study of brain development from childhood to adolescence. *J Magn Reson Imaging*, **15**, 137–43.

Kadota, T., Horinouchi, T., and Kuroda, C. (2001) Development and aging of the cerebrum: assessment with proton MR spectroscopy. *AJNR Am J Neuroradiol*, **22**, 128–35.

Keshavan, M. S., Stanley, J. A., Montrose, D. M., et al. (2003) Prefrontal membrane phospholipid metabolism of child and adolescent offspring at risk for schizophrenia or schizoaffective disorder: an in vivo ^{31}P MRS study. *Mol Psychiatry*, **8**, 316–23, 251.

Keltner, J. R., Wald, L. L., Frederick, B. D., et al. (1997) In vivo detection of GABA in human brain using a localized double-quantum filter technique. *Magn Reson Med*, **37**(3), 366–71.

Kreis, R., Ernst, T., and Ross, B. D. (1993) Development of the human brain: in vivo quantification of metabolite and water content with proton magnetic resonance spectroscopy. *Magn Reson Med*, **30**, 424–37.

Lim, K. O., Adalsteinsson, E., Spielman, D., et al. (1998) Proton magnetic resonance spectroscopic imaging of cortical gray and white matter in schizophrenia. *Arch Gen Psychiatry*, **55**, 346–52.

Mason, G. F., Behar, K. L., and Lai, J. C. (1996) The ^{13}C isotope and nuclear magnetic resonance: unique tools for the study of brain metabolism. *Metab Brain Dis*, **11**, 283–313.

McCormick, D. A. (1989) GABA as an inhibitory neuro-transmitter in human cerebral cortex. *J Neurophysiol*, **62**, 1018–27.

Miller, B. L. (1991) A review of chemical issues in ^{1}H NMR spectroscopy: *N*-acetyl-L-aspartate, creatine and choline. *NMR Biomed*, **4**, 47–52.

Moffett, J. R., Ross, B., Arun, P., et al. (2007) *N*-Acetylaspartate in the CNS: from neurodiagnostics to neurobiology. *Prog Neurobiol*, **81**, 89–131.

Moonen, C. T., Von Kienlin, M., Van Zijl, P. C., et al. (1989) Comparison of single-shot localization methods (STEAM and PRESS) for in vivo proton NMR spectroscopy. *NMR Biomed*, **2**, 201–8.

Moore, C. M., Breeze, J. L., Gruber, S. A., et al. (2000) Choline, myo-inositol and mood in bipolar disorder: a proton magnetic resonance spectroscopic imaging study of the anterior cingulate cortex. *Bipolar Disord*, **2**, 207–16.

Moore, C. M., Breeze, J. L., Kukes, T. J., et al. (1999) Effects of myo-inositol ingestion on human brain myo-inositol levels: a proton magnetic resonance spectroscopic imaging study. *Biol Psychiatry*, **45**, 1197–202.

Moore, C. M., Demopulos, C. M., Henry, M. E., et al. (2002) Brain-to-serum lithium ratio and age: an in vivo magnetic resonance spectroscopy study. *Am J Psychiatry*, **159**, 1240–2.

Moore, C. M., Frazier, J. A., Glod, C. A., et al. (2007) Glutamine and glutamate levels in children and adolescents with bipolar disorder: a 4.0-T proton magnetic resonance spectroscopy study of the anterior cingulate cortex. *J Am Acad Child Adolesc Psychiatry*, **46**, 524–34.

Moss, H. B. and Talagala, S. L. (1997) ^{31}P magnetic resonance spectroscopy of normal peripubertal children: effects of sex and fronto-occipital location. In: *Proceedings of the 5th Annual Meeting of the International Society for Magnetic Resonance in Medicine*. Vancouver, Canada, April, 1997 Vol. 2, p. 1219.

Nioka, S., Smith, D. S., Chance, B., et al. (1990) Oxidative phosphorylation system during steady-state hypoxia in the dog brain. *J Appl Physiol*, **68**, 2527–35.

Ordidge, R., Connelly, A., and Lohman, J. (1986) Image selective in vivo spectroscopy (ISIS). A new technique for spatially selective NMR spectroscopy. *J Magn Res*, **60**, 281–3.

Pettegrew, J. W., Keshavan, M. S., Panchalingam, K., et al. (1991) Alterations in brain high-energy phosphate and membrane phospholipid metabolism in first-episode, drug-naive schizophrenics. A pilot study of the dorsal prefrontal cortex by in vivo phosphorus 31 nuclear magnetic resonance spectroscopy. *Arch Gen Psychiatry*, **48**, 563–8.

Potwarka, J. J., Drost, D. J., and Williamson, P. C. (1999) Quantifying ^{1}H decoupled in vivo ^{31}P brain spectra. *NMR Biomed*, **12**(1), 8–14.

Pouwels, P. J., and Frahm, J. (1998) Regional metabolite concentrations in human brain as determined by quantitative localized proton MRS. *Magn Reson Med*, **39**, 53–60.

Pouwels, P. J., Brockmann, K., Kruse, B., et al. (1999) Regional age dependence of human brain metabolites from infancy to adulthood as detected by quantitative localized proton MRS. *Pediatr Res*, **46**, 474–85.

Provencher, S. W. (1993) Estimation of metabolite concentrations from localized in vivo proton NMR spectra. *Magn Reson Med*, **30**, 672–9.

Provencher, S. W. (2001) Automatic quantitation of localized in vivo ^{1}H spectra with LCModel. *NMR Biomed*, **14**, 260–4.

Rzanny, R., Klemm, S., Reichenbach, J. R., et al. (2003) ^{31}P-MR spectroscopy in children and adolescents with a familial risk of schizophrenia. *Eur Radiol*, **13**, 763–70.

Strauss, W. L., Unis, A. S., Cowan, C., et al. (2002) Fluorine magnetic resonance spectroscopy measurement of brain fluvoxamine and fluoxetine in pediatric patients treated for pervasive developmental disorders. *Am J Psychiatry*, **159**, 755–60.

Tyson, R. L., Sutherland, G. R., and Peeling, J. (1996) ^{23}Na nuclear magnetic resonance spectral changes during and after forebrain ischemia in hypoglycemic, normoglycemic, and hyperglycemic rats. *Stroke*, **27**, 957–64.

van der Knaap, M. S., van der Grond, J., van Rijen, P. C., et al. (1990) Age-dependent changes in localized proton and phosphorus MR spectroscopy of the brain. *Radiology*, **176**, 509–15.

Wallimann, T., Wyss, M., Brdiczka, D., et al. (1992) Intracellular compartmentation, structure and function of creatine kinase isoenzymes in tissues with high and fluctuating energy demands: the "phosphocreatine circuit" for cellular energy homeostasis. *Biochem J*, **281** (Pt 1), 21–40.

Weber, O. M., Trabesinger, A. H., Duc, C. O., et al. (1997) Detection of hidden metabolites by localized proton magnetic resonance spectroscopy in vivo. *Technol Health Care*, **5**(6), 471–91.

Wiederman, D., Schuff, N., Matson, G. B., et al. (2001) Short echo time multislice proton magnetic resonance spectroscopic imaging in human brain: metabolite distributions and reliability. *Magn Reson Imaging*, **19**(8), 1073–80.

19

Diffusion tensor imaging in developmental clinical neuroscience

Dae-Shik Kim

Introduction

Diffusion tensor imaging (DTI) based on high-resolution diffusion magnetic resonance imaging (MRI) plays an increasingly important role in adult and pediatric neuroimaging. This is thanks to diffusion MRI's exceptional capability in providing a variety of information from the same brain in a non-invasive manner: differences in mean diffusivity reflect differences in neurostructural properties, while direction-encoded DTI can reveal local and long range neuroanatomical connections. In this chapter we will introduce the basic principles of DTI, and discuss some of the important technical and conceptual issues for the future of DTI in developmental clinical neurosciences.

Magnetic resonance imaging has paved the way for accurately mapping the structural and functional properties of the brain in vivo. In particular, the intrinsic non-invasiveness of magnetic resonance (MR) methods and the sensitivity of the MR signal to subtle changes in the structural and physiological neuronal tissue fabric make it an all but ideal research and diagnostic tool for studying pediatric brains through cross-sectional and longitudinal imaging studies. For example, high-resolution volumetric MR studies have revealed that autism spectrum disorders (ASD) may result in increased cerebral gray matter volume in children aged 2–4 years, but not in ASD children aged 5–16 years (Courchesne et al., 2001), leading to the hypothesis that a period of cerebral hyperplasia in infancy might be followed by an abnormal slowing and plateauing of brain growth by age 5 years (Courchesne et al., 2001, 2003; Redcay and Courchesne, 2005). However, Herbert et al. (2003) found increased cerebral white matter, but not gray matter, in ASD boys aged 7–11 years, and Hazlett et al. (2005) found left-lateralized enlargement of cerebral gray matter, but no white matter differences, in adolescent and young adult males with ASD. Finally, Herbert et al. (2004) found that white matter enlargement in ASD boys aged 7–11 years was most pronounced in frontal, and especially prefrontal, cortex and that white matter increases were confined to the later-myelinating, radiate compartment, composed of intrahemispheric, short- and mid-range cortico-cortical connections.

While these examples demonstrate the potential utility of MRI-based volumetric analyses, their explanatory power is currently limited by the inability of anatomical MR to discriminate among the various sources of the MR signals in the observed tissue. In other words, there is no established one-to-one correspondence between a cytoarchitectonic unit, or tissue compartment, and the observed MR signal property. For example, it is not possible to deduce from volumetric studies alone whether

Neuroimaging in Developmental Clinical Neuroscience, eds. Judith M. Rumsey and Monique Ernst. Published by Cambridge University Press. © Cambridge University Press 2009.

the observed gray/white matter volume changes are due to an actual change in cell/fiber number, cell shape, and/or myelinization. Consequently and despite their ubiquitous use, the overall tissue specificity (or the "neural correlate") of structural MRI parameters remains largely elusive. This is in stark contrast to functional MRI (fMRI) where vigorous validation studies have been launched following an early realization that the observed fMRI signals correlate only indirectly with the underlying neuroelectrical activities (Heeger et al., 2000; Kim et al., 2004; Logothetis et al., 2001).

The more recently introduced technique of diffusion-based MRI promises to circumvent some of these problems by more directly evaluating the intrinsic tissue properties of the brain. For example, the sharp difference in structural characteristics between tissue properties in the central nervous system has been extensively exploited in several diffusion-weighted imaging (DWI) applications, ranging from the characterization of ischemia (Moseley et al., 1990a, 1990b), the demarcation of brain tumors (Eis et al., 1994), and the extensive investigation of fiber connectivity through the use of DTI (Basser et al., 1994; Pierpaoli et al., 1996). Finally, recent advances in compartmental-specific diffusion MRI suggest that diffusion-based MRI may also be capable of providing information about neural tissue properties at the microscopic cellular level.

Principles of diffusion tensor magnetic resonance imaging

One essential characteristic of diffusion is that a group of molecules that start at the same location will spread out over time. In a medium that is locally free of physical restrictions, each molecule experiences a series of random displacements ("Brownian motion") such that the average shape of diffusion will be more or less isotropic (see Figure 19.1 for the spherical diffusion shape in the upper panel). However, if the same molecules

"Typical" gray matter pixel: low FA

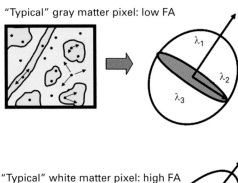

"Typical" white matter pixel: high FA

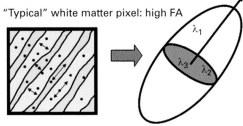

Figure 19.1. Isotropic and anisotropic diffusion. Upper panel: molecular diffusion is isotropic in the absence of limiting barriers. In typical gray matter, fractional anisotropy (FA) is low, as indicated by the circular ellipsoid indicating approximately equal diffusion in all directions. Lower panel: In typical white matter, where water molecules are "trapped" between physical barriers such as bundles of axonal fibers, the average diffusion will be "anisotropic." Here, FA is high, as indicated by the ellipsoid elongated in keeping with the fiber direction.

are "trapped" between locally oriented physical barriers, such as axonal fiber bundles or intra-axonal microtubules, then it is clear that molecular diffusion along the physical barriers will be easier than orthogonal to the barriers. Consequently, the average shape of diffusion in such restricted environments will be anisotropic, with the resulting diffusion ellipsoid being co-linear to the local physical barriers (see the lower panel in Figure 19.1). The technique of DTI attempts to reconstruct the orientation of the local physical barriers by carefully measuring the amount of direction-restricted water molecular displacement for each of the experimentally determined gradient-encoded spatial directions.

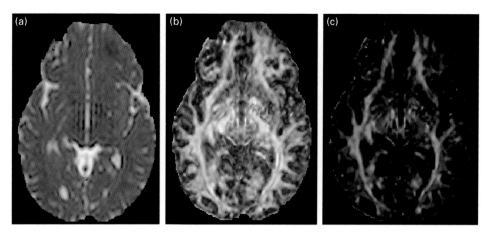

Figure 19.2. Types of maps derived from diffusion tensor imaging (DTI) studies. Different types of information can be obtained from a single DTI experiment: (a) overall diffusivity; (b) degree of anisotropy of local diffusion (brightness coding for fractional anisotropy); and (c) voxel-by-voxel color coding of the diffusion orientation. Green: diffusion ellipsoids aligned along the anterior–posterior axis; red: diffusion along left–right axis; blue: alignment along the ventral–dorsal axis.

Magnetic resonance imaging measurement of diffusion

As suggested by Stejskal and Tanner (1968), an MR image can be sensitized to diffusion in a given direction by using a couple of temporally separated gradients in the desired direction. The application of a magnetic field gradient pulse at one of the three spatial dimensions (x, y, z) dephases the protons (spin) along the respective dimension. A second pulse in the same direction, but opposite polarity ("refocusing pulse"), will *re-phase* these spins. However, such re-phasing cannot be perfect if the protons moved between the two gradient pulses. That is to say, the *signal loss*, which cannot be recovered after the application of the second gradient pulse, is a function of the local molecular motion. The amount of the molecular diffusion is known to obey Eq. 1, assuming the sample is isotropic (no directionality in water diffusion):

$$\frac{S}{S_0} = e^{-\gamma^2 G^2 \delta^2 (\Delta - \delta/3) D} \tag{1}$$

where S and S_0 are signal intensities with and without the diffusion weighting, γ is a constant (called "gyromagnetic ratio"), G and δ are gradient strength and duration, and Δ is the separation between a pair of gradient pulses. Because these parameters are all known, from the amount of signal decrease (S/S_0), diffusion constants (D) at each voxel can be derived by (Basser and Pierpaoli, 1996; Conturo et al., 1996; Westin et al., 1999):

$$b = \gamma^2 \delta^2 (\Delta - \delta/3) G^2 \tag{2}$$

$$D = \frac{1}{b} \ln \frac{S_0}{S}. \tag{3}$$

While the diffusion constant D is a scalar for conventional DWI, in the case of DTI data it is described by a 3×3 symmetrical matrix $\overline{\overline{D}} = \begin{bmatrix} D_{xx} & D_{xy} & D_{xz} \\ D_{xy} & D_{yy} & D_{yz} \\ D_{xz} & D_{yz} & D_{zz} \end{bmatrix}$. This matrix can be diagonalized to obtain $\overline{\overline{\Lambda}} = \begin{bmatrix} \lambda_1 & 0 & 0 \\ 0 & \lambda_2 & 0 \\ 0 & 0 & \lambda_3 \end{bmatrix}$ and corresponding $\overline{\overline{P}} = [\vec{p}_1 \vec{p}_2 \vec{p}_3]$, which then can in turn be used to describe the diffusivity and directionality of water diffusion within a given voxel. An important measure associated with the diffusion tensor is the so-called "trace" (see Figure 19.2a):

$$tr\{\underline{D}\} = D_{xx} + D_{yy} + D_{zz} = 3 \cdot \langle \lambda \rangle = \lambda_1 + \lambda_2 + \lambda_3. \tag{4}$$

The trace has similar values in healthy white and gray matter ($tr(D) \sim 2.1 \times 10^{-3}\,\mathrm{mm}^2/\mathrm{s}$). However, the trace value drops considerably in brain tissue affected by acute stroke (Sotak, 2002). This drop is attributed to an increase in tortuosity factor due to the shrinkage of the extracellular space (Sotak, 2002). Consequently, the trace of the diffusion tensor can be used as an early indicator of ischemic brain injury. Finally, the anisotropy of the diffusion tensor characterizes the amount of diffusion variation as a function of direction (e.g., the deviation from isotropy). Several of these anisotropy measures are normalized to range from 0 to 1. One of the most commonly used measures of anisotropy is the fractional anisotropy (FA; see Figure 19.2b) (Basser and Pierpaoli, 1996):

$$FA = \frac{1}{\sqrt{2}} \sqrt{\frac{(\lambda_1 - \lambda_2)^2 + (\lambda_2 - \lambda_3)^2 + (\lambda_3 - \lambda_1)^2}{\lambda_1^2 + \lambda_2^2 + \lambda_3^2}} \quad (5)$$

which is the ratio of the root-mean-square (RMS) of the eigenvalues' deviation from their mean normalized by the eigenvalues' Euclidian norm. Fractional anisotropy has been shown to provide the best contrast between different classes of brain tissues (Alexander et al., 2000). Finally, voxel-by-voxel color coding of the orientation of the diffusion anisotropy (Figure 19.2c) can provide a comprehensive visualization of the major physical barriers in the living cortex.

Labeling neuronal circuitry using DTI-based fiber tractography

Traditionally, connectivity patterns between cortical regions have been studied using a variety of neuro-anatomical staining techniques. The earliest developed techniques include *postmortem* methods, such as dissection of white matter, strychnine neuronography (Pribram and MacLean, 1953), and the Nauta (Whitlock and Nauta, 1956) methods of tracing neuronal degeneration following localized lesions. More recently, implantation of carbocyanine dyes, such as DiI and DiA (see, e.g., Galuske et al., 2000), has replaced the older "degeneration methods." However, in addition to the long duration

needed for passive labeling of the axonal fibers (often several weeks to months), the aforementioned *postmortem* methods inherently fail to yield a correlation to the foci of functional activation. This problem can be somewhat mitigated through the use of modern trans-neuronal markers [e.g., horseradish peroxidase (HRP), rhodamine- and fluorescein-conjugated latex microspheres (Katz and Iarovici, 1990), biocytin (Kisvarday et al., 1994), and biotinylated dextran amine (BDA)] that can be used to trace the pattern of neuronal connectivity "in vivo." Even these techniques are limited in their use by: (1) the low number of labeled tracts; (2) the short tracing distance (mostly, across two to three synapses only); (3) the need for invasive procedures, such as inserting a glass probe, and, most importantly; (d) the need to sacrifice the animal before the labeling pattern can be visualized. Naturally, such limitations invalidate the use of existing "in vivo" tracing methods for human studies.

Diffusion-based fiber tractography techniques, however, can be used to non-invasively extract the most likely fiber pathways in a given brain volume from a discrete 3D set of above-mentioned diffusion tensors. In most published fiber-tracking algorithms, tracking starts at a user-defined seeding point or region of interest. To this end, the diffusion properties are assumed to be homogeneous at the macroscopic (voxel) level. The voxel size is typically about $2 \times 2 \times 3$ mm^3 in human DTI data and $0.3 \times 0.3 \times 0.5$ mm^3 in animal data. However, particularly in humans, techniques that allow finer spatial resolution would be necessary if small-diameter fiber tracts and tract regions with a high radius of curvature are to be traced.

Mori et al. (1999) developed one of the earliest and most commonly employed algorithms: fiber assignment by continuous tracking (FACT), which is based on extrapolation of continuous vector lines from discrete DTI data. The reconstructed fiber direction within each voxel is parallel to the diffusion tensor eigenvector (ε_1) associated with the greatest eigenvalue (λ_1). Within each voxel, the fiber tract is a line segment defined by the input

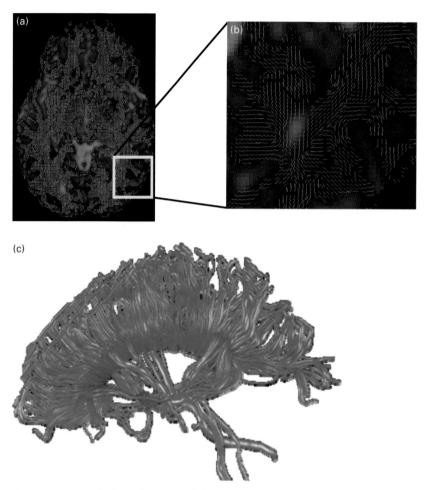

Figure 19.3. In vivo high-resolution DTI of the human corpus callosum. The left panel (a) depicts the user-defined seeding region of interest (ROI) for fiber reconstruction (see panel (b) for enlarged view of the ROI), and the panel (c) shows the result of DTI-based fiber tractography of human corpus callosum.

position, the direction of ε_1, and an output position at the boundary with the next voxel. The track is propagated from voxel to voxel. Tracking is terminated when a sharp turn in the fiber orientation occurs.

In algorithms developed by Basser (2002), Basser et al. (2000), Conturo et al. (1999) and Pajevic et al. (2002), a continuous representation of the diffusion tensor and principal eigenvector ε_1 are interpolated from the discrete voxel data. The fiber track direction at any location along the track is given by the continuous ε_1. Typically the tracking algorithm stops when the fiber radius of curvature (e.g., 60°) or the anisotropy factor (e.g., FA <0.2) falls below a threshold. With this approach, the fiber is not represented by a succession of line segments but by a relatively smooth curve that follows the local diffusion direction and is more representative of the behavior of real fibers (see Figure 19.3 for such DTI fiber tractography result).

Magnetic resonance imaging parameters

Diffusion MR images can be obtained from a variety of existing clinical and research MRI scanners at field strengths ranging from 1.5 T to 11.7 T. In practical terms, however, studies of children generally do not exceed 3 T, and the Food and Drug Administration considers field strengths greater than 4 T to pose a significant risk for infants under 1 month of age and field strengths greater than 8 T to pose a potential risk for children and adults. Diffusion tensor imaging sequences usually consist of a number of gradient direction-encoded conventional spin echo echo-planar images. In humans, ~64 DTI axial slices will cover the entire brain with the following acquisition parameters: repetition time 10 s, echo time 84 ms, flip angle 90°, matrix 128×128, field of view 256×256 mm^2, slice thickness 2 mm, no gap (2 mm isotropic voxels), 4 averages, bandwidth 1302 Hz/pixel, partial k-space 6/8, fat saturation, acquisition time 6 min and 42 s. Diffusion weighting can be performed using anywhere between 6 and 256 gradient directions (more typically, 15 independent directions are used) with a b value of ~1000 s/mm^2. A reference image is also obtained with a very low diffusion weighting (b value of ~0 s/mm^2). Using parallel imaging (e.g., SENSitivity Encoding or SENSE), each DTI dataset can be acquired within about 2 min. For high b value diffusion-weighted MRI studies (see "High b value diffusion-weighted imaging for tissue structural characterization"), multiple b values (e.g., 16 different b values) logarithmically spaced between $b = 200$ s/mm^2 and $b = 10\,000$ s/mm^2 can be used.

Conventional diffusion tensor imaging studies in adults

Diffusion tensor tractography studies in normal adults have been reported by numerous groups from a variety of cortical and subcortical locations. Most MRI systems are now routinely equipped with DTI acquisition and reconstruction techniques, and several well-designed public domain software packages are available for more detailed DT fiber tractography analyses. However, despite its great promise for non-invasive visualization of white matter connections in human brains, one has to keep in mind that DTI has important limitations. First, it is not entirely clear what is actually being measured with the diffusivity and anisotropy indices. For example, the precise contribution of these two factors, fiber density and myelination, to the anisotropy index is not completely understood. Thus, it is unclear to what degree the results of DTI correspond to the actual density and orientation of the local axonal fiber bundles. Secondly, it is also important to understand how white matter is, in general, organized. The most basic shortcoming of DTI is that it can only determine a single fiber orientation at any given location in the brain. This is clearly inadequate in regions with complex white matter architecture, where different axonal pathways crisscross through each other, such as in the pons. The crossing fibers create multiple fiber orientations within a single MRI voxel, where a voxel refers to a 3D pixel, and constitutes the individual element of the MR image. Since the DT assumes only a single preferred direction of diffusion within each voxel, an average value for that voxel may signify isotropy as the opposing fiber directions cancel each other out; DTI cannot adequately describe regions of crossing fibers or of converging or diverging fibers. Three-dimensional DTI fiber tracking techniques also founder in these regions of complex white matter architecture, since there is no well-defined single dominant fiber orientation for them to follow.

Non-conventional diffusion tensor imaging

In recent years, some of limitations of conventional DTI have been addressed by measuring the full 3D dispersion of water diffusion in each MRI voxel at high angular resolution. Thus, instead of obtaining diffusion measurements in only a few independent directions to determine a single fiber orientation as in DTI, dozens or even hundreds of uniformly distributed diffusion directions in 3D space are acquired to resolve multiple fiber orientations in high angular resolution diffusion imaging (HARDI) [e.g., diffusion spectrum imaging (DSI) or HARDI]. Each distinct fiber population can be visualized on maps of the orientation distribution function (ODF), which are computed from the 3D high angular resolution diffusion data through a projection reconstruction technique known as the Funk-Radon transform. This 3D projection reconstruction is very similar mathematically to the 2D method by which computed tomography (CT) images are calculated from X-ray attenuation data. Unlike DTI, DSI/HARDI has the advantage of being model-independent, and therefore does not assume any particular 3D distribution of water diffusion or any specific number of fiber orientations within a voxel.

Yet another way to circumvent the limitations of conventional DTI is to give up the notion of concatenated fiber reconstructions (so-called streamtubes) altogether. As explained earlier, DTI is based on the assumption that the local orientation of nerve fibers is parallel to the first eigenvector of the diffusion tensor. However due to imaging noise, limited spatial resolution, and partial volume effects, the fiber orientation cannot fundamentally be determined without the loss of certainty. Therefore, instead of attempting to reconstruct fiber pathways that are prone to fundamental uncertainties, DTI-based probabilistic mapping methods (Behrens et al., 2003; Jones and Pierpaoli, 2005; Parker et al., 2003) utilize probability density functions (PDFs) defined at each point within the brain to describe the local uncertainty in fiber orientation. Recent studies suggest PDF-based DTI holds particular utility for group comparisons between different cohorts.

Multimodal combination of diffusion tensor imaging with functional magnetic resonance imaging

One of the strengths, as well as the weakness, of DTI-based tractography is the fact that it generates "too much data." Unlike conventional neurotracing techniques (DiI, HRP, biocytin, etc.), each DTI experiment has the potential to provide the complete set of connectivity data across all imaged voxels. A key in making DTI an outstanding tool for neuroscience is therefore to develop selection criteria to determine the seeding ROI for DTI fiber tracing. Such seeding ROIs can be determined either (1) based on an anatomical ROI known a priori to the investigator, or (2) based on foci of fMRI hotspots elicited during the functional stimulation of the brain. Using the foci of fMRI activation as the "initial" and "termination" ROIs for DTI fiber tracings is a natural choice for many questions in cognitive neuroscience, since here the main interest is to elucidate the pattern of neuronal circuitry underlying the observed functional activation for a given task. To this end, we have already developed methods to combine high-resolution fMRI with DTI. Figures 19.4 and 19.5 display the methods and results, respectively, of such combined DTI and fMRI experiments.

While fiber tracking using DTI in combination with fMRI has the potential to provide crucial insights into the brain's functional architecture in vivo, major interpretative issues must be addressed before such a multimodal approach can routinely be used for clinical and basic neurosciences. Some of the critical issues are:

1. How do we extract compact seeding ROIs based on "patchy" BOLD functional activity? What is the appropriate statistical threshold for this? Should all fMRI voxels (above a certain threshold) be treated as homologous DTI seeding ROIs regardless of their p values and/or cross-correlation coefficients?

2. Areas of high BOLD activity are located within the cortical gray matter. The cortical gray matter, however, is characterized by low fractional anisotropy (because of the presence of multidirectional fibers/fiber bundles within the imaging voxels in the gray matter). Therefore, most reconstructed fibers will terminate at the gray/white matter boundary. If seeding ROIs were placed exclusively within the fMRI voxels, conventional DTI algorithms will result in zero or a low

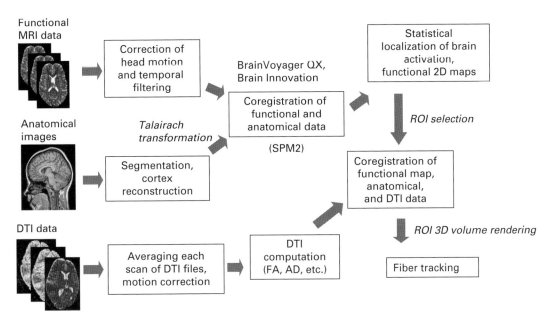

Figure 19.4. Processing strategies for multimodal fMRI-based DTI fiber tractography. Advanced image registration techniques are needed to ensure proper coregistration for these three different types of data. fMRI, functional MRI; T1W, T1 weighted; ROI, region of interest; FA, fractional anisotropy; AD, apparent diffusion.

number of reconstructed fibers. One needs therefore to develop automatic "search" algorithms (this is currently done manually) for the closest fiber termination points to be included as part of the seeding ROI.

3. One needs also to address the question how the experimentally acquired voxel size should be related to the density of the seeding points. For example, a given $1 \times 1 \times 1$ mm^3 voxel in a BOLD fMRI study may contain hundreds of thousands of physical axonal fibers. For the subsequent DTI fiber reconstruction, should the same voxel then be treated as *one* seeding point? Or should the imaging voxel be "regridded" into a multitude of sub-seeding points?

Diffusion tensor imaging studies in children

Two recent reviews of DTI studies of pediatric development (Cascio et al., 2007; Wozniak and Lim, 2006) list nearly 50 studies, many of which

deal with healthy, normative development. There is general consistency in the literature to suggest that overall diffusion decreases and anisotropy increases with age during childhood and adolescence, likely as a result of ongoing myelination and axonal pruning. However, the specific trajectories of these changes and their regional specificities are as yet unknown. Zhang et al. (2005) noted dramatic changes throughout the first 2 years of life. A normative study of 30 children, aged 0–54 months, by Hermoye et al. (2006) described rapid changes in the first year of life, which then slowed during the second year, reaching relative stability after age 2. Only a few studies have related diffusion measures to behavior and cognitive ability. Studies in healthy children have demonstrated that anisotropy in the temporal lobes increased with working memory capacity in 8- to 18-year-olds (Nagy et al., 2004), while frontal and occipitoparietal anisotropy have been related to IQ in 5- to 18-year-olds (Schmithorst et al., 2005). Three studies have shown significant positive correlations between anisotropy

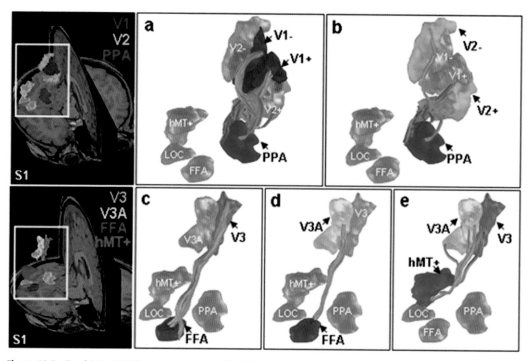

Figure 19.5. Combining DTI fiber tractography with fMRI. The explanatory power of DTI can be further increased by combining DTI fiber tractography with conventional functional imaging. Here, the areas of high fMRI activity during visual stimulation along the human ventro-temporal cortex (volume rendered in gray) are used as seeding points for DTI-based fiber reconstructions (labeled in blue). FFA, fusiform face area; LOC, lateral occipital cortex; PPA, parahippocampal place area; hMT+, human medio-temporal cortex upper visual field; V1, V2, V3, V3A, retinotopically organized primary visual areas; "+", area coding for the upper visual field; "−", area coding for the lower visual field.

in temporo-parietal white matter and reading ability (Beaulieu et al., 2005; Deutsch et al., 2005; Niogi and McCandliss, 2006). A few studies have begun to examine white matter integrity in developmental neuropsychiatric disorders using DTI, and some preliminary findings exist in attention deficit/hyperactivity disorder and autism (see Cascio et al., 2007).

A common barrier for high-resolution DTI studies in children has been the need for relatively long scan durations – especially if a large number of gradient encoding directions are used. While recent advances in parallel-imaging and high-field (3T) MRI enable a greatly accelerated acquisition of DTI data within time frames more realistic for the pediatric environment (e.g., ~5–10 min for a complete dataset), DTI in developing pediatric brains

poses a unique of set of technical and conceptual problems. First, imaging with young children is poised to be more likely to yield motion artifacts relative to studies of adults. While extensive mock training and post-processing strategies will mitigate some of the motion-related artifacts, excessive motion during image acquisition will ultimately lead to non-recoverable spatial distortions and signal dropout. Furthermore, even with perfectly motionless MR images in place, processing pediatric DTI data faces additional technical challenges: most image segmentation and coregistration algorithms have been developed for adult brains. Pediatric brains, however, exhibit unique relaxation and tissue properties, making the application of existing image-contrast-based techniques difficult.

Likewise, the lack of standard templates for developing brains poses additional difficulties for group averaging data. Finally, by definition, a pediatric brain undergoes significant changes in almost all aspects of its cytoarchitecture, gyrification, and volumetrics. Longitudinal data acquisition will therefore have to deal with the problem of detecting small, targeted differences within larger differences caused by mere brain maturation. This problem is further amplified for DTI-based fiber tractography in pediatric brains, given the large degree of fiber pruning, migration, and re-wiring during the course of postnatal development.

High *b* value diffusion-weighted imaging for tissue structural characterization

We have seen thus far how diffusion-based MRI can be used in order to assess the presence and magnitude of both the global diffusivity and the structural restrictions posed by axonal fiber bundles. More recent advances in high *b* value DWI and DTI suggest that diffusion-based MR methods can also be used to elucidate the neural tissue properties at the microscopic cellular level. This so-called compartmental-specific DTI is based on the fact that, as a result of the structural heterogeneity of tissue in a spatial scale significantly smaller than the typical image voxel size, the diffusion-weighted signals display a multiexponential dependence on diffusion weighting magnitude. The complexity of this multiexponential behavior of the signal allows a more detailed inspection of diffusion properties in matter, as proposed by Callaghan et al. (1988), and by Cory and Garroway (1990). The method, known as "*q* space imaging," is based on the acquisition of data with multiple gradient strength values *q*. When a Fourier transformation is performed pixel by pixel with respect to the variable $q = \gamma g \Delta / 2\pi$:

$$P(\vec{R}, \Delta) = \frac{1}{2\pi} \int\limits_{-\infty}^{\infty} S(\vec{q}, \Delta) \cdot \exp(-i2\pi \vec{q} \cdot \vec{R}) d\vec{q} \qquad (6)$$

the transformed dataset P represents the displacement probability of the water molecules with respect to the axis which was sensitized to diffusion, at a given diffusion time Δ. The concept of high *b* value (or *q* space) imaging has already been successfully applied in various in vitro and in vivo applications to elucidate the microscopic tissue properties of the white and gray matter (Assaf and Cohen, 1999, 2000; Assaf et al., 2000, 2002; King et al., 1994, 1997).

The other approach of using diffusion data acquired with multiple *b* values is to model the data according to a plausible model that governs the diffusion pattern in each voxel. In this approach, the data are fitted to a multiparametric model function that best represents the expected behavior of the signal with respect to *b*. The advantage of modeling the diffusion data rests with the possibility of extracting information about diffusion characteristics of water in various compartments (e.g., intra- versus extracellular diffusion) from the same dataset, and thus simultaneously obtaining volumetric and structural information about those compartments. Such compartmental-specific structural data, in turn, will enable us to address cortical volumetric questions at a level of spatial specificity not attainable using conventional anatomical images.

Conclusions

Diffusion tensor imaging can be used to elucidate the presence and orientation of macroscopic fiber bundles in adult brain. In pediatric brains, a successful application of DTI will necessitate a much more stringent control of motion and consideration of differential tissue contrasts. Combined imaging with high-resolution fMRI and high *b* value compartmental-specific diffusion imaging holds further promises for the future of pediatric DTI.

ACKNOWLEDGMENTS

Drs Itamar Ronen, Mina Kim and Susumu Mori provided crucial help in data acquisition and analyses. We also thank Mathieu Ducros, Sahil

Jain and Keun-Ho Kim for their help with DTI postprocessing. This work was supported by grants from NIH (NS44825), the Keck Foundation, and the Human Frontiers Science Program.

REFERENCES

Alexander, A. L., Hasan, K., Kindlmann, G., et al. (2000) A geometric analysis of diffusion tensor measurements of the human brain. *Magn Reson Med*, **44**, 283–91.

Assaf, Y. and Cohen, Y. (1999) Structural information in neuronal tissue as revealed by q-space diffusion NMR spectroscopy of metabolites in bovine optic nerve. *NMR Biomed*, **12**, 335–44.

Assaf, Y. and Cohen, Y. (2000) Assignment of the water slow-diffusing component in the central nervous system using q-space diffusion MRS: implications for fiber tract imaging. *Magn Reson Med*, **43**, 191–9.

Assaf, Y., Ben-Bashat, D., Chapman, J., et al. (2002) High b-value q-space analyzed diffusion-weighted MRI: application to multiple sclerosis. *Magn Reson Med*, **47**, 115–26.

Assaf, Y., Mayk, A., and Cohen, Y. (2000) Displacement imaging of spinal cord using q-space diffusion-weighted MRI. *Magn Reson Med*, **44**, 713–22.

Basser, P. J. and Pierpaoli, C. (1996) Microstructural and physiological features of tissues elucidated by quantitative-diffusion-tensor MRI. *J Magn Reson B*, **111**, 209–19.

Basser, P. J., Mattiello, J., and Lebihan, D. (1994) Estimation of the effective self-diffusion tensor from the NMR spin echo. *J Magn Reson B*, **103**, 247–54.

Basser, P. J., Pajevic, S., Pierpaoli, C., et al. (2000) In vivo fiber tractography using DT-MRI data. *Magn Reson Med*, **44**, 625–32.

Beaulieu, C., Plewes, C., Paulson, L. A., et al. (2005) Imaging brain connectivity in children with diverse reading ability. *NeuroImage*, **25**, 1266–71.

Behrens, T. E., Johansen-Berg, H., Woolrich, M. W., et al. (2003) Non-invasive mapping of connections between human thalamus and cortex using diffusion imaging. *Nat Neurosci*, **6**, 750–7.

Callaghan, P. T., Eccles, C. D., and Xia, Y. (1988) NMR microscopy of dynamic displacements: k-space and q-space imaging. *J Phys Sci Instrum*, **21**, 820–2.

Cascio, C. J., Gerig, G., and Piven, J. (2007) Diffusion tensor imaging: application to the study of the developing brain. *J Am Acad Child Adolesc Psychiatry*, **46**, 213–23.

Conturo, T. E., Lori, N. F., Cull, T. S., et al. (1999) Tracking neuronal fiber pathways in the living human brain. *Proc Natl Acad Sci U S A*, **96**, 10422–7.

Conturo, T. E., McKinstry, R. C., Akbudak, E., et al. (1996) Encoding of anisotropic diffusion with tetrahedral gradients: a general mathematical diffusion formalism and experimental results. *Magn Reson Med*, **35**, 399–412.

Cory, D. G. and Garroway, A. N. (1990) Measurement of translational displacement probabilities by NMR: an indicator of compartmentation. *Magn Reson Med*, **14**, 435–44.

Courchesne, E., Carper, R., and Akshoomoff, N. (2003) Evidence of brain overgrowth in the first year of life in autism. *J Am Med Assoc*, **290**, 337–44.

Courchesne, E., Karns, C. M., Davis, H. R., et al. (2001) Unusual brain growth patterns in early life in patients with autistic disorder: an MRI study. *Neurology*, **57**, 245–54.

Deutsch, G. K., Dougherty, R. F., Bammer, R., et al. (2005) Children's reading performance is correlated with white matter structure measured by diffusion tensor imaging. *Cortex*, **41**, 354–63.

Eis, M., Els, T., Hoehn-Berlage, M., and Hossmann, K. A. (1994) Quantitative diffusion MR imaging of cerebral tumor and edema. *Acta Neurochir Suppl (Wien)*, **60**, 344–6.

Galuske, R. A., Schlote, W., Bratzke, H., et al. (2000) Interhemispheric asymmetries of the modular structure in human temporal cortex. *Science*, **289**, 1946–9.

Hazlett, H. C., Poe, M., Gerig, G., et al. (2005) Magnetic resonance imaging and head circumference study of brain size in autism: birth through age 2 years. *Arch Gen Psychiatry*, **62**, 1366–76.

Heeger, D. J., Huk, A. C., Geisler, W. S., et al. (2000) Spikes versus BOLD: what does neuroimaging tell us about neuronal activity? *Nat Neurosci*, **3**, 631–3.

Herbert, M. R., Ziegler, D. A., Deutsch, C. K., et al. (2003) Dissociations of cerebral cortex, subcortical and cerebral white matter volumes in autistic boys. *Brain*, **126**, 1182–92.

Herbert, M. R., Ziegler, D. A., Makris, N., et al. (2004) Localization of white matter volume increase in autism and developmental language disorder. *Ann Neurol*, **55**, 530–40.

Hermoye, L., Saint-Martin, C., Cosnard, G., et al. (2006) Pediatric diffusion tensor imaging: normal database and observation of the white matter maturation in early childhood. *NeuroImage*, **29**, 493–504.

Jones, D. K. and Pierpaoli, C. (2005) Confidence mapping in diffusion tensor magnetic resonance imaging tractography using a bootstrap approach. *Magn Reson Med*, **53**, 1143–9.

Katz, L. C. and Iarovici, D. M. (1990) Green fluorescent latex microspheres: a new retrograde tracer. *Neuroscience*, **34**, 511–20.

Kim, D. S., Ronen, I., Olman, C., et al. (2004) Spatial relationship between neuronal activity and BOLD functional MRI. *NeuroImage*, **21**, 876–85.

King, M. D., Houseman, J., Gadian, D. G., et al. (1997) Localized q-space imaging of the mouse brain. *Magn Reson Med*, **38**, 930–7.

King, M. D., Houseman, J., Roussel, S. A., et al. (1994) q-Space imaging of the brain. *Magn Reson Med*, **32**, 707–13.

Kisvarday, Z. F., Kim, D. S., Eysel, U. T., et al. (1994) Relationship between lateral inhibitory connections and the topography of the orientation map in cat visual cortex. *Eur J Neurosci*, **6**, 1619–32.

Logothetis, N. K., Pauls, J., Augath, M., et al. (2001) Neurophysiological investigation of the basis of the fMRI signal. *Nature*, **412**, 150–7.

Mori, S., Crain, B. J., Chacko, V. P., et al. (1999) Three-dimensional tracking of axonal projections in the brain by magnetic resonance imaging. *Ann Neurol*, **45**, 265–9.

Moseley, M. E., Cohen, Y., Mintorovitch, J., et al. (1990a) Early detection of regional cerebral ischemia in cats: comparison of diffusion- and T2-weighted MRI and spectroscopy. *Magn Reson Med*, **14**, 330–46.

Moseley, M. E., Kucharczyk, J., Mintorovitch, J., et al. (1990b) Diffusion-weighted MR imaging of acute stroke: correlation with T2-weighted and magnetic susceptibility-enhanced MR imaging in cats. *AJNR Am J Neuroradiol*, **11**, 423–9.

Nagy, Z., Westerberg, H., and Klingberg, T. (2004) Maturation of white matter is associated with the development of cognitive functions during childhood. *J Cogn Neurosci*, **16**, 1227–33.

Niogi, S. N. and McCandliss, B. D. (2006) Left lateralized white matter microstructure accounts for individual differences in reading ability and disability. *Neuropsychologia*, **44**, 2178–88.

Pajevic, S., Aldroubi, A., and Basser, P. J. (2002) A continuous tensor field approximation of discrete DT-MRI data for extracting microstructural and architectural features of tissue. *J Magn Reson*, **154**(1), 85–100.

Parker, G. J., Haroon, H. A., and Wheeler-Kingshott, C. A. (2003) A framework for a streamline-based probabilistic index of connectivity (PICo) using a structural interpretation of MRI diffusion measurements. *J Magn Reson Imaging*, **18**, 242–54.

Pierpaoli, C., Jezzard, P., Basser, P. J., et al. (1996) Diffusion tensor MR imaging of the human brain. *Radiology*, **201**, 637–48.

Pribram, K. and MacLean, P. (1953) Neuronographic analysis of medial and basal cerebral cortex. *J Neurophysiol*, **16**, 324–40.

Redcay, E. and Courchesne, E. (2005) When is the brain enlarged in autism? A meta-analysis of all brain size reports. *Biol Psychiatry*, **58**, 1–9.

Schmithorst, V. J., Wilke, M., Dardzinski, B. J., et al. (2005) Cognitive functions correlate with white matter architecture in a normal pediatric population: a diffusion tensor MRI study. *Hum Brain Mapp*, **26**, 139–47.

Sotak, C. H. (2002) The role of diffusion tensor imaging in the evaluation of ischemic brain injury – a review. *NMR Biomed*, **15**, 561–9.

Stejskal, E. O. and Tanner, J. E. (1968) Restricted self-diffusion of protons in colloidal systems by the pulse-gradient, spin-echo method. *J Chem Phys*, **49**, 1768–77.

Westin, C. F., Maier, S. E., Khidir, B., et al. (1999) *Image Processing for Diffusion Tensor Magnetic Resonance Imaging*. New York: Springer.

Whitlock, D. G., Nauta, W. J. H. (1956) Subcortical projections from temporal neocortex in macaca mulata. *J Comp Neurol*, **106**, 183–212.

Wozniak, J. R. and Lim, K. O. (2006) Advances in white matter imaging: a review of in vivo magnetic resonance methodologies and their applicability to the study of development and aging. *Neurosci Biobehav Rev*, **30**, 762–74.

Zhang, L., Thomas, K. M., Davidson, M. C., et al. (2005) MR quantitation of volume and diffusion changes in the developing brain. *AJNR Am J Neuroradiol*, **26**, 45–9.

20

Arterial spin labeling perfusion magnetic resonance imaging in developmental neuroscience

Jiong-Jiong Wang, Hengyi Rao and John A. Detre

Introduction

Arterial spin labeling (ASL) perfusion magnetic resonance imaging (MRI) is an emerging neuroimaging technique that permits non-invasive quantitative mapping of cerebral blood flow (CBF), which is a biological parameter that is known to be regionally coupled to brain metabolism under normal conditions. Until recently, measurement of CBF involved the intravenous injection of exogenous contrast agents or radioactive tracers that are undesirable or ethically problematic in pediatric research. The development of ASL perfusion MRI now provides a non-invasive approach for measuring CBF in any age group. Arterial spin labeling is distinguished from nuclear medicine approaches and dynamic contrast MRI by its non-invasive nature, and from blood-oxygen-level-dependent (BOLD) functional MRI (fMRI) by its capability for baseline CBF mapping, independent of task activation. The above features make ASL particularly attractive for longitudinal neurodevelopmental studies. Arterial spin labeling perfusion MRI complements both structural MRI and the widely used BOLD fMRI method for measuring task-elicited changes in neural function. Here we summarize the technical aspects of performing ASL perfusion MRI in pediatric populations and discuss its current use and promises for developmental and clinical cognitive neuroscience.

Methodological background

Perfusion refers to the delivery of oxygen and nutrients to tissue by means of blood flow, which is typically measured as the volume of arterial blood (ml) delivered to 100 g of tissue per minute ($ml \cdot 100 g^{-1} \cdot min^{-1}$). The majority of methods for measuring cerebral perfusion have used tracers that differ in size (microsphere, intravascular or diffusible) and biophysical properties (e.g., radio-isotopes or contrast agents that alter MR/computed tomography signals). Arterial spin labeling perfusion MRI, one of the newest approaches, utilizes magnetically labeled endogenous arterial blood water as a tracer (Detre and Alsop, 1999; Wong, 1999). In ASL techniques, water molecules in inflowing arteries are magnetically "labeled" (MR signal manipulation through inversion or saturation) using radiofrequency (RF) pulses (Figure 20.1). Image acquisition is carried out after a delay time (~1 s) that allows this endogenous tracer (labeled blood water) to flow into the brain. During ASL scans, repeated label and control acquisitions, which differ only in whether spin labeling RF pulses are applied, are interleaved. The perfusion contrast is obtained by pairwise subtraction of the label and control acquisitions. Absolute CBF in well-characterized physiological units of $ml \cdot 100 g^{-1} \cdot min^{-1}$ is estimated by modeling expected signal changes in the brain,

Neuroimaging in Developmental Clinical Neuroscience, eds. Judith M. Rumsey and Monique Ernst. Published by Cambridge University Press. © Cambridge University Press 2009.

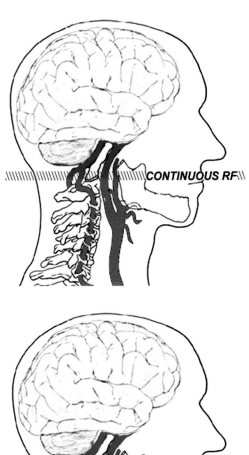

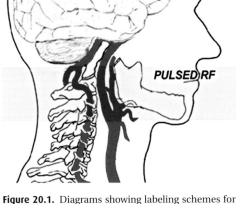

Figure 20.1. Diagrams showing labeling schemes for continuous arterial spin labeling (CASL) and pulsed arterial spin labeling (PASL).

primarily taking into account the tracer half-life determined by the T1 of blood (1–2 s).

Since ASL does not require the administration of a contrast agent or radioactive tracer, it is safer, more economical, and more convenient than

existing radionuclide or dynamic contrast MRI approaches. This is particularly appealing for the pediatric population, allowing perfusion imaging in a wide range of age groups from neonates to adolescents. Additionally, ASL scans can be repeated as many times as needed during the same scanning session without cumulative effects. During the past decade, theoretical and experimental studies have been conducted to improve the accuracy of CBF quantification using ASL by taking into account multiple parameters such as arterial transit time (Alsop and Detre, 1996; Gonzalez-At et al., 2000; Wong et al., 1998; Yang et al., 2000; Ye et al., 1997), magnetization transfer effect (Alsop and Detre, 1996), T1 (McLaughlin et al., 1997; Wang et al., 2002), labeling efficiency (Maccotta et al., 1997; Roberts et al., 1994; Utting et al., 2003), and capillary water permeability (Ewing et al., 2001; Parkes and Tofts, 2002) in models used to calculate CBF. Cerebral blood flow measurements with ASL perfusion MRI have been shown to agree with results from ^{15}O-labeled positron emission tomography (^{15}O-PET; Ye et al., 2000) and dynamic contrast MR approaches (Siewert et al., 1997). Arterial spin labeling perfusion measurements both at rest and during task activation have been demonstrated to be highly reproducible across intervals varying from a few minutes to a few days (Floyd et al., 2003; Wang et al., 2003a). In a recent concurrent ASL and fluorodeoxyglucose (FDG) PET study (Newberg et al., 2005), excellent concordance between regional CBF and metabolic rate for glucose utilization was observed during visual stimulation, adding support for ASL as a non-invasive surrogate marker of neural function and metabolism. Arterial spin labeling also has demonstrated clinical utility for detecting pathological changes in CBF and metabolism in a variety of neurological and psychiatric disorders in adults (Alsop et al., 2000; Chalela et al., 2000; Detre et al., 1998).

The primary weakness of ASL as compared to dynamic contrast MRI is the relatively small labeling effect (<1% raw signal), which may limit its sensitivity for detecting pathological perfusion changes as well as task-induced brain activation.

This small labeling effect arises from the fact that the volume of arterial blood available for labeling is only on the order of 1%–2% of total brain volume. The labeled blood further relaxes during the transit from the labeling region to the brain, resulting in a net labeling effect of less than 1% of raw MRI signal in brain tissue. Because the arterial transit time from the labeling region to the brain is comparable to the tracer half-life (blood T1), ASL techniques are very sensitive to transit effects, which can produce focal artifacts in perfusion images corresponding to intravascular signals (Detre and Alsop, 1999). Nevertheless, ASL uniquely allows arterial transit time to be quantified (Wang et al., 2003b), which may have diagnostic or mechanistic significance in cerebrovascular disorders.

Spin labeling schemes

There are two major categories of spin labeling techniques: continuous (CASL) and pulsed ASL (PASL). In CASL, arterial blood water is continuously labeled as it passes through a labeling plane, typically applied at the base of the brain. Labeling can be applied for several seconds, maximizing the effects on brain signal. In PASL, a relatively thick slab of blood inferior to the imaging slices is instantly inverted using short RF pulses (10–20 ms) (Figure 20.1). While CASL provides stronger perfusion contrast, it is more challenging for implementation and deposits a higher level of RF power into the subject compared to PASL (the direct effect is tissue heating). Therefore, CASL has to be operated or optimized according to guidelines for the specific absorption rate (SAR) of RF power, especially at high magnetic field strengths (≥ 3 T). The long labeling pulses in CASL also partially excite the imaging slices. This effect has to be balanced during the control acquisitions in order to accurately quantify CBF, and special procedures are required to permit multislice imaging (Alsop and Detre, 1998; Zaharchuk et al., 1999). In contrast, PASL methods require a well-designed slice profile for the labeling pulse with a sharp edge to eliminate

residual signal from static tissue. The image coverage is generally larger in CASL than in PASL, although both techniques are approaching the ability to image the whole brain.

One issue related to the use of CASL is that not all commercial MRI scanners allow long RF pulses for labeling (RF duty-cycle limit). However, multi-slice CASL has been successfully implemented on the MRI scanners of several major vendors. A recent development that takes advantage of the higher labeling efficiency of CASL, while circumventing hardware limitations, is a novel labeling scheme termed pseudo-CASL (Garcia et al., 2005). Pseudo-CASL utilizes a train of short RF pulses to mimic the effect of continuous spin labeling, providing improved labeling efficiency and compatibility with the standard body transmit coil and array receiver coils now widely used on commercial MRI systems.

Technical advances in arterial spin labeling imaging

The relatively low signal-to-noise ratio (SNR) is the primary limitation hampering the widespread application of ASL perfusion MRI. Performing ASL at high magnetic fields offers a natural solution to improve the image quality, as well as to reduce transit-related effects (Wang et al., 2002). The SNR is proportional to the main field strength, and there is an improved labeling effect due to prolonged tracer half-life (T1) at high fields. However, due to the shortened T2/T2*, the net SNR gain in high-field ASL is somewhat compromised. Figure 20.2a demonstrates the theoretical calculation of PASL and CASL signals as a function of the static field strength, measured in units of Tesla (T). A two-fold signal gain is readily achievable by performing PASL/CASL at 3.0 and 4.0 T, as compared to similar methods at 1.5 T (Wang et al., 2002; Yongbi et al., 2002). Higher field strengths allow improved spatio-temporal resolution, illustrated in Figure 20.2b, as well as longer post-labeling delay times to counteract delayed transit effects usually present in patient populations (Wang et al., 2005b). As 3-T

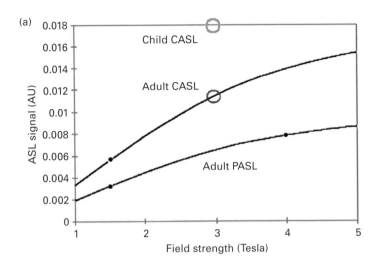

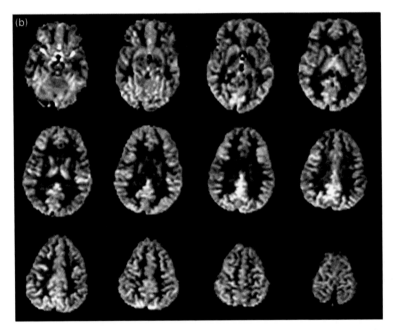

Figure 20.2a, b. (a) Theoretical ASL signal as a function of magnetic field strength as in Wang et al. (2002). The circles indicate the 3.0-T CASL signals in adults and children, respectively (reprinted with permission from Elsevier).
(b) Representative child (8-year-old boy) perfusion images acquired using CASL at 3 T, showing image quality approaching routine structural MRI (reprinted with permission from Elsevier).

MRI scanners are increasingly available at many neuroimaging centers, high-field ASL is expected to become the standard practice with widespread applications.

A complementary approach for increasing signal in ASL perfusion MRI is to use an array coil as the receiver to provide very high SNR due to its element coils. Although the signal enhancement is not homogeneous in space (higher at the cortex, lower at the center of the brain), the absolute CBF maps obtained with ASL are largely unaffected by this coil sensitivity profile effect due to several calibration steps involved in perfusion quantification (Wang et al., 2005c). Phased array coils can also be optimized for parallel imaging to shorten the image acquisition time by the acceleration factor. Although there is generally a cost in SNR for parallel imaging, in ASL perfusion MRI, much of this SNR cost can be regained through shortened echo time (TE) along with reduced distortion from susceptibility artifact (Wang et al., 2005c). This trend is noteworthy for ASL, as array receiver and parallel acquisition are becoming the standard configuration on new commercial MR systems.

Lastly, fast 3D sequences can be applied for image acquisition in ASL to further improve SNR due to the excitation of a large slab of spins and the prolonged image acquisition window. Because fast 3D sequences generally consist of a train of spin-echoes, image distortion arising from magnetic field inhomogeneity effects is reduced compared to routine 2D gradient-echo echo-planar imaging or spiral sequences. It is possible to acquire the whole brain volume within a single-shot RF excitation using ultrafast 3D sequences such as gradient- and spin-echo (GRASE) (Fernandez-Seara et al., 2005). Overall, the integrated effects of signal gains from various sources including high magnetic field (approx. 2-fold), array receiver coil (Wang et al., 2005c) (approx. 2-fold), novel spin labeling schemes (Garcia et al., 2005) (~1.5-fold) and image acquisition sequences (Fernandez-Seara et al., 2005) (~1.5-fold) have led to an impressive net SNR gain (8- to 10-fold) in ASL techniques over the past decade, bringing ASL perfusion MRI to

the frontier of practical clinical and neuroscience applications (Figure 20.3). The present ASL scan takes just a few minutes to obtain a reliable CBF measurement as part of a routine MRI exam or fMRI study.

Benefits of arterial spin labeling for developmental neuroscience

Arterial spin labeling methods may be particularly suitable for studying developmental changes in brain function in children because of the non-invasive nature of these methods and several characteristics of pediatric physiology (Wang and Licht, 2006). First, blood flow rate is higher in children compared to adults (Chiron et al., 1992). This factor enhances perfusion contrast. Similarly, water content of the brain is higher in children than in adults (Dobbing and Sands, 1973), which yields a greater concentration and half-life of the tracer. This factor increases equilibrium MR signal and spin-lattice, spin-spin relaxation times (T1, T2), improving pediatric ASL. Secondly, data from Doppler ultrasound studies suggest that blood flow velocities in carotid arteries are higher in healthy children compared to adults, with the peak velocity occurring within the age range of 5–8 years (Schoning et al., 1993). This effect can be translated into reduced arterial transit time for labeled blood to flow from the labeling region to the brain, resulting in reduced relaxation of the labeling effect and reduced transit effect (focal intravascular signal) in pediatric perfusion images.

In a feasibility study at 1.5 T, we observed that both brain T1 and equilibrium MR signal, and thereby the SNR of ASL perfusion images, all decrease linearly with age. On average, a 70% improvement in the SNR of perfusion images was observed in healthy children (1 month to 10 years) as compared to healthy adults (Wang et al., 2003c). The benefit of high-field ASL, in conjunction with the natural ASL signal gains in children, yields impressive pediatric perfusion images at 3.0 T, the quality of which approaches that of routine anatomical MR scans

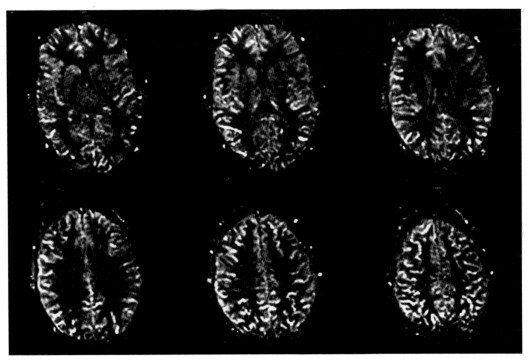

Figure 20.3. High-field, high-resolution (voxel size $= 2 \times 2 \times 5\,\text{mm}^3$) pseudo-CASL perfusion images acquired at 3 T using an eight-channel head array coil, showing excellent delineation of cortical/subcortical structures.

(Figure 20.2b). Given the unique and reciprocal benefits in terms of safety and image quality, ASL methods may be ideally suited for developmental clinical neuroscience research.

Cerebral blood flow quantification in age groups and longitudinal stability

Absolute CBF in classical units of $\text{ml}\cdot 100\,\text{g}^{-1}\cdot\text{min}^{-1}$ measured by ASL should be independent of biophysical effects and thereby provide a valuable marker for longitudinal studies to follow neurodevelopmental changes in children aged 0–18 years. Although labeled water is nominally a free diffusible tracer, because of its rapid decay (half-life $\sim$1 s), the majority of existing ASL methods employ a single-compartment perfusion model for CBF quantification (Herscovitch and Ernst, 2000), which assumes that all the labeled blood stays in the

(micro-) vasculature rather than completely exchanging into brain tissue. Potential errors in CBF quantification could arise mainly from the choice of two parameters: the T_1 of blood (T_{1a}, tracer half-life) and blood–brain partition coefficient of water (λ) (for calculation of blood magnetization). Both the blood T_1 and blood–brain partition coefficient of water are expected to be higher in children than in adults (Herscovitch and Raichle, 1985). Fortunately, the two parameters exert opposite effects on the calculation of child CBF when adult values are adopted (overestimation for T_{1a} and underestimation for λ). Currently, the adult perfusion model seems to provide a reasonable approximation for child CBF quantification, with CBF measures comparable to values obtained using radioactive methods (Chiron et al., 1992). Based on simulation, there may be a slight (10%) overestimation of perfusion in neonates, but the overall error is around

5% across the entire age span (0–18 years) (Wang et al., 2003c). It is worth noting that multi-compartment perfusion models are also being developed for accurate CBF mapping and incorporate the limited exchange of water across the blood–brain barrier (permeability) and T_2/T_2^* effects (Parkes and Tofts, 2002; St Lawrence and Wang, 2005).

The accurate determination of the labeling efficiency plays a critical role among biophysical factors that may affect the long-term test-retest repeatability of ASL. The labeling efficiency is relatively insensitive to flow velocity in PASL methods. However, the duration and volume of the tagging bolus need to be tailored/and optimized for specific clinical/age populations, such as neonates with congenital heart defects (Wang et al., 2006). In adults, we have demonstrated that robust motor cortex activation can be observed using PASL at 4 T, even when the resting state and task activation were separated by a day (Wang et al., 2003a).

Continuous ASL and pseudo-CASL methods generally require an optimal flow velocity range to reach the theoretical tagging efficiency (Maccotta et al., 1997, Wu et al., 2007). Optimization and simulation of labeling parameters may need to be carried out for pediatric perfusion imaging using CASL, as the interindividual variability of flow velocities in children may be large. The variability in the neuroanatomy of cerebral blood supply may be another factor leading to potential perfusion errors in children, since optimal labeling occurs when the artery and labeling plane are perpendicular to each other. Ideally, a fast blood and brain T_1 mapping sequence, as well as a MR angiography (to localize the major vessels for labeling), should be carried out in conjunction with CASL or pseudo-CASL scans for accurate CBF mapping. Despite these potential sources of error, excellent repeatability (with variation in values of around 10%) is readily achievable in adult studies using the existing CASL technique for scans repeated 1–2 weeks apart (Floyd et al., 2003). The accurate and absolute CBF quantification is important, not only for longitudinal studies, but also to facilitate the integration of ASL perfusion data acquired at different imaging

sites on various MRI platforms for multisite studies. For developmental studies, while PASL may seem less challenging for achieving a consistent labeling efficiency across longitudinal scans, CASL and pseudo-CASL would be preferable given the superior image quality and whole brain coverage (see below).

Perfusion magnetic resonance imaging for developmental functional magnetic resonance imaging studies

The BOLD contrast has proven successful in fMRI studies for mapping the functional neuroanatomical substrates underlying cognitive and emotional development in children. However, a major limitation of existing BOLD fMRI studies is the lack of quantitative (absolute) information concerning baseline brain function, which may affect the accurate interpretation of age-related activation effects. Arterial spin labeling perfusion MRI affords several appealing and complementary features for functional brain imaging, particularly for studies of behavioral states, mood, and pharmacological manipulations.

First, absolute CBF maps can be obtained both at rest and during activation, facilitating the comparison of fMRI results between patient and control populations, as well as across age groups. Secondly, because ASL image series show stable noise characteristics over the entire frequency spectrum ("white noise" for ASL timecourse data) (Aguirre et al., 2002; Wang et al., 2003a), it is suitable for studying sequential changes in resting or activated brain function over long periods. Perfusion fMRI has been successfully applied in visualizing persistent or gradual changes in brain function associated with sustained attention, working memory, motor learning, and psychological stress (Kim et al., 2006; Olson et al., 2006; Wang et al., 2005a). Thirdly, because the normal distribution of noise in a perfusion image series satisfies the theoretical assumption of statistical analysis (that of normality), virtually any statistical method, including both parametric and non-parametric approaches,

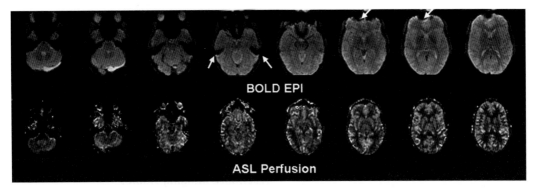

Figure 20.4. Improved coverage and reduced distortion in orbitofrontal and temporal regions (arrows) in ASL perfusion MRI acquired using spin-echo echo-planar imaging (EPI) as compared to standard blood-oxygen-level-dependent (BOLD) images acquired with gradient-echo EPI on the same subject.

can be applied in data analysis (Aguirre et al., 2005). Fourthly, given that ASL effects are not based on magnetic susceptibility, brain activation in regions of high magnetic field inhomogeneity can be measured using appropriate pulse sequences. Figure 20.4 shows improved image coverage and reduced distortion in orbitofrontal and temporal regions in ASL perfusion MRI, acquired using spin-echo echo-planar imaging (EPI), as compared to standard BOLD images acquired with gradient-echo EPI on the same subject.

Recent evidence suggests that perfusion fMRI is resilient to effects of subject motion during overt speech production (Kemeny et al., 2005) and allows certain behavioral and pharmacological manipulations to be performed outside the MR scanner so that the sustained effects on CBF can subsequently be measured using ASL (Franklin et al., 2007; Wang et al., 2003a). Perfusion fMRI, therefore, may be suitable for pediatric neuroimaging where subject motion may be an issue, as well as for many ecological paradigms that are otherwise difficult to study using the BOLD contrast. The acoustic noise level is also much lower in ASL compared to BOLD scans, because of the embedded MR quiet period (e.g., post-labeling delay time) in ASL acquisitions. Finally, ASL should provide more accurate localization of neural activation as opposed to BOLD fMRI, since the perfusion contrast mainly arises from

capillaries and brain tissue as opposed to venous structures in BOLD fMRI (Duong et al., 2001; Luh et al., 2000). However, compared to the mainstream BOLD fMRI, ASL still has the disadvantages of the need for technical sophistication, limited availability, smaller effect size of task activation, and lower temporal resolution. A head-to-head comparison of perfusion and BOLD fMRI is presented in Table 20.1.

Relationship with other neuroimaging modalities

Among in vivo neuroimaging techniques and modalities developed to elucidate the structure and function of human brain, MRI methods offer the advantages of being non-invasive, multimodal, and widely available. Structural MRI, measuring tissue volume or density, and BOLD fMRI, measuring task-induced relative brain activation, have been widely utilized in developmental and cognitive neuroscience over the last decade. However, for structural MRI, the relationship among cortical volume, functional capacity and behavior remains unknown. Blood-oxygen-level-dependent fMRI provides a useful measure of relative signal changes associated with specific tasks, but has poor sensitivity for tracking changes in brain function evolved over long time scales, thereby limiting

Table 20.1 Comparison between imaging characteristics of functional magnetic resonance imaging (fMRI) using blood-oxygen-level-dependent (BOLD) and arterial spin labeling (ASL) perfusion contrasts

	BOLD	ASL perfusion
Signal mechanism	Blood flow, blood volume, oxygenation consumption	Blood flow
Contrast parameter	T_2^*	T_1
Spatial specificity	Drain veins, venules, tissue	Tissue, capillaries, arterioles
Typical signal change	$\sim$0.5%–5%	<1%
Imaging methods	Gradient-echo	Gradient-echo
	Asymmetric spin-echo	Spin-echo
Optimal task frequency (block design)	$\sim$0.01 Hz–0.06 Hz	<0.01 Hz
Sample rate (TR)	$\sim$1–3 s per image volume	$\sim$3–8 s per image volume

its application in longitudinal studies of the developing brain.

Given the characteristics of the respective contrast mechanisms, the roles of ASL and BOLD fMRI in functional neuroimaging studies are likely to be complementary. The quantitative CBF maps obtained with ASL provide a surrogate marker for a subject's behavioral state (e.g., mood or resting states) that complements transient evoked processes visualized by BOLD fMRI. Neural states may reflect learning, affective predisposition, drug effects, or neural phenotypes. Baseline neural states assessed by ASL can also be regionally correlated with behavioral performance or other measures across cohorts, providing a complementary approach to BOLD fMRI for functional localization and structural MRI for neuroanatomical variation.

Several exploratory studies have demonstrated the feasibility of concurrent ASL and BOLD (Rao et al., 2007a), ASL and FDG-PET (Newberg et al., 2005), and ASL and electroencephalography (EEG; Rao et al., 2006). Note the combination of perfusion fMRI and EEG may be particularly useful for characterizing neural function during the sleep–wake cycle (Noeth et al., 2007), wherein BOLD fMRI is of limited use. However, our recent data using both ASL and structural MRI showed that alterations in brain anatomy and resting CBF may be dissociated in adolescents with a history of in utero cocaine

exposure, as well as in healthy adults carrying the short or long allele of the serotonin transporter promoter gene (5HTT) (Rao et al., 2007b). An unresolved question in both development and degeneration of the brain is whether functional changes drive structural changes or vice versa. Careful longitudinal correlations of structural MRI and ASL perfusion MRI should allow this question to be addressed, while concurrent ASL and FDG-PET will allow the coupling between CBF and metabolism to be explored in greater detail.

To date, the applications of ASL perfusion fMRI in cognitive neuroscience, including developmental neuroscience, are in their infancy. Below, we present some recent ASL perfusion MRI studies from our group and others in developmental cognitive/clinical neuroscience.

Normal development of cerebral blood flow from childhood to young adulthood

Human brain maturation and development are far from complete at birth. Changes in both cellular metabolism and blood flow proceed in stages through the first decade of life (Chiron et al., 1992; Epstein, 1999). For example, after birth, synaptic density increases rapidly, reaching a peak level at about 2 years of age (Huttenlocher, 1979), followed

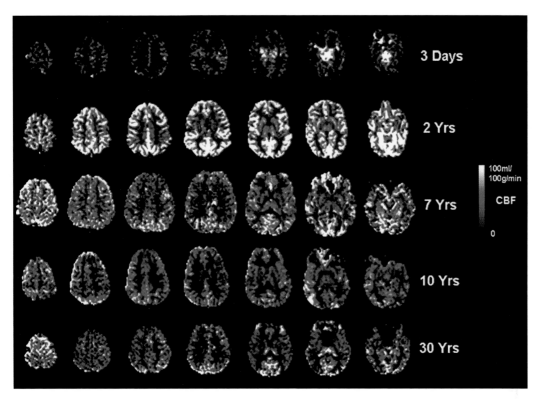

Figure 20.5. Representative CBF images across the age span from 3 days to 30 years acquired using a PASL technique at 1.5 T.

by regionally specific synaptic pruning (Webb et al., 2001). The timecourse of regional blood flow and metabolism has been inferred from the evolution of neuronal innervation and synaptic stabilization during postnatal brain growth. Although assessment of brain perfusion and its time-related changes during childhood have been addressed in infants and children with illness (Greeley et al., 1989; Greisen, 1997; Lou et al., 1979), studies of normal infants and children are relatively scarce. Several studies, using transcranial Doppler (Schoning and Hartig, 1996), single photon emission computed tomography (SPECT) (Chiron et al., 1992), PET (Takahashi et al., 1999), or contrast-enhanced cerebral CT (Wintermark et al., 2004), have attempted to catalog normative regional CBF data across age groups in children without brain abnormalities. However, conclusions from these studies are limited by small sample sizes and varied imaging protocols across age groups.

Both pulsed and continuous ASL perfusion MRI have been performed on normal children and healthy adults by the authors and other groups. As an example, Figure 20.5 shows representative PASL CBF images across the age span from 3 days to 30 years. The images clearly demonstrate an age-related increase of CBF from neonates (3 days) to toddlers (2 years), followed by tapering of CBF from childhood to adulthood. The neonatal global CBF is lower than the adult level, but regional CBF is relatively increased in sensorimotor cortex and the basal ganglia, consistent with PET imaging results in this age group (Chugani, 1998). Recent PASL data collected at 3.0 T from 29 unsedated healthy neonates replicated these findings (Miranda et al., 2006). The effect of different procedures of sedation and anesthesia on CBF is

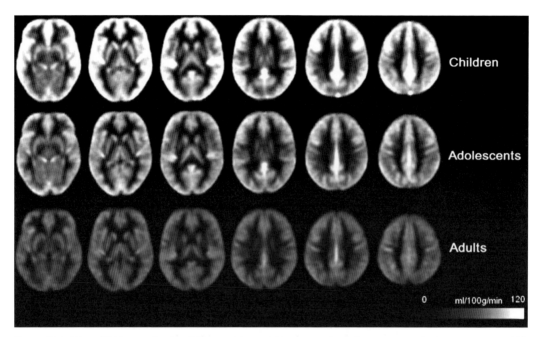

Figure 20.6. Mean CBF images for the child group (ages 5–10 years, top), adolescent group (ages 11–16 years, middle) and young adult group (ages 18–30 years, bottom).

an unsolved issue, which has just started to be investigated using ASL techniques (Kofke et al., 2007).

During the past 2 years, through a Human Brain Project supported by NIMH ("pediatric template of brain perfusion"), we collected and analyzed a large set of pediatric CBF data through collaboration with colleagues at the Children's Hospital of Philadelphia, Georgetown University Medical Center, Cincinnati Children's Hospital Medical Center, and the University of California at Davis. This project included data from 90 normal children and young adults, acquired using the same CASL perfusion MRI sequence on Siemens 3T Trio scanners. These subjects were divided into three age groups: 31 children (ages 5–10 years), 33 adolescents (ages 11–16 years), and 26 young adults (ages 18–30 years). Global CBF intensities were first calculated for each subject. The mean global CBF decreased from 81.2 ml·$100\,g^{-1}$·min^{-1} in the child group to 64.4 ml·$100\,g^{-1}$·min^{-1} in the adolescent group and to 43.5 ml·$100\,g^{-1}$·min^{-1} in the young

adult group. The CBF images of each individual subject were then normalized to a customized template by applying the parameters from the normalization of the concurrent EPI images and then averaged within each age group. Figure 20.6 shows the mean CBF maps (templates) for each group, which clearly demonstrate the overall decrease of global CBF, as well as the growing fraction of white matter and declining fraction of gray matter during brain development. After adjusting for global CBF differences, both voxel-wise and ROI analyses on these data revealed significant relative regional CBF increases with age in the frontal cortex, cingulate cortex, angular gyrus, and hippocampus (Figure 20.7). The regional CBF increases may reflect the later maturation of cortical regions associated with integrative, cognitive control and executive functions, and memory. In this study, the test-retest repeatability of ASL perfusion MRI across the participating sites was not established. The observed age-dependent absolute CBF changes

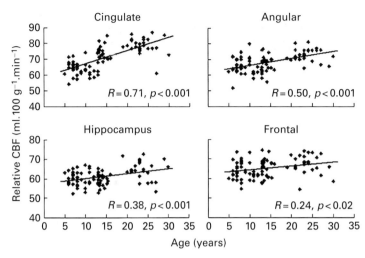

Figure 20.7. Age-related regional CBF changes in the cingulate cortex, angular gyrus, hippocampus, and frontal cortex (using Automated Anatomical Labeling toolbox in SPM). After adjusting the global CBF differences, all four regions show significant positive correlations between age and relative CBF ($p < 0.05$).

might, therefore, be confounded by technical variability across neuroimaging sites. Nevertheless, the excellent concordance of current CBF data with the literature strongly supports the feasibility and potential of ASL perfusion MRI for future longitudinal neurodevelopmental studies of the growing human brain. A recent CASL study at 1.5 T replicated these findings in 23 children and teenagers (Biagi et al., 2007).

Studies of atypical development and childhood diseases

Adolescents exposed to cocaine in utero

In utero cocaine exposure (IUCE) and its potential cognitive and psychological sequelae have been a major health-related concern during the last decade in the United States. Animal studies have clearly demonstrated certain alterations of cortical development induced by IUCE (Levitt et al., 1997). In humans, the extent to which IUCE may have an enduring impact on the neurocognitive development of children remains unclear. Using CASL perfusion MRI, we measured and compared resting CBF in two groups

of adolescents, including 24 cocaine-exposed subjects (COC, 13 male, mean age 14.4 ± 1.0 years) and 25 non-cocaine-exposed controls (CON, 12 male, mean age 13.9 ± 0.9 years) who had been followed since birth (Hurt et al., 1995; Rao et al., 2007c).

In utero cocaine exposure was associated with a significant global CBF reduction, which suggests that chronic cocaine exposure in utero may produce cerebral vasoconstriction effects in the fetus that persist into adolescence. Figure 20.8 shows the effect of IUCE on the regional absolute and relative CBF (i.e., ratio of regional to global absolute CBF). The decrease of absolute CBF in the COC group was observed mainly in the occipital lobe and thalamus, which concurs with previous findings from animal and human studies (Lee et al., 2003; Lidow, 2003). After adjusting for global CBF differences, however, a significant increase in relative CBF in the COC group was found in frontal, cingulate, and parietal regions that serve as the neural substrates for mediating attention and arousal regulation. These cortical regions also mature late (Gogtay et al., 2004). In light of reduced global CBF, these relative increases in the blood supply to frontal and parietal regions in exposed participants suggests that

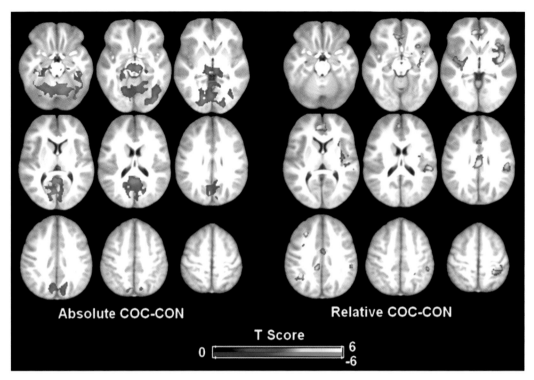

Figure 20.8. Absolute and relative CBF differences between the in utero-cocaine-exposed (COC) and non-exposed control (CON) adolescents. The threshold was set as FDR (false discovery rate) corrected p <0.05 and cluster larger than 100 voxels.

compensatory mechanisms may be triggered during neurodevelopment. The COC and CON groups did not show significant differences in the morphometric analysis of structural MRIs (voxel-based morphometry/VBM) or in behavioral testing. This suggests that perfusion fMRI may be a valuable, more sensitive approach for imaging the long-term effects of drug use during pregnancy on neurodevelopment and an important source of hypotheses concerning the mechanisms by which prenatal cocaine exposure affects brain development.

Sickle cell disease

Children with sickle cell disease (SCD) are at high risk for neurological complications including ischemia and neurocognitive impairments. Conventional diagnostic methods, such as anatomical MRI, MR angiography, and transcranial Doppler sonography, are useful for demonstrating involvement of the large cerebral vasculature in SCD, but have limited sensitivity for detecting microcirculatory occlusion and chronic regional tissue hypoxia and acidosis (Kaul and Nagel, 1993; Wang et al., 2001). Arterial spin labeling perfusion MRI has been used in SCD studies to provide additional information regarding the effects of small vessel disease and low hemoglobin concentration on brain tissue perfusion that usually are undetectable with conventional MR imaging.

In one study (Oguz et al., 2003), quantitative CBF was measured in 14 children with SCD and 7 controls using CASL perfusion MRI at 1.5 T. Sickle cell disease was associated with a significant increase of CBF in all cerebral arterial territories, which concurred with previous PET findings. Moreover, four children with SCD had relatively reduced CBF in the right side arterial territories without associated demonstrable large vessel disease on conventional MR images. These vascular territories may be at risk for ischemia given the exhausted vascular reserve in SCD children.

In another study (Strouse et al., 2006), quantitative CBF was measured in 24 children with SCD using CASL perfusion MRI at 1.5 T and correlated with their intelligence quotient (IQ) scores. Significant inverse correlations were obtained between performance IQ and anterior and right-sided absolute CBF. These correlations suggest that increased absolute CBF may be both a response to and a risk factor for cerebral hypoxia. Such results suggest that ASL perfusion MRI may be used to identify SCD children with neurocognitive impairment before damage is evident by conventional imaging.

Autism spectrum disorders

Autism is a complex and severe neurodevelopmental disorder that is associated with life-long deficits in social interaction and verbal and non-verbal communication, and with stereotyped and repetitive behaviors. The prevalence of autism may be as high as 0.6% of the male population (Hill and Frith, 2003). Both structural MRI and BOLD fMRI have been used to characterize the anatomical and physiological brain phenotypes underlying autism spectrum disorders (ASD) and have provided a wealth of data for understanding brain–behavior relationships in ASD. A range of theories and hypotheses, such as theory of mind deficits (Baron-Cohen et al., 1999; Peterson, 2005), cortical underconnectivity (Cherkassky et al., 2006; Just et al., 2004), and mirror neuron system dysfunction (Williams et al., 2001), have been proposed to explain the neurological basis and development of autistic symptoms (see Chapter 8). Functional MRI with BOLD contrast has suggested deficient activation in face processing regions (Schultz et al., 2000). However, careful studies with concurrent eye tracking suggest

that these differences may be attributable to performance factors (Dalton et al., 2005), i.e., ASD subjects not fixating normally on the presented face stimuli. Arterial spin labeling perfusion MRI can provide a surrogate index of brain function that is not confounded by performance effects and may thereby complement structural and BOLD MRI for studying the neural basis of autism.

Through the Human Brain Project supported by NIMH, we collected resting CBF data using CASL perfusion MRI from eight high-functioning male children diagnosed with ASD (mean age = 9.8 ± 2.0 years) and 12 matched control children (data acquired at Georgetown University, Center for Functional and Molecular Imaging). The global CBF showed no difference between the autistic and control children. However, both voxel-wise general linear model analysis and ROI analysis revealed significant regional relative CBF differences between the autism group and the control group. Autistic subjects showed significantly lower regional relative CBF in left inferior and medial frontal cortex, which may reflect the deficits seen in language comprehension and social cognition in autism. Autistic children also showed significantly higher regional relative CBF in right occipital cortex, thalamus, and sensorimotor cortex, which may reflect the enhanced processing of perceptual and sensory input in autism. These results are consistent with the imaging literature on autism (Cody et al., 2002; Hubl et al., 2003; Penn, 2006) and further extend our knowledge of the neural correlates of this disorder.

Potential and challenges

Magnetic resonance imaging remains the most versatile modality for studying human brain structure and function during development. A comprehensive MRI study can quantify volumes of gray and white matter, visualize fiber tracts, and demonstrate networks subserving sensorimotor and cognitive function. On the timescale of development, absolute baseline function as manifested in regional CBF and measured using ASL perfusion MRI is likely to provide critical information, reflecting functional phenotypes corresponding to genotypes, developmental stages, neurological and psychiatric disorders of development, therapeutic effects and pharmacological manipulations. While widespread applications of ASL in developmental neuroscience have previously been limited by technical factors, recent methodological advances, as described above, have dramatically improved the sensitivity and resolution of this approach. Early applications, also summarized above, clearly demonstrate the feasibility of using ASL to study brain function during development. More widespread dissemination of this approach in studies with large sample sizes may be expected to yield new insights into the mechanisms and disorders of brain development. A complete understanding of the neurochemical mechanisms (neurotransmitter and enzyme function) underlying neurodevelopmental processes, however, may require integrative approaches that combine structural, functional, and molecular neuroimaging.

ACKNOWLEDGMENTS

This work was supported by the Thrasher Research Fund, NIH MH072576 (Human Brain Project), HD049893, and DA014129. The authors are grateful to Dr Kim Cecil at the Cincinnati Children's Hospital Medical Center, Drs Thomas Zeffiro and John VanMeter at the Georgetown University Medical Center, Drs Hallam Hurt and Daniel Licht at the Children's Hospital of Philadelphia, and Dr Tony Simon at University of California, Davis.

REFERENCES

Aguirre, G. K., Detre, J. A., and Wang, J. (2005) Perfusion fMRI for functional neuroimaging. *Int Rev Neurobiol*, **66**, 213–36.

Aguirre, G. K., Detre, J. A., Zarahn, E., et al. (2002) Experimental design and the relative sensitivity of BOLD and perfusion fMRI. *NeuroImage*, **15**, 488–500.

Alsop, D. C. and Detre, J. A. (1996) Reduced transit-time sensitivity in noninvasive magnetic resonance imaging of human cerebral blood flow. *J Cereb Blood Flow Metab*, **16**, 1236–49.

Alsop, D. C. and Detre, J. A. (1998) Multisection cerebral blood flow MR imaging with continuous arterial spin labeling. *Radiology*, **208**, 410–16.

Alsop, D. C., Detre, J. A., and Grossman, M. (2000) Assessment of cerebral blood flow in Alzheimer's disease by spin-labeled magnetic resonance imaging. *Ann Neurol*, **47**, 93–100.

Baron-Cohen, S., Ring, H. A., Wheelwright, S., et al. (1999) Social intelligence in the normal and autistic brain: an fMRI study. *Eur J Neurosci*, **11**, 1891–8.

Biagi, L., Abbruzzese, A., Bianchi, M. C., et al. (2007) Age dependence of cerebral perfusion assessed by magnetic resonance continuous arterial spin labeling. *J Magn Reson Imaging*, **25**, 696–702.

Chalela, J. A., Alsop, D. C., Gonzalez-Atavalez, J. B., et al. (2000) Magnetic resonance perfusion imaging in acute ischemic stroke using continuous arterial spin labeling. *Stroke*, **31**, 680–7.

Cherkassky, V. L., Kana, R. K., Keller, T. A., et al. (2006) Functional connectivity in a baseline resting-state network in autism. *NeuroReport*, **17**, 1687–90.

Chiron, C., Raynaud, C., Maziere, B., et al. (1992) Changes in regional cerebral blood flow during brain maturation in children and adolescents. *J Nucl Med*, **33**, 696–703.

Chugani, H. T. (1998) A critical period of brain development: studies of cerebral glucose utilization with PET. *Prevent Med*, **27**, 184–8.

Cody, H., Pelphrey, K., and Pirven, J. (2002) Structural and functional magnetic resonance imaging of autism. *Int J Dev Neurosci*, **20**(3–5), 421–38.

Dalton, K. M., Nacewicz, B. M., Johnstone, T., et al. (2005) Gaze fixation and the neural circuitry of face processing in autism. *Nat Neurosci*, **8**(4), 519–26.

Detre, J. A. and Alsop, D. C. (1999) Perfusion fMRI with arterial spin labeling (ASL). In: Moonen, C. T. W., Bandettini, P. A. (eds.) *Functional MRI*. Heidelberg: Springer-Verlag, p. 47–62.

Detre, J. A., Alsop, D. C., Vives, L. R., et al. (1998) Noninvasive MRI evaluation of cerebral blood flow in cerebrovascular disease. *Neurology*, **50**, 633–41.

Dobbing, J. and Sands, J. (1973) Quantitative growth and development of human brain. *Arch Dis Child*, **48**, 757–67.

Duong, T. Q., Kim, D. K., Ugurbil, K., et al. (2001) Localized cerebral blood flow response at submillimeter columnar resolution. *Proc Natl Acad Sci U S A*, **98**, 10904–9.

Epstein, H. T. (1999) Stages of increased cerebral blood flow accompany stages of rapid brain growth. *Brain Dev*, **21**, 535–9.

Ewing, J. R., Cao, Y., and Fenstermacher, J. (2001) Single-coil arterial spin-tagging for estimating cerebral blood flow as viewed from the capillary: relative contributions of intra- and extravascular signal. *Magn Reson Med*, **46**, 465–75.

Fernandez-Seara, M. A., Wang, Z., Wang, J., et al. (2005) Continuous arterial spin labeling perfusion measurements using single shot 3D GRASE at 3T. *Magn Reson Med*, **54**, 1241–7.

Floyd, T. F., Ratcliffe, S. J., Wang, J., et al. (2003) Precision of the CASL-perfusion MRI technique: global and regional cerebral blood flow within vascular territories at one hour and one week. *J Magn Reson Imaging*, **18**, 649–55.

Franklin, T. R., Wang, Z., Wang, J., et al. (2007) Limbic activation to cigarette smoking cues independent of nicotine withdrawal: a perfusion fMRI study. *Neuropsychopharmacology*, **32**(11), 2301–9.

Garcia, D. M., de Bazelaire, C., and Alsop, D. (2005) Pseudo-continuous flow driven adiabatic inversion for arterial spin labeling. *Proc Intl Soc Magn Reson Med*, **13**, 37.

Gogtay, N., Giedd, J. N., Lusk, L., et al. (2004) Dynamic mapping of human cortical development during childhood through early adulthood. *Proc Natl Acad Sci U S A*, **101**, 8174–9.

Gonzalez-At, J. B., Alsop, D. C., and Detre, J. A. (2000) Perfusion and transit time changes during task activation determined with steady-state arterial spin labeling. *Magn Reson Med*. **43**, 739–46.

Greeley, W. J., Ungerleider, R. M., Kern, F. H., et al. (1989) Effects of cardiopulmonary bypass on cerebral blood flow in neonates, infants, and children. *Circulation*, **80**, I209–15.

Greisen, G. (1997) Cerebral blood flow and energy metabolism in the newborn. *Clin Perinatol*, **24**, 531–46.

Herscovitch, P. and Ernst, M. (2000) Functional brain imaging with PET and SPECT. In: Ernst, E., Rumsey, J. M. (eds.) *Functional Neuroimaging in Child Psychiatry*. Cambridge: Cambridge University Press, p. 3–26.

Herscovitch, P. and Raichle, M. E. (1985) What is the correct value for the brain-blood partition coefficient for water? *J Cereb Blood Flow Metab*, **5**, 65–9.

Hill, E. L. and Frith, U. (2003) Understanding autism: insights from mind and brain. *Philos Trans R Soc Lond B Biol Sci*, **358**, 281–9.

Hubl, D., Bolte, S., Feineis-Matthews, S., et al. (2003) Functional imbalance of visual pathways indicates alternative face processing strategies in autism. *Neurology*, **61**(9), 1232–7.

Hurt, H., Brodsky, N. L., Braitman, L. E., et al. (1995) Natal status of infants of cocaine users and control subjects: a prospective comparison. *J Perinatol*, **15**, 297–304.

Huttenlocher, P. R. (1979) Synaptic density in human frontal cortex – developmental changes and effects of aging. *Brain Res*, **163**, 195–205.

Just, M. A., Cherkassky, V. L., Keller, T. A., et al. (2004) Cortical activation and synchronization during sentence comprehension in high-functioning autism: evidence of underconnectivity. *Brain*, **127**, 1811–21.

Kaul, D. K. and Nagel, R. L. (1993) Sickle cell vasoocclusion: many issues and some answers. *Experientia*, **49**, 5–15.

Kemeny, S., Ye, F. Q., Birn, R., et al. (2005) Comparison of continuous overt speech fMRI using BOLD and arterial spin labeling. *Hum Brain Mapp*, **24**, 173–83.

Kim, J., Whyte, J., Wang, J., et al. (2006) Continuous ASL perfusion fMRI investigation of higher cognition: quantification of tonic CBF changes during sustained attention and working memory tasks. *NeuroImage*, **31**, 376–85.

Kofke, W. A., Blissitt, P. A., Rao, H., Wang, J., Addya, K., and Detre, J. A. (2007) Remifentanil-induced cerebral blood flow effects in normal humans: dose and ApoE genotype effects. *Anesth Analg*, **105**, 167–75.

Lee, J. H., Telang, F. W., Springer, C. S. Jr., et al. (2003) Abnormal brain activation to visual stimulation in cocaine abusers. *Life Sci*, **73**, 1953–61.

Levitt, P., Harvey, J. A., Friedman, E., et al. (1997) New evidence for neurotransmitter influences on brain development. *Trends Neurosci*, **20**, 269–74.

Lidow, M. S. (2003) Consequences of prenatal cocaine exposure in nonhuman primates. *Brain Res Dev Brain Res*, **147**, 23–36.

Lou, H. C., Skov, H., and Pedersen, H. (1979) Low cerebral blood flow: a risk factor in the neonate. *J Pediatr*, **95**, 606–9.

Luh, W. M., Wong, E. C., Bandettini, P. A., et al. (2000) Comparison of simultaneously measured perfusion and BOLD signal increases during brain activation with T(1)-based tissue identification. *Magn Reson Med*, **44**, 137–43.

Maccotta, L., Detre, J. A., and Alsop, D. C. (1997) The efficiency of adiabatic inversion for perfusion imaging by arterial spin labeling. *NMR Biomed*, **10**, 216–21.

McLaughlin, A. C., Ye, F. Q., Pekar, J. J., et al. (1997) Effect of magnetization transfer on the measurement of cerebral blood flow using steady-state arterial spin tagging approaches: a theoretical investigation. *Magn Reson Med*, **37**, 501–10.

Miranda, M. J., Olofsson, K., and Sidaros, K. (2006) Noninvasive measurements of regional cerebral perfusion in preterm and term neonates by magnetic resonance arterial spin labeling. *Pediatr Res*, **60**, 359–63.

Newberg, A. B., Wang, J., Rao, H., et al. (2005) Concurrent CBF and CMRGlc changes during human brain activation by combined fMRI-PET scanning. *NeuroImage*, **28**, 500–6.

Noeth, U., Kotajima, F., Josephs, O., et al. (2007) Brain perfusion during sleep – determination with quantitative perfusion MRI and EEG with online artefact removal. *Proc Intl Soc Magn Reson Med*, **15**, 508.

Oguz, K. K., Golay, X., Pizzini, F. B., et al. (2003) Sickle cell disease: continuous arterial spin-labeling perfusion MR imaging in children. *Radiology*, **227**, 567–74.

Olson, I. R., Rao, H., Moore, K. S., et al. (2006) Using perfusion fMRI to measure continuous changes in neural activity with learning. *Brain Cogn*, **60**, 262–71.

Parkes, L. M. and Tofts, P. S. (2002) Improved accuracy of human cerebral blood perfusion measurements using arterial spin labeling: accounting for capillary water permeability. *Magn Reson Med*, **48**, 27–41.

Penn, H. E. (2006) Neurobiological correlates of autism: a review of recent research. *Child Neuropsychol*, **12**(1), 57–79.

Peterson, C. C. (2005) Mind and body: concepts of human cognition, physiology and false belief in children with autism or typical development. *J Autism Dev Disord*, **35**, 487–97.

Rao, H., Dinges, D. F., Censits, D., et al. (2006) Simultaneous EEG and ASL perfusion fMRI during resting and mental calculation: a preliminary study. *Proc Intl Soc Magn Reson Med*, **14**, 3299.

Rao, H., Gillihan, S. J., Wang, J., et al. (2007b) Genetic variation in serotonin transporter alters resting brain function in healthy individuals. *Biol Psychiatry*, **62**(6), 600–6.

Rao, H., Wang, J., Korczykowski, M., et al. (2007c) Altered resting cerebral blood flow in adolescents with in-utero cocaine exposure revealed by perfusion functional MRI. *Pediatrics*, **120**(5), e1245–54.

Rao, H., Wang, J., Tang, K., et al. (2007a) Imaging brain activity during natural vision using CASL perfusion fMRI. *Hum Brain Mapp*, **28**, 593–601.

Roberts, D. A., Detre, J. A., Bolinger, L., et al. (1994) Quantitative magnetic resonance imaging of human brain perfusion at 1.5 T using steady-state inversion of arterial water. *Proc Natl Acad Sci U S A*, **91**, 33–7.

Schoning, M., and Hartig, B. (1996) Age dependence of total cerebral blood flow volume from childhood to adulthood. *J Cereb Blood Flow Metab*, **16**, 827–33.

Schoning, M., Staab, M., Walter, J. et al. (1993) Transcranial color duplex sonography in childhood and adolescence. Age dependence of flow velocities and waveform parameters. *Stroke*, **24**, 1305–9.

Schultz, R. T., Gauthier, I., Klin, A., et al. (2000) Abnormal ventral temporal cortical activity during face discrimination among individuals with autism and Asperger syndrome. *Arch Gen Psychiatry*, **57**(4), 331–40.

Siewert, B., Schlaug, G., Edelman, R. R. et al. (1997) Comparison of EPISTAR and T2*-weighted gadolinium-enhanced perfusion imaging in patients with acute cerebral ischemia. *Neurology*, **48**, 673–9.

St Lawrence, K. S. and Wang, J. (2005) Effects of the apparent transverse relaxation time on cerebral blood flow measurements obtained by arterial spin labeling. *Magn Reson Med*, **53**, 425–33.

Strouse, J. J., Cox, C. S., Melhem, E. R., et al. (2006) Inverse correlation between cerebral blood flow measured by continuous arterial spin-labeling (CASL) MRI and neurocognitive function in children with sickle cell anemia (SCA). *Blood*, **108**, 379–81.

Takahashi, T., Shirane, R., Sato, S., et al. (1999) Developmental changes of cerebral blood flow and oxygen metabolism in children. *Am J Neuroradiol*, **20**, 917–22.

Utting, J. F., Thomas, D. L., Gadian, D. G., et al. (2003) Velocity-driven adiabatic fast passage for arterial spin labeling: results from a computer model. *Magn Reson Med*, **49**, 398–401.

Wang, J. and Licht, D. J. (2006) Pediatric perfusion MRI with arterial spin labeling. *Neuroimaging Clin N Am*, **16**, 149–67.

Wang, J., Aguirre, G. K., Kimberg, D. Y., et al. (2003a) Arterial spin labeling perfusion fMRI with very low task frequency. *Magn Reson Med*, **49**, 796–802.

Wang, J., Alsop, D. C., Li, L., et al. (2002) Comparison of quantitative perfusion imaging using arterial spin labeling at 1.5 and 4.0 Tesla. *Magn Reson Med*, **48**, 242–54.

Wang, J., Alsop, D. C., Song, H. K., et al. (2003b) Arterial transit time imaging with flow encoding arterial spin tagging (FEAST). *Magn Reson Med*, **50**, 599–607.

Wang, J., Licht, D. J., Jahng, G. H., et al. (2003c) Pediatric perfusion imaging using pulsed arterial spin labeling. *J Magn Reson Imaging*, **18**, 404–13.

Wang, J., Licht, D. J., Silverstre, D. W., et al. (2006) Why perfusion in neonates with congenital heart defects is negative – technical issues related to pulsed arterial spin labeling. *Magn Reson Imaging*, **24**, 249–254.

Wang, J., Rao, H., Wetmore, G. S., et al. (2005a) Perfusion functional MRI reveals cerebral blood flow pattern under psychological stress. *Proc Natl Acad Sci U S A*, **102**, 17804–9.

Wang, J., Zhang, Y., Wolf, R. L., et al. (2005b) Amplitude modulated continuous arterial spin labeling perfusion MR with single coil at 3T-feasibility study. *Radiology*, **235**, 218–28.

Wang, W., Enos, L., Gallagher, D., et al. (2001) Neuropsychologic performance in school-aged children with sickle cell disease: a report from the Cooperative Study of Sickle Cell Disease. *J Pediatr*, **139**, 391–7.

Wang, Z., Wang, J., Connick, T. J., et al. (2005c) Continuous ASL perfusion MRI with an array coil and parallel imaging at 3T. *Magn Reson Med*, **54**, 732–7.

Webb, S. J., Monk, C. S., and Nelson, C. A. (2001) Mechanisms of postnatal neurobiological development: implications for human development. *Dev Neuropsychol*, **19**, 147–71.

Williams, J. H., Whiten, A., Suddendorf, T., and Perrett, D. I. (2001) Imitation, mirror neurons and autism. *Neurosci Biobehav Rev*, **25**(4), 287–95.

Wintermark, M., Lepori, D., Cotting, J., et al. (2004) Brain perfusion in children: evolution with age assessed by quantitative perfusion computed tomography. *Pediatrics*, **113**, 1642–52.

Wong, E. C. (1999) Potential and pitfalls of arterial spin labeling based perfusion imaging techniques for MRI. Moonen, C. T. W., Bandettini, P. A., (eds.) In: *Functional MRI*. Heidelberg: Springer-Verlag, pp. 63–9.

Wong, E. C., Buxton, R. B., and Frank, L. R. (1998) Quantitative imaging of perfusion using a single subtraction (QUIPSS and QUIPSS II). *Magn Reson Med*, **39**, 702–8.

Wu, W. C., Fernandez-Seara, M., Wong, E. C., et al. (2007) A theoretical and experimental investigation of the tagging efficiency of pseudocontinuous arterial spin labeling. *Proc Intl Soc Magn Reson Med*, **15**, 375.

Yang, Y., Engelien, W., Xu, S., et al. (2000) Transit time, trailing time, and cerebral blood flow during brain activation: measurement using multislice, pulsed spin-labeling perfusion imaging. *Magn Reson Med*, **44**, 680–5.

Ye, F. Q., Berman, K. F., Ellmore, T., et al. (2000) H(2)(15)O PET validation of steady-state arterial spin tagging cerebral blood flow measurements in humans. *Magn Reson Med*, **44**, 450–6.

Ye, F. Q., Mattay, V. S., Jezzard, P., et al. (1997) Correction for vascular artifacts in cerebral blood flow values measured by using arterial spin tagging techniques. *Magn Reson Med*, **37**, 226–35.

Yongbi, M. N., Fera, F., Yang, Y., et al. (2002) Pulsed arterial spin labeling: comparison of multisection baseline and functional MR imaging perfusion signal at 1.5 and 3.0T: initial results in six subjects. *Radiology*, **222**, 569–75.

Zaharchuk, G., Ledden, P. J., Kwong, K. K., et al. (1999) Multislice perfusion and perfusion territory imaging in humans with separate label and image coils. *Magn Reson Med*, **41**, 1093–8.

21

Neuroimaging of treatment effects in developmental neuropsychiatric disorders

Steven R. Pliszka and David C. Glahn

Introduction

As has been reviewed in Section 2 of this volume, neuroimaging studies are beginning to delineate those brain systems involved in a wide range of developmental psychopathology. Regions involved in a functional domain (attention, impulse control, mood regulation, etc.) are found to be altered in children with a psychiatric disorder relative to a control group. From a treatment perspective, the question is whether such differences constitute a "biomarker," a biological characteristic that is objectively measured and serves as an indicator of disease processes or responses to a therapeutic intervention (Frank and Hargreaves, 2003; Wise and Tracey, 2006) (see Table 21.1). Importantly, both pharmacological and psychosocial interventions can alter biomarkers.

Biomarkers need not relate directly to etiology of the disease. Take for example the clinical use of a chest radiograph. A patient with pneumonia presents with fever, shortness of breath, and infiltrates on the radiograph. The pneumonia might be viral, bacterial or due to a toxic substance. The infiltrate represents the location and seriousness of the compromised lung function independent of etiology. Treatment of the underlying condition will lead to resolution of the infiltrate, thus making the radiograph valuable in monitoring treatment effects.

It is likely that the first clinical uses of neuroimaging in pediatric mental disorders will parallel this example in which the utility is tied to monitoring treatment effects. Pediatric neuroimaging studies implicate many of the same brain regions in multiple disorders. Frontal and anterior cingulate areas are involved in both attention deficit/hyperactivity (ADHD) as well as in affective disorders (Castellanos and Tannock, 2002; Mayberg, 2003). Striatal areas are likely to be altered in both ADHD and tic disorders (Peterson, 2001; Peterson et al., 1999; Semrud-Clikeman et al., 2006). Furthermore, children show functional impairments of attention, impulse control, and mood regulation across a variety of disorders. This makes it unlikely that a neuroimaging signature for a specific disorder (i.e., ADHD vs. bipolar) (type 0 biomarker) will be discovered soon. Nevertheless, many effective treatments are available for ADHD, affective disorders, and psychotic disorders. The effects of these treatments can be studied presently using neuroimaging approaches, without knowing the specific etiology of the disorders. The neuroimaging markers might predict response to a given treatment (type 1 biomarker), document improvement or long-term effects on brain function, or uncover adverse events not observable by clinical interview or behaviorally (type 2 biomarker).

Neuroimaging in Developmental Clinical Neuroscience, eds. Judith M. Rumsey and Monique Ernst. Published by Cambridge University Press. © Cambridge University Press 2009.

Table 21.1 Key definitions in clinical trials research

Biomarker	A characteristic that is objectively measured and evaluated as an indicator of normal biological processes, pathogenic processes, or pharmacological responses to a therapeutic regimen
Type 0	Markers of natural history of disease that correlate longitudinally with symptoms and long-term outcome
Type 1	Markers that capture the effects of an intervention in accordance with mechanism of the intervention (psychological or pharmacological). Marker need not be related to disease process
Type 2 (surrogate endpoint)	Change in marker predicts clinical benefits. Reliable and valid to the point that it can be used in place of subjective clinical endpoint
Clinical endpoint	A measure of improvement of symptoms in a clinical trial. Currently assessed by subjective clinical judgment or rating scales in psychiatric studies
Phase I trial	A proof of concept study. A treatment is administered to a small number of healthy controls or patients to determine basic safety, pharmacokinetics, and "signal" of treatment effect
Phase II trial	A small double-blind placebo-controlled trial of the treatment in patients used to confirm presence of treatment effect. Effect size determined to find sample size for Phase III study
Phase III (pivotal) trial	Large-scale double-blind, placebo-controlled trials. FDA uses data to determine approval for clinical use
Phase IV (post marketing)	Once approved and in general clinical use, a treatment is further studied for safety and/or use in special populations (children, elderly, those with comorbid conditions)

Methodological issues

Neuroimaging investigation in psychiatric disorders has employed a wide range of neuroimaging techniques: positron emission tomography (PET), single photon emission computed tomography (SPECT), structural and functional magnetic resonance imaging (MRI, fMRI), magnetic resonance spectroscopy (MRS), and advanced electroencephalographic (EEG) techniques such as event-related potentials (ERP). PET and SPECT studies have been performed mainly in adults, due to ethical concerns regarding the exposure of children (particularly healthy controls) to even very low levels of radiation. PET also requires insertion of intravenous and, at times, arterial lines, to which children often will not assent. As a result, fMRI, MRS, and EEG techniques have been used to study treatment effects of psychotropic medication in children, but PET studies in adults have also contributed to the understanding of the effects of agents used in children.

One of the principal challenges in neuroimaging treatment studies is to match the study design to the timecourse of the drug or psychotherapy treatment. The Food and Drug Administration (FDA) classifies clinical trials into four phases, and this terminology is widely used outside the agency (Table 21.1). Three major study designs can be considered: (1) obtaining the imaging studies both pre- and post-treatment in a group of patients suffering from the disorder (this corresponds to a Phase II trial); (2) a crossover design in which the same subject undergoes two imaging studies, once on placebo and again on an active agent (ideally these should be done in the same session, but this is rarely possible); and (3) a parallel groups design in which subjects are scanned at baseline, then randomized in a double-blind fashion to either active drug or placebo (or psychotherapy vs. wait list or attention control) and then scanned at follow-up. This last design is the most rigorous, corresponding to a Phase III pivotal trial. Each of these designs has advantages or feasibility issues depending on the therapeutic agent studied.

Neuroimaging a group of patients before and after treatment has the advantages of both feasibility and simplicity. Patients suffering from a disorder are often eager to contribute to the understanding of their condition, and their treatment is not delayed or altered from the usual standard of care. Such a design cannot control for either placebo effects or the effects of time or practice on the task. The latter issue is sometimes addressed by having a control group undergo the imaging studies twice as well (with no treatment). This also yields data regarding the reliability of the neuroimaging method. If the sample of patients is large enough, differences on the neuroimaging measures between responders and non-responders to the treatment can be explored.

Crossover (repeated measure) designs with a placebo control are useful for studying acute effects of a single dose of drug. Such designs are ideal for the study of the effect of stimulant drugs using PET, fMRI, and MRS, as the therapeutic effects of these agents have a timecourse that very much parallels their pharmacokinetic properties. It is feasible (although difficult) to have children with ADHD perform an imaging task one day on placebo and another day on a stimulant (Pliszka et al., 2004; Vaidya et al., 1998). Given the long safety record of stimulants and low risk of a single dose of stimulant, some studies have administered the medication to healthy controls (Vaidya et al., 1998). Such studies provide insight as to whether a treatment has the same effect on the brain in both patients and controls or whether there are unique effects in the patient group. Unfortunately, stimulants are the only psychotropic treatment whose pharmacokinetic properties permit this design. Antidepressants, antipsychotics, and mood stabilizers require several weeks to achieve their therapeutic effects. While the effects of a single dose of these agents on brain function would be of interest, such effects are unlikely to be relevant to their mechanism of therapeutic action. Such agents also require too long of a washout to make a crossover design feasible.

For agents requiring several weeks of administration for therapeutic effect, only a parallel groups design is feasible. An alternative design is to randomize patients to different treatment modalities (medication vs. psychotherapy). Neuroimaging is obtained at baseline and at the end of the study. In some instances, imaging at other time points is added to distinguish the physiological effects of the drug from those resulting from downstream clinical improvements. Ideally, a matched control group of healthy controls is scanned multiple times as well. This design has the advantage of controlling both for expectations and time, and responders and non-responders to both active treatment and placebo can be studied (Mayberg et al., 2002). These designs raise feasibility issues – patients may object to being on placebo; and subject drop-out can reduce power since a between-groups analysis is required. Inter-subject variability on the neuroimaging variables may outweigh the small, but possibly critical, change in brain function related to the treatment effect. Nonetheless such studies will be key in the development of a type 2 biomarker, one that can stand in for the usual clinical endpoint (Table 21.1).

At present, no type 0 or 1 biomarker has been validated for any neuropsychiatric disorder of childhood, so naturally no type 2 biomarker is available. All neuroimaging studies of the treatment of children and adolescents with developmental neuropsychiatric disorders have been done in patients with ADHD, bipolar disorder, and obsessive-compulsive disorder (OCD). Where possible, it will be determined which findings suggest an avenue for development of a type 1 biomarker for treatment of the conditions.

Effects of placebo and cognitive therapy on brain function

Neuroimaging biomarkers may be sensitive to psychological interventions, including placebo. Placebo analgesia to pain can be blocked by administration of opiate antagonists (Levine et al., 1978);

fMRI studies have shown that administration of a placebo during a painful stimulus leads to μ-opiate receptor activation in several areas of the brain key to pain perception and response including the dorsolateral prefrontal cortex (DLPFC), anterior cingulate, nucleus accumbens, and insula (Zubieta et al., 2005). Given the high placebo response rate (50%–66%) in clinical trials of antidepressants in children and adolescents (Hughes et al., 2007), the biological mechanism of placebo response in affective disorder will be of great interest for neuroimaging treatment studies of depression in this population.

Mayberg (2003) reviewed PET studies examining brain glucose metabolism in depressed adults and healthy controls, who were studied both in a baseline neutral state and during induced sadness. She proposed a model of both normal sadness and severe depression as states characterized by prefrontal hypometabolism and limbic hypermetabolism (particularly in the insula and subgenual anterior cingulate gyrus). Mayberg (2003) further proposed that antidepressant treatment reverses this pattern and that pretreatment hypermetabolism in the rostral anterior cingulate predicted a positive response to antidepressants. Mayberg and colleagues (2002) specifically examined the placebo effect by conducting PET in eight subjects (four on fluoxetine, four on placebo), all of whom were treatment responders after a 6-week double-blind placebo-controlled medication trial. Changes in glucose metabolism that were common to both placebo and fluoxetine included increases in prefrontal (Brodmann's area, BA 46), anterior cingulate (BA 24b), inferior parietal (BA 40), posterior cingulate (BA 31/23), and posterior insula, as well as decreases in subgenual cingulate (BA 25), thalamus, and parahippocampus. Changes that were unique to fluoxetine included decreases in glucose metabolism in caudate, putamen, hippocampus, and anterior insula. All of the metabolism changes were more robust with fluoxetine.

It would be reasonable to hypothesize that psychotherapy would have effects similar to placebo and that the pattern of cortical increases and subcortical decreases of brain metabolism would represent a final common pathway of depression treatment, but this is not the case. Cognitive therapy has been associated with a quite different response pattern, which is nearly the reverse of that seen with drug or placebo; that is, cognitive therapy has been associated with dorsolateral and medial frontal decreases and hippocampal increases (Benedetti et al., 2005). In children, neuroimaging of responses to cognitive therapy has been conducted in patients with OCD (Gilbert et al., 2000; Rosenberg et al., 2000b). Despite very dissimilar designs and methodologies (MRS in children, PET in adults) but similar improvements in symptoms, differential effects of medication and psychotherapy on several brain regions were noted.

Treatment responses in developmental neuropsychiatric disorders

Stimulant and non-stimulant effects in attention deficit/hyperactivity disorder

Positron emission tomography studies

Positron emission tomography has been used to examine elements of the dopaminergic system in ADHD and its response to medications, as well as to measure regional cerebral glucose utilization and regional cerebral blood flow.

Shown in Figure 21.1a are the dopaminergic and norepinephrinergic projections from subcortical nuclei to the cortex thought to be involved in the effects of stimulants and non-stimulants, such as atomoxetine, used to treat ADHD. Both dopamine neurons in the ventral tegmental area and substantia nigra and norepinephrine neurons in the locus ceruleus send projections to the frontal cortex, where the moderate release of these neurotransmitters enhances executive function, while very low or very high release impairs such functions (Arnsten, 2006). Norepinephrinergic axons project throughout the entire cortex, including the prefrontal cortex, which is of particular importance for ADHD theories. Prefrontal norepinephrine

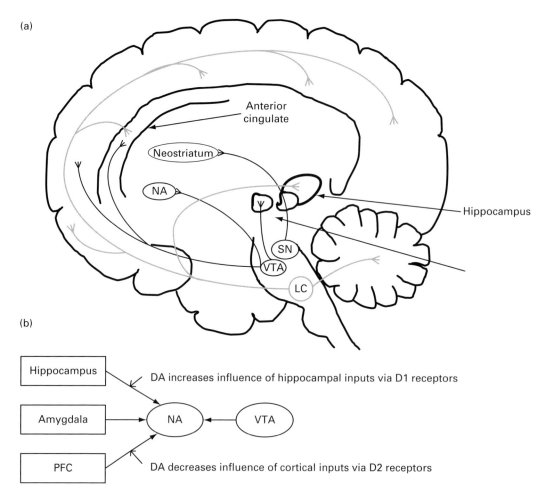

Figure 21.1. Integrating hypotheses regarding catecholamine function in attention deficit/hyperactivity disorder (ADHD). NA, nucleus accumbens; PFC, prefrontal cortex; DA, dopamine (black axons). (a) Noradrenergic axons project widely throughout the PFC; DA neurons project to PFC, NA and neostriatum. Both neurotransmitters modulate executive function, while DA plays a major role in response to reward via the NA circuitry. (b) The NA integrates input from hippocampus, amygdala, and PFC. Dopamine release into NA shifts the dominance of its input from cortical to limbic afferents (Grace et al., 2007).

release has been shown to enhance information processing (Pliszka, 2005). Dopamine axon projections are more limited and include three major contingents: mesolimbic (striatum and amygdala), mesocortical (prefrontal cortex), and tuberoinfundibular. In the neostriatum (e.g., caudate nucleus), dopamine plays a role in modulating motor responses, while in the nucleus accumbens it is involved in motivational processes, including the processing of information related to the presence of reward and punishment (Grace, 2001).

Grace et al. (2007) have reviewed the effects of dopamine on processing by nucleus accumbens of information from both cortical and limbic areas. As summarized in Figure 21.1b, by acting through D2 receptors, dopamine decreases input from the prefrontal cortex, and D1 stimulation enhances hippocampal input. To understand the functional implications of this, one must understand that dopamine exists in both a phasic and a tonic pool, as shown in Figure 21.2a. Tonic dopamine exists in low concentration around the terminals of neurons, activating autoreceptors which actually dampen dopamine release when an action potential arrives (Grace et al., 2007). It is the tonic dopamine, acting through D2 autoreceptors, that suppresses cortical input. In contrast, a larger phasic pool is released upon neuronal stimulation; this dopamine stimulates postsynaptic receptors and is cleared by the dopamine transporter (which in turn can be labeled in SPECT and PET studies). Phasic dopamine potentiates limbic input via the D1 receptors. Grace et al. (2007) suggests that the cortical and limbic inputs thus compete for control of information and that stimulant medications can affect this balance.

As shown in Figure 21.2b, stimulants block the reuptake of both dopamine and norepinephrine by binding to the transporters. The binding rate of stimulants, particularly methylphenidate (MPH), has been estimated using PET and SPECT, which provides a measure of transporter concentration. Studies have been inconsistent as to whether adults with ADHD have increased or decreased dopamine transporter (DAT) concentration relative to controls, indicating that the number of DAT has not been established as a type 0 biomarker for ADHD (Volkow et al., 2007a). Raclopride is a molecule that binds to postsynaptic D2 receptors. While giving a measure of D2 receptor concentration, this radioligand has been used to also assay dopamine release using displacement paradigms. For example, the administration of stimulant enhances (either through excess release or blockade of dopamine transporters) synaptic dopamine, which binds to D2 receptors and displaces the bound radioligand raclopride, reducing the measure of D2 receptors. This reduction in raclopride binding indexes the amount of dopamine released by stimulants. Such binding can be used as a measure of dopamine release by assessing raclopride binding at baseline,

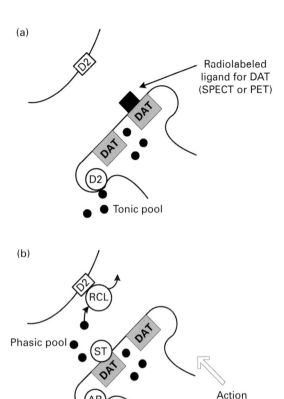

Figure 21.2. Dopamine neurotransmission and stimulant effects. (a) Dopamine transmission is segregated into two dissociable compartments: (1) an intrasynaptic fast-acting, spatially restricted "phasic" compartment and (2) a slowly changing, extrasynaptic "tonic" compartment. Because DAT are primarily intrasynaptic, only dopamine transmission is heavily influenced by DAT. When a radioligand such as raclopride (RCL) is introduced, it binds to D2 receptors, thus yielding a measure of D2 receptor concentration. (b) When stimulants (ST) are introduced, the DAT is blocked and the increased DA in the cleft displaces RCL from the D2 receptor. This reduction in bound raclopride thus indexes the amount of dopamine released by stimulants.

and then administering an intervention. If the intervention causes a release of dopamine, the newly released dopamine will push the raclopride off the receptor, leading to a decline in binding.

When healthy control adults were administered 60 mg MPH, there was a decline in raclopride binding relative to placebo, indicating that MPH had increased dopamine release (Volkow et al., 2001). This effect of MPH on synaptic dopamine was more pronounced when subjects were performing a mathematics task than during a less attention-demanding task (Volkow et al., 2004). Rosa-Neito et al. (2005) performed a PET raclopride study with nine adolescents with ADHD. The subjects were scanned both on placebo and again on a therapeutic dose of MPH. Methylphenidate was associated with a decrease in raclopride binding (indicating increased dopamine release) relative to placebo and the magnitude of this decrease correlated with MPH-induced improvements on cognitive testing. This adolescent study had no control group, but Volkow et al. (2007b) were able to use a similar methodology in adults with ADHD and a matched control group. Volkow et al. (2007b) found that adults with ADHD showed a blunted DA response in the caudate, hippocampus and amygdala in response to intravenous MPH relative to controls. ADHD subjects rated themselves as liking the drug more, and these ratings correlated with both the blunted MPH-induced dopamine release and ratings of inattention. Interestingly, Volkow et al. (2007b) also found that, during placebo administration, subjects with ADHD had lower D2/D3 receptor availability rather than the up-regulation one might expect if subjects with ADHD had chronically lower levels of dopamine release.

These studies indicate that the ultimate effect of stimulants on catecholamine release is more complex than the induction of a simple increase of neurotransmitter in the synaptic cleft. Rather, stimulants affect the balance between levels of extrasynaptic dopamine (tonic dopamine) and levels of intrasynaptic dopamine (phasic dopamine), which determines the end-point behavioral effect of dopamine (Grace, 2001). This balance is determined by a number of factors including the influence of other critical brain regions. For example, tonic dopamine, which exerts an inhibitory effect on phasic dopamine, is regulated by afferents from prefrontal and amygdalar regions (Arnsten, 2006).

Positron emission tomography has also been used to measure regional cerebral glucose utilization and cerebral blood flow (CBF) as a measure of neuronal activity. The effect of MPH/dextroamphetamine on glucose utilization has been examined in adults with ADHD, but this measure has not been shown to be sensitive to stimulant effects, even when stimulant treatment induces robust improvements in attentional performance (Ernst et al., 1994; Matochik et al., 1993, 1994). Mehta et al. (2000) administered MPH or placebo to healthy adults while PET was obtained during a spatial memory task; MPH improved task performance and decreased CBF in dorsolateral frontal cortex, the temporal poles and visual cortex. Increases in CBF were seen only in the right cerebellum. Resting state PET ($H_2^{15}O$) to measure CBF was obtained in ten medication-free adult men with ADHD who then underwent an open trial of MPH (Schweitzer et al., 2003). Resting state PET was repeated after 3 weeks of treatment and led to increases (relative to baseline) in cerebellar blood flow but decreases to the precentral gyrus and caudate. The same sample was later studied on and off MPH with PET, this time while they performed an executive function task (Paced Auditory Serial Addition Task) (Schweitzer et al., 2004). Methylphenidate improved task performance and decreased blood flow in the right prefrontal cortex (BA 9/10), while increases were noted in the right thalamus and precentral gyrus.

Single photon emission computed tomography can also be used to assess CBF as a proxy for neuronal activity, but only one such study (Kim et al., 2001) has examined the effects of MPH on blood flow in a rigorously controlled fashion. Significant increases in CBF after MPH treatment were found in bilateral prefrontal cortex, caudate, and thalamus of ADHD subjects who were MPH responders. The lack of similar data in healthy controls means that this measure lacks validity as a type 0 biomarker, and insufficient data exist to validate SPECT as a type 1 biomarker, despite its commercial availability

(Castellanos, 2002). Most recent treatment research in ADHD has utilized MRI and ERP or EEG methods. These studies are summarized in Table 21.2.

Anatomic magnetic resonance imaging

The long-term effect of stimulant treatment of ADHD on brain development is of great public interest. In a large case–control study conducted at the National Institute of Mental Health (Castellanos et al., 2002), anatomic MRI was obtained on 152 children with ADHD and 139 matched controls. At both baseline and follow-up, children with ADHD had decreased cerebral and cerebellum volumes (both gray and white matter) relative to controls, but these reductions in brain volume were most pronounced for those children with ADHD who had never been treated with medication. Recently, the anterior cingulate volume was found to be reduced in treatment-naive children with ADHD, compared to healthy controls and children with ADHD who had received chronic stimulant treatment (Semrud-Clikeman et al., 2006). In contrast, caudate volume was reduced in children with ADHD regardless of treatment history. These two studies suggest that long-term stimulant treatment of ADHD has a normalizing effect on brain growth in some areas but not others, though the mechanisms of such changes is unknown.

Functional magnetic resonance imaging

While limited in number, fMRI studies of medication effects in children with ADHD are emerging. No-go trials on a go/no-go task increase caudate activation relative to go trials; this difference is enhanced by MPH in children with ADHD (Newcorn et al., 2006; Vaidya et al., 1998), but decreased by MPH in controls (Vaidya et al., 1998). Both atomoxetine, a specific norepinephrine reuptake inhibitor, and MPH increase activation of the anterior cingulate and ventrolateral prefrontal cortex (Newcorn et al., 2006; Pliszka et al., 2004), while MPH appears unique in producing more caudate activation than atomoxetine, perhaps due to MPH's greater effect

on dopamine. In contrast, atomoxetine has been shown to increase activity on no-go trials relative to go trials in parietal, temporal, and posterior cingulate cortex, where norepinephrinergic are denser than dopaminergic projections (Newcorn et al., 2006). As noted in Table 21.2, the samples in these studies are limited. Shafritz et al. (2004) performed a double-blind, placebo-controlled crossover study of MPH in 15 adolescents with ADHD. Functional MRI was obtained in both sessions while subjects performed a divided attention task. There was no effect of MPH on task performance and minimal effects on brain activation. Only the left caudate showed a stronger activation on MPH relative to placebo.

Pliszka et al. (2004) recruited 26 children aged 8–15 years of age to participate in a 5-week double-blind placebo-controlled crossover study of MPH to determine the optimal dose of MPH for reducing symptoms of ADHD. Subjects then underwent fMRI twice, once on placebo and once on their optimal dose of MPH, with order of medication counterbalanced. During fMRI, children performed the stop-signal task, a measure of inhibitory control. Brain-oxygen-level-dependent (BOLD) signal during inhibitory trials (both successful and unsuccessful inhibitions) was contrasted with signal during go trials.

This study illustrates several of the challenges related to fMRI treatment research in children. Seventeen subjects completed the double-blind trial, but only 13 agreed to continue with the two follow-up fMRI scans. Of these, only eight subjects had both a placebo and MPH scan that was suitable for analysis due to motion artifact or technical problems with the scanner. Relative to placebo, MPH increased activation of the anterior cingulate cortex (ACC) and right dorsolateral prefrontal cortex (DLPFC) during unsuccessful inhibitions, while it decreased activation in the bilateral ventrolateral prefrontal cortex (VLPFC) during successful inhibitions (see Figure 21.3). These results suggested that MPH may improve cognitive control in ADHD by enhancing the function of error processing networks in the ACC and VLPFC, as well as

Table 21.2 SPECT, fMRI and aMRI studies of treatment effects in ADHD

Study	Method	Study design paradigm	Sample	Findings
Kim et al. (2001)	SPECT ^{99}Tc HMPAO	Pre-post design, scanned at baseline and after an 8-week open trial of MPH	32 boys age 7–14 years with ADHD, no LD or comorbid disorder	Significant increases in cerebral blow flow after MPH were found in bilateral prefrontal cortex, caudate, and thalamus of subjects who were MPH responders
Castellanos et al. (2002)	aMRI	Longitudinal follow-up. Subjects scanned up to 4 times each over 10 years. Subjects mean age ~9.0 years at first scan	152 children with ADHD and 139 matched controls. 49 ADHD subjects had never been treated with medication	At both baseline and follow-up, children with ADHD had decreased cerebral and cerebellum volumes relative to controls; reductions most pronounced for those children with ADHD who had never been treated
Semrud-Clikeman et al. (2006)	aMRI	Comparison of children with and without a history of stimulant treatment to controls	30 children with ADHD (16 treated long-term with stimulants, 14 treatment naive) and 21 controls, aged 9–15 years	ACC volume was found to be reduced in treatment-naive children with ADHD, compared to healthy controls and children with ADHD who had received chronic stimulant treatment. Caudate volume was reduced in children with ADHD regardless of treatment history
Vaidya et al. (1998)	fMRI	DBPC crossover trial of MPH in both children with ADHD and controls, fMRI performed twice (placebo and MPH) using the go/no-go task	10 children with ADHD, 10 healthy controls, mean age 9.0 years	During fast-paced version of CPT: MPH increased striatal activity in children with ADHD but decreased it in controls. MPH increased frontal lobe activity relative to placebo in both groups. During slow-paced CPT: No effect of MPH vs. placebo in either group

352

Pliszka et al. (2004)	fMRI	DBPC crossover trial of MPH in children with ADHD, fMRI with the stop-signal task obtained twice (placebo & MPH)	8 children with ADHD, aged 9–15 years, mean age 12.0	Relative to placebo, MPH increased activation of the ACC and right DLPFC during unsuccessful inhibitions, while it decreased activation in the bilateral VLPFC during successful inhibitions
Shafritz et al. (2004)	fMRI	DBPC crossover trial of MPH	15 adolescents with ADHD, mean age 15 years	Relative to placebo, MPH activated left caudate
Newcorn et al. (2006)	fMRI	Pre-post design. Both open and controlled trials of atomoxetine and MPH, fMRI obtained at baseline and on medication. Subjects performed the go/no-go task	16 children with ADHD treated with either MPH ($n = 8$) or atomoxetine ($n = 8$) for 6 weeks	No-go trials increased brain activity in several regions relative to go trials; both atomoxetine and MPH increased this difference relative to baseline in the VLPFC and ACC. MPH produced greater activations than atomoxetine in the caudate nuclei, medial frontal areas and right motor and inferior temporal cortex. Atomoxetine produced more activation than MPH in bilateral parietal cortex, insula, posterior cingulate gyrus, and superior temporal gyri
Epstein et al. (2007)	fMRI	DBPC crossover trial of MPH, subjects performed go/no-go task	15 parent-child dyads, both with ADHD	Youths showed increases in brain activation on MPH in ACC, PFC, caudate, cerebellum. Adults showed decreased activation in PFC and parietal areas

Notes:

ACC, anterior cingulate cortex; ADHD, attention deficit/hyperactivity disorder; aMRI, anatomical magnetic resonance imaging; CPT, Continuous Performance Tasks; DBPC, double-blind, placebo-controlled; DLPFC, dorsolateral prefrontal cortex; fMRI, functional magnetic resonance imaging; LD, learning disorder; MPH, methylphenidate; PFC, prefrontal cortex; SPECT, single photon emission computed tomography; VLPFC, ventrolateral prefrontal cortex.

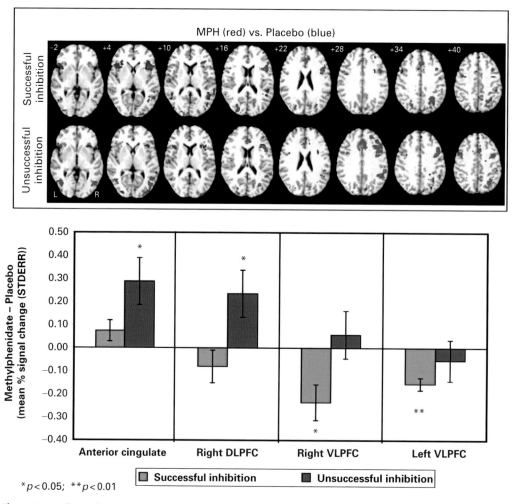

Figure 21.3. Effects of methylphenidate (MPH) versus placebo in children with attention deficit/hyperactivity disorder (ADHD), detected with functional magnetic resonance imaging (fMRI). Upper panel: regions of significantly greater activity during stop trials of the stop-signal task versus go trials on MPH (red) and placebo (blue). Lower panel: difference in activity between MPH and placebo conditions in four regions of interest as a function of trial type (successful versus unsuccessful inhibitions).

by improving the action of right prefrontal mechanisms that support inhibitory control.

Most recently, Epstein et al. (2007) obtained fMRI on parent-child dyads recruited from the study population of the Multimodal Treatment of ADHD (MTA) study (Jensen et al., 2007). Subjects were 7–9 years of age when they entered the study and were older adolescents (mean age 17 years) at the time of

fMRI study. Fifteen parent-child dyads in which the child's parent also had ADHD were scanned while performing a go/no-go task on placebo and MPH in a double-blind fashion. In both age groups, MPH improved task performance. In the youths, MPH activated the anterior cingulate, PFC, caudate, inferior parietal lobule, and cerebellum; there were no areas where MPH decreased activity. In contrast, the parents with ADHD showed activation with MPH only in the cerebellum, caudate, and hippocampus, while MPH decreased activation in PFC and the inferior parietal lobule.

Event-related potentials

In ERP, the subject performs a repetitive cognitive task and an EEG is obtained during each trial. The EEG is then averaged over many trials, canceling out random brain activity and producing a waveform that represents the brain's response to each class of stimuli in the task. In studies of ADHD, oddball auditory ERP tasks and inhibitory tasks such as the Continuous Performance Task (CPT) have been most utilized (Barry et al., 2003). In the auditory oddball task, the subject must detect rare ("oddball") tones among a long string of common tones. Healthy controls produce a larger P300 wave to the oddball tones than to the common tones, but this difference is markedly reduced in children with ADHD (Barry et al., 2003). The meaning of this difference in the P300 is debated. The P300, which in these tasks is most prominent over the parietal areas, possibly reflects activity related to evaluation of the stimuli, but may also represent the amount of mental effort that is invested in the task (Kok, 1997). In most treatment studies using oddball tasks with ERP, MPH enhances the P300 response to the rare stimuli (Jonkman et al., 2000; Klorman, 1991).

On the CPT, the child must respond to target stimuli and avoid responding to other stimuli. In the go/no-go task, the child responds to go stimuli the majority of the time but must refrain from responding when the no-go stimuli are presented. No-go or stop signals on an inhibitory task are associated with a right lateralized N200, which may signal the triggering of prefrontal inhibitory processes (Kok, 1986). No-go stimuli of inhibitory tasks also elicit a fronto-central P300, which is thought to be generated by the anterior cingulate (Schmajuk et al., 2006) and which differs from the parietal P300 discussed above. Two ERP studies of the effects of MPH in ADHD children showed the stimulant to enhance both the N200 and no-go P300 (Pliszka et al., 2007; Seifert et al., 2003).

Potential type 1 biomarkers

Functional MRI activity in the anterior cingulate while performing inhibitory or Stroop-like tasks may be a potential biomarker of pharmacological response to ADHD medications. Stimulants and atomoxetine appear to increase activity in this area on complex or inhibitory trials relative to simple trials. Further studies are needed to determine if these changes in the cingulate correlate with treatment response as assessed by behavior rating scales and clinician assessment of outcome. The fact that anterior cingulate volume in children with ADHD was related to treatment history (Semrud-Clikeman et al., 2006) supports the hypothesis that this area may be a mediator of treatment effects. Another key issue from a developmental perspective is that while MPH is associated with increased activation in PFC in children with ADHD, adults may show the opposite pattern. Yet, these divergent shifts with drug treatment may both be associated with improved task performance.

Mood stabilizers and antipsychotics in bipolar disorder

Magnetic resonance spectroscopy

Magnetic resonance spectroscopy has been more widely used than other imaging modalities in the study of treatment effects in bipolar disorder, in part because of its ability to assess *myo*-inositol (mI) concentration. Inositol forms part of the phosphatidylinositol second messenger system and lithium, a principal treatment for bipolar disorder, interacts with this system (Silverstone et al., 2005). Thus it is hoped that direct effects of lithium can be assessed with MRS. N-Acetylaspartate (NAA) concentrations can also be assessed; this compound is viewed as a marker of neuronal integrity; choline (Cho) and creatine (Cr) may reflect energy processes in the brain as well as the breakdown of neuronal membranes.

Magnetic resonance spectroscopy studies of the effect of mood stabilizer treatment on mI in adults with bipolar disorder have been inconsistent. *Myo*-inositol has been shown to be decreased with lithium treatment in bipolar depressed adults in one study (Moore et al., 1999), but mI was increased in another study of all subtypes of bipolar patients treated with valproic acid (Friedman et al., 2004). In one study, the lack of difference in mI between patients with bipolar on mood stabilizers and controls was interpreted as a normalizing effect of the drugs, even though no pre- and post-medication studies were preformed in the bipolar patients (Silverstone et al., 2002). An added issue in these studies is whether the change in mood state alters the neurochemistry, as opposed to reflecting the effects of the mood stabilizer per se. Lithium treatment increased NAA in two studies of adults with bipolar disorder (Brambilla et al., 2005; Moore et al., 2000b). This effect has been less consistently seen with other mood stabilizing agents (Friedman et al., 2004).

Most treatment studies in children and adolescents with bipolar disorder have focused on MRS and fMRI. Davanzo et al. (2001) treated 11 bipolar children and adolescents (age 7–17 years) with lithium for 3 weeks; only the mI/Cr ratio showed a decrease with lithium treatment. No effect of lithium on mI concentrations in depressed bipolar adolescents was found in a study with a larger sample and a more rigorous design than any of the adult treatment studies (Patel et al., 2006). Eight weeks of lamotrigine treatment increased left DLPFC NAA/Cr and mI/Cr ratios in depressed bipolar adolescents (Gallelli et al., 2005). In contrast, these ratios were unchanged from baseline in 10 children of bipolar parents with severe mood dysregulation who underwent 12 weeks of treatment with divalproex (Chang et al., 2006b). See Table 21.3.

In the largest, most well controlled MRS treatment study to date in adolescents, 20 patients who were admitted for the first time to an inpatient psychiatric hospital for bipolar I disorder (mixed or manic type) underwent a 4-week trial of olanzapine monotherapy. Magnetic resonance spectroscopy was obtained at baseline, day 7 and day 28 (Del Bello et al., 2006). The patients had no history of prior psychotropic medication treatment. *N*-Acetylaspartate, mI, choline, creatine, and glutamate/glutamine were assessed in the ventral medial prefrontal cortex. A matched group of healthy controls scanned at the same time points showed no changes in concentrations of any metabolites over time. For the bipolar group as a whole, there were also no changes in the concentration of NAA, but patients who remitted showed a significant increase (6%) in NAA concentration, while non-responders showed a significant decrease (–4%). Changes in NAA correlated strongly ($r = 0.68$) with reductions in manic symptoms. While change in Cho did not correlate with change in symptoms, those who remitted had higher Cho concentrations at baseline. The different patterns of NAA response in remitters and non-remitters is of considerable significance, as it suggests that NAA changes are related to clinical status as opposed to a direct pharmacologic effect of the olanzapine on brain physiology.

Functional and anatomic magnetic resonance imaging

An adult study examining the effects of age and treatment history on brain volume is informative in terms of the developmental effects of medication. Blumberg et al. (2006) found an age by diagnosis by medication interaction effect on ventral prefrontal cortex volume (VPFC). Overall, VPFC volume declined with age in all subjects, but the effect was more pronounced in patients with bipolar disorder. Treatment with psychotropic medication appeared to attenuate this decline. This suggests neurotropic effects of medications for bipolar disorder, consistent with other findings in adults that have shown increases in NAA and gray matter with mood stabilizer treatment (Moore et al., 2000a; Sassi et al., 2002, 2005). Amygdalar volume was larger in children with bipolar disorder who had been on lithium or valproate relative to those who had not, although the length of time the subjects had been on a mood stabilizer was not reported (Chang et al., 2005).

Chang et al. (2006a) obtained fMRI on the same sample of depressed bipolar adolescents studied by Gallelli et al. (2005). Treatment with lamotrigine decreased amygdala activation to negative pictures from the International Affective Pictures System (IAPS), and this decrease correlated with clinicians' rating of improvement in depressed mood. In a study of children of bipolar parents (Chang et al., 2006b), increased amygdalar activation to negative pictures seen in the patients relative to controls at baseline was not present after 12 weeks of divalproex treatment. However, right DLPFC activation to negative pictures was increased in patients relative to controls post treatment.

Potential type 1 biomarkers

At present, MRI work has not progressed to the point that any biomarker of affective disorder treatment in children and adolescents using this technique can be proposed. The increase in NAA in adolescents with mania who respond to olanzapine is intriguing (Del Bello et al., 2006). This would suggest neurotropic effects of olanzapine in these patients, but it needs to be borne in mind that the responders experienced a mood shift from manic/mixed to euthymic. Longitudinal MRS studies of NAA concentrations in bipolar adolescents will be needed to distinguish changes related to medication from changes in mood related to cycling.

Selective serotonin reuptake inhibitors and cognitive behavior therapy in obsessive-compulsive disorder

Functional MRI and PET have been used to study the effects of both antidepressant and psychotherapeutic modalities in the treatment of obsessive-compulsive (OCD) and other anxiety disorders. Two PET studies (Baxter et al., 1992; Schwartz et al., 1996) have shown that both cognitive behavioral therapy and fluoxetine decrease metabolism

in the right orbitofrontal cortex and right caudate of adults with OCD. Activity in these two areas was correlated at baseline, but not after treatment. In contrast to the PET studies of depression treatment (Mayberg, 2003), there were no brain regions where treatment increased metabolism. Nakao et al. (2005) obtained fMRI in patients with OCD during phobic symptom provocation before and after antidepressant or cognitive behavioral therapy. Due to low statistical power, data from six patients treated with cognitive behavioral therapy and four treated with fluvoxamine had to be combined. Post treatment, there were decreases in activation of the orbitofrontal cortex, DLPFC, and anterior cingulate in response to threat, but the effects of psychotherapy and drug treatment cannot be disentangled.

Treatment studies of adult phobia have also shown similar effects of psychotherapy and medication, though with a slightly different pattern of brain regions involved than in OCD: decreases of activation during symptom provocation were found in paralimbic areas (Furmark et al., 2002), right DLPFC and parahippocampal areas (Paquette et al., 2003), and bilateral insula and anterior cingulate (Straube et al., 2006). Changes in orbitofrontal metabolism seen in OCD treatment were not evident in phobia treatment. In panic disorder, Prasko et al. (2004) studied the effects of treatment on adult patients by obtaining FDG-PET before and after treatment with cognitive behavioral therapy ($n = 6$) or antidepressants ($n = 6$). Again, both treatment groups showed a similar pattern of changes: decreases in metabolism in right frontal and temporal regions and increases in left frontal metabolism.

Cognitive behavioral therapy and paroxetine appear to have different effects on brain functioning in children and adolescents with OCD (see Table 21.4). Paroxetine treatment for 12 weeks reduced glutamate/glutamine caudate concentrations as well as thalamic volume to levels seen in controls (Gilbert et al., 2000; Rosenberg et al., 2000b), effects not seen in a separate sample of children treated successfully for OCD with cognitive therapy alone (Benazon et al., 2003; Rosenberg

Table 21.3 Neuroimaging studies of treatment effects in child and adolescent bipolar disorder

Study	Method	Study design paradigm	Sample	Findings
Davanzo et al. (2001)	MRS	Pre-post. Subjects with BP treated with open trial of lithium for 11 weeks, ACC assessed by MRS at baseline and after 1 week of treatment	11 children with BP; in mixed or manic state; 11 controls also assessed by MRS	NAA/Cr, mI/Cr, Glx/Cr, and Cho/Cr ratios were assessed; only the mI/Cr ratio showed a decrease with lithium treatment. None of the ratios in patients differed from controls, either before or after treatment
Patel et al. (2006)	MRS	Pre-post. Subjects treated with open trial of lithium for 6 weeks	28 adolescents with BP (aged 12–18 years) in depressed phase	No change in mI concentration post treatment with lithium
Chang et al. (2005)	aMRI	Baseline comparison of subjects with BP to healthy controls	20 children with BP euthymic at time of scan; 20 age-matched healthy controls	The children with BP had reduced amygdala volume relative to controls overall, but the amygdala volume of those patients on lithium or valproate not differ from controls
Gallelli et al. (2005)	MRS	Pre-post. Subjects had MRS at baseline and after 8 weeks of treatment with lamotrigine	11 adolescents with BP, all in depressed phase	NAA/Cr and mI/Cr ratios increased at 8 weeks over baseline
Chang et al. (2006a)	fMRI	Pre-post. Subjects viewed positive or negative valence pictures; BOLD signal in amygdala contrasted to neutral picture	As above	Decrease in clinician ratings of the child's depression post treatment correlated with a decrease in amygdala activation to negative pictures
Chang et al. (2006b)	MRS fMRI	Pre-post. Subjects with mood dysregulation underwent a 12-week open trial of divalproex for severe mood dysregulation. Healthy controls studied at baseline only	10 children of BP parents with mood dysregulation compared to 10 healthy controls at baseline. Mean age 11.4 years	MRS: no changes found in NAA/Cr or mI/Cr ratios from baseline to post treatment fMRI: controls had greater amygdalar activation when viewing negative pictures than children with mood dysregulation at baseline. Post treatment with divalproex, no difference between controls and patients in amygdala activation, but patients now showed greater right DLPFC activation to negative pictures

Del Bello et al. (2006)	MRS	Pre-post. 4-week inpatient open trial of olanzapine monotherapy, MRS obtained at baseline, days 7 and 28	20 adolescents with BP I, manic or mixed (age 12–18 years), no prior medications	NAA, mI, Cho, Cr, and Glx were assessed in the ventral medial prefrontal cortex. There were no overall changes in concentration of NAA; remitted showed a significant increase (6%) in NAA concentration, while non-responders showed a significant decrease (−4%). Changes in NAA correlated strongly ($r = 0.68$) with reductions in manic symptoms, while remitters showed higher Cho concentrations at baseline

Notes:

ACC, anterior cingulate cortex; aMRI, anatomical magnetic resonance imaging; BOLD, blood-oxygen-level-dependent; BP; bipolar disorder; Cho, choline; Cr, creatine; DLPFC, dorsolateral prefrontal cortex; Glx, glutamine/glutamate; MRS, magnetic resonance spectroscopy; mI, *myo*-inositol; NAA, *N*-acetylaspartate.

Table 21.4 Neuroimaging studies of treatment effects in child and adolescent OCD

Study	Method	Study design paradigm	Sample	Findings
Rosenberg et al. (2000b)	MRS	Pre-post. Subjects treated with a 12-week open trial of paroxetine (10–60 mg/day). Glutamate/glutamine assessed before and after treatment	11 treatment-naive children with OCD, aged 8–17 years, 11 matched controls	Glutamate/glutamine caudate concentrations were greater in children with OCD at baseline compared to healthy controls, but these declined to levels indistinguishable from controls post treatment
Benazon et al. (2003)	MRS	Pre-post. Subjects treated with 12 weeks of CBT, treatment was very effective	21 treatment-naive children with OCD aged 6–16 years	No changes in any neurochemical measure
Rosenberg et al. (2000a)	aMRI	Pre-post. Subjects treated with 12 weeks of CBT, treatment was very effective	11 treatment-naive children with OCD aged 8–17 years	No change in thalamic volume
Gilbert et al. (2000)	aMRI	Pre-post, thalamic volume assessed before and after 12 weeks of treatment with paroxetine	10 treatment-naive children with OCD, matched controls	OCD children had greater thalamic volume than controls at baseline. Thalamic volume decreased to levels similar to controls post treatment
Szeszko et al. (2004)	aMRI	Pre-post, amygdala volume assessed before and after 16 weeks of treatment with paroxetine	11 treatment-naive children with OCD, mean age 12.8, 11 matched healthy controls	At baseline, children with OCD had increased asymmetry of amygdala relative to controls. No asymmetry post treatment
Lazaro et al. (2006)	aMRI	Pre-post, assess DLPFC volume in children with OCD before and after 6 months of paroxetine treatment	15 children aged 9–17 years and 15 age-matched controls	OCD patients had less gray matter bilaterally in the DLPFC at baseline relative to controls, but after 6 months of paroxetine treatment there was a trend for gray matter volume to be increased in the patient group

Notes:

aMRI, anatomical magnetic resonance imaging; CBT, cognitive behavioral therapy; DLPFC, dorsolateral prefrontal cortex; OCD, obsessive-compulsive disorder.

et al., 2000a). As noted in Table 21.4, paroxetine has been shown to alter amygdala volume symmetry (Szeszko et al., 2004) and a trend has been found toward increasing gray matter volume in the DLPFC (Lazaro et al., 2006).

Potential type 1 biomarkers

The studies of Rosenberg and colleagues are notable for their homogeneous samples of children with OCD and the use of treatment-naive subjects, increasing their power to detect differences despite relatively small samples. In two studies, paroxetine appeared to reduce thalamic volume as well as glutamate/glutamine concentrations. Baxter (1999) have hypothesized that OCD is related to overactivity of a corticothalamic circuit involving the orbito-frontal cortex. Perhaps paroxetine treatment leads to reduction in excitatory neurotransmission in this circuit (reflected by decreased glutamate) which ultimately reduces thalamic volume, both correlates of symptomatic reduction.

Conclusions and future directions

Several major challenges in the neuroimaging of treatment effects lie ahead. The reliability of neuroimaging techniques, particularly of fMRI and MRS, must be clearly established, as well as the relationship of neuroimaging findings to symptoms. This is particularly true for MRS, where neurochemistry is assessed in multiple brain regions and many metabolites (as well as their ratios to Cr) are studied. The problem of type 1 error is serious. Clinical samples are precious, the imaging suite must be available at the time the patient is appropriately medicated, and the environment must be child friendly. Since it is rarely ethical to withdraw children from medication prior to scanning, great efforts must be made to recruit children who are treatment naive for the disorder under scrutiny. In the study of ADHD, this means the capacity to scan young children (age 5–7) as this is increasingly the age at which children present for

treatment. For mood disorders, the neuroimaging signatures of mood shifts need to be differentiated from the pharmacodynamic effects of medications or the direct effects of cognitive therapies on brain activity. Once these challenges are met, pediatric neuroimaging may allow the field to predict more accurately the most beneficial treatment for a child and to ensure that such treatment is fostering healthy brain development. Indeed, perhaps the most hopeful sign emerging from the work to date is early evidence that our current treatments (stimulants, mood stabilizers) have normalizing or neurotropic effects on brain volume and chemistry (Castellanos et al., 2002; Del Bello et al., 2006; Semrud-Clikeman et al., 2006).

REFERENCES

Arnsten, A. F. (2006) Fundamentals of attention-deficit/hyperactivity disorder: circuits and pathways. *J Clin Psychiatry*, **67** [Suppl 8], 7–12.

Barry, R. J., Johnstone, S. J., and Clarke, A. R. (2003) A review of electrophysiology in attention-deficit/hyperactivity disorder: II. Event-related potentials. *Clin Neurophysiol*, **114**, 184–98.

Baxter, L. R. (1999) Functional imaging of brain systems mediating obsessive-compulsive disorder. In: Charney, D. S., Nestler, E. J., and Bunney, B. S. (eds.) *Neurobiology of Mental Illness.* New York: Oxford University Press, pp. 534–47.

Baxter, L. R., Schwartz, J. M., Bergman, K. S., et al. (1992) Caudate glucose metabolic rate changes with both drug and behavior therapy for obsessive-compulsive disorder. *Arch Gen Psychiatry*, **49**, 681–9.

Benazon, N. R., Moore, G. J., and Rosenberg, D. R. (2003) Neurochemical analyses in pediatric obsessive-compulsive disorder in patients treated with cognitive-behavioral therapy. *J Am Acad Child Adolesc Psychiatry*, **42**, 1279–85.

Benedetti, F., Mayberg, H. S., Wager, T. D., et al. (2005) Neurobiological mechanisms of the placebo effect. *J Neurosci*, **25**, 10390–402.

Blumberg, H. P., Krystal, J. H., Bansal, R., et al. (2006) Age, rapid-cycling, and pharmacotherapy effects on ventral prefrontal cortex in bipolar disorder: a cross-sectional study. *Biol Psychiatry*, **59**, 611–18.

Brambilla, P., Stanley, J. A., Nicoletti, M. A., et al. (2005) ^{1}H magnetic resonance spectroscopy investigation of the dorsolateral prefrontal cortex in bipolar disorder patients. *J Affect Disord*, **86**, 61–7.

Castellanos, F. X. (2002) Proceed, with caution: SPECT cerebral blood flow studies of children and adolescents with attention deficit hyperactivity disorder. *J Nucl Med*, **43**, 1630–3.

Castellanos, F. X. and Tannock, R. (2002) Neuroscience of attention deficit/hyperactivity disorder: the search for endophenotypes. *Nature Rev Neurosci*, **3**, 617–28.

Castellanos, F. X., Lee, P. P., Sharp, W., et al. (2002) Developmental trajectories of brain volume abnormalities in children and adolescents with attention-deficit/hyperactivity disorder. *J Am Med Assoc*, **288**, 1740–8.

Chang, K., Karchemskiy, A., Barnea-Goraly, N., et al. (2005) Reduced amygdalar gray matter volume in familial pediatric bipolar disorder. *J Am Acad Child Adolesc Psychiatry*, **44**, 565–73.

Chang, K. D., Karchemskiy, A., Garrett, A., et al. (2006a) Effects of lamotrigine on brain chemistry and function in adolescents with bipolar disorder. Presented at the 53rd Annual Meeting of the American Academy of Child and Adolescent Psychiatry, 24–29 October 2006, San Diego, CA.

Chang, K. D., Wagner, C., Karchemskiy, A., et al. (2006b) Effects of divalproex on brain chemistry, morphometry and function in adolescents at risk for bipolar disorder. Presented at the Annual Meeting of the American College of Neuropharmacology, 24–29 December, Fort Lauderdale, FL.

Davanzo, P., Thomas, M. A., Yue, K., et al. (2001) Decreased anterior cingulate myo-inositol/creatine spectroscopy resonance with lithium treatment in children with bipolar disorder. *Neuropsychopharmacology*, **24**, 359–69.

Del Bello, M. P., Cecil, K. M., Adler, C. M., et al. (2006) Neurochemical effects of olanzapine in first-hospitalization manic adolescents: a proton magnetic resonance spectroscopy study. *Neuropsychopharmacology*, **31**, 1264–73.

Epstein, J. N., Casey, B. J., Tonev, S. T., et al. (2007) ADHD- and medication-related brain activation effects in concordantly affected parent-child dyads with ADHD. *Journal of Child Psychology and Psychiatry*, **[On-Line]** Available at: http://www.blackwell-synergy.com/doi/abs/10.1111/j.1469-7610.2007.01761.x, accessed 4 August, 2008.

Ernst, M., Zametkin, A. J., Matochik, J. A., et al. (1994) Effects of intravenous dextroamphetamine on brain metabolism in adults with attention-deficit hyperactivity disorder (ADHD). Preliminary findings. *Psychopharmacol Bull*, **30**, 219–25.

Frank, R. and Hargreaves, R. (2003) Clinical biomarkers in drug discovery and development. *Nature Rev Drug Discov*, **2**, 566–80.

Friedman, S. D., Dager, S. R., Parow, A., et al. (2004) Lithium and valproic acid treatment effects on brain chemistry in bipolar disorder. *Biol Psychiatry*, **56**, 340–8.

Furmark, T., Tillfors, M., Marteinsdottir, I., et al. (2002) Common changes in cerebral blood flow in patients with social phobia treated with citalopram or cognitive-behavioral therapy. *Arch Gen Psychiatry*, **59**, 425–33.

Gallelli, K. A., Wagner, C., Spielman, D., et al. (2005) Lamotrigine treatment and neurometabolite changes in adolescents with bipolar depression. Presented at the 53rd Annual Meeting of the American Academy of Child and Adolescent Psychiatry, 24–29 October, San Diego, CA.

Gilbert, A. R., Moore, G. J., Keshavan, M. S., et al. (2000) Decrease in thalamic volumes of pediatric patients with obsessive-compulsive disorder who are taking paroxetine. *Arch Gen Psychiatry*, **57**, 449–56.

Grace, A. A. (2001) Psychostimulant actions on dopamine and limbic system function: relevance-related behavior and impulsivity. In: Solanto, M. V., Arnsten, A. F. T., and Castellanos, F. X. (eds.) *Stimulant Drugs and ADHD: Basic and Clinical Neuroscience.* New York, NY: Oxford University Press, pp. 134–57.

Grace, A. A., Floresco, S. B., Goto, Y., et al. (2007) Regulation of firing of dopaminergic neurons and control of goal-directed behaviors. *Trends Neurosci*, **30**, 220–7.

Hughes, C. W., Emslie, G. J., Crismon, M. L., et al. (2007) Texas Children's Medication Algorithm Project: update from Texas Consensus Conference Panel on Medication Treatment of Childhood Major Depressive Disorder. *J Am Acad Child Adolesc Psychiatry*, **46**, 667–86.

Jensen, P. S., Arnold, L. E., Swanson, J. M., et al. (2007) 3-year follow-up of the NIMH MTA study. *J Am Acad Child Adolesc Psychiatry*, **46**, 989–1002.

Jonkman, L. M., Kemner, C., Verbaten, M. N., et al. (2000) Attentional capacity, a probe ERP study: differences between children with attention-deficit hyperactivity disorder and normal control children and effects of methylphenidate. *Psychophysiology*, **37**, 334–46.

Kim, B. N., Lee, J. S., Cho, S. C., et al. (2001) Methylphenidate increased regional cerebral blood flow in subjects with attention deficit/hyperactivity disorder. *Yonsei Med J*, **42**, 19–29.

Klorman, R. (1991) Cognitive event-related potentials in attention deficit disorder. *J Learn Disabil*, **24**, 130–40.

Kok, A. (1986) Effects of degradation of visual stimuli on components of the event-related potential (ERP) in go/no-go reaction tasks. *Biol Psychol*, **23**, 21–38.

Kok, A. (1997) Event-related-potential (ERP) reflections of mental resources: a review and synthesis. *Biol Psychol*, **45**, 19–56.

Lazaro, L., Bargallo, N., Castro-Fornieles, J., et al. (2006) Brain changes in children and adolescents with obsessive compulsive disorder. Presented at the 53rd Annual Meeting of the American Academy of Child and Adolescent Psychiatry, 24–29 October, San Diego, CA.

Levine, J. D., Gordon, N. C., and Fields, H. I. (1978) The mechanisms of placebo analgesia. *Lancet*, **2**, 654–7.

Matochik, J. A., Liebenauer, L. L., King, A. C., et al. (1994) Cerebral glucose metabolism in adults with attention deficit hyperactivity disorder after chronic stimulant treatment. *Am J Psychiatry*, **151**, 658–64.

Matochik, J. A., Nordahl, T. E., Gross, M., et al. (1993) Effects of acute stimulant medication on cerebral metabolism in adults with hyperactivity. *Neuropsychopharmacology*, **8**, 377–86.

Mayberg, H. S. (2003) Modulating dysfunctional limbic-cortical circuits in depression: towards development of brain-based algorithms for diagnosis and optimised treatment. *Br Med Bull*, **65**, 193–207.

Mayberg, H. S., Silva, J. A., Brannan, S. K., et al. (2002) The functional neuroanatomy of the placebo effect. *Am J Psychiatry*, **159**, 728–37.

Mehta, M. A., Owen, A. M., Sahakian, B. J., et al. (2000) Methylphenidate enhances working memory by modulating discrete frontal and parietal lobe regions in the human brain. *J Neurosci*, **20**, RC65.

Moore, G. J., Bebchuk, J. M., Hasanat, K., et al. (2000b) Lithium increases *N*-acetyl-aspartate in the human brain: in vivo evidence in support of bcl-2's neurotrophic effects? *Biol Psychiatry*, **48**, 1–8.

Moore, G. J., Bebchuk, J. M., Parrish, J. K., et al. (1999) Temporal dissociation between lithium-induced changes in frontal lobe *myo*-inositol and clinical response in manic-depressive illness. *Am J Psychiatry*, **156**, 1902–8.

Moore, G. J., Bebchuk, J. M., Wilds, I. B., et al. (2000a) Lithium-induced increase in human brain grey matter. *Lancet*, **356**, 1241–2.

Nakao, T., Nakagawa, A., Yoshiura, T., et al. (2005) Brain activation of patients with obsessive-compulsive disorder during neuropsychological and symptom provocation tasks before and after symptom improvement: a functional magnetic resonance imaging study. *Biol Psychiatry*, **57**, 901–10.

Newcorn, J. H., Schulz, K. P., and Fan, J. (2006) FMRI measures of mechanisms and response to methylphenidate (MPH) and atomoxetine (ATX). Presented at the 53rd Annual Meeting of the American Academy of Child and Adolescent Psychiatry, 24–29 October, San Diego, CA.

Paquette, V., Levesque, J., Mensour, B., et al. (2003) "Change the mind and you change the brain": effects of cognitive-behavioral therapy on the neural correlates of spider phobia. *NeuroImage*, **18**, 401–9.

Patel, N. C., DelBello, M. P., Cecil, K. M., et al. (2006) Lithium treatment effects on *myo*-inositol in adolescents with bipolar depression. *Biol Psychiatry*, **60**, 998–1004.

Peterson, B. S. (2001) Neuroimaging studies of Tourette syndrome: a decade of progress. *Adv Neurol*, **85**, 179–96.

Peterson, B. S., Skudlarski, P., Gatenby, C., et al. (1999) An fMRI study of Stroop word-color interference: evidence for cingulate subregions subserving multiple distributed attentional systems. *Biol Psychiatry*, **45**, 1237–58.

Pliszka, S. R. (2005) The neuropsychopharmacology of attention-deficit/hyperactivity disorder. *Biol Psychiatry*, **57**, 1385–90.

Pliszka, S. R., Glahn, D. C., Liotti, M., et al. (2004) Neuroimaging of inhibitory control in ADHD, combined type. Presented at the 51st Annual Meeting of the American Academy of Child and Adolescent Psychiatry, 19–24 October, Washington, DC.

Pliszka, S. R., Liotti, M., Bailey, B.Y., et al. (2007) Electrophysiological effects of stimulant treatment in children with attention deficit hyperactivity disorder (ADHD). *J Child Adolesc Psychopharmacol*, **17**, 356–66.

Prasko, J., Horacek, J., Zalesky, R., et al. (2004) The change of regional brain metabolism (^{18}FDG PET) in panic disorder during the treatment with cognitive behavioral therapy or antidepressants. *Neuroendocrinol Lett*, **25**, 340–8.

Rosa-Neto, P., Lou, H. C., Cumming, P., et al. (2005) Methylphenidate-evoked changes in striatal dopamine correlate with inattention and impulsivity in adolescents with attention deficit hyperactivity disorder. *NeuroImage*, **25**, 868–76.

Rosenberg, D. R., Benazon, N. R., Gilbert, A., et al. (2000a) Thalamic volume in pediatric obsessive-compulsive disorder patients before and after cognitive behavioral therapy. *Biol Psychiatry*, **48**, 294–300.

Rosenberg, D. R., MacMaster, F. P., Keshavan, M. S., et al. (2000b) Decrease in caudate glutamatergic concentrations in pediatric obsessive-compulsive disorder patients taking paroxetine. *J Am Acad Child Adolesc Psychiatry*, **39**, 1096–103.

Sassi, R. B., Nicoletti, M., Brambilla, P., et al. (2002) Increased gray matter volume in lithium-treated bipolar disorder patients. *Neurosci Lett*, **329**, 243–5.

Sassi, R. B., Stanley, J. A., Axelson, D., et al. (2005) Reduced NAA levels in the dorsolateral prefrontal cortex of young bipolar patients. *Am J Psychiatry*, **162**, 2109–15.

Schmajuk, M., Liotti, M., Busse, L., et al. (2006) Electrophysiological activity underlying inhibitory control processes in normal adults. *Neuropsychologia*, **44**, 384–95.

Schwartz, J. M., Stoessel, P. W., Baxter, L. R., Jr. et al. (1996) Systematic changes in cerebral glucose metabolic rate after successful behavior modification treatment of obsessive-compulsive disorder. *Arch Gen Psychiatry*, **53**, 109–13.

Schweitzer, J. B., Lee, D. O., Hanford, R. B., et al. (2003) A positron emission tomography study of methylphenidate in adults with ADHD: alterations in resting blood flow and predicting treatment response. *Neuropsychopharmacology*, **28**, 967–73.

Schweitzer, J. B., Lee, D. O., Hanford, R. B., et al. (2004) Effect of methylphenidate on executive functioning in adults with attention-deficit/hyperactivity disorder: normalization of behavior but not related brain activity. *Biol Psychiatry*, **56**, 597–606.

Seifert, J., Scheuerpflug, P., Zillessen, K. E., et al. (2003) Electrophysiological investigation of the effectiveness of methylphenidate in children with and without ADHD. *J Neural Transm*, **110**, 821–9.

Semrud-Clikeman, M., Pliszka, S. R., Lancaster, J., et al. (2006) Volumetric MRI differences in treatment-naive vs chronically treated children with ADHD. *Neurology*, **67**, 1023–7.

Shafritz, K. M., Marchione, K. E., Gore, J. C., et al. (2004) The effects of methylphenidate on neural systems of attention in attention deficit hyperactivity disorder. *Am J Psychiatry*, **161**, 1990–7.

Silverstone, P. H., McGrath, B. M., and Kim, H. (2005) Bipolar disorder and myo-inositol: a review of the magnetic resonance spectroscopy findings. *Bipolar Disord*, **7**, 1–10.

Silverstone, P. H., Wu, R. H., O'Donnell, T., et al. (2002) Chronic treatment with both lithium and sodium valproate may normalize phosphoinositol cycle activity in bipolar patients. *Human Psychopharmacol*, **17**, 321–7.

Straube, T., Glauer, M., Dilger, S., et al. (2006) Effects of cognitive-behavioral therapy on brain activation in specific phobia. *NeuroImage*, **29**, 125–35.

Szeszko, P. R., MacMillan, S., McMeniman, M., et al. (2004) Amygdala volume reductions in pediatric patients with obsessive-compulsive disorder treated with paroxetine: preliminary findings. *Neuropsychopharmacology*, **29**, 826–32.

Vaidya, C. J., Austin, G., Kirkorian, G., et al. (1998) Selective effects of methylphenidate in attention deficit hyperactivity disorder: a functional magnetic resonance study. *Proc Natl Acad Sci*, **95**, 14494–9.

Volkow, N. D., Wang, G., Fowler, J. S., et al. (2001) Therapeutic doses of oral methylphenidate significantly increase extracellular dopamine in the human brain. *J Neurosci*, **21**, RC121.

Volkow, N. D., Wang, G. J., Fowler, J. S., et al. (2004) Evidence that methylphenidate enhances the saliency of a mathematical task by increasing dopamine in the human brain. *Am J Psychiatry*, **161**, 1173–80.

Volkow, N. D., Wang, G. J., Newcorn, J., et al. (2007a) Brain dopamine transporter levels in treatment and drug naive adults with ADHD. *NeuroImage*, **34**(3), 1182–90.

Volkow, N. D., Wang, G. J., Newcorn, J., et al. (2007b) Depressed dopamine activity in caudate and preliminary evidence of limbic involvement in adults with attention-deficit/hyperactivity disorder. *Arch Gen Psychiatry*, **64**, 932–40.

Wise, R. G. and Tracey, I. (2006) The role of fMRI in drug discovery. *J Magn Reson Imag*, **23**, 862–76.

Zubieta, J. K., Bueller, J. A., Jackson, L. R., et al. (2005) Placebo effects mediated by endogenous opioid activity on mu-opioid receptors. *J Neurosci*, **25**, 7754–62.

22

Functional alleles, neuroimaging and intermediate phenotypes in the deconstruction of complex behavioral variation

David Goldman, Beata Buzas and Ke Xu

Introduction

Heritability plays a critical role in behavioral variation and psychiatric disease. Mounting evidence suggests that most psychiatric disorders are moderately to highly heritable. For example, the heritability of major depression is approximately 0.37 (Sullivan et al., 2000), the heritability of schizophrenia is approximately 0.80 (Sullivan et al., 2003), and the heritability of ten addictive disorders ranges from 0.27 to 0.65 (Goldman et al., 2005).

Roles for multiple genes and gene–environment interactions are often invoked to explain complex psychiatric diseases. A common hypothesis is that multiple genes, each with only a moderate effect, combine to produce risk for psychiatric diseases. By evaluating gene effects in the context of stress exposures that are permissive for the actions of certain alleles, investigators have clarified the effects and effect sizes of certain functional polymorphisms affecting behavior. For example, a common serotonin transporter polymorphism with a significant, but quantitatively modest, effect on neuroticism has large effects on depressive ideation and suicidality in stress-exposed individuals (Roy et al., 2007). Similarly, the effect of a functional monoamine oxidase A (*MAOA*) polymorphism in behavioral dyscontrol is far more manifest in stress-exposed individuals (Caspi et al., 2002). In another context, the same low-expression *MAOA* genotype was shown to be permissive for the action of higher testosterone levels to predict dyscontrol behaviors in males, with no effect of testosterone in the absence of the low-expression genotype (Sjöberg et al., 2008). These findings raise the question of whether the action of genetic variants in complex behavior is generally to be understood only in the context of exposures. However, this conclusion may be modified by the ability to more directly capture gene effects on the brain by accessing and measuring intermediate neurobiological changes.

Concepts of genetics

Gene, allele, haplotype, and functional variant

Traditionally a gene was defined as the physical unit of heredity, which occupies a specific locus on a chromosome. Biochemically, a gene is a segment of deoxyribonucleic acid (DNA) that carries all the necessary information for the manufacture of a protein and for its own regulation. As knowledge of the genome has grown, it has become increasingly difficult to define a gene. Overlapping transcripts, variations in splicing, expressed sequences within introns [non-coding regions spliced out during messenger ribonucleic acid (mRNA) processing] of other genes, and

Neuroimaging in Developmental Clinical Neuroscience, eds. Judith M. Rumsey and Monique Ernst. Published by Cambridge University Press. © Cambridge University Press 2009.

promoter and enhancer regions in adjacent genes that may alter gene expression have made the definition much more fuzzy. Increasingly, the importance of ribonucleic acid (RNA) gene products as direct effectors has become recognized. RNAs that exert direct actions without translation into protein include the transfer and ribosomal RNAs that are involved in protein translation, nucleolar RNAs (involved in the synthesis and processing of RNAs), and microRNAs (miRNAs; small 21- to 23-nucleotide RNAs) that regulate gene expression.

Most genes have characteristic regulatory DNA sequences called promoters that enable enhanced and controlled expression. These regions often contain an enrichment of **c**ytosine-**p**hospho**g**uanine (CpG) dinucleotides. The methylation status of cytosines of CpG dinucleotides influences the regulation of gene expression. DNA does not exist as a naked molecule but is wound around nucleosome protein cores comprising evolutionarily conserved histone proteins. The chromatin structure of genes changes according to transcriptional status. To accomplish these changes, the histones are chemically modified or substituted, leading to tighter packing of the DNA or opening of the DNA to transcription by RNA polymerase enzymes. This structural configuration directly affects transcription, which is reflected in the level of gene expression.

Those gene variants (alleles) that become common in populations (frequency >1%) are termed polymorphisms. Most of the 11 million human polymorphisms appear to have no direct effect on function (or fitness of the organism) and are therefore termed neutral. On chromosomes, combinations of alleles are known as haplotypes. Haplotypes are a combination of alleles at multiple, linked loci that are inherited as a unit. This phenomenon, known as linkage disequilibrium, occurs because blocks of DNA may not be broken up frequently enough to achieve a random assortment by recombination which occurs during meiosis. In addition, certain haplotypes confer a selective evolutionary advantage and are preferentially maintained. Haplotypes

can be used to track the effects of unknown alleles that reside on the haplotype, reflecting the ancient chromosomal segment on which a mutation originally occurred.

Using either alleles (gene variants) or haplotypes (DNA segments including several alleles), one important aim of genetic studies is to identify the relatively small fraction of alleles that are functional, or non-neutral. Neuronal cell structure is a manifestation of gene expression and of complex adjustments in gene expression during the differentiation and migration of neurons. These adjustments are in response to hormonal, nutritional and trophic signals and also result from the interactions of neurons with neighboring cells, including glia, and interactions with other neurons in networks. Gene mutations can affect any of these processes or cause no effect. The study of neurogenetic variation can help us better understand individual differences in behavior and disease, and the processes leading to complex behavioral phenotypes.

Genetic models of psychiatric diseases

The genetic complexity of psychiatric diseases is poorly understood in part because the etiology of psychiatric diseases is poorly understood. For this reason, genetic analysis is usually performed on a heterogeneous, clinically defined entity. Questions regarding the number of genes involved in any specific psychiatric disease or whether the genetic variations act alone or in combination to produce vulnerability are the object of both research and speculation. There are many origins of genetic complexity including lack of penetrance of allele effects, variable expressivity, genetic heterogeneity, and polygenicity (Goldman et al., 2005). In genetic heterogeneity, different risk genes and alleles individually may be sufficient to lead to the same phenotype. In polygenicity, a combination of risk alleles is required for the manifestation of the phenotype. In addition, the interaction between these different risk alleles (epistasis) complicates the identification of the genetic cause of a given phenotype. Certain psychiatric diseases, such as autism or schizophrenia, have high monozygotic (MZ)/dizygotic (DZ) twin concordance ratios, on the order of 10:1 for autism and 8:1 for schizophrenia indicative of polygenicity. However, several other psychiatric diseases, including the addictive disorders, show MZ/DZ ratios that are remarkably close to 2:1 (Goldman et al., 2005).

Gene identification approaches

Genetic linkage and genetic association are classic approaches for identifying locations of genes influencing phenotypes, e.g., psychiatric diseases. *Classic genetic linkage* is locus-based linkage that relies on the observation of either co-segregation (coupling) or repulsion between an allele and a trait in related individuals. Genetic association is actually a variant of genetic linkage. The *genetic association* approach is allele-based linkage that provides a measure of the relationship between an allele, genotype (combination of alleles), haplotype or diplotype (combination of haplotypes) and the phenotype. This is usually accomplished by identifying a difference in the frequency of these genetic markers between patients and controls, but can also be accomplished in studies of family pedigrees by evaluating allele transmissions from heterozygous parents to affected offspring for deviations from a 50:50 transmission using a transmission disequilibrium test.

Recently several landmark whole genome association studies have been reported for complex diseases. For example, the Wellcome Trust Case Control Consortium, a collaboration of leading human genetics laboratories, is analyzing DNA from 2000 patients, each with one of seven complex diseases, including bipolar disorder, and 3000 healthy controls (Wellcome Trust Case Control Consortium, 2007). Using a panel of 550 000 genetic markers, a total of 24 genome-wide significant loci were detected. These loci account for only a small portion of the total variance for the diseases. For example, a single genome-wide significant locus of modest effect size (odds ratio 1.9) was identified

on chromosome 16 for bipolar disorder. In this analysis, all of the alleles that were genotyped or imputed had frequencies of >5%, and no locus with an effect on the odds ratio of less than 1.5 was significant at the genome-wide level. Perhaps many of the alleles influencing these common diseases are rare. Alternatively, many may have smaller effect sizes, requiring either larger samples or samples in which intermediate phenotypes that are more responsive to allele effects have been measured.

Intermediate phenotypes and endophenotypes

An intermediate phenotype is a trait or characteristic that is associated with disease and linked to the action of functional alleles. Studies using intermediate phenotypes involve the deconstruction of complex phenotypes and create opportunities to identify, on one hand, more homogeneous subgroups, and, on the other hand, gene effects across clinically defined diseases that share causation. Certain intermediate phenotypes reflecting relevant neurobiological changes are likely to be more directly influenced by genetic variation.

An "endophenotype" is a heritable intermediate phenotype. In 1966, John and Lewis coined this term in the context of insect biology. Gottesman and Shields introduced "endophenotype" to the field of biological psychiatry in 1972. Frequently this term is applied to any trait assessed by a laboratory-based method including functional neuroimaging, electrophysiology, biochemistry, neuroanatomy, and neuropsychology. However, the Gottesman criteria for an endophenotype are: (1) association with disease; (2) heritability; (3) state-independence (i.e., a trait); (4) co-segregation (of both endophenotype and disease) within families; and (5) an increased ratio of affected/unaffected phenotypes in relatives of probands over that in the general population (Gottesman and Gould, 2003).

Determination of the heritability of intermediate phenotypes can require extensive and costly research.

Table 22.1 Intermediate phenotypes in psychiatric disorders

Intermediate phenotype	Heritability	Phenotype relationship	Gene
Alcohol response	Yes	Alcoholism risk	GABAA α 6
Alcohol-induced flushing	Yes	Alcoholism protective	ALDH2
Neurotransmitter metabolites			
5HIAA	Yes	Impulsivity/depression	
HVA	Yes		
MHPG	Yes	Parkinson's	
CRH, cortisol, NPY	Yes	Depression, anxiety	NPY
EEG alpha	Yes	Alcoholism, anxiety	
EEG beta	Yes	Alcoholism	Chr 4 GABA$_A$
P300 ERP power and latency	Yes	Alcoholism, schizophrenia	
Frontal cognitive function	Yes	Alcoholism, schizophrenia	COMT
MRI structure			
Hippocampus	?	Episodic memory	BDNF
Amydgala	?	Emotion	HTT
Frontal	Yes	Impulsivity	MAOA
Task-associated fMRI			
Frontal	?	Frontal cognition/schizophrenia	COMT
Amygdala	?	Emotion/depression	HTT, COMT, MAOA
Imaging with specific ligands			
Mu opioid receptor	?	Pain threshold/affect	COMT, NPY
D2, D3 receptor	?	Reward	
5HT1A receptor	?	Depression	HTT (?)
5HT and DA transporters	?	Depression	HTT

Notes:
ALDH2, aldehyde dehydrogenase 2 gene; BDNF, brain-derived neurotrophic factor; COMT, catechol-*O*-methyltransferase; CRH, corticotropin releasing hormone; ERP, event-related potential; fMRI, functional magnetic resonance imaging; GABA$_A$, gamma-aminobutyric acid receptor A; HTT, serotonin transporter; HVA, homovanillic acid; MAOA, monoamine oxidase A; MHPG, 3-methoxy-4-hydroxyphenylglycol; NPY, neuropeptide Y; 5H1AA, 5-hydroxyindole acetic acid.

Mechanistically relevant phenotypes associated with heritable psychiatric disease are often assumed to be heritable and called endophenotypes. Furthermore, intermediate phenotypes that are not associated with a disease can nevertheless be valuable for identifying a gene for the disease. This can occur when the intermediate phenotype identifies a small, otherwise unrecognized, disease subgroup or when the intermediate phenotype is protective. An intermediate phenotype that identifies the action of an environmental factor could help sort out the action of genes by exclusion or interaction.

Thus far, more than a dozen intermediate phenotypes have been identified for schizophrenia, mood disorders and addictions, and several have been linked to the action of functional alleles (see Table 22.1). Intermediate phenotypes that are indirectly accessed using behavioral measures, e.g., personality questionnaires and neuropsychological test batteries, yield linkage results that are often inconsistent, with relatively modest effect sizes of the allele. In contrast, neuroimaging and electrophysiological techniques accessing brain responses associated with unique sensory-motor, cognitive or affective processes have shown stronger and more consistent associations with functional alleles and thus are increasingly used in genetic studies.

Imaging genetics and its principles

The new field of "imaging genetics" bridges imaging technology and genetics, enabling correlations between neural function and genotype. A recent review of imaging genetic studies by Goldman and Ducci (Goldman and Ducci, 2007) reported a median effect size of 3.9 as compared to 1.9 for 24 complex disease gene loci identified by the Wellcome Trust Case Control Consortium using whole genome association and much larger samples (2000 for each complex disease, and 3000 controls). Thus, imaging genetic studies yielding replicated results are often adequately powered with relatively small samples (e.g., 30), the required samples size also being a function of allele frequency.

Several critical design considerations have emerged from early imaging genetic studies:

1. Selection of candidate functional alleles: ideally, functional alleles should be abundant in the study population. However, an alternative is to stratify the dataset such that rarer genotypes are represented. For example, the frequency of the *COMT Met158* allele varies between 28% and 50% in different ethnic populations. This is a high allele frequency leading to relatively high abundances of the *Val158/Val158*, *Val158/Met158* and *Met/Met158* genotypes, but the rarer *Met158/Met158* homozygotes can be selectively studied as well.

2. Careful matching across genotypic groups: gender, age, ethnicity, IQ, socioeconomic status, treatment, and diagnosis may be matched. Key variables are those that predict genotype or that predict the intermediate phenotype. Effects of certain variables on intermediate phenotypes can be covaried. Ethnic factor scores computed from ancestry informative markers (AIMs) represent an important approach to determining the existence of ethnic covariance of intermediate phenotypes and correcting for these effects (Enoch et al., 2006).

3. Task selection for imaging genetics studies: tasks that consistently engage discrete brain circuits, produce robust signals, and reveal reproducible interindividual variations are the best candidates for probing relationships between imaging phenotypes and specific genotypes.

Next we describe validated associations of two functional loci to complex behaviors via the imaging genetics approach.

Catechol-*O*-methyltransferase, cognitive function, and emotional resilience

The catechol-*O*-methyltransferase gene and its inherited variations

Located on chromosome 22q11, the catechol-*O*-methyltransferase (*COMT*) gene spans 27 kilobases (kb) and includes six exons (Figure 22.1a). The COMT enzyme, discovered by Axelrod and Vesell (Axelrod and Vesell, 1970), metabolizes dopamine and other catecholamine neurotransmitters. Two forms of COMT are found: soluble (S-COMT) and membrane-bound (M-COMT). S-COMT predominates in peripheral tissues, while M-COMT predominates in the central nervous system.

In 1977, Weinshilboum and Raymond observed that red blood cell COMT activity was bimodally distributed in the general population (Weinshilboum and Raymond, 1977). In 1996, Lachman et al. identified a guanine-to-adenine (G to A) single nucleotide polymorphism (SNP) located at codon 158 of M-COMT, resulting in an amino acid substitution: valine-to-methionine (SNP designation *Val158Met*, dbSNP ID: rs4680[1]) (Lachman et al., 1996). The *Val* and *Met* alleles are co-dominant in action, such

---

[1] The single nucleotide polymorphism database (dbSNP) is a public domain archive available through the National Library of Medicine in which each SNP is identified by an ID number.

(a)

(b)

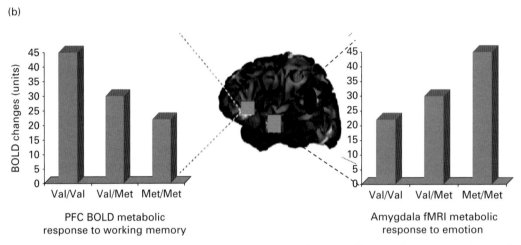

PFC BOLD metabolic
response to working memory

Amygdala fMRI metabolic
response to emotion

Figure 22.1. Catechol-*O*-methyltransferase (*COMT*) gene, polymorphisms, and functional imaging. (a) There are both membrane-bound and soluble forms of COMT: M-COMT and S-COMT, respectively, which differ by 50 amino acids. The common missense variant *Val158Met* (rs4680) alters COMT enzyme activity. The activity of Val-COMT is three- to four-fold higher than that of Met-COMT, but the difference is smaller in vivo (see text). The four-locus haplotypes containing rs6269, rs4633, rs4818, and rs4680 have been shown to affect COMT protein expression by altering mRNA secondary structure and translatability. (b) During a working memory test, *Val/Val* homozygotes show greater activations in prefrontal cortex (PFC) and poorer performance (left). *Met* allele carriers show greater activations of the amygdala following exposure to negative emotional stimuli. Thus, the *Met* allele appears beneficial to cognitive function under many circumstances, but appears to lead to stronger emotional responses and reduced stress resiliency.

that the enzyme activity of *Val/Met* heterozygous individuals is intermediate. Although the *Val* allele was observed to be three- to four-fold higher in activity as compared to the *Met* enzyme in vitro, more recent data indicate that at 37°C (body temperature), the *Val* allele is only up to 40% more active (Chen et al., 2004).

In addition to *Val158Met*, other *COMT* loci modify the activity of the enzyme. A promoter SNP, rs737865 A to G substitution has a minor effect on COMT activity (Chen et al., 2004). The SNP rs165599 has also been shown to correlate with differential allele expression (Bray et al., 2003). Recently, it was found that different *COMT* SNPs on the same haplotype can interact to alter expression of *COMT* by modifying mRNA structure and translatability (Nackley et al., 2006). Three SNPs (rs6269 G to A, rs4633 C to T, rs4818 G to C) and *Val158Met* (rs4680 G to A) form three common haplotypes that lead to different mRNA structures altering the expression and activity of *COMT* (Figure 22.1a). These three functional haplotypes were previously associated with differing pain thresholds (Diatchenko et al., 2005).

COMT Val158Met *and frontal cognition*

COMT regulates dopamine level in prefrontal cortex

Both human and animal studies have shown that dopamine modulates prefrontal cortex (PFC) activity during working memory (Mattay et al., 1996; Sawaguchi and Goldman-Rakic, 1991; Seamans et al., 1998). The synaptic action of dopamine is terminated either by dopamine transporter reuptake, which can be followed by metabolism by monoamine oxidase (MAO), metabolism by COMT, or diffusion. Because the dopamine transporter is expressed at low levels in PFC, COMT plays a major role in modulating dopaminergic neurotransmission in this region, which has a paucity of dopamine transporters, as shown by pharmacological studies (Karoum et al., 1994). In COMT knockout mice, dopamine levels are increased in PFC but not in striatum (Gogos et al., 1998), and working memory performance is remarkably enhanced. In humans, performance on cognitive tasks that engage the PFC is improved by augmenting dopamine neurotransmission. Dopamine receptor agonists improve attention and working memory in healthy individuals (Kimberg et al., 1997). Stimulants, which increase dopamine release, are useful in the treatment of individuals with attention deficit/hyperactivity disorder (ADHD) (Paule et al., 2000). After traumatic brain injury, cognitive performance improves significantly following a single dose of bromocriptine (McAllister et al., 2006), a dopamine receptor 1 agonist.

Val158Met *predicts frontal cognitive performance in varied groups*

Healthy adults

Val/Val homozygotes have the highest enzyme activity and presumably the lowest PFC dopamine levels and they show the poorest performance on working memory tasks. *Met/Met* homozygotes tend to perform better (Egan et al., 2001), and heterozygotes perform within an intermediate range,

displaying an allele dosage effect. The *Val158Met* genotype predicts response accuracy on the n-back working memory task (Bertolino et al., 2006; Goldberg et al., 2003). In healthy controls, *Met/Met* individuals were 5% more accurate than *Val/Val* homozygotes on the 1-back task, but 15% more accurate on the more demanding 2-back task. In controls, *Val158Met* appears to account for approximately 4% of the variance in performance on working memory tasks.

Adult patients and their relatives

It is well known that schizophrenia and brain injury are associated with poorer performance on cognitive tests thought to engage PFC. The unaffected siblings of schizophrenic patients tend to be intermediate in performance. A few studies have investigated the relationship between *Val158Met* and working memory in patients with schizophrenia (including those on antipsychotic medications), the well siblings of schizophrenic patients, and patients with brain injury (Bertolino et al., 2006; Bruder et al., 2005; Egan et al., 2001; Mattay et al., 2003; McAllister et al., 2006; McIntosh et al., 2006; Weickert et al., 2004). These studies are remarkably congruent in showing similar *Val158Met* genotype effects as in controls, despite impairments on test performance observed in these clinical and at-risk groups.

Healthy children

Dopamine levels in PFC may also alter cognitive function in the developing brain. However, it is important to note that, as compared to adults, the relationship of COMT to cognitive function in children is more complicated and controversial, due to developmental changes in the dopamine system itself. Nevertheless, in one study of healthy children, *Val158Met* genotype predicted performance on the Dots-mixed task, which measures visual working memory and inhibitory functions of the PFC (Diamond et al., 2004). Among 39 12-year-old children, *Met/Met* homozygotes performed significantly better than *Val/Val* homozygotes (Diamond et al., 2004), consistent with findings in adults.

In view of the developmental plasticity of dopaminergic neurotransmission, it will be of great interest to see if these findings are replicable.

Children with ADHD or the chromosome 22q11 deletion syndrome

A meta-analysis revealed no significant association of *Val158Met* to ADHD in children aged 4–16 years old (Cheuk and Wong, 2006). However, the *COMT* genotype was associated with working memory in ADHD children, but in a direction opposite to that seen in controls (Bellgrove et al., 2005). In contrast to the findings in healthy children, *Met/Met* homozygote and *Met/Val* heterozygote children with ADHD performed significantly worse than *Val/Val* homozygotes on two tasks that engage working memory: walk/don't walk and the sky search dual task. The cognitive impairment in *Met/Met* or *Met/Val* ADHD children may be due to excessively high dopamine function following treatment with amphetamine.

The role of the *COMT* genotype has also been studied in the context of a severe behavioral syndrome in which one copy of the *COMT* gene is frequently deleted. *COMT* is located in the chromosome 22q11 region, which is deleted in children with velocardiofacial syndrome (VCFS). Children with VCFS have one chromosome carrying the *COMT Met* or *Val* allele and a second chromosome in which *COMT* is deleted. It is unclear whether *COMT Val* genotype worsens deficits in PFC found in children with VCFS. Two studies reported that *Met/deletion* children with VCFS performed better on cognitive tests thought to engage PFC as compared to *Val/deletion* children (Bearden et al., 2004; Shashi et al., 2006). However, another study reported that the *COMT Met/deletion* genotype was associated with poorer performance on tests of expressive language and verbal working memory (Baker et al., 2005). Potentially, *Met*-allele-associated impairments, which seem paradoxical, may be explained by the inverted "U" relationship of dopamine concentration to PFC function, whereby performance is suboptimal if dopamine is either deficient or excessive (Mattay et al., 2003). The

VCFS *Met/deletion* children have extremely low COMT activity and therefore may have excessively high dopamine levels in the PFC, impairing cognitive function. This idea is supported by the finding that amphetamine, through its dopamine-augmenting action, has a differential ability to improve the frontal cognitive performance of normal individuals with different *COMT* genotypes. *Met/Met* homozygotes suffered impairment at higher levels of task difficulty for the n-back task, indicating that they may already have had frontal dopamine levels optimal for this task difficulty (Mattay et al., 2003). More studies are needed to better understand the relationship between *COMT* and cognitive function in children with VCFS.

Effects of COMT on functional neuroimaging measures

Val158Met and *COMT* haplotype predict measures of frontal function

More than a dozen neuroimaging studies have consistently shown that *COMT Val158Met* predicts PFC activation during executive cognitive tasks. Under conditions of equivalent performance on the 2-back working memory task, *Val/Val* homozygotes had greater activations of blood-oxygen-level-dependent (BOLD) signal changes in dorsolateral prefrontal cortex (DLPFC) and cingulate cortex (Figure 22.1b left). The increased activation has been interpreted as reflecting the compensatory effort required for *Val/Val* individuals, who recruit more neuronal activity, to accomplish the task (Egan et al., 2001).

This cortical inefficiency also ties the *Val* allele to schizophrenia. In schizophrenia patients, activation of DLPFC during equivalent performance on the 2-back task and other frontal tasks is higher than in healthy controls and even greater activation is predicted by the *Val* allele (Bertolino et al., 2006). Also, compared to *Met/Met* schizophrenia patients, *Val/Val* schizophrenia patients showed reduced volumes of the left anterior cingulate cortex and the right middle temporal gyrus (Ohnishi et al., 2006). Such within-disorder studies offer potential clues to the etiologic heterogeneity of schizophrenia and potentially explain variation in treatment response. Interestingly, Bertolino et al. reported that schizophrenia patients with *Met* genotypes tended to show more cognitive improvement during treatment with olanzapine (Bertolino et al., 2004). Studies of at-risk individuals offer the opportunity to capture effects of genotype without potential interactions of treatment and other clinical history. In unaffected 16- to 25-year-old subjects at high familial risk for schizophrenia, the *Val* allele was associated with reduced gray matter density in the anterior cingulate cortex (McIntosh et al., 2006). In addition, these *Val* carriers showed increased activations of DLPFC and anterior cingulate as difficulty increased on a sentence completion test requiring response initiation and response suppression shown to be sensitive to frontal lesions (Burgess and Shallice, 1996).

Polymorphisms such as *Val158Met* are likely to be informative for neural function within the context of many diseases, as well as in behaviors within the normal range. The *Val* allele predicted stronger bilateral frontal and basal ganglia activations in individuals with traumatic brain injury who were challenged with the n-back task (McAllister et al., 2006). In these brain-injured patients, a similar effect of the *Val* allele on dorsal cingulate activity was also seen during a task requiring increasing levels of attentional control.

COMT haplotype predicts cortical efficiency
As mentioned earlier, the haplotype level of analysis is capturing information on additional functional loci at *COMT*. However, thus far, haplotype-based imaging genetic studies are few in number, principally because of the complexity of dealing with two to five common haplotypes, as may be typically found. This problem can be addressed via the functional grouping of haplotypes, much as certain alleles [e.g., the *MAOA* 3.5 and 4 repeat low-expressing alleles (Sabol et al., 1998) and the HTTLPR *S* and *L_G*

low-expressing alleles (Hu et al., 2006) have been grouped on the basis of functional equivalence. In Meyer-Lindenberg's study, a haplotype including the rs737865 guanine to adenine promoter locus and *Val158Met* accounted for more variation in PFC function than *Val158Met* alone (Meyer-Lindenberg et al., 2006). The diplotype *A-Val*, which carries two *A-Val* haplotypes and is predicted to yield the highest *COMT* expression, was associated with worse PFC function. Later, an additional SNP (rs165599) located at the 3′ end of *COMT* was added to the analysis, forming the same haplotype, *G-Val-G*, previously linked to schizophrenia in a large sample of Israelis of Ashkenazi descent (Shifman et al., 2002). The *G-Val-G* risk haplotype strongly predicted PFC inefficiency during working memory, with a maximum effect in ventrolateral PFC. It should be noted that the haplotypes used in these studies do not follow the functional four-locus *COMT* haplotype as mentioned earlier (Diatchenko et al., 2005). Thus, potentially this haplotype or other combinations of *COMT* genetic variants might explain still more variability in cognition mediated by PFC.

COMT Val158Met *and stress resiliency*

Val158Met balancing selection: warrior versus worrier

The proposed negative effect of the *Val158* allele on cognitive performance and its possible role as a risk factor in schizophrenia raises the question of why it has been maintained at high frequencies in populations worldwide. The *Met* allele, although associated with better cognitive performance, may be linked to diminished stress resiliency, higher levels of anxiety, and disorders whose risk is increased by higher anxiety or lower stress resiliency (Zubieta et al., 2003). Thus, the mechanism could be a form of natural selection that works to maintain multiple alleles within a population ("balancing selection"), favoring behavioral diversity in cognition and emotion. This "worrier/warrior" balancing selection model (Zhu et al., 2004) is illustrated in Figure 22.2.

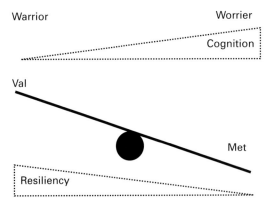

Figure 22.2. The "warrior/worrier" balancing selection model for *COMT Val158Met*. As compared to *Met/Met* homozygotes, *Val/Val* homozygotes tend to benefit with respect to stress resiliency but tend to have poorer cognitive function, and *Val/Met* heterozygotes tend to be intermediate.

COMT Val158Met *is associated with stress resiliency*

Met/Met homozygosity was associated with higher levels of anxiety among women in two populations (Enoch et al., 2003). The *COMT Met* allele may also be associated with obsessive-compulsive disorder (OCD), which is characterized by anxiety-producing intrusive thoughts and the performance of anxiety-reducing rituals. The *Met* allele has been shown to be more abundant in OCD patients, particularly in male patients (Karayiorgou et al., 1997, 1999). A family-based study also found that OCD patients were more likely to be *Met/Met* homozygotes (Schindler et al., 2000). Addictions have also been linked to *COMT*, with some studies of later-onset, anxious alcoholics implicating the *Met* allele.

Anxiety and dysphoric mood are frequently triggered by stressful life experiences. Individuals who are anxious and dysphoric are less tolerant of stress and pain. Therefore, it was predicted that the *Met* allele would be associated with lower pain thresholds and impaired brain opioid responses to painful, stressful stimuli that correlate with pain thresholds (Zubieta et al., 2003). There was a correlated effect on pain threshold and negative affective response to pain. In a replication study of experimental pain in women, the four-locus *COMT* haplotype containing the *Met* allele was again correlated with lower pain threshold, as mentioned, an important clue that preceded molecular studies showing epistatic interactions of SNPs within the *COMT* haplotype to alter *COMT* expression (Diatchenko et al., 2005).

COMT Val158Met, *stressful stimuli and brain activation*

Using fMRI, the effect of *COMT* genotype on the neural processing of unpleasant stimuli is most pronounced in the prefrontal cortex (Drabant et al., 2006; Smolka et al., 2005), the same region previously implicated in the effect of *COMT* on cognition. The number of *Met* alleles is positively correlated with BOLD response in the PFC. *Met/Met* individuals show the greatest changes in these cortical responses to unpleasant pictures, and *Val/Val* individuals have the smallest responses to the same unpleasant pictures.

The *Val158Met* genotype appears to explain 13%–38% of inter–individual variance in PFC response to unpleasant stimuli (Drabant et al., 2006; Smolka et al., 2005). *Met158* also predicts stronger activation by unpleasant stimuli of different parts of subcortical structures, including the hippocampus, amygdala, and thalamus (Figure 22.1b, right). In *Met/Met* homozygotes, there is increased functional coupling of limbic and PFC regions (Drabant et al., 2006). In an imaging genetics study using the μ opioid receptor ligand ^{11}C carfentanil, the *Met* allele predicted impaired endorphin release following a painful stimulus. In other words, the *Met* allele was associated with a reduction of the displacement of the exogenous radioligand by endogenous endorphins, whose release is activated by painful stimuli. This effect was found in the thalamus and at various locations within the limbic system (Zubieta et al., 2003).

In summary, *COMT* genetic variation alters cognitive and emotional processing via both local and distributed neural effects. The pleiotropy of *COMT* effects enables an evolutionarily relevant counter-balance of emotion and cognitive function.

The serotonin transporter gene and emotional circuits

Serotonin pathway and emotion

Converging evidence from animal and human studies indicates that serotonin (also known as 5-hydroxy tryptamine or 5HT) is a critical regulator of emotional behaviors, including depression, anxiety, fear, and aggression (Ressler and Nemeroff, 2000). Serotonin neurons in the midbrain raphe complex modulate neuronal function in the amygdala, orbitofrontal cortex, hippocampus, brainstem, and hypothalamus. The amygdala is a primary neuro-anatomical region implicated in anxiety disorders (Rauch et al., 2003). Functional measures of amygdala responses to negative emotional stimuli predict anxiety (Hariri et al., 2002).

The serotonin transporter (5HTT) terminates 5HT's action in the synaptic cleft by reuptake into the presynaptic terminal. The 5HTT protein is abundantly expressed in structures of the limbic network including the amygdala. Changes in sero-tonin transporter function have been connected to anxiety and dysphoria using both pharmaco-behavioral and genetic paradigms. For example, selective serotonin reuptake inhibitors (SSRIs, e.g., fluoxetine) block the serotonin transporter and are often associated with relief of anxiety and depression. More direct evidence comes from neuroimaging studies (positron emission tomography, PET) show-ing higher serotonin transporter binding in depres-sive patients compared with healthy subjects (Meyer et al., 2004). These studies suggest that patients with mood disorder have either a greater density of trans-porters or reduced amounts of endogenous 5HT that is able to bind to the transporter (Ichimiya et al., 2002). In rodents, 5HTT gene knockout mice show anxiety-like behaviors (Holmes et al., 2003; Murphy et al., 2001) and have an enhanced cortico-sterone response to stress (Tjurmina et al., 2002).

Functional serotonin transporter (5HTT) promoter alleles

A functional promoter polymorphism (5HTTLPR)

The human $5HTT$ ($SLC6A4$), located on chromosome 17q11.1-q12, has a variable repeat polymorphism located in the promoter region (5HTTLPR) (Heils et al., 1996; Lesch et al., 1994). As originally described, there are two common alleles. The long allele (L) has 16 copies of a 22-bp imperfect repeat sequence, and the short allele (S) has 14 copies (Lesch and Mossner, 1998) (Figure 22.3a). The S allele drives 5HTT tran-scription less efficiently and leads to lower trans-porter expression in vitro (Heils et al., 1996) and in living brain (Heinz et al., 2000). Homozygous L/L lymphoblasts express 60% more 5HTT than homo-zygous S/S lymphoblasts.

A third common, functional HTTLPR allele

More recently, Hu et al. (2006) discovered that $5HTTLPR$ is functionally tri-allelic. They found that a common single nucleotide substitution adenine (A) to guanine (G) within the L allele creates a func-tional transcription suppressor site and leads to low levels of transcription, equivalent to the S allele (Figure 22.3a). At the transcriptional level, the three common alleles (L_A, L_G, and S) act co-dominantly, and, taken together, account for more of the inter-individual variations in expression than S or L alone. This raises the possibility that the role of HTT in behavior can be better refined by the improved information captured with tri-allelic genotyping. In fact, the highest 5HTT binding was observed in the putamen of eight L_A/L_A homozy-gotes compared to individuals with genotypes leading to reduced serotonin transporter expres-sion (Praschak-Rieder et al., 2007). However, later in vivo PET studies have failed to detect an effect of 5HTTLPR on serotonin transporter binding in either healthy individuals or medication-free major depression patients, regardless of whether the

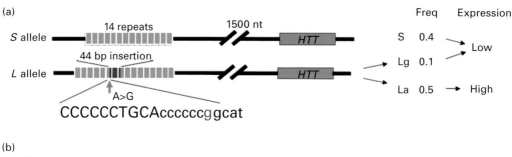

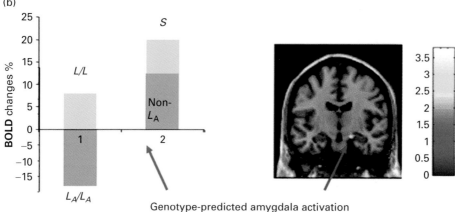

Genotype-predicted amygdala activation

Figure 22.3. *HTTLPR* alters amygdala activations by emotional stimuli. (a) The *S* allele decreases *5HTT* transcription and is associated with lesser 5HTT function (Lesch et al., 1994). A common A to G SNP located within the *L* allele leads to a similar reduction-of-function (by creating an AP2 transcriptional suppressor site) (Hu et al., 2006). The L_A /L_A genotype predicts greatest *5HTT* transcription and greatest *5HTT* function. (b) Individuals with genotypes comprised of *S* or L_G alleles show greatest amygdala activations following exposure to negative emotional stimuli.

genotype was bi-allelic or tri-allelic (Oquendo et al., 2007; Parsey et al., 2006; Shioe et al., 2003).

HTTLPR predicts anxiety and dysphoria

The low transcribing *S* allele has been implicated in human behavior, especially in anxiety. However, the findings are inconsistent, with extensive cross-study heterogeneity. On average, a single copy of the *S* allele leads to only a 0.1 standard deviation increase in trait anxiety, explaining the need for large population studies to generate sufficient power for linkage to the complex behavior. Thus, the effects on anxiety are much more modest than the effects on the neural responses associated with

anxiety. Allele effects are also much larger following severe stress exposure. In a prospectively studied cohort from New Zealand, the effect of the *S* allele in increasing depressive ideation and suicidality was observed only in those who had undergone severe stress (Caspi et al., 2003). In a retrospectively studied population of African American substance abuse patients with very high expected rates of suicidality, the *S* allele effect on suicidality risk was again seen only in individuals exposed to high levels of emotional trauma or neglect (Roy et al., 2007). A similar observation was made in non-human primates in whom early-life trauma exposures were systematically manipulated. Rhesus macaque monkeys carrying an *S* allele were more likely to drink heavily if they had undergone stressful rearing and had elevated cortisol responses (Barr et al., 2004).

Functional imaging of HTTLPR

HTTLPR predicts stronger amygdala activations and volume

As already mentioned, the amygdala is a central node of the limbic circuit that regulates fear, anxiety, and other emotional responses. Hariri et al. used fMRI to test the relationship between *HTTLPR* genotypes and amygdala response to fear stimuli (angry and fearful facial expressions) (Hariri et al., 2002). The *S/S* and *S/L* genotypes were grouped together due to small sample size and the view, at the time, that the *S* allele was functionally dominant. Two comparison groups (*L/L* versus *[L/S* and *S/S]*) each contained fourteen individuals matched by age, gender, IQ, and education. During fMRI scanning, the subjects performed a task that required them to match the expression of fearful and angry human faces. Consistent with the role of the *S* allele in anxiety, BOLD fMRI percentage signal change in the right amygdala was significantly greater in individuals with *S* allele genotypes compared to subjects homozygous for the *L* allele (Figure 22.3b). This finding suggests that individuals with the *S* allele, which is associated with increased anxiety, show amygdalar hyper-reactivity to stressful environmental stimuli. Thus, amygdala activation may provide an intermediate phenotype for emotional responses that are more proximal to the genotype, compared to the behavioral manifestations of anxiety. These results were replicated in a larger cohort of 92 individuals and in other studies (Bertolino et al., 2005; Hariri et al., 2005; Heinz et al., 2005, 2007; Pezawas et al., 2005). In Furmark et al., both amygdala blood flow (^{15}O-PET) and emotional responses to a public speaking paradigm were predicted by *HTTLPR* genotype (Furmark et al., 2004). In more recent studies, *HTTLPR* was associated with increased amygdala activation following emotional stimuli

in patients with major depression (Dannlowski et al., 2007).

HTTLPR genotypes alter amygdala/frontal cortex coupling

Greater amygdala reactivity to negative emotion-related stimuli may underlie the increased frequency of depressive ideation and anxiety in individuals carrying the *S* allele (or the reduction-of-function L_G allele), but the mechanism is unclear. The *S*-allele-dependent enhancement of amygdala activation may be driven by responses to negative stimuli or responses to ambiguous stimuli (Canli et al., 2005; Heinz et al., 2007). Using negative, neutral, or positive words as stimuli, Canli et al. (2005) found that amygdala activation was relatively greater in *S* allele carriers when contrasting responses to negative words with responses to neutral words. However, when responses to neutral words were compared to visual fixation, *S* allele carriers actually decreased activity in both right and left amygdalae. In fact, *S* carriers and *L/L* homozygotes did not differ on their response to negative stimuli versus the visual fixation condition. This observation was replicated in a study using standard affectively unpleasant, pleasant, and neutral picture stimuli (Heinz et al., 2007). In *S* allele carriers, both right and left amygdala activations were greater for aversive than neutral stimuli, but again in *S* allele carriers the visual fixation condition revealed deactivation compared to the neutral pictures. Thus, *S* allele carriers may be more responsive not only to clearly aversive stimuli, but also to situations in which there is uncertainty, which can be stressful.

S allele is associated with increased coupling between amygdala and frontal cortex under stress

HTTLPR's modulatory effects on the amygdala may be secondary to actions of the serotonin transporter in other brain regions whose activations correlate with modulation of amygdala responses.

The right posterior fusiform gyrus, implicated in the processing of face images, is also distinctly activated in individuals with different *HTTLPR* genotypes. Similar distributed effects of *HTTLPR* genotypes in regions that modulate the amygdala were found in two other studies involving passive viewing of emotional stimuli. These stimuli were either standardized pictures from the International Affective Picture System (Heinz et al., 2007) or emotionally negative words (Canli et al., 2005).

Most interestingly, the *S* allele appears to influence the functional coupling of amygdala and prefrontal regions. Serotonin modulates incoming projections from the prefrontal cortex onto the basolateral amygdala. The *S* allele has also been associated with volume reductions of both prefrontal cortex and amygdala (Pezawas et al., 2005). Higher emotion-induced activity and reduced functional coupling of the ventromedial prefrontal cortex (VMPFC) to the amygdala were associated with the *S* allele. The VMPFC has been shown to be important in emotion processing and to be functionally altered in depressive disorders. In contrast to the Heinz et al. finding, but again implicating frontal–amygdala coupling, the *S* allele predicted decoupling of the perigenual anterior cingulate and amygdala, correlating with depression (Pezawas et al., 2005). The cingulate cortex displays the highest density of 5HTT terminals within the human cortex (Varnas et al., 2004) and is a target zone of dense projections from the amygdala. These genotype effects on functional connectivity may also explain the association of OCD with rare and common serotonin transporter alleles that result in a "gain of function" involving an increase in transcription (Hu et al., 2006; Ozaki et al., 2003). In OCD, deep brain stimulation of the anterior limb of the internal capsule appears to have at least short-term benefit (Greenberg et al., 2006) indicating that functional coupling of brain regions is important in OCD. Through its role in modulating functional connectivity, the serotonin transporter influences the coding of negative emotions relative to positive emotion, highly relevant for many psychiatric disorders. In addition, based on functional changes

of the amygdala over child and adolescent development, the pattern of activation seen in adults is expected to differ from that seen in minors. These differences may contribute to behaviors associated with distinct maturational periods (e.g., novelty seeking in adolescence), as well as to discrete periods of vulnerability to psychiatric disorders known to peak in adolescence. An understanding of the interaction of the genetics with developmental changes and environmental influences will be critical for elucidating normal trajectories of affective cognitive development in health and disease.

Conclusions and future directions

Intermediate phenotypes informative for brain activity can be used to deconstruct complex end-point behavioral phenotypes and may ultimately lead to a more etiologically based redefinition of disease. Neuroimaging provides tools to quantitatively assess reliable intermediate phenotypes that are closer to gene effects. The convergence of imaging technology and genetics has created the rapidly growing field of imaging genetics. Via this approach, two genes, *COMT* and *5HTT*, have been linked to specific neural circuits associated with cognition and emotion.

A functional polymorphism in *COMT*, *Val158Met* (rs4680), partially accounts for variation in responses to working memory tasks in PFC and emotion regulation in the amygdala. The *Met* allele that leads to lower enzyme activity and predicts higher dopamine levels is associated with improved working memory, but is also correlated with high anxiety. Thus, the *Met* allele appears to be beneficial for cognitive function, at least under conditions of low or moderate stress, while the *Val* allele may benefit emotional resiliency.

In the domain of serotonergic neurotransmission, the functional HTTPLR has also been related to emotion. The short allele (*S*) reduces *HTT* transcription, as does the L_G allele which has recently been recognized. The *S* allele has been associated with decreased amygdala activation by neutral stimuli, but enhanced activation by aversive or ambiguous cues. The *S* allele effect appears partly mediated by a decrease in connectivity between amygdala and prefrontal/cingulate cortices, and other changes in both structure and function, suggesting the possibility of effects on developing neurocircuitry in children, studies of which will require repeated analyses across developmental stages. The *HTTLPR* and *COMT* effects appear additive, presaging a future in which effects of various functional alleles are integrated with each other and with environmental predictors.

REFERENCES

Axelrod, J. and Vesell, E. S. (1970) Heterogeneity of *N*- and *O*-methyltransferases. *Mol Pharmacol*, **6**, 78–84.

Baker, K., Baldeweg, T., Sivagnanasundaram, S., et al. (2005) COMT Val108/158 Met modifies mismatch negativity and cognitive function in 22q11 deletion syndrome. *Biol Psychiatry*, **58**, 23–31.

Barr, C. S., Newman, T. K., Shannon, C., et al. (2004) Rearing condition and rh5-HTTLPR interact to influence limbic-hypothalamic-pituitary-adrenal axis response to stress in infant macaques. *Biol Psychiatry*, **55**, 733–8.

Bearden, C. E., Jawad, A. F., Lynch, D. R., et al. (2004) Effects of a functional COMT polymorphism on prefrontal cognitive function in patients with 22q11.2 deletion syndrome. *Am J Psychiatry*, **161**, 1700–2.

Bellgrove, M. A., Domschke, K., Hawi, Z., et al. (2005) The methionine allele of the COMT polymorphism impairs prefrontal cognition in children and adolescents with ADHD. *Exp Brain Res*, **163**, 352.

Bertolino, A., Arciero, G., Rubino, V., et al. (2005) Variation of human amygdala response during threatening stimuli as a function of 5′HTTLPR genotype and personality style. *Biol Psychiatry*, **57**, 1517–25.

Bertolino, A., Caforio, G., Blasi, G., et al. (2004) Interaction of COMT Val108/158 Met genotype and olanzapine treatment on prefrontal cortical function in patients with schizophrenia. *Am J Psychiatry*, **161**, 1798–805.

Bertolino, A., Caforio, G., Petruzzella, V., et al. (2006) Prefrontal dysfunction in schizophrenia controlling for COMT Val158Met genotype and working memory performance. *Psychiatry Res*, **147**, 221–6.

Bray, N. J., Buckland, P. R., Williams, N. M., et al. (2003) A haplotype implicated in schizophrenia susceptibility is associated with reduced COMT expression in human brain. *Am J Hum Genet*, **73**, 152–61.

Bruder, G. E., Keilp, J. G., Xu, H., et al. (2005) Catechol-*O*-methyltransferase (COMT) genotypes and working memory: associations with differing cognitive operations. *Biol Psychiatry*, **58**, 901–7.

Burgess, P. W. and Shallice, T. (1996) Response suppression, initiation and strategy use following frontal lobe lesions. *Neuropsychologia*, **34**, 263–72.

Canli, T., Omura, K., Haas, B. W., et al. (2005) Beyond affect: a role for genetic variation of the serotonin transporter in neural activation during a cognitive attention task. *Proc Natl Acad Sci U S A*, **102**, 12224–9.

Caspi, A., McClay, J., Moffitt, T. E., et al. (2002) Role of genotype in the cycle of violence in maltreated children. *Science*, **297**, 851–4.

Caspi, A., Sugden, K., Moffitt, T. E., et al. (2003) Influence of life stress on depression: moderation by a polymorphism in the 5-HTT gene. *Science*, **301**, 386–9.

Chen, J., Lipska, B. K., Halim, N., et al. (2004) Functional analysis of genetic variation in catechol-*O*-methyltransferase (COMT): effects on mRNA, protein, and enzyme activity in postmortem human brain. *Am J Hum Genet*, **75**, 807–21.

Cheuk, D. K. and Wong, V. (2006) Meta-analysis of association between a catechol-*O*-methyltransferase gene polymorphism and attention deficit hyperactivity disorder. *Behav Genet*, **36**, 651–9.

Dannlowski, U., Ohrmann, P., Bauer, J., et al. (2007) Serotonergic genes modulate amygdala activity in major depression. *Genes Brain Behav*, **6**(7), 672–6.

Diamond, A., Briand, L., Fossella, J., et al. (2004) Genetic and neurochemical modulation of prefrontal cognitive functions in children. *Am J Psychiatry*, **161**, 125–32.

Diatchenko, L., Slade, G. D., Nackley, A. G., et al. (2005) Genetic basis for individual variations in pain perception and the development of a chronic pain condition. *Hum Mol Genet*, **14**, 135–43.

Drabant, E. M., Hariri, A. R., Meyer-Lindenberg, A., et al. (2006) Catechol *O*-methyltransferase val158met genotype and neural mechanisms related to affective arousal and regulation. *Arch Gen Psychiatry*, **63**, 1396–406.

Egan, M. F., Goldberg, T. E., Kolachana, B. S., et al. (2001) Effect of COMT Val108/158 Met genotype on frontal lobe function and risk for schizophrenia. *Proc Natl Acad Sci U S A*, **98**, 6917–22.

Enoch, M. A., Shen, P. H., Xu, K., et al. (2006) Using ancestry-informative markers to define populations and detect population stratification. *J Psychopharmacol*, **20**, 19–26.

Enoch, M. A., Xu, K., Ferro, E., et al. (2003) Genetic origins of anxiety in women: a role for a functional catechol-*O*-methyltransferase polymorphism. *Psychiatr Genet*, **13**, 33–41.

Furmark, T., Tillfors, M., Garpenstrand, H., et al. (2004) Serotonin transporter polymorphism related to amygdala excitability and symptom severity in patients with social phobia. *Neurosci Lett*, **362**, 189–92.

Gogos, J. A., Morgan, M., Luine, V., et al. (1998) Catechol-*O*-methyltransferase-deficient mice exhibit sexually dimorphic changes in catecholamine levels and behavior. *Proc Natl Acad Sci U S A*, **95**, 9991–6.

Goldberg, T. E., Egan, M. F., Gscheidle, T., et al. (2003) Executive subprocesses in working memory: relationship to catechol-*O*-methyltransferase Val158Met genotype and schizophrenia. *Arch Gen Psychiatry*, **60**, 889–96.

Goldman, D. and Ducci, F. (2007) Deconstruction of vulnerability to complex diseases: enhanced effect sizes and power of intermediate phenotypes. *Sci World J*, **7**, 124–30.

Goldman, D., Oroszi, G., and Ducci, F. (2005) The genetics of addictions: uncovering the genes. *Nat Rev Genet*, **6**, 521–32.

Gottesman, I. I. and Gould, T. D. (2003) The endophenotype concept in psychiatry: etymology and strategic intentions. *Am J Psychiatry*, **160**, 636–45.

Greenberg, B. D., Malone, D. A., Friehs, G. M., et al. (2006) Three-year outcomes in deep brain stimulation for highly resistant obsessive-compulsive disorder. *Neuropsychopharmacology*, **31**, 2384–93.

Hariri, A. R., Drabant, E. M., Munoz, K. E., et al. (2005) A susceptibility gene for affective disorders and the response of the human amygdala. *Arch Gen Psychiatry*, **62**, 146–52.

Hariri, A. R., Tessitore, A., Mattay, V. S., et al. (2002) The amygdala response to emotional stimuli: a comparison of faces and scenes. *NeuroImage*, **17**, 317–23.

Heils, A., Teufel, A., Petri, S., et al. (1996) Allelic variation of human serotonin transporter gene expression. *J Neurochem*, **66**, 2621–4.

Heinz, A., Braus, D. F., Smolka, M. N., et al. (2005) Amygdala-prefrontal coupling depends on a genetic variation of the serotonin transporter. *Nat Neurosci*, **8**, 20–1.

Heinz, A., Jones, D. W., Mazzanti, C., et al. (2000) A relationship between serotonin transporter genotype and in vivo protein expression and alcohol neurotoxicity. *Biol Psychiatry*, **47**, 643–9.

Heinz, A., Smolka, M. N., Braus, D. F., et al. (2007) Serotonin transporter genotype (5-HTTLPR): effects of neutral and undefined conditions on amygdala activation. *Biol Psychiatry*, **61**(8), 1011–14.

Holmes, A., Lit, Q., Murphy, D. L., et al. (2003) Abnormal anxiety-related behavior in serotonin transporter null mutant mice: the influence of genetic background. *Genes Brain Behav*, **2**, 365–80.

Hu, X. Z., Lipsky, R. H., Zhu, G., et al. (2006) Serotonin transporter promoter gain-of-function genotypes are linked to obsessive-compulsive disorder. *Am J Hum Genet*, **78**, 815–26.

Ichimiya, T., Suhara, T., Sudo, Y., et al. (2002) Serotonin transporter binding in patients with mood disorders: a PET study with $[^{11}C](+)McN5652$. *Biol Psychiatry*, **51**, 715–22.

Karayiorgou, M., Altemus, M., Galke, B. L., et al. (1997) Genotype determining low catechol-*O*-methyltransferase activity as a risk factor for obsessive-compulsive disorder. *Proc Natl Acad Sci U S A*, **94**, 4572–5.

Karayiorgou, M., Sobin, C., Blundell, M. L., et al. (1999) Family-based association studies support a sexually dimorphic effect of COMT and MAOA on genetic susceptibility to obsessive-compulsive disorder. *Biol Psychiatry*, **45**, 1178–89.

Karoum, F., Chrapusta, S. J., and Egan, M. F. (1994) 3-Methoxytyramine is the major metabolite of released dopamine in the rat frontal cortex: reassessment of the effects of antipsychotics on the dynamics of dopamine release and metabolism in the frontal cortex, nucleus accumbens, and striatum by a simple two pool model. *J Neurochem*, **63**, 972–9.

Kimberg, D. Y., D'Esposito, M., and Farah, M. J. (1997) Effects of bromocriptine on human subjects depend on working memory capacity. *NeuroReport*, **8**, 3581–5.

Lachman, H. M., Papolos, D. F., Saito, T., et al. (1996) Human catechol-*O*-methyltransferase pharmacogenetics: description of a functional polymorphism and its potential application to neuropsychiatric disorders. *Pharmacogenetics*, **6**, 243–50.

Lesch, K. P. and Mossner, R. (1998) Genetically driven variation in serotonin uptake: is there a link to affective spectrum, neurodevelopmental, and neurodegenerative disorders? *Biol Psychiatry*, **44**, 179–92.

Lesch, K. P., Balling, U., Gross, J., et al. (1994) Organization of the human serotonin transporter gene. *J Neural Transm Gen Sect*, **95**, 157–62.

Mattay, V. S., Frank, J. A., Santha, A. K., et al. (1996) Whole-brain functional mapping with isotropic MR imaging. *Radiology*, **201**, 399–404.

Mattay, V. S., Goldberg, T. E., Fera, F., et al. (2003) Catechol *O*-methyltransferase val158-met genotype and individual variation in the brain response to amphetamine. *Proc Natl Acad Sci U S A*, **100**, 6186–91.

McAllister, T. W., Flashman, L. A., McDonald, B. C., et al. (2006) Mechanisms of working memory dysfunction after mild and moderate TBI: evidence from functional MRI and neurogenetics. *J Neurotrauma*, **23**, 1450–67.

McIntosh, A. M., Baig, B. J., Hall, J., et al. (2006) Relationship of catechol-*O*-methyltransferase variants to brain structure and function in a population at high risk of psychosis. *Biol Psychiatry*, **61**, 1127–34.

Meyer, J. H., Houle, S., Sagrati, S., et al. (2004) Brain serotonin transporter binding potential measured with carbon 11-labeled DASB positron emission tomography: effects of major depressive episodes and severity of dysfunctional attitudes. *Arch Gen Psychiatry*, **61**, 1271–9.

Meyer-Lindenberg, A., Nichols, T., Callicott, J. H., et al. (2006) Impact of complex genetic variation in COMT on human brain function. *Mol Psychiatry*, **11**, 867–77.

Murphy, D. L., Li, Q., Engel, S., et al. (2001) Genetic perspectives on the serotonin transporter. *Brain Res Bull*, **56**, 487–94.

Nackley, A. G., Shabalina, S. A., Tchivileva, I. E., et al. (2006) Human catechol-*O*-methyltransferase haplotypes modulate protein expression by altering mRNA secondary structure. *Science*, **314**, 1930–3.

Ohnishi, T., Hashimoto, R., Mori, T., et al. (2006) The association between the Val158Met polymorphism of the catechol-*O*-methyl transferase gene and morphological abnormalities of the brain in chronic schizophrenia. *Brain*, **129**, 399–410.

Oquendo, M. A., Hastings, R. S., Huang, Y. Y., et al. (2007) Brain serotonin transporter binding in depressed patients with bipolar disorder using positron emission tomography. *Arch Gen Psychiatry*, **64**, 201–8.

Ozaki, N., Goldman, D., Kaye, W. H., et al. (2003) Serotonin transporter missense mutation associated with a complex neuropsychiatric phenotype. *Mol Psychiatry*, **8**, 933–6.

Parsey, R. V., Hastings, R. S., Oquendo, M. A., et al. (2006) Effect of a triallelic functional polymorphism of the serotonin-transporter-linked promoter region on expression of serotonin transporter in the human brain. *Am J Psychiatry*, **163**, 48–51.

Paule, M. G., Rowland, A. S., Ferguson, S. A., et al. (2000) Attention deficit/hyperactivity disorder: characteristics, interventions and models. *Neurotoxicol Teratol*, **22**, 631–51.

Pezawas, L., Meyer-Lindenberg, A., Drabant, E. M., et al. (2005) 5-HTTLPR polymorphism impacts human cingulate–amygdala interactions: a genetic susceptibility mechanism for depression. *Nat Neurosci*, **8**, 828–34.

Praschak-Rieder, N., Kennedy, J., Wilson, A. A., et al. (2007) Novel 5-HTTLPR allele associates with higher serotonin transporter binding in putamen: a [(11)C] DASB positron emission tomography study. *Biol Psychiatry*, **62**, 327–31.

Rauch, S. L., Shin, L. M., and Wright, C. I. (2003) Neuroimaging studies of amygdala function in anxiety disorders. *Ann N Y Acad Sci*, **985**, 389–410.

Ressler, K. and Nemeroff, C. (2000) Role of serotonergic and noradrenergic systems in the pathophysiology of depression and anxiety disorders. *Depress Anxiety*, **12**, 2–19.

Roy, A., Hu, X. Z., Janal, M. N., et al. (2007) Interaction between childhood trauma and serotonin transporter gene variation in suicide. *Neuropsychopharmacology*, **32**, 2046–52.

Sabol, S. Z., Hu, S., and Hamer, D. (1998) A functional polymorphism in the monoamine oxidase A gene promoter. *Hum Genet*, **103**, 273–9.

Sawaguchi, T. and Goldman-Rakic, P. S. (1991) D1 dopamine receptors in prefrontal cortex: involvement in working memory. *Science*, **251**, 947–50.

Schindler, K. M., Richter, M. A., Kennedy, J. L., et al. (2000) Association between homozygosity at the COMT gene locus and obsessive compulsive disorder. *Am J Med Genet*, **96**, 721–4.

Seamans, J. K., Floresco, S. B., and Phillips, A. G. (1998) D1 receptor modulation of hippocampal-prefrontal cortical circuits integrating spatial memory with executive functions in the rat. *J Neurosci*, **18**, 1613–21.

Shashi, V., Keshavan, M. S., Howard, T. D., et al. (2006) Cognitive correlates of a functional COMT polymorphism in children with 22q11.2 deletion syndrome. *Clin Genet*, **69**, 234–8.

Shifman, S., Bronstein, M., Sternfeld, M., et al. (2002) A highly significant association between a COMT haplotype and schizophrenia. *Am J Hum Genet*, **71**, 1296–302.

Shioe, K., Ichimiya, T., Suhara, T., et al. (2003) No association between genotype of the promoter region of serotonin transporter gene and serotonin transporter binding in human brain measured by PET. *Synapse*, **48**, 184–8.

Sjöberg, R. L., Ducci, F., Barr, C. S., et al. (2008) A non-additive interaction of a functional MAO-A VNTR and testosterone predicts antisocial behavior. *Neuropsychopharmacology*, **33**(2), 425–30.

Smolka, M. N., Schumann, G., Wrase, J., et al. (2005) Catechol-*O*-methyltransferase val158met genotype affects processing of emotional stimuli in the amygdala and prefrontal cortex. *J Neurosci*, **25**, 836–42.

Sullivan, P. F., Kendler, K. S., and Neale, M. C. (2003) Schizophrenia as a complex trait: evidence from a meta-analysis of twin studies. *Arch Gen Psychiatry*, **60**, 1187–92.

Sullivan, P. F., Neale, M. C., and Kendler, K. S. (2000) Genetic epidemiology of major depression: review and meta-analysis. *Am J Psychiatry*, **157**, 1552–62.

Tjurmina, O. A., Armando, I., Saavedra, J. M., et al. (2002) Exaggerated adrenomedullary response to immobilization in mice with targeted disruption of the serotonin transporter gene. *Endocrinology*, **143**, 4520–6.

Varnas, K., Halldin, C., and Hall, H. (2004) Autoradiographic distribution of serotonin transporters and receptor subtypes in human brain. *Hum Brain Mapp*, **22**, 246–60.

Weickert, T. W., Goldberg, T. E., Mishara, A., et al. (2004) Catechol-*O*-methyltransferase val108/158met genotype predicts working memory response to antipsychotic medications. *Biol Psychiatry*, **56**, 677–82.

Weinshilboum, R. M. and Raymond, F. A. (1977) Inheritance of low erythrocyte catechol-*O*-methyltransferase activity in man. *Am J Hum Genet*, **29**, 125–35.

Wellcome Trust, Case Control Consortium (2007) Genome-wide association study of 14,000 cases of seven common diseases and 3,000 shared controls. *Nature*, **447**, 661–78.

Zhu, G., Lipsky, R. H., Xu, K., et al. (2004) Differential expression of human COMT alleles in brain and lymphoblasts detected by RT-coupled 5′ nuclease assay. *Psychopharmacology (Berl)*, **177**, 178–84.

Zubieta, J. K., Heitzeg, M. M., Smith, Y. R., et al. (2003) COMT val158met genotype affects mu-opioid neurotransmitter responses to a pain stressor. *Science*, **299**, 1240–3.

Progress and future directions

Introduction to Section 5

In this section (Chapter 23), Rumsey and Ernst highlight the value of a developmental perspective, progress to date in understanding normal and atypical neurodevelopment through the use of neuroimaging, and technological advances that hold promise for continued scientific progress. They identify critical needs in pediatric neuroimaging and discuss potential translational applications. Also discussed are current limitations and impediments to progress. Finally, the emerging field of neuroethics is addressed.

Neuroimaging in developmental clinical neuroscience today

Judith M. Rumsey and Monique Ernst

Introduction

Since the initial publication in 2000 of our volume entitled *Functional Neuroimaging in Child Psychiatry* (Ernst and Rumsey, 2000), there has been a burgeoning of neuroimaging research in pediatric populations focused on understanding both normal development and childhood disorders. This trend stems in large part from technological advances, in particular the advent of safe, non-invasive magnetic resonance (MR) imaging methods in the 1980s and 1990s, their increasing diversity, and improved image processing tools. The application of these methods to pediatric populations has begun to yield a wealth of information concerning the neurocircuitry underlying both healthy development and developmental neuropsychiatric disorders. This literature has benefited from the increasing sophistication and precision with which adult studies have mapped the neural substrates of executive functions, face and object perception, emotion processing, language, and other critical domains of functioning. Moreover, these advances in imaging have coincided with changing conceptualizations of mental illnesses, conditions now recognized to be largely disorders of abnormal brain development. This changing conceptualization, in tandem with technological advances in imaging, sets the stage for future work to address major questions on development through an integrative neuroscience approach, relying heavily on brain imaging. Going forward, the realization of the full potential of neuroimaging approaches in developmental clinical neuroscience will require a greater focus on longitudinal research and the integration of clinical, systems-level research with basic neuroscience research at the genetic and cellular-molecular levels.

In this final chapter, we discuss the rationale for a developmental approach, highlight selected contributions of neuroimaging in developmental clinical neuroscience, and describe key technological advances enabling us to expand our knowledge base. We then outline some critical needs and future research directions, including translational applications. Finally, we briefly consider ethical issues that have emerged in parallel with the exponential increase in neuroimaging research.

A developmental perspective

Developmental neuroimaging research seeks to identify age-related changes in brain structure, function, and chemistry. Its goals are to understand critical developmental transitions and establish normative trajectories against which to evaluate deviations associated with the risk and manifestation of disease. Why is this important? Such knowledge would help constrain theories of the

Neuroimaging in Developmental Clinical Neuroscience, eds. Judith M. Rumsey and Monique Ernst. Published by Cambridge University Press. © Cambridge University Press 2009.

neurobiology of childhood-onset disorders and inform the development of novel treatments.

The earliest neurobiological events leading to disability may occur at critical stages in neural development, long before the emergence of symptoms. The delayed expression of early pathology may stem from the limited behavioral repertoire of infants and young children, as well as from the cascading effects of early neuropathology on later-developing circuitry (Muller and Courchesne, 2000; Schultz, 2005). For example, the protracted development of prefrontal systems may not permit executive function deficits to be fully manifest until adolescence, a time when these systems are coming on line and environmental demands for self-organization are increasing. Alternatively, subcortical neuropathology may, over the course of development, alter the normal development of circuits involving prefrontal cortex (Krasnova et al., 2007). Plasticity may result in compensatory neural changes that can be adaptive or maladaptive (Muller, 2007; Pascual-Leone et al., 2005).

Furthermore, disabilities that strike in childhood may derail a variety of neurodevelopmental processes important for functioning throughout life. Early neuropathology impacts brain–behavior development, not only via direct biological effects, but also through its effects on experience-dependent brain organization. For example, limbic-based pathophysiology in autism may result in gaze aversion (Schultz, 2005), which in turn may preclude the development of expertise in face perception and limit the specialization of the fusiform gyrus for face processing. Such an early developmental failure may impede the development of social cognition (Pelphrey and Perlman, Chapter 5). Understanding the normal trajectories of structural and functional brain development, including experience-dependent cortical specialization, circuit formation, connectivity, and neurochemistry, will provide a basis for identifying risk factors and early biomarkers of disease.

For many diseases, early identification and treatment are key in delaying, slowing or halting the progression of illness. Furthermore, identification of the earliest precursors along a causal pathway to disease can elucidate pathophysiology and contribute to the development of biomarkers with which to predict illness, monitor health status and treatment effects, and ultimately prevent disease. While many adult-onset psychiatric disorders are thought to have a developmental component, the study of adults will frequently capture the effects of treatments (e.g., psychotropic medications), as well as the effects of chronic disease, on the brain. The study of development in disease offers unique opportunities for identifying specific neuroimaging correlates of primary pathophysiology, independent of these potential confounds. Furthermore, although not yet realized, a developmental approach may permit interventions to be targeted toward specific developmental epochs.

Deviations in developmental trajectories associated with neuropsychiatric disease have only begun to be explored, but hold promise for answering important questions. Given neural risk factors for disease, what events (neural or environmental) transform vulnerability into disease? What factors are protective against symptom onset? In the case of early-onset disorders [e.g., attention deficit/ hyperactivity disorder (ADHD), autism], what are the neural predictors and correlates of good versus poor outcomes? Given the heterogeneity associated with many childhood-onset disorders, can phenotypes be isolated based on neural and genetic correlates? Which deviations in neuroimaging profiles reflect primary pathophysiology versus compensatory adaptations?

Contributions of neuroimaging in developmental clinical neuroscience

Normal developmental processes

As highlighted by Lu and Sowell (Chapter 1), the use of longitudinal designs and advances in image analysis are yielding more precise and comprehensive descriptions of structural brain development than heretofore possible. Methods of whole-brain image analysis now include volumetric image analyses, voxel-based morphometry, cortical pattern matching, measures of cortical thickness, and deformation methods for identifying differences in the shape of various structures. Findings include curvilinear trajectories of gray matter maturation across many cortical regions, the continued development of peri-Sylvian asymmetries between childhood and adulthood, and great inter-subject variability in inferior parietal/posterior temporal regions. Although few studies have related structural brain maturation to cognitive or behavioral development, a recent longitudinal study (Lu et al., 2007) documenting a relationship between left inferior gyral cortical thickening and improved phonological skill suggests that regional structural changes may indeed relate to cognitive-behavioral changes in a meaningful way.

Functional magnetic resonance imaging (fMRI) studies have begun to identify separable developmental trajectories for specific cognitive and affective processes in healthy children and adolescents. Critical to many of these processes are prefrontal and limbic regions, thought to be involved in many psychiatric disorders. The relatively high spatial resolution of fMRI is allowing researchers to distinguish prefrontal subregions involved in discrete aspects of cognition and to identify changes in functional connectivity that underlie maturing cognitive abilities. For example, the strengthening of prefrontal–parietal and prefrontal–subcortical (e.g., striatal) connections appears to be critical to improvements in cognitive control seen over middle childhood and adolescence (Bunge and Crone, Chapter 2). Work in emotion regulation (Gotlib and Joormann, Chapter 3) has begun to distinguish the involvement of dorsal prefrontal cortex in cognitive reappraisal, a regulatory strategy for dealing with affect, from ventral prefrontal substrates associated with more automatic affective responses. An emerging developmental literature suggests that, while the neural substrates of primary emotions are similar in adults and children, children show greater, more diffuse activation of prefrontal cortex when trying to regulate affect,

highlighting the mutual dependence of emotion regulation and cognitive control on the maturation of prefrontal cortex.

As reviewed by McNealy et al. (Chapter 6), early perceptual competencies set the stage for expressive language development. While findings in school-age children suggest qualitative similarities to the adult model of language organization in the brain, the neural mechanisms supporting language appear to undergo a fine tuning as they mature. While hemispheric specialization for language may be present during the preschool years, the contributions of the two hemispheres may shift over development as a function of both age and the linguistic task. Higher-level linguistic functions, such as discourse comprehension and the integration of prosodic information, may continue to engage bilateral regions well into adolescence and beyond, with the nature and degree of lateralization dependent on the specific task.

New areas of investigation include social cognitive development and goal-directed behavior, as reflected in decision making. Pelphrey and Perlman (Chapter 5) propose a working model of social cognitive development which holds that early-developing social perceptual processes involving eye gaze and facial emotion provide building blocks for the development of social cognition (the understanding of others' goals and intentions, "theory of mind"). Ernst and Hardin (Chapter 4) outline a heuristic model of goal-directed behavior in which decision making is modulated by a functional balance involving the amygdala, ventral striatum, and prefrontal cortex, each following distinct, but interactive, developmental trajectories.

Selected findings in developmental neuropsychiatric disorders

Although most neuroimaging research on disorders has thus far involved cross-sectional designs and small samples, some interesting findings and hypotheses have begun to emerge. In ADHD, reductions are seen in whole brain volumes, with the caudate nucleus and prefrontal cortex comprising regions that show the most consistent volumetric reductions (Epstein, Chapter 7). Functional imaging studies suggest hypoactivation of fronto-striatal regions, consistent with findings implicating perturbed dopamine transmission in ADHD. Furthermore, neuroimaging studies suggest that stimulants may help to normalize fronto-striatal brain function and that the neuroimaging profiles of stimulant and non-stimulant treatments (atomoxetine) may differ (Pliszka and Glahn, Chapter 21).

Despite many inconsistent findings in autism, abnormally large increases in total brain volume within the first few years of life is emerging as a compelling finding (Zilbovicius et al., Chapter 8). Functional imaging studies examining social-cognitive neural networks are reporting abnormalities detected using a variety of tasks involving facial expression, voice perception, eye gaze perception, biological motion, and mentalizing. In fragile X syndrome, deficient synthesis of the fragile X mental retardation gene 1 (*FMR1*) protein (FMRP) results in impaired cortical dendritic spine maturation and a failure of normal synaptic pruning, leading to structural and functional irregularities that can be visualized using neuroimaging (Rivera and Reiss, Chapter 13). These include a "dose–response" effect of FMRP on brain activation, reduced fractional anisotropy in fronto-striatal and parietal sensorimotor tracts, enlarged caudate nuclei, hypoplasia of the cerebellar vermis, and enlarged fourth and lateral ventricles.

In Tourette syndrome (TS) and obsessive-compulsive disorder (OCD), two genetically and phenotypically related disorders, findings suggest disturbances in fronto-striatal systems subserving inhibitory control (Marsh et al., Chapter 12). The striatal portions of these circuits are primarily affected in TS, whereas their prefrontal portions are implicated in OCD. Enlarged prefrontal volumes seen in children with TS are associated with reduced tic severity and are hypothesized to result from a compensatory process associated with the continuous need to suppress tics.

In schizophrenia, findings reported early in disease, i.e., in first-episode patients, include reductions in whole brain volumes, temporal, and prefrontal gray matter and increased ventricular volumes; hypoactivation of prefrontal cortex; and reduced *N*-acetylaspartate (NAA), a putative marker of neuronal integrity, in prefrontal and hippocampal regions. Some of these findings may be present in unaffected relatives, thus reflecting genetic risks (Keshavan et al., Chapter 9). Superimposed on these apparent developmental differences, neurodegenerative changes may evolve with the progression of symptoms.

In bipolar disorder, neuroimaging data implicate the involvement of a cortico-limbic neural system subserving emotion regulation and motivation. Kalmar et al. (Chapter 10) propose a developmental model that suggests that mesial temporal lobe (amygdala, hippocampus) and ventral striatum abnormalities are present by puberty, whereas those seen in ventral prefrontal cortex progress over adolescence and may not be fully expressed until early adulthood. Studies of adolescent bipolar disorder reporting differences between remitters and non-remitters treated with olanzapine suggest that changes in NAA are related to clinical status rather than to drug, highlighting the need to distinguish drug effects from the effects of mood shifts (Pliszka and Glahn, Chapter 21).

The new field of imaging genetics (Goldman et al., Chapter 22) illustrates the promise of an intermediate phenotype approach for deconstructing complex behavioral phenotypes, including schizophrenia and mood and anxiety disorders. Two genes, *COMT* and *5HTT*, have been linked to cognition and emotion, respectively, with *COMT* allelic variants accounting for variations in prefrontal cortical responses to cognitive challenges, and *5HTT* allelic variants associated with amygdala reactivity to emotional stimuli in adults. Importantly, the sample sizes required for establishing these associations are much lower than those required by classic linkage and association approaches. While these genes have been studied in relationship to task performance in children, to date few studies have examined their effects on brain function in children.

Technological advances

Imaging modalities

Technological advances have provided a variety of complementary neuroimaging techniques that together will offer considerable insight into pediatric brain maturation. Their common reliance on conventional MRI scanners, improvements in hardware, software, and the speed of image acquisitions now allow data to be collected in a fairly efficient manner. The integration of data from these techniques will make it possible to achieve a better understanding of normal development, disease vulnerability and onset, and response to therapeutics.

As reviewed by Bandettini (2007), since its beginning in 1991, blood-oxygen-level-dependent (BOLD) fMRI, the functional technique most widely used in pediatric neuroimaging, has seen a number of improvements in hardware (coils, higher field strength magnets), methods (e.g., pulse sequences, post-processing, multimodal integration techniques), signal interpretability, and paradigm designs (e.g., event-related). There is now confidence that BOLD fMRI signal changes reflect meaningful increases and decreases in underlying neuronal activity. Importantly, there is an increasing sophistication with respect to the practical, technical, methodological, and analytic issues involved in scanning pediatric populations (Levita et al., Chapter 17).

Single-voxel magnetic resonance spectroscopy (MRS) is now available on most clinical scanners, but is limited in spatial coverage, resolution, and spatial localization (Silveri et al., Chapter 18). Chemical shift imaging (CSI) permits an improved evaluation of the spatial distribution of metabolites. Although limited by low resolution and sensitivity, MRS is the only modality that directly yields information on brain chemistry without the use of ionizing radiation, permitting its widespread use in pediatric populations. Efforts toward enhancements for measuring neurotransmitters implicated in mental illnesses, such as gamma-aminobutyric acid (GABA) and glutamate, continue, with an eye

toward an increasing role in drug development (Mason and Krystal, 2006; Novotny et al., 2003).

Diffusion tensor imaging (DTI) makes it possible to segment white matter into smaller anatomic units and to monitor its status in a regional, tract-specific manner (Kim, Chapter 19). By directly evaluating the intrinsic tissue properties of the brain's white matter, DTI offers quantitative measures that reflect the macroscopic three-dimensional architecture of white matter. The assessment of white matter tracts appears particularly important in development, due to evidence of robust developmental changes in the nature of the brain's connections. Given that myelination and pruning continue through adolescence, DTI offers a technique for assessing brain maturation throughout the developmental period. As many psychiatric disorders (e.g., schizophrenia, autism) are hypothesized to involve perturbations in white matter development and connectivity (Muller, 2007; Segal et al., 2007), DTI can be used to investigate the timing, nature, and phenotypic correlates of deviations in white matter development.

While BOLD fMRI holds advantages for cognitive activation studies (e.g., larger signal changes, better temporal resolution, efficient whole brain coverage) over arterial spin labeling (ASL) perfusion fMRI, ASL may nevertheless offer unique and complementary advantages over BOLD (Wang et al., Chapter 20). Although not widely available, ASL provides absolute, quantitative measures of regional cerebral blood flow (rCBF), greater capillary specificity relative to BOLD (which may improve localization), and potentially greater long-term temporal stability (given the minimal signal drift). As such, ASL may offer advantages for longitudinal studies of children. Given its limited temporal resolution, ASL may be better suited for capturing changes associated with longer time scales, such as sustained pharmacological effects, changes in mood states, changes in baseline, e.g., resting states, or certain behavioral manipulations with sustained effects. While widespread application of ASL has been limited by technical factors, recent advances have improved its sensitivity and resolution.

As interest in functional studies of children younger than school age grows, investigators are increasingly looking to the development of imaging techniques that might be applied in naturalistic environments, as opposed to a scanning suite. Although lacking the spatial resolution of fMRI, near-infrared spectroscopy (NIRS) can detect changes in cortical concentrations of hemoglobin associated with neural activity (Taga et al., 2007). Its portability allows it to be used in naturalistic settings, including those in which children can engage in social interaction (Obrig and Villringer, 2003). With continued improvements, NIRS may assume a more active role in developmental research.

Neuroinformatics

The wealth of information produced by today's research enterprise has created a need for computational models and analytic tools directed toward knowledge management and integration. Neuroinformatics, a field that stands at the intersection of neuroscience and information technology, has evolved to address these needs (Huerta et al., 2006). This emerging field provides tools (e.g., querying and data mining approaches), resources (e.g., databases), and computational analysis techniques (e.g., statistical or probabilistic tools), and optimally provides for their interoperability.

A number of large-scale neuroinformatics initiatives related to neuroimaging are being sponsored by the National Institutes of Health (NIH) to meet these needs (Huerta et al., 2006). Several have adopted psychiatric disorders as driving biological projects to ensure the clinical relevance of the tools and infrastructure developed. The Neuroimaging Informatics Tools and Resources Clearinghouse (NITRC) (www.nitrc.org) is meant to catalog all fMRI tools and related resources, as well support communications concerning them (e.g., software reviews, workshop announcements). Several centers, each composed of multidisciplinary teams including medical investigators, computer scientists, and software engineers, are working on a wide variety of projects, an adequate description of which is beyond the scope of this chapter. Interested readers are encouraged to visit the centers' websites (listed below) to familiarize themselves with the range of projects supported.

The National Alliance for Medical Image Computing (NA-MIC) (http://na-mic.org) develops computational tools for the analysis and visualization of imaging data (e.g., DTI, structural MRI, fMRI) and oversees the dissemination of these tools and related training workshops. The Center for Computational Biology (CCB) (http://cms.loni.ucla.edu/CCB) develops methods directed toward the quantitative integration of brain anatomical, physiological, and genetic data. The CCB has developed the computational brain atlas for managing multidimensional data (Toga et al., 2006; see www.loni.ucla.edu/Atlases/). Its digital nature allows algorithms to be applied to imaging data, making the atlas comparable to a dynamic database capable of generating statistical data on cortical cytoarchitecture, receptor distributions, white matter tracts and projections, and fMRI activations for various populations.

The Biomedical Informatics Research Network (BIRN) (www.nbirn.net) supports the development of computational tools and infrastructure for a digital framework to enable cross-site collaborations and data sharing. A BIRN Coordinating Center develops and implements technology to support federated databases, allowing for data and computational resources distributed across sites to be integrated, as an alternative to centralized data repositories. Three test beds focus on structural neuroimaging (Morphometry BIRN), functional neuroimaging (Function BIRN), and multi-scalar data from a molecular to whole brain level (Mouse BIRN). The BIRN is developing robust imaging paradigms and calibration methods to achieve cross-site data comparability to enable multi-site studies, critical for efforts requiring large patient samples.

Specifically relevant to developmental clinical neuroscience are the National Database for Autism Research (NDAR) (http://ndar.nih.gov), now under development using the BIRN infrastructure, and the NIH-Pediatric MRI Data Repository (www. NIH-PediatricMRI.org), a centralized database of longitudinal imaging data from normal children, that has recently become available to researchers and which will be integrated into BIRN. The NIH-Pediatric MRI Data Repository contains data from a multi-site study that has collected structural MRI, ancillary DTI and MRS, and correlated clinical/behavioral data from approximately 500 children, ages newborn through late adolescence/ early adulthood (Almli et al., 2007; Evans et al., 2006; Waber et al., 2007). Data in the repository have been de-identified to comply with federal regulations and are freely available to qualified biomedical and biobehavioral researchers whose institutions are covered by a Federalwide Assurance (FWA) (see Hermes, Chapter 16) and who agree to the terms of a Data Use Certification designed to protect human subjects. See Appendix B for additional information on this and other neuroinformatics resources.

Critical needs and future directions

Understanding normal development

An in-depth understanding of normal development will be critical for elucidating the pathophysiology of risk, onset, and progression of disease. This will require the use of longitudinal designs, the development of robust cognitive and behavioral assays for use across ages, data acquisition and analysis methods with demonstrated stability, and the integration of data across domains (behavioral, physiological), imaging modalities, and scales (cellular-molecular, systems level). Longitudinal, within-subject designs used with structural imaging studies (e.g., Lenroot et al., 2007) have increased the statistical power with which to discern developmental trajectories within the high degree of inter-subject variability seen in brain anatomy. Such studies require the monitoring and control of machine drift and careful evaluation of the impact of software upgrades (Han et al., 2006; Jovicich et al.,

2006). There remains a need to validate structural imaging measures (e.g., cortical thickness) against histological measures. Although advances in structural image analysis are now capable of producing a wealth of data describing regional maturational changes, relationships among these changes have yet to be explored in detail. How do changes in various volumetric measures, cortical thickness, and the shape and size of various structures relate to each other, to total brain volumes, and to demographic, cognitive, and behavioral measures in normal children? Answers to such questions will be facilitated by the availability of a shared normative longitudinal dataset (Evans et al., 2006).

Blood-oxygen-level-dependent fMRI studies seek to identify the neural networks that underlie cognitive or affective processes, as captured by task-specific activations. Given that children develop at different rates, immature networks in pediatric populations will likely prove to be more variable than mature networks found in adults, making longitudinal studies critical for identifying maturational changes. The challenges presented by such studies are great and, to date, very few functional imaging studies have used longitudinal designs (but see Durston et al., 2006).

Longitudinal studies require measures with good stability, such that changes seen with age can be attributed to development, unconfounded by other potential sources of variance. Machine drift (Friedman and Glover, 2006), age-related anatomical differences (Lu and Sowell, Chapter 1), physiological variables (e.g., arousal, emotional state; Gotlib and Joormann, Chapter 3), time of day, site differences (for multi-site studies) (Friedman et al., 2008), and other contextual variables all may obscure true brain maturational differences. Few functional imaging studies report test-retest reliabilities of any sort. Research in adults has shown moderate reliability for BOLD-fMRI-detected levels of amygdala activations to fearful faces over an 8-week period, but lower reliability for neutral faces or fearful versus neutral contrasts (Johnstone et al., 2005). Test-retest reliabilities for adults performing a sensorimotor task with BOLD fMRI over 2 days, reported by the Function BIRN (Friedman et al., 2008), ranged from 0.21 to 0.80, depending on the region of interest and its size, value used (e.g., median versus maximum percent signal change), field strength (better for 3 T than at 1.5 T), and the number of runs averaged. Whether children and adolescents will yield less reliable functional imaging data is unknown. There is some suggestion in the literature that patterns of activation (Berl et al., 2006; Voyvodic, 2006) may offer better reliability than levels of activation for detecting age-related changes. However, patterns of activation may be more variable in children than in adults.

Changes seen with functional neuroimaging may be qualitative or quantitative. Johnson (2003) proposes three alternative models of functional brain development. New regions may *come on line,*" allowing new behavioral abilities to appear. Changes in the interactions between regions that were previously partially active (*"interactional specialization"*) may support the appearance of new capabilities. Certain regions may become active with the acquisition of new skills and decrease their activity as expertise increases (*"skill learning"*). Similarly, Berl et al. (2006) propose several models of developmental alterations in brain functional organization. *Discontinuous transitions* are seen when new regions come on line; there is little overlap between immature and mature networks. With *progressive specialization or focalization,* diffuse, widespread, low-intensity activation becomes more intense and focal, increasingly resembling the adult model. The *regional weighting model* holds that the degree of engagement of regions within a network changes systematically with development. Vertical (subcortical-cortical) or horizontal (anterior-posterior, left-right) weightings may change.

Any or all of these models may apply to specific domains of functioning and/or specific developmental transitions. Further complicating matters are the existence of normal variants (e.g., right-lateralized language function) and adaptive or atypical variants, e.g., persistence of immature networks, use of novel strategies, and alternative networks. Attention to individual differences and the development and validation of robust paradigms for capturing these differences are discussed by Berl et al. (2006) and Bush et al. (2003). A number of creative approaches for measuring individual variability are reviewed by Berl et al. (2006).

The need to control age-related performance differences may create a dilemma when studying development. Levita et al. (Chapter 17) describe a number of rigorous approaches, behavioral and statistical, to this problem. Nonetheless, designing tasks that tap the same construct across different age groups remains a challenge for developmental studies. Equal accuracy or reaction times do not guarantee equal cognitive effort or the use of comparable strategies. Simple, basic sensorimotor or psychophysical tasks may help to anchor, track, and index processes with potential cascading influences on higher-level processes. Parametric task designs incorporating several levels of difficulty with which to generate "cognitive-dose-response curves" (Berl et al., 2006) appear promising, but increase the length of data collections, presenting additional challenges for pediatric neuroimaging studies. Training paradigms involving the teaching of new skills may be informative for determining whether the neural circuitry for acquiring new knowledge, skills, or expertise is in place and how well it is functioning. As skill or learning increases, corresponding neural changes can be examined and alternative models of functional brain development tested.

Studies utilizing DTI, ASL, and MRS are beginning to provide a picture of age-related changes in white matter architecture, regional cerebral blood flow, and brain neurochemistry, respectively (Cascio et al., 2007; Silveri et al., Chapter 18; Wang et al., Chapter 20). As with anatomic and functional BOLD MRI, longitudinal data are needed. Much remains to be learned from these approaches, which may provide absolute quantitative measures, and their integration with each other and the more widely used anatomic and BOLD fMRI techniques. As discussed above, each technique offers unique information, advantages, and limitations. Given the methodological complexities presented by BOLD fMRI studies of development, ASL may provide a useful adjunct to or alternative to BOLD depending on the questions being asked. Resting perfusion may be correlated with diagnosis or behavioral measures collected outside the scanner and offer greater stability for longitudinal studies. Coupling BOLD fMRI with perfusion, MRS, and anatomic MRI may yield more generalizable knowledge than BOLD alone, as well as further improving our understanding of BOLD signals.

Understanding atypical development

Studies of children near the time of onset of disease will allow the effects of diseases to be evaluated free of the confounds associated with treatment and disease progression. Studies of at-risk populations may permit predictors of disease to be identified. Within disorders with heterogeneous outcomes, imaging might be helpful in mapping the trajectories associated with the persistence versus remission of symptoms and impairments. The integration of data across imaging modalities will offer a more comprehensive picture of the imaging correlates of disease. Studies are needed to determine which modalities will be most helpful for specific diseases or purposes.

Studies of childhood disorders raise concerns about cohort effects, comorbidities, and greater potential for confounds associated with differences in cognitive strategies, performance, and anxiety or mood states. Cross-sectional studies comparing different age cohorts run the risk of confounds when older subjects differ in important ways, such as severity or persistence, from younger subjects with the same disorder. Longitudinal studies can avoid such confounds and may better distinguish primary pathophysiology from secondary compensatory changes. Comorbidities can be expected to increase variability in patient populations. As patients are more likely than controls to have cognitive deficits and functional imaging paradigms are likely to focus on domains of impairment, equating performance and controlling non-specific sources of variability may be difficult.

Where research goals require multi-site studies for adequate sample sizes, the ability to standardize data acquisitions and to ensure data comparability across time and sites presents even greater challenges. The adoption of core protocols across sites is but a first step in establishing comparability. Systematic cross-site analyses performed in the Function BIRN have demonstrated site differences exceeding inter-subject differences (Friedman et al., 2008). Nonetheless, calibration methods developed in the Function BIRN (use of a special phantom, use of a breath hold "task" as a surrogate for maximum BOLD response), standardization of response devices, and analytic strategies that hold promise for reducing inter-site sources of variance are evolving (Thomason et al., 2007).

Resting perfusion studies may provide a stable, quantitative marker to complement conventional BOLD fMRI activation paradigms and minimize variability across scanners and sites. Such studies may reveal default networks, thought to reflect the brain's intrinsic (non-evoked) activity. These networks, composed of regions that consistently show decreased activity in response to a variety of tasks, are hypothesized to maintain a state of preparedness, facilitate responses to stimuli, and potentially sculpt communication pathways in the brain (Raichle and Snyder, 2007). Although investigations of adult psychopathology have begun to suggest abnormalities in default mode connectivity (Garrity et al., 2007), it remains to be seen whether resting states will yield data that are sensitive to pediatric psychopathology. However, a recent hypothesis, proposed by Sonuga-Barke and Castellanos (2007), holds that default-mode processing protrudes into task-related processing in ADHD, disrupting attention and goal-directed activity and potentially accounting for increased variability in task performance.

While this volume has focused on the use of human neuroimaging across development, the correlational nature of imaging data necessitates their integration with basic neuroscience in order to achieve a deeper understanding of causal relationships. Thus, studies in rodents and non-human primate models directed toward understanding development at the cellular-molecular level and toward building translational bridges to human systems-level research are critical. To date, relatively little research in animals has focused on developmental transitions. Studies directed toward understanding to the fullest extent possible the formation of neurocircuitry subserving various processes and domains of functioning will deepen our understanding of the signals detected through the use of human neuroimaging. The development of parallel behavioral paradigms (e.g., response inhibition, fear conditioning, reward learning) and imaging protocols for use in animals holds translational potential. Further, little is known about the ways in which gene expression changes over the course of development. Preliminary human neuroimaging studies involving twin samples suggest changing genetic and environmental contributions to brain morphology over the course of development (Lenroot et al., 2007). Basic research might help elucidate the underlying mechanisms of such changes.

Biomarker and therapeutics development

Despite unprecedented progress, there is as yet no role for neuroimaging in guiding the clinical diagnosis of developmental neuropsychiatric disorders or therapeutic decision making. At this point in time, neuroimaging remains a research tool in child psychiatry and psychiatry in general. Major translational research applications involve the mapping of neurocircuitry and elucidation of pathophysiology toward the goals of dissecting the heterogeneity characteristic of disorders, developing biomarkers, predicting disease, developing novel interventions, and individualizing treatments.

Importantly, clinical applications demand the ability to reliably distinguish patients with a disorder from controls and from patients with other disorders, to predict treatment responses at the level of the individual, or to monitor biomarkers robustly associated with disease or treatment response (Bush et al., 2005). Although nuclear

medicine techniques (PET, SPECT) allow the direct assessment of neurotransmitter function and quantification of drug effects on specific receptors (Meyer et al., 2004), their use in children is limited by ethical constraints (Arnold et al., 2000). Findings in adults cannot be generalized to children, given structural, functioning, and neurochemical differences in the developing brain (Lu and Sowell, Chapter 1; Chugani, 2004; Durston et al., 2001). Thus, strategies for dealing with the limitations in mapping brain neurochemical development in children include the use and further development of MR-based imaging techniques such as MRS and fMRI (BOLD, ASL) for therapeutics and biomarker development.

Pharmacologic MRI (phMRI) combines the administration of drugs with MR techniques to examine drug effects on brain structure, metabolism, and function. Recent interest in this approach on the part of the pharmaceutical industry suggests an expanding role for phMRI in drug discovery (Honey and Bullmore, 2004; Wise and Tracy, 2006). By providing an imaging assay of drug effects on relevant brain function, fMRI holds potential for biomarker development. New compounds can be evaluated in small proof-of-concept studies to yield initial data on efficacy and mechanisms of action. Although fMRI does not provide direct information about molecular drug action at the level of receptors, appropriate experimental design and controls, together with corroborating evidence, may yield reasonable inferences about effects on receptor systems (Wise and Tracy, 2006). Limitations to the interpretation of the BOLD response present significant challenges, but might be addressed via integration with information obtained from ASL, conventional electroencephalography or evoked potential measures (Matthews et al., 2006). Functional MRI might provide a surrogate endpoint for treatment trials, following successful validation of specific fMRI markers. A recent review (Honey and Bullmore, 2004) has highlighted the importance of genotype and cognitive phenotype as conditions determining drug effects on brain activation, thus opening up possibilities for tailoring treatments to specific groups or individuals.

Given the safety of these techniques, MR-based neuroimaging may in the future assume a significant role in pediatric drug trials. The ability to study children repeatedly may allow investigators to predict treatment responses, identify the circuits involved in therapeutic effects, dissociate the immediate and long-term effects of medications from improvements in symptoms, and detect potential adverse effects on the brain across developmental epochs.

Little is known about the similarities versus differences in the effects of pharmacological versus psychotherapeutic interventions on brain mechanisms. Data in adults suggest that these two approaches work via different pathways in depression, while engaging similar circuitry in obsessive-compulsive disorder (Linden, 2006). However, preliminary data suggest that the two approaches may impact different aspects of brain functioning in pediatric obsessive-compulsive disorder (Benazon et al., 2003). Thus, another promising avenue for research involves the investigation of psychotherapeutic interventions on brain structure, function, and neurochemistry (Linden, 2006). Increasing interest in novel interventions targeted toward altering the function of specific neural circuits (e.g., cognitive training, Westerberg and Klingberg, 2007; non-invasive brain stimulation, Fregni and Pascual-Leone, 2007) may make use of neuroimaging measures both for baseline assessments and for monitoring and evaluating treatment effects.

Other potential therapeutic applications involve the use of real-time BOLD fMRI in interventions themselves. Recent advances in acquisition techniques, computational power and algorithms now permit data to be processed immediately, continuously updated, and displayed in real time, allowing its experimental use by investigators or subjects as neurofeedback to alter activity in specific brain regions (Weiskopf et al., 2007). As an adjunctive therapeutic tool, it might also be used, possibly in conjunction with immersive virtual reality systems, to evaluate the effects of cognitive behavioral therapeutic approaches or other interventions on neural activity in targeted brain regions.

Similarly, other emerging technologies, such as near-infrared spectroscopy (NIRS), hold potential for use in neurofeedback applications conducted in naturalistic environments.

Neuroethics

The increasing use of neuroimaging and other neuroscience research approaches, and our growing knowledge of the neural bases of behavior, personality, consciousness, and morality have led to the development of a specialized field of bioethics termed "neuroethics" (Illes and Bird, 2006). Neuroethics encompasses both practical safety and ethical issues encountered in the conduct of neuroimaging and other neuroscience research and broader societal issues associated with the use of neuroscience findings.

Among the practical issues impacting research are how to deal with incidental findings discovered on brain scans during the course of research (Illes and Bird, 2006) and how to ensure MRI safety and the protection of human subjects in non-clinical (e.g., university) research settings, where medical and emergency personnel are frequently not readily available (NIMH Council Workgroup on MRI Research Practices, 2007; see Appendix B). Of particular relevance to pediatric neuroimaging is the use of sedation, as children, particularly young or impaired children, are frequently less able than adults to lie motionless in a scanner without sedation. Other issues include the potential uses of in utero neuroimaging (Illes and Raffin, 2005), child discomfort, the prediction of intelligence and talent, and tendencies on the part of the media and the public to overinterpret group-averaged brain images as meaningful and causative at the level of the individual (Hinton, 2002; Illes and Raffin, 2005). Specific ethical problems are raised by research using nuclear medicine methods, such as positron emission tomography (PET). While offering potential for mapping the development of specific neurotransmitter systems, thus informing the identification of molecular targets and developmental windows for intervention (Chugani, 2004), safety issues concerning exposure to radiation limit the use of PET in children. For a discussion of ethical issues concerning the use of nuclear medicine techniques with children, see Munson et al. (2006).

Beyond the basic safety and ethical issues lie broader societal, legal, and philosophical implications that stem from our ability to monitor and potentially alter brain function using pharmacological, neurophysiological, or other interventions. For example, while clinical treatments currently target cognitive deficits associated with disorders, will our increasing knowledge base support cognitive enhancements beyond what is considered normal for reasons of academic or other competition? Would this be desirable in the larger scheme of things? Within the legal arena, structural brain imaging research demonstrating continued age-related changes throughout adolescence has been used to support judicial arguments for the differential treatment of adolescents (versus adults) in the criminal justice system (Beckman, 2004; Levin, 2006). How should information gleaned from the ability to evaluate the neural basis of traits such as empathy and moral capacity that might predict problematic behavior, such as delinquency, be used? From a philosophical perspective, our growing knowledge of the neural bases of decision making is fueling discussions concerning core issues of personal responsibility and human nature (see Farah, 2005; Farah and Heberlein, 2007). Such issues will doubtlessly be increasingly debated in legal, bioethics, and neuroscience publications. See Appendix B for a listing of selected neuroethics resources.

Conclusions

Substantial progress has been made toward the development of neuroimaging methods applicable to the investigation of human brain maturation. A developmental approach offers substantial benefits for understanding both normal maturational trajectories and those associated with disease,

holding potential for the early identification of disease and risk factors, the prediction of treatment response, and treatment development. Progress has been made in furthering our understanding of developmental neuropsychiatric disorders and pharmacological treatment responses. Although barely explored in pediatric populations, imaging genetics holds promise for elucidating genetic contributions to variations in brain maturational trajectories.

Advances in imaging technology are offering opportunities for more comprehensive assessments of the developing brain. Both the integration of data across imaging modalities and an improved understanding of the unique strengths of each modality promise to yield a more complete picture of healthy brain maturation and disease. Greater integration with basic neuroscience approaches, including animal models, is needed to achieve a deeper elucidation of the cellular-molecular mechanisms contributing to neuroimaging signals. Although neuroimaging modalities in child and adolescent psychiatry (and psychiatry in general) remain research tools, potential clinical applications, e.g., in treatment development, are on the horizon. The development of a neuroinformatics infrastructure with which to support multi-center studies and integrate data from diverse sources will be instrumental in accomplishing these aims.

ACKNOWLEDGMENT

The authors wish to thank Daniel Pine, M.D., for his helpful review of this chapter.

REFERENCES

Almli, C. R., Rivkin, M. J., McKinstry, R. C., et al. (2007) The NIH MRI study of normal brain development (objective 2): newborns, infants, toddlers, and preschoolers. *NeuroImage*, **35**, 308–24.

Arnold, L. E., Zametkin, A. J., Caravella, L., et al. (2000) Ethical issues in neuroimaging research with children.

In: Ernst, M., Rumsey, J. M. (eds.) *Functional Neuroimaging in Child Psychiatry.* Cambridge: Cambridge University Press. Chapter 6, pp. 99–109.

Bandettini, P. (2007) Functional MRI today. *Int J Psychophysiol,* **63,** 138–45.

Beckman, M. (2004) Crime, culpability, and the adolescent brain. *Science,* **305,** 596–9.

Benazon, N. R., Moore, G. J., and Rosenberg, D. R. (2003) Neurochemical analyses in pediatric obsessive-compulsive disorder in patients treated with cognitive-behavioral therapy. *J Am Acad Child Adolesc Psychiatry,* **42,** 1279–85.

Berl, M. M., Vaidya, C. J., and Gaillard, W. D. (2006) Functional imaging of developmental and adaptive changes in neurocognition. *NeuroImage,* **30,** 679–91.

Bush, G., Shin, L. M., Holmes, J., et al. (2003) The multi-source interference task: validation study with fMRI in individual subjects. *Mol Psychiatry,* **8,** 60–70.

Bush, G., Valera, E. M., and Seidman, L. J. (2005) Functional neuroimaging of attention deficit/hyperactivity disorder: a review and suggested future directions. *Biol Psychiatry,* **57,** 1273–84.

Cascio, C. J., Gerig, G., and Piven, J. (2007) Diffusion tensor imaging: application to the study of the developing brain. *J Am Acad Child Adolesc Psychiatry,* **46,** 213–23.

Chugani, D. C. (2004) Serotonin in autism and pediatric epilepsies. *Ment Retard Dev Disabil Res Rev,* **10,** 112–16.

Durston, S., Davidson, M. C., Tottenham, N., et al. (2006) A shift from diffuse to focal cortical activity with development. *Dev Sci,* **9,** 1–8.

Durston, S., Hulshoff, P. H. E., Casey, B. J., et al. (2001) Anatomical MRI of the developing brain: what have we learned? *J Am Acad Child Adolesc Psychiatry,* **40,** 1012–20.

Ernst, M. and Rumsey, J. M. (eds.) (2000) *Functional Neuroimaging in Child Psychiatry.* Cambridge: Cambridge University Press.

Evans, A. C. and the Brain Development Cooperative Group (2006) The NIH MRI Study of Normal Brain Development (2006). *NeuroImage,* **30**(1), 184–202.

Farah, M. (2005) Neuroethics: the practical and the philosophical. *Trends Cogn Sci,* **9,** 34–40.

Farah, M. and Heberlein, A. S. (2007) Personhood and neuroscience: naturalizing or nihilating? *Am J Bioeth,* **7,** 37–48.

Fregni, F. and Pascual-Leone, A. (2007) Technology insight: noninvasive brain stimulation in neurology – perspectives on the therapeutic potential of rTMS and tDCS. *Neurology,* **3,** 383–93.

Friedman, L. and Glover, G. H. (2006) Report on a multicenter fMRI quality assurance protocol. *J Magn Reson Imaging,* **23,** 827–39.

Friedman, L., Stern, H., Brown, G. G., et al. (2008) Test-retest and between-site reliability in a multicenter fMRI study. *Hum Brain Mapp,* **29**(8), 958–72.

Garrity, A. G., Pearlson, G. D., McKiernan, K., et al. (2007) Aberrant "default mode" functional connectivity in schizophrenia. *Am J Psychiatry,* **164,** 450–7.

Han, X., Jovicich, J., Salat, D., et al. (2006) Reliability of MRI-derived measurements of human cerebral cortical thickness: the effects of field strength, scanner upgrade and manufacturer. *NeuroImage,* **32,** 180–94.

Hinton, V. J. (2002) Ethics of neuroimaging in pediatric development. *Brain Cogn,* **50,** 455–68.

Honey, G. and Bullmore, E. (2004) Human pharmacological MRI. *Trends Pharmacol Sci,* **25,** 366–74.

Huerta, M. F., Liu, Y., and Glanzman, D. L. (2006) A view of the digital landscape for neuroscience at NIH. *Neuroinformatics,* **4,** 131–8.

Illes, J. and Bird, S. J. (2006) Neuroethics: a modern context for ethics in neuroscience. *Trends Neurosci,* **29,** 511–17.

Illes, J. and Raffin, T. A. (2005) No child left without a brain scan? Toward a pediatric neuroethics. *Cerebrum,* **7,** 33–46.

Johnson, M. H. (2003) Development of human brain function. *Biol Psychiatry,* **54,** 1312–16.

Johnstone, T., Somerville, L. H., Alexander, A. L., et al. (2005) Stability of amygdala BOLD response to fearful faces over multiple scan sessions. *NeuroImage,* **25,** 1112–23.

Jovicich, J., Czanner, S., Greve, D., et al. (2006) Reliability in multi-site structural MRI studies: effects of gradient non-linearity correction on phantom and human data. *NeuroImage,* **30,** 436–43.

Krasnova, I. N., Betts, E. S., Dada, A., et al. (2007) Neonatal dopamine depletion induces changes in morphogenesis and gene expression in the developing cortex. *Neurotoxicol Res,* **11,** 107–30.

Lenroot, R. K., Gogtay, N., and Greenstein, D. K. (2007) Sexual dimorphism of brain developmental trajectories during childhood and adolescence. *NeuroImage,* **36,** 1065–73.

Lenroot, R. K., Schmitt, J. E., Ordaz, S. J., et al. (2007) Differences in genetic and environmental influences on the human cerebral cortex associated with development during childhood and adolescence. *Hum Brain Mapp,* epub ahead of print.

Levin, A. (2006) Imaging studies guide policy on crime and punishment. *Psychiatric News,* **41,** 8.

Linden, D. E. J. (2006) How psychotherapy changes the brain – the contribution of functional imaging. *Mol Psychiatry*, **11**, 528–38.

Lu, L. H., Leonard, C. M., Thompson, P. M., et al. (2007) Normal developmental changes in inferior frontal gray matter are associated with improvement in phonological processing: a longitudinal MRI analysis. *Cereb Cortex*, **17**, 1092–9.

Mason, G. F. and Krystal, J. H. (2006) MR spectroscopy: its potential role for drug development for the treatment of psychiatric disease. *NMR Biomed*, **19**, 690–701.

Matthews, P. M., Honey, G. D., and Bullmore, E. T. (2006) Applications of fMRI in translational medicine and clinical practice. *Nat Rev Neurosci*, **7**, 732–44.

Meyer, J. H., Wilson, A. A., Sagrati, S., et al. (2004) Serotonin transporter occupancy of five selective serotonin reuptake inhibitors at different doses: an [11-C]DASB positron emission tomography study. *Am J Psychiatry*, **161**, 826–35.

Muller, R. A. (2007) The study of autism as a distributed disorder. *Ment Retard Dev Disabil Rev*, **13**, 85–95.

Muller, R. A. and Courchesne, E. (2000) The duplicity of plasticity: a conceptual approach to the study of early lesions and developmental disorders. In: Ernst, M., Rumsey, J. M. (eds.) *Functional Neuroimaging in Child Psychiatry*. Cambridge: Cambridge University Press, Chapter 20, pp. 335–65.

Munson, S., Eshel, N., and Ernst, M. (2006) Ethics of PET research in children. In: Charron, M. (ed.) *Practical Pediatric PET Imaging*. New York: Springer-Verlag, Chapter 7, pp. 72–91.

NIMH Council Workgroup on MRI Research Practices (2007) MRI Research Safety and Ethics: Points to Consider. Available at: http://www.nimh.nih.gov/about/advisory-boards-and-groups/namhc/reports/mri-research-safety-ethics.pdf, accessed 5 August 2008.

Novotny, E. J., Fulbright, R. K., Pearl, P. L., et al. (2003) Magnetic resonance spectroscopy of neurotransmitters in human brain. *Ann Neurol*, **54** [Suppl. 6], 525–31.

Obrig, H. and Villringer, A. (2003) Beyond the visible – imaging the human brain with light. *J Cereb Blood Flow Metab*, **23**, 1–18.

Pascual-Leone, A., Amedi, A., Fregni, F., et al. (2005) The plastic human brain cortex. *Annu Rev Neurosci*, **28**, 377–401.

Raichle, M. E. and Snyder, A. Z. (2007) A default mode of brain function: a brief history of an evolving idea. *NeuroImage*, **37**, 1083–90.

Schultz, R. (2005) Developmental deficits in social perception in autism: the role of the amygdala and fusiform face area. *Int J Dev Neurosci*, **23**, 125–41.

Segal, D., Koschnick, J. R., Slegers, L. H., et al. (2007) Oligodendrocyte pathophysiology: a new view of schizophrenia. *Int J Neuropsychopharmacol*, **10**, 503–11.

Sonuga-Barke, E. J. S. and Castellanos, F. X. (2007) Spontaneous attentional fluctuations in impaired states and pathological conditions: a neurobiological hypothesis. *Neurosci Biobehav Rev*, **31**, 977–86.

Taga, G., Homae, F., and Watanabe, H. (2007) Effects of source-detector distance of near infrared spectroscopy on the measurement of cortical hemodynamic response in infants. *NeuroImage*, **38**, 452–60.

Thomason, M. E., Foland, L. C., and Glover, G. H. (2007) Calibration of BOLD fMRI using breath holding reduces group variance during a cognitive task. *Hum Brain Mapp*, **28**, 59–68.

Toga, A. W., Thompson, P. M., Mori, S., et al. (2006) Towards multimodal atlases of the human brain. *Nat Rev Neurosci*, **7**, 952–66.

Voyvodic, J. T. (2006) Activation mapping as a percentage of local excitation: fMRI stability within scans, between scans and across field strengths. *Magn Reson Imaging*, **24**, 1249–61.

Waber, C. P., deMoor, C., Forbes, P. W., et al. (2007) The NIH MRI study of normal brain development: performance of a population based sample of healthy children aged 6 to 18 years on a neuropsychological battery. *J Int Neuropsychol Soc*, **13**, 729–46.

Weiskopf, N., Sitaram, R., Josephs, O., et al. (2007) Real-time functional magnetic resonance imaging: methods and applications. *Magn Reson Imaging*, **25**, 989–1003.

Westerberg, H. and Klingberg, T. (2007) Changes in cortical activity after training in working memory–a single subject analysis. *Physiol Behav*, **92**, 186–92.

Wise, R. G. and Tracey, I. (2006) The role of fMRI in drug discovery. *J Magn Reson Imaging*, **23**, 862–76.

Appendix A: Functional MRI educational resources and software

Educational resources

Jody Culham's guide for fMRI imaging, experimental design and analysis, fMRI4 Newbies. A Crash Course in Brain Imaging

http://psychology.uwo.ca/fmri4newbies/, accessed 5 August 2008

Douglas C. Noll's Primer on MRI and Functional MRI

http://www.eecs.umich.edu/~dnoll/primer2.pdf, accessed 5 August 2008

Joseph P. Hornak's Web Tutorial, The Basics of MRI

http://www.cis.rit.edu/htbooks/mri/, accessed 5 August 2008

Magnetic Resonance – Technology Information Portal

http://www.mr-tip.com/serv1.php?type=welcome, accessed 5 August 2008

Current software platforms

Analysis of Functional NeuroImages (AFNI)

http://afni.nimh.nih.gov/afni/, accessed 5 August 2008

AFNI is a set of C programs for processing, analyzing, and displaying functional MRI (fMRI) data. It runs on Unix+X11+Motif systems, including SGI, Solaris, Linux, and Mac OS X. It is available for free (in C source code format, and some pre-compiled binaries) for research purposes.

Analyze

http://www.analyzedirect.com/, accessed 5 August 2008

Analyze was developed by the Biomedical Imaging Resource (BIR) at Mayo Clinic. Analyze is a powerful software package developed by the Biomedical Imaging Resource (BIR) at Mayo Clinic for multi-dimensional display, processing, and measurement of multi-modality biomedical images. It is a commercial program and is used for medical tomographic scans from magnetic resonance imaging, computed tomography, and positron emission tomography.

BrainVISA

http://brainvisa.info/doc/brainvisa/en/processes/aboutBrainVISA.html, accessed 5 August 2008

BrainVISA is a software platform for visualization and analysis of multi-modality brain imaging data, including structural and functional MRI, EEG, and MEG. BrainVISA is the result of collaborative work in France and is available free. The goal of the software was to provide users with a friendly unifying interface to simplify exchange of software between different laboratories that are a part of the Institut Fédératif de Recherche (IFR49).

BrainVoyager

http://www.brainvoyager.com/

BrainVoyager QX is a commercially available cross-platform neuroimaging tool designed for advanced analysis and visualization of structural and functional MRI data. It can run on all major computer platforms, including Linux/Unix, Windows and Mac OS X. TMS Neuronavigator provides the software to navigate a TMS coil to the desired anatomical or functionally defined brain regions. Turbo-BrainVoyager is a multi-platform program specialized for real-time fMRI data analysis that allows researchers to observe a subject's or patient's brain activity during an ongoing fMRI scanning session. BrainVoyager Brain Tutor is an educational program, which is available for free download.

FMRIB Software Library (FSL)

http://www.fmrib.ox.ac.uk/fsl/, accessed 5 August 2008

FSL is a comprehensive library of analysis tools for fMRI, MRI, and DTI brain imaging data, written mainly by members of the Analysis Group, FMRIB, Oxford, UK. FSL runs on Apple, PCs (Linux and Windows) and Sun workstations and is easy to install. Major new features in FSL 4.0 include: tensor independent components analysis (ICA) for multi-subject model-free fMRI analysis; subcortical segmentation of structural images; crossing-fiber modeling for diffusion tractography; structural, histological, and white matter probabilistic brain atlases; atlas integration and 3D rendering in FSLView.

FreeSurfer

http://surfer.nmr.mgh.harvard.edu/, accessed 5 August 2008

FreeSurfer is a set of automated tools for reconstruction of the brain's cortical surface from structural MRI data and for overlay of functional MRI data onto the reconstructed surface.

MRIcro

http://www.sph.sc.edu/comd/rorden/mricro.html, accessed 5 August 2008

MRIcro allows Windows and Linux computers to view medical images. It is a stand-alone program, but includes tools to complement SPM (software that allows researchers to analyze MRI, fMRI, and PET images). MRIcro allows efficient viewing and exporting of brain images. In addition, it has the capacity to identify regions of interest (ROIs, e.g., lesions). MRIcro can create Analyze format headers for exporting brain images to other platforms.

NeuroLens

http://www.neurolens.org/NeuroLens/Home.html, accessed 5 August 2008

NeuroLens is an integrated environment for the analysis and visualization of functional neuroimages. All processing operations in NeuroLens are built around a plug-in architecture that allows for easy extension of its functionality. NeuroLens runs on Apple computers based on the G4, G5, or Intel chipsets and running Mac OS X 10.4 (Tiger) or later. It is available for free for academic and non-profit research use.

Statistical Parametric Mapping (SPM)

http://www.fil.ion.ucl.ac.uk/spm/, accessed 5 August 2008

The SPM software package has been designed for the analysis of brain imaging data. The data can be a series of images from different cohorts or time-series from the same subject. The current release is designed for the analysis of fMRI, PET, SPECT, EEG and MEG. AutoSPM: is an automated SPM software package for surgical planning.

Appendix B: Neuroinformatics and neuroethics resources

Databases

NIH Pediatric MRI Data Repository (www.NIH-PediatricMRI.org, accessed 5 August 2008)

The NIH Pediatric MRI Data Repository is a US government-sponsored research resource for the scientific community. The repository contains neuroimaging and clinical/behavioral data from the NIH MRI Study of Normal Brain Development, a longitudinal, multi-site study of healthy pediatric brain development from ages newborn to late adolescence/early adulthood, sponsored by four NIH institutes. Using comparable acquisition protocols and an epidemiologic sampling strategy, data were acquired at six pediatric study centers, and quality controls implemented through a clinical coordinating center and a data coordinating center. Children were scanned at three or more time points with T1 and T2/PD sequences. Those aged 4 ½ years to 18 years at the time of enrollment were scanned at approximately 2-year intervals, taking the oldest subjects into early adulthood (age 24). Children aged 7 days to 4 years and 5 months at the time of enrollment were scanned at shorter intervals; the youngest, at 3-month intervals. Where feasible, structural MR acquisitions were supplemented with diffusion tensor imaging and magnetic resonance spectroscopy. Clinical/behavioral measures include biospecimens for hormonal measures, neuropsychological data, behavioral ratings, psychiatric interviews, and neurological examinations.

A first data release involving cross-sectional structural MRI and clinical/behavioral data for ages 4 ½ and older occurred in June of 2007. Upcoming data releases will make longitudinal data from all subjects and all modalities available. Data, which are de-identified to comply with HIPAA and other federal regulations, is available to qualified biomedical and biobehavioral researchers whose institutions are covered by a Federalwide Assurance (FWA) and who agree to the terms of a Data Use Certification designed to protect human subjects.

National Database for Autism Research (http://ndar.nih.gov)

The National Database for Autism Research (NDAR) is a collaborative biomedical informatics system being created by the NIH to provide a national resource to support and accelerate research in autism. Components include clinical assessment functionality, data repository services, data mining, analysis resources, and neuroimaging support. The data repository will ultimately contain neuroimaging, genomic, epidemiological and clinical/behavioral data. For more information, see: http://ndar.nih.gov.

Clearinghouse

Neuroimaging Informatics Tools and Resources Clearinghouse (NITRC) (www.nitrc.org)

This clearinghouse is meant to catalog all fMRI tools and resources and provide downloads or pointers to downloadable tools and resources. NITRC offers peer-based ratings and comments to help neuroimagers decide whether a tool or resource will be valuable to them. The website also lists meetings of interest to the neuroimaging community, as well as training workshops relevant to the available resources.

Tools and infrastructure

Biomedical Informatics Research Network (BIRN) (www.nbirn.net/index_ie6.shtm)

Launched in 2001 as an initiative of the National Institutes of Health's National Center for Research Resources (NCRR), the Biomedical Informatics Research Network (or BIRN) is building an infrastructure of networked high-performance computers, data integration standards, and other emerging technologies to provide a collaborative environment for biomedical research. The BIRN approach allows for data and computational resources distributed across sites to be integrated, as an alternative to centralized data repositories.

A coordinating center oversees the networking, distributed storage, and software development needs of three neuroimaging test beds. The Function BIRN test bed employs functional neuroimaging to investigate the neural causes of schizophrenia and to subsequently assess the impact of novel interventions on brain. The Morphometry test bed focuses on the pooling of data across neuroimaging sites to investigate brain anatomy in depression, early Alzheimer's disease and mild cognitive impairment. The Mouse BIRN uses multimodal, multi-scale imaging data from mouse models of neurological and psychiatric disorders.

BIRN offers a diverse array of software tools, including image analysis tools, and some datasets. BIRN is working on issues of data comparability across sites, a prerequisite for multi-site collaborative efforts. Both the NIH Pediatric Data Repository and the National Database for Autism Research will utilize the BIRN infrastructure.

National Centers for Biomedical Computing (www.bisti.nih.gov/ncbc/)

Center for Computational Biology (CCB) (cms.loni.ucla.edu/CCB/)

The CCB applies developments in mathematics, computational science, computer graphics, image processing, and informatics to brain imaging data to achieve an integrated understanding of anatomy, physiology, and genetics. It is developing a new form of software infrastructure, the computational atlas, to manage multidimensional data spanning many scales and modalities. It provides software tools and training in computational biology.

National Alliance for Medical Image Computing (NA-MIC) (www.na-mic.org)

The National Alliance for Medical Imaging Computing (NA-MIC) is a multi-institutional, interdisciplinary team of computer scientists, software engineers, and medical investigators who develop computational tools for the analysis and visualization of medical image data. NA-MIC provides the infrastructure and environment for the development of computational algorithms and open source technologies and oversees the training and dissemination of these tools to the medical research community. The driving biological projects for this effort include schizophrenia, autism, and velocardiofacial syndrome. However, the methods and infrastructure being developed are applicable to other diseases as well. The tools and technologies developed are being applied to the study of brain anatomy and connectivity. Imaging modalities include structural MRI, diffusion MRI, quantitative EEG, and metabolic and receptor PET, as well as microscopic and genomic data.

NA-MIC provides training workshops to introduce researchers to image analysis tools such as FreeSurfer and Slicer – see https://surfer.nmr.mgh.harvard.edu/fswiki/FreeSurferBeginnersGuide

Software tools available through the National Institutes of Health

ImageJ (http://rsb.info.nih.gov/ij/)

This freely available, public domain, Java-based image processing program developed at the National Institutes of Health was designed with an open architecture that provides extensibility via plug-ins and recordable macros. Downloadable distributions are available for a number of platforms. ImageJ can be used for the analysis of numerous image type and modalities (e.g., MRI, CT, XRay, Microscopy, etc.).

Medical Imaging Processing Analysis and Visualization (MIPAV) (http://mipav.cit.nih.gov/)

MIPAV is a general purpose, extensible image processing and visualization program that executes on multiple platforms to assist researchers in extracting quantitative information from various imaging modalities including but not limited to: PET, MRI, CT, SPECT, X-ray, microscopy. Inherently MIPAV provides numerous segmentation, registration (linear, non-linear, automatice, semi-manual), quantification algorithms to support 2D, 3D, and 4D image analysis of many imaging modalities. Through the use of MIPAV's plug-in infrastructure, it is possible to develop complex user-specific image processing or analysis solutions. In addition, MIPAV supports many real-time 3D renderering techniques enabling interactive visual investigation of volume datasets.

National Library of Medicine Insight Segmentation and Registration Toolkit (ITK) (http://www.itk.org/)

This is an open-source software system for performing registration and segmentation, developed to support the Visible Human Project, which created a 3D representation of the normal male and female bodies using CT, MRI, and cryosection images. Because the software is an open-source product, developers can extend the software. Communication among members of the ITK community helps to manage the evolution of the software.

Neuroethics

NIMH Council Workgroup on MRI Research Practices (2007) *MRI Research Safety and Ethics: Points to Consider.* Available at http://www.nimh.nih.gov/about/advisory-boards-and-groups/namhc/reports/mri-research-safety-ethics.pdf, accessed 5 August 2008.

The move of MRI research from clinical into non-clinical settings, such as university neuroscience or psychology departments engaged in basic cognitive and affective neuroscience research, has increased the burden on the investigator for establishing comprehensive subject screening, staff training, operating, and emergency procedures. This, along with the lack of any comprehensive guidance on such issues led the National Institute of Mental Health to sponsor a Council work group to address basic ethical and safety issues concerning MRI screening (including for pregnancy and other exclusions); training, operating, and emergency procedures; physical facilities; scanning and participant health variables (e.g., subject monitoring, hearing protection, repeated scanning, implants, and devices); and context-specific considerations (university versus medical settings). Available through the NIMH website, this "points to consider" document provides a resource for interested investigators and their institutions.

Websites of interest

http://neuroethics.stanford.edu/, accessed 5 August 2008
Hosted by the Stanford Center for Biomedical Ethics Program in Neuroethics.
    http://neuroethics.upenn.edu/, accessed 5 August 2008

Hosted by Martha J. Farah and Melissa K. R. Hozik of the Center for Cognitive Neuroscience at the University of Pennsylvania.

http://www.aei.org/events/eventID.1072/event_detail.asp, accessed 5 August 2008

The New Neuromorality, W. H. Brady Program in Culture and Freedom Conference. Available on this website are video webcasts and papers by several leading neuroscientists addressing the application of neurobiological findings to a wide range of social issues (e.g., addiction, criminal behavior, etc.) and the legal and moral agency issues that arise out of the new neuroscience.

Glossary

This glossary consists of two parts: Part I contains terms related to neuroimaging, and, to a lesser extent, to electrophysiology. Part II contains genetic terms.

PART I IMAGING TERMS

Notations that appear following the definitions indicate the modalities to which the terms apply as follows: **MRI** signifies that the term applies to two or more MRI techniques (aMRI, ASL, BOLD fMRI, DTI, and MRS); **aMRI**, anatomic MRI; **ASL**, arterial spin labeling perfusion MRI; **BOLD fMRI**, functional MRI based on BOLD; **DTI**, diffusion tensor imaging; **EEG**, electroencephalography; **ERP**, event-related potential; **MRS**, magnetic resonance spectroscopy; **fMRI**, ASL and/or BOLD fMRI; **NIRS**, near-infrared spectroscopy; **PET**, positron emission tomography; **SPECT**, single photon emission computed tomography. Terms and definitions without these notations are either self-explanatory or not specific to neuroimaging or electrophysiology, e.g., statistical terms, terms from cognitive psychology.

acquisition Process of acquiring and storing imaging signals. (MRI, PET/SPECT)

anisotropic diffusion/diffusion anisotropy Diffusion that is biased in some direction (as opposed to isotropic diffusion). Ordered physical barriers, such as the extracellular myelin sheaths of axons and intracellular microtubuli, will cause the diffusion in white matter to be anisotropic. (DTI)

arterial spin labeling (ASL) perfusion MRI Non-invasive neuroimaging technique that uses magnetically labeled arterial blood water as an endogenous tracer to provide a quantitative mapping of regional cerebral blood flow in physiological units of ml· $100 \, \text{g}^{-1} \cdot \text{min}^{-1}$. (ASL)

arterial transit time The time for the labeled water to flow from the labeling region to the imaging slice. In cerebrovascular diseases such as stenosis, arterial transit time may be prolonged due to the narrowing of blood vessels, resulting in *delayed transit effects* (focal intravascular signals) in ASL perfusion images. (ASL)

***b* value** Factor of diffusion-weighted sequences; summarizes, among other things, the influence of the gradient duration and amplitude on the diffusion-weighted images. The higher the *b* value, the stronger the diffusion weighting. (DTI)

B_0 A conventional symbol for the static main magnetic field of an MRI scanner, which is aligned along the bore of the magnet and normally never switched off during the lifetime of the scanner. This main field can be linearly altered in space by the application of time-variable gradient fields. (MRI)

B_{max} The total number of receptors per unit volume of tissue (i.e., concentration). The term is an adaptation of the classical V_{max} (maximal reaction velocity) used in equilibrium enzyme kinetics. Because a basal level of receptor occupation by endogenous neurotransmitter is believed to be present at all times, an estimate of B_{max} is rarely, if ever, the end result a PET study analysis. (PET)

B'_{max} The total number of available receptor sites per unit volume (as if counted at steady state). Because of the omnipresence of endogenous neurotransmitter, B'_{max} is usually the receptor density parameter that is estimated in PET. (PET)

binding potential (BP)   Index of receptor binding activity. Originally introduced into the PET literature by Mintun in 1984. Due to limitations of parameter identifiability, BP is often the only parameter that can be reliably estimated from dynamic PET data. $BP = B_{max}/K_d$ (or B'_{max}/K_d) where B_{max} is a measure of the total number of receptors (B'_{max} is a measure of the number of receptors available to be bound, i.e., not occupied by endogenous ligands), and K_d is the equilibrium constant of dissociation. Note, a change in BP cannot be assigned to a change in either B'_{max} or K_d without additional a priori knowledge. (PET)

blocked designs   Experimental designs used in fMRI in which each condition is presented continuously for a period of time (generally 10–120 s), allowing the hemodynamic response to reach maximal levels. The time between blocks allows the hemodynamic response to return to baseline during rest, providing a reference signal from which the BOLD magnitude differences between conditions can be computed. Blocked designs can be used in ASL as well. (fMRI)

blood–brain partition coefficient of water   The ratio between the tissue and blood water concentrations at equilibrium; generally expressed in units of ml/g. (ASL)

BOLD contrast   fMRI provides an indirect measure of brain activity by measuring regional changes in oxygenated hemoglobin. The magnetic resonance (MR) signal of blood differs slightly depending on the level of oxygenation; deoxygenated hemoglobin attenuates the signal. Neural activity causes an increase in demand for oxygen, and the vascular system actually overcompensates for this, increasing the amount of oxygenated hemoglobin relative to deoxygenated hemoglobin. This leads to an MR signal increase that is related to the neural activity. This difference in MR signal is referred to as the blood-oxygen-level-dependent (BOLD) contrast, where higher BOLD signal intensities arise from an increase in the concentration of oxygenated hemoglobin relative to a decrease in the concentration of deoxygenated hemoglobin. (BOLD fMRI)

BOLD effect (blood-oxygen-level-dependent effect)   The change in T_2^* that is induced by changes in the amount of oxygenated hemoglobin (Hb) in the venous circulation of the brain. Because oxygenated Hb has a much smaller *magnetic susceptibility* than deoxygenated Hb, and because neural activity alters the amount of oxygenated Hb in the venous blood, the susceptibility of the blood decreases, T_2^* increases, and therefore the intensity in T_2^*-*weighted* images increases. The BOLD signal is thought to mainly reflect inputs into a brain region and local processing within that region, rather than neuronal spiking and axonal signals. (BOLD fMRI)

Brownian motion   Random displacements of suspended particles in, or molecules of, a liquid. (DTI)

capillary water permeability   The ability of the capillary wall to transport water molecules into tissue. Water permeability in the systemic circulation is an order of magnitude higher than that of the blood–brain barrier in the central nervous system. (ASL)

carbon-13 (^{13}C) MRS   Technique that detects carbon-13, the low natural abundance of carbon, which makes it possible to administer and detect labeled compounds using MRS. Offers the opportunity to study the flux of metabolic pathways, including the tricarboxylic acid cycle. (MRS)

cerebral blood flow (CBF)   Strictly, blood flow is in mL · min^{-1}. By convention, however, CBF refers to perfusion in ml · min^{-1} · g^{-1} tissue (i.e., blood flow per mass of tissue). (ASL, PET/SPECT)

cerebral metabolic rate of glucose (CMRGlc)   Rate of glucose utilization in the brain per unit of mass (μmol · min^{-1} · g^{-1}). (PET)

chemical shift   Changes in resonance frequency of various MR-visible compounds due to their immediate chemical environments, such as the type and number of chemical bonds, that permit assessment of distinct molecules or groups of molecules. For example, hydrogen nuclei in water and hydrogen nuclei in fat experience different magnetic fields and therefore have different Larmor frequencies. These frequencies are used to encode position in MRI, and make possible the differentiation of different molecular compounds and different sites within the molecules in high-resolution NMR spectra. The amount of the shift is proportional to magnetic field strength and is usually specified in parts per million (ppm) of the resonance frequency relative to a standard. (MRS)

chemical shift imaging (CSI) [= spectroscopic imaging (SI)]   Multi-voxel MRS method in which spectral and positional (structural image) data are encoded together and acquired at the same time. Signal is collected from a wide region of tissue in a single acquisition, the volume covered may include an entire brain slice or volume, and the data are later processed as individual spectra from each of the voxels. (MRS)

chronometry (= mental chronometry; chronometrics)   The study of the temporal sequencing of information processing in the human brain or a measurement of psychological processes; a core paradigm of experimental and cognitive psychology that has been applied in cognitive neuroscience. Reaction times are measured and used with subtraction techniques to determine the time needed to perform various mental processes and the stages associated with them. In functional neuroimaging, mental chronometry has been used by having subjects perform tasks based on reaction time data to determine which brain regions are involved in the cognitive processes.

coil   Single or multiple loops of wire (or other electrical conductors and components such as capacitors and diodes) encased in rigid or flexible plastic housing. Coils are designed either to produce a magnetic field from current flowing through the wire (e.g., the *magnet coil*, an RF *transmission coil*, or a *gradient coil*), or to detect a changing magnetic field by voltage induced in the wire (RF *receiver coils*, such as *phased-array coils*), or both.

An *RF coil* is a device that transmits radiofrequency (RF) waves into the tissue or receives the emitted RF signal from the tissue. *RF coils* that transmit RF electromagnetic energy are used to flip the magnetization vector and to receive the RF signal generated by the RF pulse. Transmission and reception can be obtained by either a single coil or two separate coils. The reason for using separate coils is that the requirements are different. A *transmit coil* needs to be able to send its RF energy uniformly anywhere in the body. The *receive coil* may only need to pick up signal from a small portion of the body. For example, a *surface coil* is sometimes used to receive the signal. It has a better signal-to-noise ratio near its center than a receiver coil designed to detect signals uniformly throughout a larger portion of the body. Surface coils can be used for localization of sites for measurement of chemical shift spectra, especially of phosphorus, and blood flow studies. An *array coil* is composed of separate multiple smaller coils that can be used individually or simultaneously to transmit and receive the RF signal. Array coils are useful for imaging large areas and provide good signal-to-noise ratio. *Phased array coils* consist of multiple non-interacting coil elements that simultaneously receive MR signals. They are used to achieve an optimized signal-to-noise ratio over a large volume, while allowing faster acquisitions with consequent reductions in motion artifact. The data from the individual coil elements have to be integrated by special software to produce a single image.

A *gradient coil* is used to provide magnetic field gradients. Inside the main MRI magnet are three gradient coils that produce the desired gradient (magnetic) fields Gx, Gy, and Gz to alter the influence of the static field on the imaged object through which selective spatial excitation and spatial encoding is achieved. The labels "x," "y," and "z" indicate the orientation of the magnetic fields that the coils generate. In the case of an axially oriented brain image, the "z" gradient coil is used for slice selection, the "x" gradient coil is normally used for frequency encoding of the image, and the "y" gradient coil is normally used for phase encoding. Variations are possible, and, in general, combinations of these coils are utilized in order to permit imaging in planes other than the axial plane (such as coronal, sagittal, and arbitrary oblique planes). The gradient coils are driven by extremely high currents, producing forces and subsequent vibrations of the wires in the coil, which cause the typical acoustic noise of an MRI scanner.

The *magnet coil* generates the main magnetic field. The magnet is usually superconducting and has to be cooled. It is the most expensive part of an MRI scanner. (MRI)

coil loading   Interaction between placement of the subject's head and the radiofrequency (RF) coil, which can cause shifts of the resonance frequency, thereby influencing the detection and quality of the spectral data obtained. (MRS)

continuous arterial spin labeling (CASL) perfusion imaging   ASL technique that uses long radio-frequency (RF) pulses (1–3 s) applied at a thin plane to label blood water flowing through that plane. Provides stronger perfusion contrast and generally provides greater brain coverage than pulsed ASL, but deposits a higher level of RF power into the subject. (ASL)

contrast   Relative difference of MRI signal intensities and the associated image brightness in different tissues. In anatomical imaging, brain regions are distinguished from each other only by the contrast created when the signal emanating from

protons of water molecules differ. MRI can be tailored to create additional contrasts, for example based on diffusion and perfusion weighting, temperature sensitivity, etc. In functional imaging, contrast can also refer to the dependence of the signal on the stimulus ("dynamic contrast"). (MRI)

contrast agents   Compounds, generally injected, used to improve the visibility or contrast in biological imaging. In MRI, contrast agents are often gadolinium-based and work by altering the magnetic properties of nearby hydrogen nuclei. (aMRI)

convolution   A mathematical method of combining two signals to form a third signal. Convolution is a critical technique in digital signal processing and is mostly used together with Fourier transformations for MRI signal/image processing. In fMRI, convolution is used to model the hemodynamic response based on a given hemodynamic response function and the stimulus function, which defines the timing of the stimuli. If responses in rapid event-related designs are so close that they are overlapping, under certain conditions *deconvolution* can be used to estimate the responses from the observed signal. (MRI)

cortical pattern matching (CPM)   An image analysis technique that uses sulcal patterns to achieve improved matching of cortical surface anatomy across subjects. (aMRI)

cortical thickness   A measure of distance from the border defined by the contrast between gray matter and white matter to the border of gray matter in the cortical rim with cerebrospinal fluid (outer boundary of cortex). Given that this measure relies on MR contrast, the precision of this measure depends on the contrast characteristics of the particular scan, which can vary across MR sequences and scanners, partial voluming artifacts, and other such considerations. (aMRI)

creatine/phosphocreatine (Cr)   Metabolite pool detectable as a single resonance with proton MRS. Because the total concentration of creatine and phosphocreatine is similar in many brain regions (although slightly higher in cerebral cortex than in white matter), it is often used as a reference standard. (MRS)

cytosolic choline compounds (Cho) Most of brain choline is incorporated in the membrane lipid phosphatidylcholine and is largely invisible to in vivo MRS. The Cho peak primarily reflects phosphocholine and glycerophosphocholine. Choline is considered an indicator of membrane activity, since glycerophosphocholine is a product of cellular breakdown. (MRS)

deconvolution Operation to reverse the effects of convolution on recorded data. In fMRI image processing, deconvolution algorithms are used to parse apart interwoven or overlapping hemodynamic responses resulting from the use of rapid event-related designs that do not allow the hemodynamic response to return to baseline between events or trials. Conditions necessary to make it work are a linear dependence of the hemodynamic response on the stimulus and an experimental design with variable time intervals between stimuli. (fMRI)

default network Network of brain regions that show preferentially greater activity during passive states such as rest, as compared to active tasks. This network is thought to reflect the organized, intrinsic activity of the brain and may reflect neural functions that consolidate the past, stabilize brain ensembles, and prepare the individual for the future. Marcus Raichle brought attention to what he termed the "default network" in 2001 when he noticed that tasks used with quantitative neuroimaging techniques, such as PET, elicited deactivations (e.g., decreases in cerebral blood flow) in addition to increases, or activations. Recent research suggests that these networks are active even during anesthesia and may be relevant to understanding clinical conditions.

dephasing; dephasing spins The transverse proton spin components (in the plane perpendicular to the z axis) go out of phase when the radiofrequency pulse is switched off. *Dephasing* refers to the loss of net magnetization in the transverse plane due to the fact that the individual nuclei are precessing (or wobbling) at different rates. Whereas the rate at which the magnetic orientation of the individual nuclei returns to the longitudinal direction is relatively slow ("T_1"), the rate at which the transverse magnetization disappears is relatively rapid (T_2^*). *Rephasing* returns the nuclei to their original position, which can be accomplished using a $180°$ pulse. (MRI)

depth-resolved surface coil spectroscopy (DRESS) MRS method that relies on the characteristics of a surface coil or phased array of coils to excite a region of tissue at a certain depth within the brain or body. Often used to collect ^{31}P MRS data. (MRS)

diffusion A spontaneous process by which molecules move about due to random thermal (Brownian) motion. (DTI)

diffusion coefficient Value representing the Brownian motion (or random thermal motion) of water molecules. If a diffusion coefficient is determined voxelwise from diffusion-weighted images, as obtained in DWI and DTI, it is called "apparent diffusion coefficient" (ADC). (DTI)

diffusion tensor A 3×3 symmetrical matrix used to characterize the three-dimensional shape of diffusion in a voxel. A mathematical operation called "diagonalization" yields three *eigenvalues* and *eigenvectors* describing the strength and direction of diffusion, respectively, in space. These values can be used to calculate measures of *anisotropy*, such as *fractional anisotropy*. (DTI)

diffusion tensor imaging (DTI) Introduced in the mid 1990s, this MR imaging technique attempts to reconstruct the orientation of the local physical barriers in white matter (i.e., axonal tissue) by measuring the amount of direction-restricted water molecular displacement for a predefined set of spatial directions. DTI sequences usually consist of spin echo sequences with embedded diffusion gradients, which probe the sample for the different diffusion directions.

A significant limitation of the technique is the fact that it assumes a single preferred direction of diffusion within each voxel; it does not resolve intravoxel inhomogeneities such as crossing fibers. DTI does not measure a single aspect of white matter development. In addition to being sensitive to myelin status, it reflects other aspects of axonal integrity and the organization/architecture of axons and fibers in white matter and may be differentially sensitive to various aspects at different stages of brain development. For example, during the neonatal period, changes seen on DTI may reflect overall changes in brain water content and the physical growth of axons. Later, in childhood, changes are more likely to reflect myelination. (DTI)

diffusion-weighted imaging (DWI) Diffusion MRI in which no directional dependence of diffusion is computed. In most cases, only one to three diffusion gradients are applied. DWI displays an "apparent diffusion coefficient" (ADC) for each voxel and is of significant utility for detecting acute stroke and other cortical tissue damage. (DTI)

high-*b*-value DWI If different compartments contribute to diffusion in a single voxel, it cannot be assumed any more that signal loss is a simple exponential function of a single diffusion coefficient. Rather, different diffusion compartments, for example intra- and extracellular diffusion, cause a multi-exponential signal loss, which can be detected using a range of b *values.* To obtain significant contributions from compartments with small diffusion constants, the used b values have to be larger than the ones normally used in DWI, requiring strong diffusion gradients or long scan times. In clinical imaging, high-*b*-value DWI is used for some diagnostic scans, usually with a fixed b value. (MRI)

Doppler ultrasound A non-invasive procedure used to evaluate blood flow and pressure by bouncing high-frequency sound waves (ultrasound) off red blood cells. It employs the *Doppler effect*, the change in frequency and wavelength of a wave as

perceived by an observer moving relative to the source of the waves.

echo In MRI, "echo" refers to the regrowth of the transverse component of magnetization after it has disappeared due to dephasing and subsequent rephasing. This echo is, in fact, the NMR signal that is normally recorded and analyzed in MRI. (MRI)

echo time (TE) Represents the time between the application of the excitation radiofrequency pulse and the peak of the echo signal. (MRI)

echo-planar imaging (EPI) A technique of imaging in which a complete planar image is obtained from a single excitation pulse. The signal is measured repeatedly while rapidly switching the readout gradients (*Cartesian acquisition*) or letting readout and phase gradients oscillate (*spiral acquisition*). The *Fourier transform* of the resulting echo train can be used to produce an image of the excited plane. EPI requires gradient power supplies of greater strength and speed than are needed for more conventional imaging strategies. Its advantage is speed, making (gradient-echo) EPI still the predominant method in fMRI. (BOLD fMRI)

eddy currents Persistent electric currents in a conductor after initially being induced by a changing magnetic field or motion of the conductor through a magnetic field. MRI conductors such as a patient or the metal shielding of portions of the equipment may create eddy currents that can degrade image quality. (DTI)

eigenvalues (of a diffusion tensor) Three direction-dependent diffusion coeffcients, obtained from the "diagonalization" of the diffusion tensor. Diagonalization is a mathematical operation on the diffusion tensor that yields diffusion coeffcients independent of the used coordinate system (e.g., orientation of the head). (DTI)

eigenvector (of a diffusion tensor) Like the eigenvalues, the eigenvectors are obtained from a diffusion tensor by diagonalization. The three eigenvectors define a local coordinate system that is oriented on the principal and two minor diffusion directions. In homogeneous tissues across the voxel, the principal eigenvector indicates the local fiber orientation. (DTI)

emission computed tomography Tomography using emissions from radionuclides and a computer algorithm to reconstruct the image. (PET/SPECT)

epoch In fMRI, this term is often used to refer to a portion of a single fMRI *run* during which the stimulus presentation type and/or response task type is unchanged. (fMRI)

Ernst angle The *flip angle* that yields the most signal for a given TR and TE. (MRI)

event-related fMRI fMRI that uses an event-related experimental design to present stimuli in arbitrary sequences, mixing different types of trials. The trials can be sorted following the experiment to derive fMRI measures related to different types of trials or responses (e.g., trials on which a subject performed accurately versus error trials). (BOLD fMRI)

event-related potential (ERP) (= **evoked response**) An electrical potential recorded from a human or animal following the presentation of a stimulus, as distinct from spontaneous potentials detected by electroencephalograms or electromyograms. Because evoked potential amplitudes tend to be low (<1 μv to several microvolts), signal averaging is required. The stimulus is presented repeatedly, and the signal is time-locked to it, whereas most of the "noise" is random and, therefore, averaged out. This yields a well-defined waveform, whose components are defined by their temporal characteristics. Naming conventions signify whether the component is positive or negative and the number of milliseconds (ms) following stimulus onset at which they occur, e.g., P300 is a positive-going wave that occurs approximately 300 ms following a stimulus. (ERP)

ferromagnetic Property of a substance, such as iron, that has a large positive magnetic susceptibility (ability to become magnetized). Once magnetized, ferromagnetic materials stay magnetized for some time and become "magnets." (MRI)

fiber assignment by continuous tracking (FACT) One of earliest and most commonly employed algorithms for DTI tractography. The algorithm uses the greatest eigenvalue to define the fiber orientation and terminates when a sharp turn in fiber orientation occurs or the diffusion anisotropy becomes low. (DTI)

field of view (FOV) The area or volume of the object that can be "seen" by an imaging system. In PET/SPECT it defines the volume from which emitted activity may be detected. In MRI, its dimensions are independently controlled by the frequency-encode and phase-encode gradients. (MRI, PET/SPECT)

field strength In MRI, the strength of a main static magnetic field. For MRI scanners in clinical use, this can range from 0.2 to 3 Tesla. Research scanners may have much higher field strengths. The field strength improves the signal-to-noise ratio, but may increase certain artifacts, such as susceptibility and motion artifacts. (MRI)

filtering A mathematical operation used extensively on nuclear medicine images to reduce random noise or enhance specific image features such as edges. It is also used in the reconstruction of tomographic images. Filtering can be performed in either the spatial domain or frequency space. (MRI, PET/SPECT)

flip angle Amount of rotation of the magnetization vector produced by a radiofrequency (RF) pulse, with respect to the direction of the static magnetic field. (MRI)

fluorine-19 (^{19}F) MRS Technique used to monitor the uptake and metabolism of fluorinated anesthetics (e.g., halothane, isoflurane) and other drugs, including selective serotonin reuptake inhibitor (SSRI) antidepressants. Fluorine-19 has high MRS sensitivity; however, ^{19}F MRS requires exposure of the subject to fluorinated compounds, either as an infusion during the scanning session or as part of a pharmacological treatment regimen. (MRS)

fluorodeoxyglucose [= **2-fluoro-2-deoxy-D-glucose (FDG)**] An analog of glucose used with PET to obtain regional brain measures of the rate of glucose utilization. In PET, FDG is labeled with the positron-emitting radioactive isotope fluorine-18 to produce ^{18}F-FDG. (PET)

fMRI relaxometry Novel functional magnetic resonance imaging procedure for deriving steady-state (e.g., resting) blood flow measures and testing for enduring medication effects in specific regions of the brain. Regional blood flow is regulated to match perfusion with ongoing metabolic demand, and the deoxyhemoglobin concentration becomes constant between regions in the steady state. Regions with greater continuous activity are perfused at a greater rate and over time receive a greater volume of blood and greater number of deoxyhemoglobin molecules per volume of tissue. (fMRI)

Fourier analysis Mathematical procedure to decompose a signal into its spatial or temporal frequency components, or to compute a signal from its temporal or spatial frequency components (inverse Fourier transform). The inverse Fourier transform is the fundamental algorithm to compute images from the acquired data in MRI. In MRS, it permits the analysis of a mixture of signals at different frequencies into their component frequencies (specified by the frequency-encode gradient). (MRI)

fractional anisotropy (FA) A normalized measure of the degree of diffusion anisotropy. Values range from 0 to 1. An FA of 0 corresponds to isotropic diffusion, i.e., diffusion is equal in all directions. A value of 1 denotes an ideal linear diffusion, i.e., diffusion is completely restricted along one direction. Well-defined fiber tracts generally have an FA larger than 0.20, but the exact value depends on the signal-to-noise ratio of the experiment, determined by, for example, the main field strength and number of diffusion gradients used, and can reach almost 1.0. These values are frequently represented as brain maps. FA generally increases with brain maturation as a result of changes in axonal

membranes and myelination. However, myelin is not necessary for fibers to have diffusion anisotropy. The precise contribution of fiber density and myelination to this and other anisotropy indices is not completely understood. (DTI)

free induction decay (FID) Transient nuclear signal induced in the MRI coil after a radiofrequency pulse has excited the nuclear spin system. This is referred to as the free induction decay signal because the signal is induced by the free precession of the nuclear spins around the static field after the radiofrequency pulse has been turned off. A plot of this signal as a function of time looks like an exponentially damped sinusoid at the Larmor frequency. In MRS, the FID can be converted to a series of peaks (spectrum) by Fourier analysis. (MRI)

frequency encoding Encoding of the distribution of sources of MR signals along a direction in space with different frequencies to provide information on spatial position. Refers to the use of a magnetic field gradient to cause different rates of precession (*Larmor frequencies*) along the direction of the gradient during the data acquisition time. Thus, the frequency composition of the collected data (as determined by Fourier analysis) will correspond to different spatial locations. (MRI)

functional magnetic resonance imaging (fMRI) Non-invasive technique used to detect brain activity utilizing the fact that the MR signal intensity is correlated with cerebral blood flow, which is in turn correlated with neural activity (e.g., BOLD effect). fMRI can localize brain regions that respond to various stimuli presented during the course of an experiment. Although most fMRI is performed using BOLD methods, fMRI can also be performed using arterial spin labeling (ASL). (fMRI)

Funk-Radon transform A mathematical technique used in *q-ball* imaging. It is used to compute the *orientation distribution function* (ODF) of fibers from the diffusion-weighted images, which then can be used to characterize crossing fibers, etc. (DTI)

gamma-amino-butyric acid (GABA) Major inhibitory neurotransmitter in the mammalian brain. GABA is synthesized in neurons from glutamine, with glutamate as an intermediate step, via glutamic acid decarboxylase (GAD). GABA is near the lower limit of detection using ^{1}H MRS (~1.0 mm), due in part to overlapping metabolite peaks of higher concentrations, especially the creatine spectral peak. Contributes to the Glx peak. Detectable using ^{1}H MRS. (MRS)

general linear model (GLM) An approach to functional imaging analysis that compares signals under different task conditions to each other on a voxel-wise or region-of-interest basis. The GLM allows for the inclusion of nuisance parameters such as motion and physiological signals. (MRI)

glutamate Major excitatory neurotransmitter found in all brain cell types, with a greater concentration generally being observed in neurons. Only a small fraction of brain glutamate participates in neurotransmission (~10%). Detectable using ^{1}H MRS. (MRS)

Glx resonance Resonance arising from the combination of glutamate, glutamine, and GABA signals, generally difficult to individually resolve in brain spectra at field strengths of 1.5 T or lower and thus often reported as a combined "Glx" resonance intensity. Methods for resolving these resonance lines are under development and will be important for identifying neurotransmitter function/dysfunction. Detectable using ^{1}H MRS. (MRS)

glycerophosphocholine (GPC) A phospholipid catabolite that contributes to the ^{1}H MRS resonance of choline-containing compounds (Cho) and to the ^{31}P phosphodiester PDE resonance. (MRS)

glycerophosphoethanolamine (GPE) A phospholipid catabolite that contributes to the ^{31}P PDE resonance. (MRS)

gradient The amount and direction of the rate of change in space of some quantity, such as magnetic field strength. Also commonly used to refer to the coil producing a magnetic field gradient. (MRI)

gradient coil See *coil*.

gradient field A linearly varying magnetic field used to encode the spatial location of the measured signal. (MRI)

gradient pulse Briefly applied *magnetic field gradient*. (MRI)

GRASE (gradient and spin echo) sequence An efficient MR sequence that acquires MR signals using combined gradient and spin echoes; ultrafast sequence. Advantages over echo-planar imaging include increased signal-to-noise ratio and absence of distortions and artifacts due to magnetic field inhomogeneities. (MRI)

gray matter density The proportion of signal intensity that is segmented as gray matter by automated segmentation software within a predefined sphere. Given that this measure relies on MR contrast, the precision of this measure depends on the contrast characteristics of the particular scan, which can vary across MR sequences and scanners, partial volume artifacts, and other such considerations. (MRI)

gyromagnetic ratio (γ) Ratio of the angular frequency of precession of a spin to the magnetic field strength in which the spin-carrying nucleon resides; constant and specific for a given isotope. Also called magnetogyric ratio and defined by the Larmor equation. (MRI)

hemodynamic response (HR) and HR function (HRF) In BOLD fMRI, the HR is the time course of the BOLD response to an event. The response rises and peaks approximately 5–6 s after stimulus onset and then returns to baseline. The slow prolonged nature of the HR limits the temporal resolution of fMRI. The HRF is a stimulus-independent model response function that normally is kept constant throughout the brain and across subjects. Given the HRF and an arbitrary *stimulus function*, the HR can be computed as the *convolution* of both. This way, fMRI data can be conveniently analyzed for arbitrary experimental settings. Similar concepts apply for modeling the cerebral blood flow in ASL. (fMRI)

high angular resolution diffusion imaging (HARDI) Diffusion imaging data acquisition method that attempts to resolve multiple fiber orientations (e.g., crossing fibers) within a single voxel. HARDI obtains diffusion measurements in each voxel at high angular resolution in many diffusion directions and does not assume any particular 3D distribution of water diffusion or specific number of fiber orientations within a voxel. HARDI provides datasets used for *q-space* and *q-ball imaging*. (DTI)

high field Term used to describe the *comparative* magnetic field strength of MRI magnets. As magnets increase in strength, what is considered "high field" changes. However, currently in human imaging, "high field" is often used to refer to magnets ranging in strength from 3 T and up. (MRI)

homogeneity (of a magnetic field) Homogeneity (uniformity) of the static magnetic field is an important criterion of the quality of the magnet. It can be improved by the use of *shim coils*. Homogeneity of the RF field of RF coils is of equal importance. Homogeneity requirements for MRI are generally lower than the homogeneity requirements for MRS. (MRI)

image registration See *registration*.

image selected in vivo spectroscopy (ISIS) An MRS subtractive technique that uses ±180° selective pulses in all possible combinations on the three axes such that the final signal is only from the required voxel. A volume-selective technique that is quite sensitive to subject motion. Often used with ^{31}P MRS. (MRS)

imaging genetics Relatively new field of science that uses anatomical or physiological imaging technologies as phenotypic assays to evaluate genetic variation.

inhomogeneity effects (= magnetic field inhomogeneity effects) Non-uniformities or intensity variations in MR images that complicate quantitative analyses. Inhomogeneity is produced by imperfections in the MRI magnet, magnetic susceptibility differences (for example, between bone or air and tissue) or the presence of external metallic objects. Techniques used to reduce these effects include *shimming* to fine-tune the magnetic field. (MRI)

inorganic phosphate (Pi) A high-energy phosphorus metabolite that reflects intracellular metabolism, detectable using ^{31}P MRS. When examined in conjunction with PCr, the ratio of PCr relative to Pi reflects phosphorylation potential. (MRS)

interpulse time Time between successive radiofrequency pulses used in pulse sequences. Particularly important are the inversion time (TI) in inversion recovery, and the time between a 90° pulse and the subsequent 180° pulse to produce a spin echo, which will be approximately one-half the spin-echo time (TE). The time between repetitions of pulse sequences is the *repetition time* (TR). (MRI)

inversion An excited spin state in which the net magnetization vector is in a direction opposite to that of the main field. (MRI)

inversion recovery Rate of recovery as the nuclei return to equilibrium magnetization (after their magnetization was inverted by the radiofrequency pulse). The rate of recovery depends upon the longitudinal relaxation time T_1. (MRI)

inversion recovery sequence MRI pulse sequence in which the magnetization is inverted by means of a 180° radiofrequency (RF) pulse, and the recovery from this inversion is monitored by means of a 90° RF pulse applied after a time TI. This sequence is commonly used for measurement of T_1. (MRI)

inversion time (TI) In inversion recovery, time between middle of inverting (180°) RF pulse and middle of the subsequent exciting (90°) pulse to detect amount of longitudinal magnetization. (MRI)

isotropic diffusion; isotropy Diffusion that is equal in all directions, suggesting a lack of physical barriers (e.g., cell membranes, axons, myelin, etc.)

to restrict the diffusion. In DTI, a major barrier is thought to be the myelin sheath of axons. (DTI)

isotropic voxel A volume element with equal dimensions in x, y, and z.

k-space The spatial frequency information in two or three dimensions of an object being scanned in MRI; the image is constructed from k-space by the inverse Fourier transform. (MRI)

labeling efficiency The degree of spin labeling as reference to a perfect inversion (180°) of magnetization (100%). For example, flipping the spin by 90° represents a 50% labeling efficiency. (ASL)

lactate Marker of anaerobic metabolism that can be detected with proton MRS. (MRS)

Larmor equation States that the frequency of precession of the nuclear magnetic moment is proportional to the applied magnetic field. The Larmor equation is $\omega = \gamma B$, where ω is the Larmor frequency in radians per second, γ is the gyromagnetic ratio, and B is the magnetic field strength. The latter is the combined strength of the main magnetic field and the applied gradient field. (MRI)

Larmor frequency (ω) Precession frequency of a given nucleus with a spin in a magnetic field of a given strength. It corresponds to the resonance frequency at which magnetic spins can be excited and is proportional to the field strength and given by the Larmor equation. Expressed in Hz (f) or radians per second (ω); $\omega = 2\pi f$. By varying the magnetic field across the body with a gradient magnetic field, the corresponding variation of the Larmor frequency can be used to encode position. For protons (hydrogen nuclei), the Larmor frequency is 42.58 MHz/T. (MRI)

lattice Environment with which nuclei exchange energy, causing longitudinal relaxation. (MRI)

lithium-7 (^{7}Li) MRS Technique used to monitor the concentration of the predominant lithium isotope in the human brain in persons who take lithium therapeutically. (MRS)

longitudinal relaxation (= spin lattice relaxation) Gradual recovery of the net magnetization (M) due to the main magnetic field (B_0) after an RF excitation pulse has flipped the longitudinal magnetization by some angle. (MRI)

longitudinal relaxation time (T_1) Characterizes the rate at which excited nuclei reach equilibrium by exchanging energy with the environment (lattice). The magnetization in the z-direction will grow after a 90° excitation from 0 to about 63% of its final thermal equilibrium value in a time of T_1. (MRI)

magnetic field gradient Magnetic field gradients are used to change the strength of the magnetic field B_0 in a certain direction. Gradients are used in MR imaging with selective excitation to select a region for imaging, to encode the location of MR signals received from the object being imaged, to read out the signal, to rephase spins (in gradient echo imaging), or to generate certain contrast mechanisms such as in the case of diffusion gradients. (MRI)

magnetic field strength The main magnetic field B_0 of an MRI system is produced by the superconducting electric coils, and its strength is measured in *Tesla* (T; metric unit system) which equals 10 000 Gauss (G; outdated non-metric unit system). Earth's magnetic field is 0.5 G, and a refrigerator magnet is a few Gauss. (MRI)

magnetic moment Measure of the magnetic properties of an object or particle (nucleus) that cause it to align with the static magnetic field. It is also responsible for the precession of spins around the z-axis. (MRI)

magnetic resonance Absorption or emission of electromagnetic energy by nuclei in a (static) magnetic field after excitation by suitable radiofrequency (RF) waves: the frequency of resonance is given by the Larmor equation. (MRI)

magnetic resonance imaging (MRI) Creation of images with an MRI scanner. An MRI scanner utilizes the nuclear magnetic resonance (NMR) phenomenon. The image contrast usually depends on both the spin density and the relaxation times, with their relative importance determined by the particular imaging technique employed. MRI is a highly versatile technique that can probe many different tissue properties and physiological processes. It is still rapidly developing, with no end in sight yet. Because it does not involve ionizing radiation, it is considered safe for use in children. (MRI)

magnetic resonance spectroscopy (MRS) [= nuclear magnetic resonance spectroscopy (NMRS)] Technique to detect species of the atomic nuclei in a sample (e.g., brain), and to identify the compounds in which they are bound. The technique is based on the detection of resonance radiofrequency signals emitted characteristically by certain magnetically susceptible atomic nuclei. This occurs when a sample is placed in a strong magnetic field (B_0) and excited by a second pulse of electromagnetic energy B_1 of appropriate frequency and orientation. The resonance frequency of the emitted radiation is characteristic of the atomic nucleus and the nature of its immediate chemical environment; its exact value is proportional to the magnetic field, B_0. In NMRS, small shifts in resonance frequency are observed. The *chemical shifts* are characteristic of molecular bonding patterns adjacent to the susceptible nucleus, yielding information such as the presence of ion complexes. Not all atoms give rise to NMR signals, due to their non-magnetic nature. MR-visible nuclei that are considered of biological importance include ^1H, ^{19}F, ^{31}P, ^{13}C, and ^{23}Na, listed in order of decreasing NMR sensitivity. (MRS)

magnetic susceptibility Denotes the intensity of the magnetization produced in a substance by an applied magnetic field. Paramagnetic substances have positive susceptibility, whereas diamagnetic substances have negative susceptibilities. (MRI)

magnetization transfer Selective saturation of a particular spin within a multispin system. (MRI)

mean diffusivity Equals one-third of the *trace* of the diffusion tensor and is used as an indicator of direction-independent average water molecular displacement. Mean diffusivity has been used to infer cell size and integrity in adult and developing brains. It decreases with maturation while fractional anisotropy increases. (DTI)

morphometry Measurement of brain structures, generally for group comparisons. (MRI)

MR sensitivity (= MR or NMR visibility) MR sensitivity reflects the intrinsic resonance signal *from a nucleus, based its gyromagnetic ratio*, which is influenced by the natural abundance of the nucleus being studied, noise and magnetic field strength. For example, ^1H has the highest gyromagnetic ratio, and therefore has the highest relative sensitivity. (MRS)

multi-compartment perfusion models Detailed models that take into account arterial, capillary, tissue, and venous compartments in quantifying perfusion using ASL methods. A two-compartment perfusion model includes capillary and tissue compartments. (ASL)

***myo*-inositol (mI)** Metabolite detectable with proton MRS that is involved in phospholipid metabolism and the maintenance of osmotic equilibrium. mI-containing phospholipids participate in a number of cell signaling patterns. Detectable using ^1H MRS. (MRS)

***N*-acetylaspartate (NAA)** Metabolite found primarily in neurons that contributes the largest signal to water-suppressed cerebral spectra; often viewed as a neuronal viability marker, although its exact role remains unknown. Detectable using ^1H MRS. (MRS)

near-infrared spectroscopy (NIRS) A spectroscopic method utilizing the near infrared region of the electromagnetic spectrum. As an imaging method, NIRS can non-invasively monitor brain function through an intact skull in humans by detecting changes in blood hemoglobin concentrations associated with neural activity, particularly at the cortical level. NIRS can be used at bedside and is starting to be used in pediatric critical care and developmental cognitive research. (NIRS)

nuclear magnetic resonance (NMR) The absorption and emission of electromagnetic energy tuned to the *Larmor frequency* of a nucleus precessing in a magnetic field (B_0). The frequency ω_0 of the magnetic resonance is the same as the frequency of the Larmor precession of the nuclei in the magnetic field and is proportional to the strength of the field. Thus, $\omega_0 = \gamma B_0$, where γ is a characteristic constant, called the *gyromagnetic ratio*, for a given nucleus. Freely moving, or mobile, nuclei in molecules become aligned when placed in a magnetic field. An additional resonant magnetic field, applied at the correct frequency, can disturb alignment with the magnetic field until this second field is removed. The nuclei then realign with the static magnetic field, back to their original state, emitting energy in the process. The strength of the emitted signal is proportional to the local concentration of the mobile metabolites. These signals are then used, usually with some type of computer processing, to produce an image (MRI, based on protons) or to identify the spectrum of chemical substances. (MRI)

nuclear magnetic resonance imaging See *magnetic resonance imaging*. (MRI)

nuclear magnetic resonance spectroscopy See *magnetic resonance spectroscopy*.

nuclear spin An atomic nucleus (with an odd number of nucleons, i.e., protons and neutrons) can be thought of as a spinning charged body, which acts like a tiny magnet. (MRI)

nucleoside triphosphate (NTP) Metabolite detectable using ^{31}P MRS, primarily reflecting adenosine triphosphate (ATP) in brain. This resonance includes signals from β NTP, α NTP, and γ NTP. (MRS)

orientation distribution function (ODF) Distribution function of diffusion directions. The ODF can resolve multiple intravoxel fiber orientations, for example crossing, bending, and "kissing" fibers, and does not require a model of the diffusion process. It can be computed from *high-angular-resolution diffusion* data. (DTI).

outer volume suppression (OVS) A technique used in MRS that selectively reduces the contribution of signals from outside the region of interest, or voxel, e.g., lipids, which can interfere with metabolite signal detection. (MRS)

parallel imaging MRI method based on the use of multiple RF coils; reduces acquisition time by using multiple coil elements and is applicable to all pulse sequences. In contrast to the use of array coils to cover separate anatomical regions, parallel imaging uses the multiple coils to simultaneously measure the same region, e.g., the brain. (MRI)

paramagnetic A substance with a positive magnetic susceptibility (magnetizability). The addition of a small amount of paramagnetic substance may greatly reduce the relaxation time of water. (MRI)

parcellation Segmentation, or subdivision, of an image into component regions based on the recognition of similar patterns of intensity, contrast, functional characteristics, or other features; required for most types of morphometry; may be manual, semi-automatic or fully automated. (MRI)

partial volume effect The alteration in voxel values that occurs when the structure being imaged has a spatial extent that is similar or smaller than the resolving capabilities of the imaging device. For example, a voxel appearing like gray matter may contain white matter and cerebrospinal fluid. The result is that the value in a particular voxel reflects a mixture of tissues. In DTI, crossing fibers also can lead to a partial volume effect in that the anisotropy may appear to be less than it actually is.

phase encoding Use of a magnetic field gradient to cause different rates of precession for a brief period of time, resulting in phase differences across space in the direction of the gradient. Phase encoding is used in combination with frequency encoding and slice selection to encode the position in space. (MRI)

phosphocreatine (PCr) High-energy phosphate resource pool that acts as a buffer to maintain brain ATP levels (β NTP levels). PCr provides a shuttle for energy from sites of production to sites of utilization. Detectable using ^{31}P MRS. (MRS)

phosphodiesters (PDE) Metabolite detectable using ^{31}P MRS, reflective of breakdown products of membrane phospholipids. Proton decoupling and the use of higher field strength scanners (4.0 T and higher) can be used to resolve the PDE peak into constituent components, glycerophosphocholine (GPC) and glycerophosphoethanolamine (GPE). (MRS)

phosphomonoesters (PME) Metabolite detectable using ^{31}P MRS, reflective of phospholipid membrane anabolites, or precursors, necessary for the building of new membranes. Proton decoupling and the use of higher field strength scanners (4.0 T and higher) can be used to resolve the PME peak into constituent components, phosphocholine (PC) and phosphoethanolamine (PE). (MRS)

phosphorus-31 (^{31}P) MRS Technique that provides a means of detecting high-energy phosphate metabolites and constituents of membrane synthesis, indicating cellular bioenergetic state and the integrity and function of cell membranes. Detectable metabolites include PCr, NTP, Pi, PME (PC and PE), and PDE (GPC and GPE). Intracellular pH can also be measured based on the chemical shift of Pi. (MRS)

planes of the brain The *sagittal* plane divides the brain into right and left halves; the *midsagittal* plane does this at the midline, through the corpus callosum; *parasagittal* slices run parallel to the midsagittal plane on either side of it. The *axial, transverse, or horizontal* plane is parallel to the floor when the patient is standing and divides the brain into slices that are more or less superior or inferior; the *coronal or frontal* plane passes it at a right angle to the floor when the patient is standing and divides the brain into slices that are more or less anterior, or rostral, and posterior, or caudal.

point-resolved spectroscopy (PRESS) A single-voxel spatial localization technique, based on spin-echo sequence, used to define the anatomical location of an MRS voxel, so that the MRS signal can be measured more precisely without contamination from other spatial positions. MRS sequence that provides twice the signal-to-noise ratio of STEAM when spectra are obtained at the same echo time and thus recommended for acquisitions at longer echo times, e.g., lactate. Used to collect both single-voxel and chemical shift imaging (CSI) ^{1}H MRS data. (MRS)

positron emission tomography (PET) A procedure used for the study of regional tissue physiology and biochemistry. It is based on the in vivo detection and imaging of positron-emitting radioisotopes that are introduced as tracer elements into the physiological systems of interest. The tracer tag should not perturb the behavior of the molecule. (PET)

precession Slow gyration (wobbling) of the axis of a spinning body (e.g., nuclei) so as to trace a cone; caused by the application of a perpendicular force (torque) that tends to change the direction of the rotation axis. Similar to the effect of gravity on the motion of a spinning top or gyroscope. (MRI)

probabilistic brain atlas A brain atlas that includes *probability maps* that depict the likelihood of a particular feature. Used to show how frequently (what percentage of times) the feature is present in a particular population or to reflect the degree of uncertainty, e.g., of fiber direction as in DTI. The major application thus far has been to depict the probabilistic position and borders of cortical areas based on large samples within a population.

probability density functions (PDFs) A continuous probability distribution follows from an integral over the PDF.

probability distribution A probability distribution describes the values and probabilities that a random event can take. The values must cover all of the possible outcomes of the event, while the total probabilities must sum to exactly 1, or 100%. A *discrete distribution* contains a countable numbers of discrete outcomes with positive probabilities. A *continuous distribution* describes the probability of events over a continuous range of infinitely dense real numbers. Here, although the probability of an event occurring at any one real number in the range is technically zero, the total probability of events occurring over the entire range is one.

proton decoupling MRS technique that can increase MR sensitivity by producing line-narrowing effects or by minimizing the contribution of ^{1}H nuclei on other (e.g., ^{31}P) nuclear spins to which they are bound. For instance, in ^{31}P MRS, the PME and PDE resonances are influenced by proton-containing compounds. Thus decoupling improves the resolution of PC and PE within the *PME peak* and GPC and GPE within the *PDE peak*. A special head coil, dually tuned for both ^{1}H and the nucleus of interest (e.g., ^{31}P nuclei), is required for proton decoupling. Proton decoupling is also essential for ^{13}C MRS. (MRS)

proton density (ρ) (= spin density) In the context of MRI, the number of hydrogen atoms per unit volume, a principal determinant of strength of the MRI signal from a region. Images based on proton density are generated using a long TR and a short TE (e.g., spin-echo, TR = 1800 ms, TE = 20–40 ms). A spin-density-weighted image usually has a low contrast, since hydrogen content differences between tissues are small, but may be useful to detect lesions. (MRI)

proton (^{1}H) MRS Technique that permits the quantification of important amino acids and neurotransmitters, including NAA, Cr, Cho, mI, lactate, GABA, glutamate and glutamine (Glx). (MRS)

pseudo-continuous ASL An imaging technique that uses a train of short RF pulses to simulate the effect of continuous ASL (CASL). (ASL)

180° pulse (π pulse) Inversion pulse. Radio-frequency (RF) pulse designed to rotate the macroscopic magnetization vector 180° in space as referred to the rotating frame of reference. If the spins are initially aligned with the magnetic field, this pulse will produce inversion. (MRI)

90° pulse (π/2 pulse) Saturation pulse. Radiofrequency (RF) pulse designed to rotate the macroscopic

magnetization vector 90° in space as referred to the rotating frame of reference, usually about an axis at right angles to the main magnetic field. If the spins are initially aligned with the magnetic field, this pulse will produce transverse magnetization and a free induction decay (FID). (MRI)

pulse sequence A program that controls the MRI scanner or the process defined by such a program. It determines the timing, duration, and strength of radiofrequency (RF) and magnetic field gradient pulses. Pulse sequence programming can be quite complex due to many interrelations between scanning parameters, the complexity of the pulse trains themselves, the complexity of the signal location procedure in k-space, and safety issues such as keeping the specific absorption ratio (SAR) within bounds. (MRI)

pulse sequence programming Specification of the signals being sent to the RF transmit coil, slice-selection coil, frequency-encoding coil, and phase-encoding coil. This programming is complex. For example, the RF excitation pulse cannot be at a single frequency if it is to work in conjunction with the slice-selection gradient to define a slice of tissue that will be excited. In particular, the Fourier spectrum of the excitation pulse should be tuned to the range of frequencies that correspond to the different Larmor frequencies created by the slice-select gradient, over the spatial extent desired for imaging. (MRI)

pulsed arterial spin labeling (PASL) perfusion imaging Imaging technique that uses almost instant RF pulses (10–20 ms) to label a thick slab of blood water. (ASL)

q-ball imaging A diffusion imaging data acquisition and post-processing technique that requires considerably less diffusion gradient applications than *q-space* imaging with the benefit of reduced scan time. The *orientation distribution function* (ODF) of diffusion, which can resolve multiple intravoxel fiber orientations and does not require a particular diffusion model, is computed using the concept of a *Funk-Radon transform*. (DTI)

q-space imaging   Diffusion imaging using a range of multiple b values and a large number of diffusion gradient directions to better define white matter structure in a spatial scale significantly smaller than the typical image voxel size. (DTI)

quantitative   electroencephalography   (qEEG) Measurement of electrical patterns (brainwaves) at the surface of the scalp and characterization of the amount of power in various frequency bands (delta, theta, alpha, beta), reflecting the degree of activation and arousal. These data can be converted to digital form, presented in graphical, topographic maps, and compared with data from normal individuals to reveal group differences. (EEG)

raclopride   A synthetic compound that acts on a subset of dopamine receptors (D2 receptors) as an antagonist and which can be radiolabeled and used in PET to assess the degree of dopamine binding. At the doses used for PET, no physiological effects are expected. (PET)

radiofrequency (RF)   Electromagnetic wave frequency below the infrared range. The RF used in MRI studies is commonly in the megahertz (MHz) range. The principal effect of RF fields on the body is power deposition in the form of heating, mainly at the surface; this is a principal area of concern for safety limits. (MRI)

radiofrequency (RF) pulse   A short burst of radiofrequency (RF) electromagnetic energy delivered by an RF transmitter. If the RF frequency is at the Larmor frequency, the result is a rotation ("flip") of the magnetization vector. (MRI)

radioisotope   A chemical element that has an unstable nucleus and emits radiation during its decay to a stable form. (PET/SPECT)

radioligand   A radioactive biochemical substance (a ligand) used with nuclear medicine techniques, such as PET or SPECT, that binds to proteins, such as receptors or transporters, or is degraded by enzymes. Mathematical models are then used to measure the behavior of these radioligands

(e.g., binding potential, metabolic rates). (PET/ SPECT)

receiver coil   See *coil.*

region of interest (ROI)   Outlined area on a computer-processed image, defined automatically or manually, for data analysis. The purpose is to test predictions based solely on a selected region (e.g., amygdala) or regions and thus to reduce the stringency of correction for multiple tests done for each unit of volume (voxel) being examined. (PET/ SPECT, MRI)

In MRS, a region or volume of interest in single voxel acquisition signifies the region targeted for data acquisition. (MRS)

registration   The process of transforming different sets of data into one coordinate system, which then allows one to compare data across individual datasets. Because human brains differ in size and shape, as part of spatial normalization, they are warped to fit a standard space, typically employing a three-dimensional transformation model. (MRI, PET/SPECT)

relative anisotropy (RA)   A normalized representation of the degree of anisotropy, similar to fractional anisotropy. RA and FA can be translated into each other, making RA, which in addition has worse signal-to-noise properties, redundant. (DTI)

relaxation (= recovery)   The process by which the nuclei realign themselves with the static magnetic field (T_1 relaxation is within the longitudinal axis and T_2 is within the transverse axis). The energy that is released during relaxation emits a measurable signal, which decays over time, referred to as *free induction decay (FID).* (MRI)

relaxation rates   Reciprocals of the relaxation times, T_1 and T_2 ($R_1 = 1/T_1$ and $R_2 = 1/T_2$). (MRI)

relaxation times   After excitation, nuclei tend to return to their equilibrium position in accordance with these time constants. On cessation of the radiofrequency (RF) excitation pulse, the longitudinal magnetization M_z returns toward the equilibrium value M_0 at a rate characterized by the time constant T_1, and any transverse magnetization decays towards zero with a time constant T_2 (or T_2^* in the real situation). (MRI)

T_1 ("T-one") Spin-lattice or longitudinal relaxation time; the time constant that describes the recovery of the longitudinal component of the net magnetization over time. Starting from zero magnetization in the z direction, the z magnetization will grow after excitation from zero to a value of about 63% of its final value in a time of T_1. The T_1 time is a contrast-determining tissue parameter. Fat nuclei have a short T_1, whereas water nuclei do not give up their energy to the lattice (surrounding tissue) as quickly as fat, and therefore take longer to regain longitudinal magnetization, resulting in a long T_1 time. (MRI)

T_2 ("T-two") Spin-spin or transverse relaxation time; is a tissue-specific time constant that describes the decay of the transverse component of net magnetization due to accumulation of phase differences caused by *spin-spin* interactions. Starting from a non-zero value of the magnetization in the xy plane, the xy magnetization will decay so that it loses 63% of its initial value in a time T_2. Fat has a very efficient energy exchange and therefore it has a relatively short T_2. Water is less efficient than fat in the exchange of energy and therefore it has a long T_2. (MRI)

T_2^* ("T-two-star") The time constant that describes the decay of the transverse component of the net magnetization due to both accumulated phase differences and local field inhomogeneities. T_2^* is shorter then T_2. BOLD-contrast fMRI relies on T_2^* contrast. (MRI)

repeated free induction decay Another term for *saturation recovery*. A sequence in which 90° pulses are repeated for excitation and measurement. It results in partial saturation if the period between the 90° pulses is of the order of T_1 or less and gives a T_1-weighted signal. Generally the term is only applied where the signal is detected as a free induction decay (FID) (and not as an echo). (MRI)

repetition time (RT) The period between an RF excitation pulse and the next pulse that excites the same volume of tissue. (MRI)

rephase; rephasing pulse The process of returning out of phase magnetic moments back into phase coherence. Caused either by reversing a magnetic gradient (gradient echo, "GE") or by applying a 180° RF pulse (spin echo, "SE"). In the spin-echo pulse sequence, this action effectively cancels out field-inhomogeneity effects from the signal. (MRI)

rephasing gradient Gradient magnetic field applied for a brief period after a selective excitation pulse, in the direction opposite to the gradient used for the selective excitation. The result of the gradient reversal is a rephasing of the spins (which will have gotten out of phase with each other along the direction of the selection gradient), forming an echo by "time reversal," and improving the sensitivity of imaging after the selective excitation process. (MRI)

resonances Distinct metabolite peaks whose area is proportional to the concentration of molecules that contribute to the resonance; used to derive tissue concentration estimates for brain chemicals. (MRS)

run In the context of fMRI, a run refers to a single, continuous collection of images. (fMRI)

saturation Non-equilibrium state in MRI in which there is no usable net magnetization. After exposure to a single radiofrequency (RF) pulse, if T_2 is much shorter than T_1, then the net transverse magnetization will disappear before significant repolarization of the spins occurs. During this time the sample is said to be saturated. (MRI)

saturation recovery (SR) A partial saturation pulse sequence in which the preceding pulses leave the spins in a state of saturation, so that recovery at the time of the next pulse has taken place from an initial condition of no magnetization. A sequence not often used, employing a 90° RF excitation pulse, with a very long repetition time. T_1 times can be measured faster with this technique than with inversion recovery pulse sequences. (MRI)

seeding region of interest (seeding ROI) A region selected on an image to begin a calculation, e.g., correlations with values in other regions for mapping structural or functional connectivity or for fiber tractography. When combining fMRI with DTI, a seeding point is frequently selected to reflect a "hot spot" of activation. (MRI)

sequence delay time   Time between the last pulse of a pulse sequence and the beginning of the next identical pulse sequence. It is the time allowed for the nuclear spin system to recover its magnetization and is equal to the sum of the acquisition delay time, data acquisition time, and the waiting time. (MRI)

shim coils   Coils carrying a relatively small current that are used to provide auxiliary magnetic fields to compensate for inhomogeneities in the main magnetic field of an MRI system. (MRI)

shimming   Process of adjusting field gradients to optimize the magnetic homogeneity of the static magnetic field, B_0, of an MR imager produced by imperfections in the magnet or the presence of external ferromagnetic objects. Shimming may involve changing the configuration of the magnet or adding shim coils or small pieces of steel. *Passive shimming* involves the placement of metal within the bore of magnet or on the outer surface of the magnet to correct distortions in the magnetic field and create a more homogenous field. *Active shimming* uses coils in the magnet through which current is passed to generate an auxiliary, corrective field with which to improve homogeneity. (MRI)

signal attenuation   In diffusion imaging, the decrease of signal due to the application of diffusion gradients. The amount of signal attenuation depends on the diffusion properties of the tissue and the chosen b value. (DTI)

signal averaging   Technique to improve signal-to-noise ratio by averaging repeated scans over a few minutes through the same region of interest. The noise tends to decrease because of its random nature, whereas the signal reinforces itself. (generally MRI)

signal loss   Loss of MRI signal that occurs when protons dephase between excitation and signal readout. Unrecoverable intravoxel dephasing can be caused by strong magnetic field inhomogeneities, for example close to bone, metallic implants, or air cavities. It is most pronounced in gradient-echo EPI sequences such as in fMRI applications. (MRI)

signal-to-noise ratio (SNR)   The ratio between the signal intensity of the object and standard deviation of the background noise. A higher magnetic field produces strong MR signals, with the SNR theoretically increasing linearly with the field, although other factors may offset these gains. (MRI)

single photon emission computed tomography (SPECT)   Imaging technique that uses radioisotopes that allow for measurements of regional cerebral blood flow. The resolution available with SPECT is inferior to that of positron emission tomography. (SPECT)

single-shot RF excitation   A high-speed MRI method. Traditional MRI acquisition methods collect one signal (echo) for each radiofrequency (RF) excitation pulse. Fast MRI breaks the one-excitation one-signal link. Instead, subsecond, ultrafast MRI acquires all image data after a single RF excitation pulse. This is referred to as "single-shot" imaging. (MRI)

single-voxel MRS   MRS that collects data from single, predefined tissue volumes (voxels). Most single-voxel MRS studies collect data from one to three voxels per imaging session. (MRS)

slice selection   The region whose electromagnetic vector will be flipped can be limited to a slice by applying a slice-selection gradient in the orthogonal direction while the RF excitation pulse is presented. (MRI)

smoothing   Process whereby the values of voxels or pixels are weighted or averaged to a small number of surrounding voxels or pixels to increase the signal-to-noise ratio and decrease the influence of variable anatomy, for example gyral/sulcal patterns, across subjects. Generally performed early in image analysis (as part of "preprocessing" prior to comparisons of groups).

The purpose of smoothing is to enhance an image by reducing high-frequency phenomena, such as statistical noise, while preserving the overall form of the data. However, since edges within an image or large gradients in a curve are dominated by high-frequency components, the effect of smoothing is to reduce or "average out" such features. Another price to pay for increasing the signal-to-noise ratio by smoothing is an increase in bias (bias-variance dilemma). Smoothing can be applied successfully to both images (spatial representation) and curves (temporal representation) from radionuclide images. For example, in many "dynamic" studies, temporal (i.e., between corresponding pixels in several frames) or spatial (i.e., between adjacent pixels within a frame) smoothing is usually implemented prior to viewing the data in cine mode. It is important to understand that if the image data have been smoothed, then subsequent data extraction or display may be modified by the smoothing algorithm.

sodium-23 (^{23}Na) MRS   Technique that has been used to probe the concentration gradient of intracellular versus extracellular sodium, providing important information about cellular processes such as the generation of neuronal impulses and regulation of cell volume. Abnormalities or alterations in the sodium resonance have been associated with cerebral ischemia. (MRS)

spatial frequency filtering   Technique for eliminating in an image the higher spatial frequencies, assumed to be noise. Employs Fourier transform analysis. (MRI)

spatial normalization   An image processing step involving image registration (MRI, PET, SPECT).

spatial resolution   Ability of an instrument to image two separate line or point sources of signal as separate entities. The smaller the distance between the two sources, the better the spatial resolution. A measure of spatial resolution is the point spread function (PSF) or the line spread function (LSF) and the derived system transfer function (STF).

In MRS, spatial localization techniques are often used to more accurately identify the origin of the MRS signal without contamination from other spatial positions. Thus anatomical images with high spatial resolution, to distinguish between brain structures, are used in the placement of MRS voxels. (All imaging modalities)

spatial smoothing   See *smoothing*.

specific absorption rate (SAR)   A measure of energy deposited by an RF pulse which causes heating in the patient, expressed in watts per kilogram (W/kg). Used to evaluate safety. (MRI)

spectra   See *magnetic resonance spectroscopy*.

spectroscopic imaging   Acquisition of MRS spectra from multiple volumes simultaneously (each contiguous volume of 1–2 ml each); typically takes about 20–40 min to complete. See *chemical shift imaging*. (MRS)

spin   Intrinsic angular momentum of an elementary particle or system of particles, such as a nucleus, that is responsible for the magnetic moment. (MRI)

spin axes   The direction of a spin, which is a pseudovector. (MRI)

spin coupling   The diffusion of magnetic moments due to actual movement of the associated molecule and/or chemical exchange. (MRI)

spin density   Density of resonating spins in a given region. It is a principal determinant of the strength of the MRI signal from that region. For water, there are about 1.1×10^5 mol of hydrogen per m^3, or 0.11 mol of hydrogen/cm^3. True spin density is normally not imaged directly, but influences signals received with different pulse sequences. Proton-density-weighted images show an intensity approximately related to the spin density. (MRI)

spin echo   The reappearance of an MRI signal, arising from refocusing or rephasing the various components of magnetization in the *xy* plane. This usually results from the application of the 180° (inversion) pulse after decay of the initial free induction decay (FID). It can be used to determine T_2 without contamination from inhomogeneous effects of the magnetic field. Example of spin-echo sequences: TR: 1800–2500 ms, TE: 80–120 ms. Most ^{1}H MRS use a spin echo with a long TE (>100 ms), which permits the measurement of three to four compounds present in the brain at relatively high concentration (e.g., NAA, Cr and phosphocreatine, Cho). Short TE (<20 ms) permit the observation and quantitation of other metabolites (e.g., glucose, *myo*-inositol, glutamate, glutamine, GABA). (MRI)

spin-lattice relaxation time (= T_1, **or longitudinal relaxation time**) Characteristic time constant for spins to align themselves with the external static magnetic field. The exponential time constant that characterizes the growth or decay of the component of magnetization parallel to the external field. The physical mechanism involved in this process is the interaction of the nucleus with its entire surroundings (lattice). (MRI)

spin-quantum number Refers to a physical characteristic of each MR-visible nucleus, indicating the number of possible energy levels for a given nucleus in a fixed magnetic field [(2 × spin-quantum number) + 1 = number of energy levels]. Nuclei with an odd mass number or an unpaired electron, such as ^{1}H and ^{31}P, have a spin-quantum number of ½, and, accordingly, two possible energy levels (– ½ and + ½). (MRS)

spin-spin broadening Increased line width of the MRS spectra due to interactions between neighboring dipoles. The line width of a peak due to its intrinsic T_2 is typically <1 Hz, whereas the line width from field inhomogeneity may be from 5 Hz to 10 Hz. The term T_2* refers to the combined effect of the intrinsic T_2 of the peak and the magnetic field inhomogeneity that affects the field width. (MRI)

spin-spin relaxation time (= T_2, **or transverse relaxation time**) Characteristic time constant for loss of phase coherence among spins oriented at an angle in the main magnetic field owing to interactions between the spins. The exponential time constant that characterizes the decay of confinement of magnetization perpendicular to the external field. This decay results from interaction at the nuclei with its immediate neighboring nuclei. Also called transverse relaxation time. (MRI)

spin-spin splitting Splitting in the lines of an MRS spectrum arises from the interaction of the nuclear magnetic moment with those of neighboring nuclei. (MRS)

spiral sequences/imaging MR sequences and imaging technique for fast image acquisition that uses sinusoidally changing gradients to trace a spiral trajectory through k-space. An advantage to the use of spiral sequences in fMRI is reduced susceptibility artifacts in orbitofrontal brain regions, relative to echo-planar scan sequences based on a cartesian data acquisition grid. (MRI)

stereotaxic, stereotaxic space Relating to stereotaxis or sterotaxy, i.e., to a three-dimensional arrangement. In neuroimaging research, images from individuals are generally warped, using anatomical landmarks, to fit into a common stereotaxic (coordinate) space, e.g., *Talairach* space, such that each voxel coordinate is anatomically comparable across subjects for purposes of group comparisons. *Stereotaxic normalization* is a term used to describe this process.

stimulated echo acquisition (STEAM) A single-voxel spatial localization technique used to define the anatomical location of an MRS voxel, so that the MRS signal can be measured more precisely without contamination from other spatial positions. Widely used MRS sequence for single-voxel spectroscopy. Provides less signal than PRESS, but is useful for acquiring spectra with short TE values, e.g., Glx. (MRS)

stream-tubes Simple methods of fiber tractography used in DTI, such as through concatenation of vectors along the principal diffusion axis. (DTI)

subtraction technique   Image analysis technique that statistically assesses the difference in activity between conditions or groups in each voxel or region of interest.

susceptibility artifacts   Artifacts created when homogeneity of the magnetic field is degraded by differences between the magnetic properties of adjoining tissues or tissue–air boundaries. (MRI)

T_1-weighted MRI   An MR image is generated using imaging parameters that cause contrast to be primarily based on differences in T_1 times for different tissues. (Short TR and short TE are used for T_1-weighting such as TR: 400–600 ms, TE: 10–30 ms.) Tissues with short T_1 are bright in T_1-weighted images. T_1 contrast is best to delineate anatomical structures, differentiate between white and gray matter, and detect subacute hemorrhage. Thus, T_1 scans are sometimes described as "anatomical scans." In T_1-weighted images, cerebrospinal fluid is generally dark, water-based tissues are mid-gray, and fat-based tissues are very bright. (MRI)

T_2-weighted MRI   An MR image generated using imaging parameters that cause contrast to be primarily based on differences in T_2 times for different tissues. (Long TR and long TE are used for T_2-weighting, such as TR: 1800–2500 ms, TE: 80–120 ms.) Tissues with long T_2 are bright in T_2-weighted images. T_2-weighted images have strong contrast between normal brain tissue and tissues with high water content, namely CSF and pathological processes (i.e., tumors, inflammation, cysts, demyelinating processes). High iron content decreases MR signal, especially on T_2-weighted images. Gradient-echo sequences do not allow one to obtain true T_2 weighting (because of the absence of a refocusing RF pulse), but rather T_2^* contrast, which corresponds to spontaneous signal decay along the transverse plane, and depends on magnetic field inhomogeneities. T_2^* contrast is widely used in functional MRI. (MRI)

tagging efficiency   See *labeling efficiency*. (ASL)

Talairach coordinates   A system for specifying locations in individual brains. It yields three coordinates (x, y, z) based on a rigid rotation of the brain to an orientation specified by anatomical landmarks, followed by a piecewise linear transformation of the brain in six sections that preserves continuity. It is the most widely used system for comparing brains between individuals. It was first developed and presented by J. Talairach in 1967.

TE (= time to echo)   The time interval between an excitation pulse and data acquisition. The time (milliseconds) between presentation of the saturating RF pulse that flips the longitudinal magnetization by 90° and the time that an echo is detected (because of a refocusing pulse). The total echo time (TE) affects the peak's intensity as a function of T_2. (MRI)

temporal resolution   The shortest amount of time that can be measured between two events in an imaging study. Also, the sampling time in an fMRI experiment. (MRI, PET/SPECT)

Tesla   Unit of magnetic flux density or field strength. One Tesla is equal to 10 000 Gauss (a non-metric unit) and is 20 000 times the Earth's magnetic field. (MRI)

TI (= time to inversion)   The time (milliseconds) between the center of the first inversion pulse and the middle of the saturating (90°) pulse in an inversion recovery pulse sequence. (MRI)

tissue segmentation   Subdivision of an image into component regions based on the recognition of similar patterns of intensity, contrast, functional characteristics, or other features; required for most types of morphometry; may be manual, semi-automatic or fully automated. (MRI)

TR (= time to repetition)   TR is the pulse sequence repetition time. It is delineated by initiating the first RF pulse of the sequence, then repeating the same RF pulse at a time t. Variations in the value of TR have an important effect on the control of image contrast characteristics. The time (milliseconds) between presentation of the RF pulses that flip the longitudinal magnetization to generate an NMR signal. The repetition time (TR) affects a peak's intensity as a function of T_1. (MRI)

trace (*D*) A scalar measure of mean diffusivity within a voxel. Trace has similar values in white and gray matter. Trace maps depict the mean diffusivity in each voxel. The "apparent diffusion coefficient" is defined as ADC = trace(*D*)/3. (DTI)

tracer Substance used for tracking some biological process.

    endogenous tracer Naturally occurring substance occurring within the body that can be used to track some biological process. Examples: MRI uses blood-water protons as an endogenous tracer and fMRI relies on the endogenous tracer deoxyhemoglobin. (PET/SPECT, MRI)

tractography Technique for non-invasively extracting the most likely fiber pathways in a given brain volume from a discrete 3D diffusion tensor, *q-space*, or *q-ball* imaging dataset. In most algorithms, tracking starts at a user-defined seeding point or region of interest. Resolution is limited by the voxel size, but the apparent resolution can be much higher by starting in different seed points within a voxel. (DTI)

transcranial Doppler sonography (TCD) Non-invasive technique that measures the velocity of blood flow through the brain's blood vessels using sound waves from an ultrasound probe. The speed of the blood flow increases or decreases the frequency of the sound and these changes are recorded and analyzed (*Doppler effect*).

transcranial magnetic stimulation (TMS) A non-invasive method that creates a brief and intense magnetic field by placing a coil against the scalp to induce in the brain electric currents capable of depolarizing neurons or modulating neuronal activity. TMS can be used to assess brain, especially cortical, activity, and to modulate (prime or suppress) neural activity in specific, targeted neural circuits. When used with neuroimaging, it can create a transient "virtual lesion" by disrupting activity in a specific region, allowing one to determine whether a region is essential or critical to task

performance. Potential therapeutic applications of TMS are being investigated.

transverse magnetization Component (M_{xy}) of the net magnetization vector orthogonal to the direction of the main field (longitudinal, B_0), whose precession, at the Larmor frequency, generates the NMR response signal. In the absence of externally applied RF energy. M_{xy} decays to zero with a characteristic time constant T_2, or more strictly T_2^*. (MRI)

transverse relaxation time (spin-lattice relaxation time) Time constant, T_2, characterizing the rate at which nuclei reach equilibrium, or go out of phase, with each other, i.e., gradual loss of the measured magnetization in the plane perpendicular to B_0 due to dephasing of the individual nuclei. (MRI)

voxel A volume element in an image; the three-dimensional counterpart of a two-dimensional pixel (= picture element). In MRI, a single number represents some process or characteristic (e.g., blood flow, MR signal, etc.) in a voxel. Voxels can also contain multiple numbers such as a diffusion tensor.

voxel-based morphometry (VBM) A technique for measuring brain structure or function on a voxel-by-voxel basis. Initially used to evaluate functional imaging data, this approach has recently begun to be applied to structural imaging data as well. VBM entails automated spatial normalization of imaging data into a standard coordinate space and scaling of images to achieve anatomical comparability of the voxels across subjects. (MRI)

voxelwise On a voxel-by-voxel basis. In image analysis, the calculation of a statistic within each voxel in order to provide a fine-grained image of statistical differences, such as those between an experimental and control task or between two groups.

PART II. GENETIC TERMS

activation ratio The ratio of affected/unaffected activated X chromosomes. In fragile X syndrome, this is highly correlated with amount of the fragile X mental retardation protein (FMRP) produced.

actively translating ribosomes Ribosomes that are assembled on a messenger RNA molecule and are engaged in translating the mRNA into protein.

adenine One of the four bases in DNA that make up the letters ATGC, adenine is the "A." The others are "G" for guanine, "C" for cytosine, and "T" for thymine. Adenine always pairs with thymine. Cytosine always pairs with guanine.

allele Alternative form of a gene found at the same locus on homologous chromosomes. A single allele for each locus is inherited separately from each parent (e.g., at a locus for eye color the allele might result in blue or brown eyes). A dominant allele controls the phenotype even in a single copy, i.e., heterozygous form. The phenotype of a recessive allele is only apparent when two copies of the allele are present, in a homozygous form.

amino acids Building blocks of proteins. Vertebrates have 20 amino acids. In a gene, each amino acid is encoded by a sequence of three nucleotides (a "triplet") that instructs the cell to insert that amino acid in a specific position of a protein as it is assembled. No triplet encodes for more than one amino acid, but different triplets may encode for the same amino acid.

ancestry informative markers (AIMs) Polymorphisms that exhibit substantially different frequencies between groups of descendants derived from inbred ancestral groups (e.g., races). By using AIMS, one can estimate the ancestral (racial) proportion of an individual.

association The occurrence of a particular allele in a group of patients more often than can be accounted for by chance.

balancing selection Form of natural selection that works to maintain genetic polymorphisms (multiple alleles) within a population.

base One of four nitrogenous substances in nucleic acid molecules (A = adenine, C = cytosine, G = guanine, T = thymine).

base pair (bp) Unit of complementary DNA bases in a double-stranded DNA molecule. Two nitrogenous bases [adenine and thymine (A-T) or guanine and cytosine (C-G)] are held together by hydrogen bonds. Base pairing ensures that the genetic information, the sequence of bases in the DNA, is passed from generation to generation with high fidelity in a process called DNA replication.

chromosome The self-replicating genetic structure of cells containing the cellular DNA that bears in its nucleotide sequence the linear array of genes. In *prokaryotes* (organisms without a defined nucleus), chromosomal DNA is circular, and the entire genome is carried on one chromosome. *Eukaryotic* (higher level organisms with membrane-bound nucleus) genomes consist of a number of chromosomes; each chromosome consists of one DNA strand tightly wound around nucleosomes, which are further condensed with the help of additional histone proteins. In somatic cells, chromosomes are found in pairs (diploid cells); human beings typically have 23 pairs of chromosomes.

chromosome 22q11 deletion syndrome (= velocardiofacial syndrome) A syndrome resulting from the loss of chromosomal material on the q (lower) arm of chromosome 22 that is characterized by developmental delays, learning disabilities, social/emotional symptoms, heart defects, and immune deficiency. Present in 1 of every 2000–4000 live births, affected individuals are at risk for the development of schizophrenia, bipolar disorder, and attention deficit/hyperactivity disorder.

co-causality Determination of a phenotype by more than one cause, e.g., by both genetic and environmental components or by more than one allele.

co-dominant Alleles that are both phenotypically expressed when they occur together in a heterozygous state (e.g., A and B alleles in the ABO blood group system).

codon A sequence of three adjacent nucleotides (bases) that encode a specific amino acid.

complex disease (= genetically complex disease) Diseases whose genetic causes involve more than a single gene. Most psychiatric diseases are genetically complex and poorly understood.

concordance State of having the same trait, e.g., when both twins in a pair share a trait, they are *concordant*. If only one has the trait, they are *discordant*.

concordance ratio Ratio of concordance of two individuals having the same trait; most frequently applied to twin data in which the concordance for monozygotic (MZ) versus dizygotic (DZ) twins is compared to estimate the heritability of a trait. A high MZ/DZ ratio suggests strong heritability.

cytosine (C) One of the four main nucleobases found in the nucleic acids DNA and RNA. In DNA it is paired with guanine, but it is unstable and can change into uracil. This can lead to a point mutation if not repaired by DNA repair enzymes. Other bases are adenine, guanine, and thymine.

demethylation Enzymatic process resulting in the removal of a methyl group from a molecule, for example DNA.

deoxyribonucleic acid (DNA) A linear sequence of deoxyribonucleotides. DNA is a double-stranded molecule held together by weak bonds between base pairs of nucleotides, and encodes genetic information. The four nucleotides in DNA contain the bases adenine (A), guanine (G), cytosine (C), and thymine (T).

diplotype Combination of haplotypes on a chromosome pair; analogous to genotype.

dizygotic twin Fraternal twin; a twin who shares only as much genetic material (50%) with his/her co-twin as would a sibling; a twin resulting from a fertilization event that is separate from his/her co-twin.

endophenotype A heritable intermediate phenotype.

enhancer Regulatory DNA sequence that increases the transcription of a related gene, independently of position and orientation, by interacting with specific transcription factors.

epistasis Interaction whereby one gene masks or interferes with the phenotypic expression of one or more genes at other loci. The gene whose phenotype is expressed is *epistatic*. The gene or genes whose phenotype is altered or suppressed is *hypostatic*.

exon Region of gene that contains the code for producing the gene's protein. Each exon codes for a specific portion of the complete protein. In some species (including humans), most gene's exons are separated by long regions of DNA called *introns*. Introns are excised during RNA splicing, and exons are "stitched" together to form mature mRNA that will be translated.

expressivity (= penetrance) Variation in the expression of a trait given the same allele. If an allele is highly penetrant, most individuals will express (display) the trait; if it has low penetrance, the allele will produce the trait less frequently.

FMR1 **gene** Gene on the X chromosome, a mutation in which is responsible for fragile X syndrome. A DNA segment, known as the CGG (cytosine-guanine-guanine) triplet repeat, is expanded from the number normally seen (fewer than 55). When the number of repeats exceeds 200, the expansion is called a "*full mutation*" and the expression of the *FMR1* gene is turned off, leading to a deficiency in FMR protein and causing the fragile X syndrome. When the number falls between 55 and 200 repeats, a "*premutation*" exists which leads to RNA toxicity by a gain-of-function mechanism.

FMR1 protein (FMRP) A protein found in both dendrites and synapses of neurons associated with actively translating ribosomes during protein synthesis. Its absence or underproduction results in developmental changes at the neuronal level and results in fragile X syndrome.

full mutation A mutation of sufficient magnitude that it causes disease. Example: in fragile X, the full mutation consists of more than 200 CGG (triplet) repeats, which involve methylation, which stops the synthesis of the FMR1 protein (FMRP), thereby causing the fragile X syndrome.

functional polymorphism (functional allele) Genetic variants that affect protein function, gene transcription, regulation, or translation.

gain of function Mutations that result in a protein product that is increased in quantity or has a new function. Compare with *loss of function* in which alteration results in a non-functional protein product.

gene Functional and physical unit of heredity passed from parent to offspring; ordered sequence of nucleotides located in a particular position on a particular chromosome that encodes a specific functional product (i.e., a protein or RNA molecule).

gene dose–response curve Relationship between degree of genetic abnormality and the disease phenotype; may account for severity of disease and its phenotypic manifestations. Example: severity within the fragile X spectrum shows a correspondence to the number of trinucleotide triplet repeats.

gene expression The process by which information coded by genes is converted into the structures present and operating in the cell. Expressed genes include those that are transcribed and in most but not all (e.g., microRNA, ribosomal and transfer RNAs) cases translated into protein.

genetic heterogeneity Ability of more than one allele to cause the same trait or disease. Phenomenon by which different risk genes and alleles individually may be sufficient to lead to the same phenotype.

genetic loci Genetic position on the linkage map or on the chromosome.

genetic markers Alleles of genes, or DNA polymorphisms, used as experimental probes to keep track of an individual, a tissue, a cell, a nucleus, a chromosome, or a gene. Stated another way, a genetic marker is any character that acts as a signpost or signal of the presence or location of a gene or hereditary characteristic in an individual in a population. There are four common chromosome changes that are inherited from generation to generation, and commonly used as markers in genetic studies:

1. **indels** are insertions or deletions of the DNA at particular locations on the chromosome. An example is the YAP (Y chromosome alu polymorphism).
2. **SNPs** are single nucleotide polymorphisms in which a particular nucleotide is changed (e.g., A is changed to G). There is about 1 SNP per 300 bp in the human genome.
3. **microsatellites** are short sequences of nucleotides (typically 2–5 core base pairs, example: ATCG) which are repeated multiple times in tandem. Over time changes in copy number do occur; thus the number of repeats may increase or decrease.
4. **minisatellites** are longer sequences of nucleotides (typically 9–80 core base pairs, example: TAAGGGCCA) which are repeated multiple times in tandem. Over time changes sometimes do occur and the number of repeats may increase or decrease.

genome Entire DNA sequence of an organism or species. Complete hereditary information of an organization; a term coined in 1920 by Hans Winkler, a Professor of Botany at the University of Hamburg, Germany.

genome-wide scan Association or linkage studies which search the whole genome for single nucleotide polymorphisms (SNPs) or linkage regions that occur more frequently in people with a particular disease than in people without the disease. These scans typically assess tens or hundreds of thousands of SNPs using array-based genotyping methods and are often used to identify novel risk genes in complex disorders.

genomics Study of an organism's entire genome, the function and interactions of all an organism's genes, as opposed to the study of single genes (*genetics*).

genotype   Genetic constitution of an individual; set of specific alleles at a gene locus in an individual.

guanine   One of the four bases in DNA that make up the letters ATGC, guanine is the "G." Guanine always pairs with cytosine. These letters are used as shorthand for the sequences of fragments of DNA, e.g., CCAAGTAC.

haplotype   A set of closely linked genetic markers (DNA polymorphisms) that are inherited as a unit.

heritability   The proportion of total variation of a trait attributable to genetic factors; a statistical estimate of the genetic component of liability, ranging from zero to one.

heterozygote   An individual who possesses two different alleles at a particular locus on a pair of homologous chromosomes.

heterozygous   State of carrying two different alleles at a locus on homologous chromosomes.

histone   Type of protein rich in lysine and arginine found in association with DNA in chromosomes; protein core around which DNA is wound in a chromosome.

homozygosity   State of having identical alleles at one or more loci in homologous chromosomes.

homozygote   An individual with two identical alleles at a particular locus on a pair of homologous chromosomes.

hybridization   Base pairing of two single strands of DNA or RNA. Formation of double helix from complementary strands of DNA or RNA derived from different sources.

imaging genetics   A relatively new field of science investigating the relationship between genetic variations and variations in neural structure or function as assessed by imaging.

intermediate phenotype   A trait or characteristic that is associated with disease and linked to the action of functional alleles; may include personality traits, neuropsychological profiles, neuroimaging profiles or any aspect of a syndrome and may more readily be linked to a gene than the disease diagnosis itself.

intron (= **intervening sequence**)   Non-coding region of DNA found between two exons that is spliced out during RNA processing.

kilobase (kb)   1000 DNA base pairs (bp).

knockout mice   Animal (mouse) model in which a specific gene has been genetically disabled.

liability   Tendency toward expression of a multifactorial trait, consisting of a combination of genetic and non-genetic factors.

linkage   Association of genes and/or markers located near each other on a chromosome. Given their proximity, linked genes and markers tended to be inherited together, i.e., transmitted together during meiosis in gamete formation.

linkage disequilibrium (LD)   Non-random association of alleles at linked loci in populations; greater co-occurrence of two or more alleles at closely linked loci in a population more frequently than would be expected by chance for independent markers.

locus   Location of a specific gene or other marker on a chromosome (plural loci).

lymphoblasts   Immature (precursor) cells, normally found in bone marrow, that typically differentiate to form mature lymphocytes (white blood cells in the immune system).

marker   An identifiable physical location on a chromosome (e.g., restriction enzyme cutting site, SNP, microsatellite) whose inheritance can be monitored.

meiosis   Cell division in which haploid gametes are formed from diploid germ cells.

messenger RNA (mRNA)   The ribonucleic acid (RNA) transcribed from the DNA of a gene in the cell nucleus. The messenger RNA is then translocated into the cytoplasm, where it serves as a template for protein synthesis.

metabotropic glutamate receptor 5 (mGluR) theory of fragile X syndrome (FXS) Theory that posits that exaggerated signaling in mGluR pathways may underlie many of the cognitive, behavioral, and neurological symptoms of FXS.

methylation Addition of methyl groups to cytosine bases in cytosine-phosphoguanine (CpG) dinucleotides, forming 5-methylcytosine. Methylation is usually correlated with reduced transcription.

microarray Grid of multiple DNA fragments fixed to a glass slide for hybridization of RNA or DNA. Microarrays may be used for gene expression profiling or genotyping. The technique is based on the selectivity of DNA-DNA or DNA-RNA hybridization and fluorescence-based detection.

microRNAs (miRNA) A class of small (21–23 nucleotides), non-coding RNA molecules that have emerged as critical and pervasive regulators of eukaryotic gene expression.

missense variant (= missense mutation) A point mutation (affecting a single base pair), which results in a change in an amino acid sequence.

monozygotic Denoting genetically identical twins derived from a single zygote, or fertilized egg.

mosaic Existence of two or more genetically different cells lines in an individual; an individual with two different cell lines derived from a single zygote. Example: within an individual with fragile X, some cells express the FMR1 protein and others do not.

mutation Alteration in the sequence of DNA at a genetic locus. Change in genetic material, either of a single nucleotide or in the copy number of DNA segments or structure of the chromosome. A mutation that occurs in the gametes is inherited; one which occurs in somatic cells is not. Mutations most frequently have no effect or cause harm, but may occasionally confer an advantage or benefit.

neutral polymorphisms Polymorphisms which have no direct effect on function. Most human polymorphisms fall into this category.

non-synonymous polymorphism A variant that specifies a different amino acid and which can have a significant effect on the biological actions of the protein.

nucleolar RNAs [small nucleolar RNAs (snoRNAs)] Located with the nucleolus, involved in synthesis and processing of RNAs within the nucleolus; class of small RNA molecules that guide chemical modifications (e.g., methylation) of ribosomal RNA and other RNA genes.

nucleolus Small rounded mass within the cell nucleus where the production and assembly of ribosome components occurs.

nucleosome protein core Four pairs (H2A, H2B, H3, H4) of histone proteins around which DNA winds.

nucleotide A building block of DNA or RNA consisting of a nitrogenous base (adenine, guanine, thymine, or cytosine in DNA; adenine, guanine, uracil, or cytosine in RNA), a phosphate molecule, and a sugar molecule (deoxyribose in DNA and ribose in RNA).

nucleus The cellular organelle in eukaryotes that contains the genetic material. The nucleus is a membrane-bound structure, where the entire DNA, packaged in chromosomes, is contained.

odds ratio The ratio of the odds of an event occurring in one group to the odds of it occurring in another group (or other dichotomous classification). An odds ratio of 1 signifies equal probability in both groups; an odds ratio greater than 1 indicates a greater likelihood of some event or trait in the first group. The odds ratio must be greater than or equal to zero. In linkage studies an odds ratio of greater than 3 is considered evidence of linkage. Odds ratio is not the same as relative risk.

penetrance Characteristic phenotypic effect of a genotype. The proportion of individuals carrying a particular variation of a gene who also express a particular trait. If a gene is highly penetrant, most individuals will display the trait; if it has low penetrance, the gene will produce the trait less frequently.

phenotype Appearance or observed characteristics (physical, biochemical, physiological) of an individual resulting from the interaction of the environment and the genotype.

pleiotropy The genetic effect of a single gene on multiple phenotypic traits; diverse effects, for example different phenotypes, produced by the same allele.

polygenicity Combined additive effects of multiple genes on a trait, needed to produce a certain phenotype.

polymorphism Frequently occurring (>1% in the population) variation in a nucleotide sequence. Polymorphisms may have an effect on gene transcription, regulation, and translation. Some polymorphisms are associated with an increased risk for disease.

premutation Existence of a gene in an unstable form which can undergo a further mutational event to cause disease; sequence variation that predisposes to mutation, usually moderate expansion of a triplet that leads to further expansion, such as in fragile X syndrome. In fragile X, the "premutation" is defined by ~50–200 CGG repeats. Individuals affected with the premutation can transmit a full mutation to their offspring, but are not generally affected with the full-blown fragile X syndrome. Instead they may have cognitive or behavioral problems, premature ovarian failure in female carriers, and an adult-onset neurological disorder causing tremor and difficulties with balance and gait.

proband Affected family member in genetic studies involving families and larger pedigrees. Synonymous with propositus (if male), proposita (if female) and *index case.*

promoter A DNA sequence at the 5′ end of genes which contain recognition sequences for RNA polymerase and regulatory transcription factors.

protein A large molecule composed of one or more chains of amino acids in a specific order; the order is determined by the base sequence of nucleotides in the gene coding for the protein. Proteins are required for the structure, function, and regulation of the body's cells, tissues, and organs, and each protein has unique functions.

recombination Cross over between two linked loci; occurrence among offspring of new combinations of alleles, resulting from crossovers that occur during parental meiosis. Likelihood of recombination increases with increasing physical distance between genes or gene segments.

repeat number polymorphism Trinucleotide sequences, micro- or mini satellites that vary in their copy number in the genome.

ribonucleic acid (RNA) The nucleic acid which is found mainly in the nucleolus and ribosomes; single-stranded molecule that consists of a sugar (ribose), phosphate group, and a series of bases (adenine, cytosine, guanine, and uracil). *Messenger RNA* transfers genetic information from the nucleus to the ribosomes in the cytoplasm and also acts as a template for the synthesis of polypeptides. *Transfer RNA* transfers activated amino acids from the cytoplasm to messenger RNA.

ribosome Cellular structure involved in translation of mRNA into protein.

RNA polymerase enzyme Enzyme that binds to a promoter site and synthesizes RNA from a DNA template.

RNA toxicity A mechanism by which expansion of trinucleotide repeats causes disease when expressed as RNA by altering protein function with loss- or gain-of-function effects.

Example: Fragile X-associated tremor/ataxia syndrome (FXTAS) is a late-onset progressive neurological disorder reported in some older men with the fragile X premutation, whose pathogenesis is thought to result from overexpression and toxicity of FMR1 mRNA.

single nucleotide polymorphisms (SNPs; pronounced "snips") Polymorphisms that result from variation at a single nucleotide. Common variations that occur in human DNA at a frequency of 1 per every 300 bases. Currently it is the most commonly used genetic marker used to track inheritance in families and in association studies.

susceptibility gene A gene which has alleles that increase the risk of disease, but which alone is insufficient to cause the disease.

thymine One of the four bases in DNA that make up the letters ATGC, thymine is the "T." Thymine always pairs with adenine. These letters are used as shorthand for the sequences of fragments of DNA, e.g., CCAAGTAC. These sequences are the code for genetic information.

transcription Process whereby an RNA sequence is synthesized from a DNA template.

transcription factors Proteins that bind to DNA to influence and regulate transcription.

transfer RNA (tRNA) Small RNA chain (73–93 nucleotides) that transfers a specific amino acid to a growing polypeptide chain at the ribosomal site of protein synthesis during translation; essential in protein synthesis.

translation Process whereby genetic information from messenger RNA is translated into protein.

transmission disequilibrium test A statistical test to determine the association of an allele with a trait by analyzing non-random segregation of the allele in families with an affected offspring.

trinucleotide repeat expansion (= triplet repeat expansion) Type of mutation in which a gene segment containing multiple copies of a triplet of bases is expanded in length. Increase in the number of copies of triplet repeat sequences responsible for mutations in a number of single-gene disorders.

velocardiofacial syndrome (VCFS) See *chromosome 22q deletion syndrome*.

X chromosome A chromosome that is involved in sex determination. In humans, females have two X chromosomes; males have one X and one Y chromosome.

X inactivation Process by which one of the X chromosomes becomes inactive. It occurs early in female embryonic development. Process is random as to whether the X chromosome from the father or mother is inactivated.

X-inactivation ratio Fraction of cells with the same (maternally or paternally derived) X chromosome active.

ACKNOWLEDGMENTS

The editors wish to thank Henning U. Voss, Ph.D., Radiology Citigroup Biomedical Imaging Center, Weill Medical College of Cornell University for his invaluable contributions to the imaging glossary. Thanks are also due the authors of Chapters 1, 6, 13, 17, 18, 19, 20 and 22 for their contributions to this glossary.

Index

Note: National Commission for the Protection of Human Subjects of Biomedical and Behavioral Research is abbreviated to Commission

22q11 deletion syndrome 372, 429

5HTT genotype *see* serotonin transporter gene

acquisition 407
activation ratio 220, 428
actively translating ribosomes 218, 429
additive factors logic 285
adenine 429
adenosine triphosphate (ATP) 298, 300
ADHD *see* attention deficit/hyperactivity disorder
adolescent alcohol use disorder (AUD)
    cognitive deficits 235
    comorbid disruptive behavior disorders 234
    dangers of lifetime dependence 232
    diagnosis of, problems with 234
    functional brain abnormalities 236–8
    leading to abuse of other drugs 232–4
    pathophysiology of 232
    structural brain abnormalities 236
alcohol 229
    prenatal exposure
        behavioral and cognitive consequences 229–30
        morphological consequences 230–2
    see also adolescent alcohol use disorder (AUD)
allele 429
amino acids 429
amygdala (AMY) 81
    anxiety and depressive disorders 187, 188–9
    approach-avoidance behaviors 58–9
    autism studies 131
    bipolar disorder 172–3
    emotion regulation 39–40, 45–6

face processing 82–4, 189, 191–3
fear/threat response 185–6
fragile X syndrome 224
and *HTTLPR* genotypes 377–8
and intrinsically dangerous stimuli 186
schizophrenia 152–3
social phobia 189
Analysis of Functional NeuroImages (AFNI), software 401
Analyze, software platform 401
ancestry informative markers (AIMs) 369, 429
animal studies
    alcohol withdrawal 235
    cocaine exposure in utero, effects of 337
    developmental changes 395–6
    dopamine, prefrontal cortex and working memory 371
    drug use, causative relationship 233–4
    eating behaviors, role of neuropeptides 249
    emotional modulation of behavior 184–6
    goal-directed behaviors 55
    stimulus-response learning 210
anisotropic diffusion 317, 407
anorexia nervosa (AN) 245
    functional neuroimaging studies
        receptor imaging studies 250–2
        task-activation studies 250
    phenotype of 245
    risk factors
        congenital risks 248
        during development 248–9
        genetic influences 247–8
        serotonin and dopamine levels 247
        trait behaviors 248
    structural brain imaging studies 249

anxiety and depressive disorders 183
　electrophysiology 194–5
　functional magnetic resonance imaging 188
　　data in adults 188–91
　　data in juveniles 191–4
　goal of brain imaging research 184–5
　magnetic resonance spectroscopy (MRS) 194
　morphometry 187
　　data in adults 187–8
　　data in juveniles 188
　neural circuits and behavior in anxiety and
　　　depressive disorders 186–7
　neural circuits in emotionally modulated behavior 185
　　brain response to rewards 186
　　brain response to threats 185–6
　relevant conditions 183
　　anxiety disorders 183–4
　　depressive disorders 184
　　identifying at-risk juveniles 184
　see also depression; stress reactivity
approach and avoidance behavior 54–5
　frontal EEG asymmetry model 44
　role of amygdala 58–9
　role of prefrontal cortex (PFC) 59–60
　role of striatum 57–8
　see also goal-directed behavior
arithmetic processing, effect of decreased fragile X mental
　　　retardation gene protein (FMRP) 223
array coils 330, 409
arterial spin labeling (ASL) perfusion MRI 326, 407
　atypical development and childhood diseases, studies of
　　　adolescents exposed to cocaine in utero 337–8
　　autism spectrum disorders 339
　　sickle cell disease 338–9
　benefits of ASL for developmental neuroscience 330–1
　　cerebral blood flow (CBF) quantification in age
　　　groups and longitudinal stability 331–2
　　perfusion MRI for developmental fMRI studies 332–3
　methodological background 326–8
　normal development of CBF from childhood to
　　　young adulthood 334–7
　potential and challenges of 339–40
　relationship with other neuroimaging
　　　modalities 333–4
　spin labeling schemes 328
　technological advances 330, 391
arterial transit time 328, 330, 407
ASDs see autism spectrum disorders (ASDs)
assent see child assent (informed consent)

association, genetic 367, 429
attention 120–1
　ADHD functional neuroimaging studies 121–3
　executive attention 121, 122–3
　selective attention 22, 120, 121–2
　sustained attention 120, 121
attention deficit/hyperactivity disorder (ADHD) 113
　clinical phenomenology and epidemiology 113
　functional imaging 114–15
　　attention 120–1
　　　developmental considerations 123
　　　functional neuroimaging results 121–3
　　response inhibition 115
　　　developmental considerations 119
　　　functional neuroimaging results 115–19
　　resting state studies 124
　　studies of other activation tasks
　　　motor overflow 123
　　　reward processing 124
　　　spatio-visual attention 123
　　　timing deficits 123–4
　　working memory (WM) 119
　　　functional neuroimaging results 119–20
　future research directions 126
　methylphenidate (MPH) effects 395
　neurobiology of 113–14, 124–5
　　altered anatomical connectivity 125
　　neurochemical perturbation 125
　　structural anomalies 125, 390
　results of studies, challenges of integrating 125–6
　structural imaging 114, 351
AUD see adolescent alcohol use disorder (AUD)
auditory oddball paradigm, ERP 92–3, 355
authorization, written release for pediatric research 271
autism spectrum disorders (ASDs) 130
　anatomical imaging studies 314
　　diffusion tensor imaging (DTI) 132–3
　　structural MRI 131–2
　association with fragile X syndrome 218, 225
　brain overgrowth 131, 142, 390
　functional brain imaging
　　activation studies 137
　　　exceptional abilities 142
　　　language stimulation studies 137–8
　　　mirror neuron system 140–2
　　　social perception and mentalizing studies
　　　　eye gaze perception 140
　　　　face perception 138–9
　　　　mentalizing 140

theory of mind 139–40
voice perception 139
ASL perfusion MRI 339
resting state studies 136–7
future directions for research 143
magnetic resonance spectroscopy (MRS) 133–6
social impairments, neurobiology of 143
avoidance behavior *see* approach and avoidance behavior

B_0 407
B_{max} 407–8
B'_{max} 408
b value 319, 323, 407, 412
balancing selection 373, 429
basal ganglia
    effect of psychotropic medication 175
    excess glutamatergic activity in bipolar disorder 175, 310
    inconsistent findings in major depressive
        disorder (MDD) 187–8
    less activation in attention deficit/hyperactivity
        disorder (ADHD) 121–2
    smaller in obsessive-compulsive disorder (OCD) 205–6
base, genetic term 429
base pair (bp) 429
Belmont Report (National Commission) 274–5
binding potential 408
biological motion 73, 74, 81, 82
biomarkers 344
    potential type 346
        olanzapine for bipolar disorder 357
        paroxetine treatment for obsessive-compulsive
            disorder (OCD) 361
        stimulants for attention deficit/hyperactivity
            disorder (ADHD) 355
    and therapeutics development 396–7
Biomedical Imaging Resource (BIR) 401
Biomedical Informatics Research Network (BIRN) 392, 404
bipolar disorder (BD) 161
    clinical features of 161–71
    future research directions 178–9
    glutamatergic functioning 310
    mood-stabilizing medication, effects of 176–7
    neurodevelopmental cortico-limbic model 171–2,
        178, 390
    amygdala dysfunction 172–3
    hippocampus 174
    striatum abnormalities 174–5
    ventral prefrontal cortex (VPFC) dysfunction 175–6
    overlapping disorders, distinguishing from 178

block designs 191, 284, 408
blood–brain partition coefficient of water 331, 408
B_{max} 407–8
BOLD (blood-oxygen-level-dependent) contrast
    281–2, 408
BOLD (blood-oxygen-level-dependent) effect 408
BOLD fMRI 281
    combining with diffusion tensor imaging (DTI) 320–1
    compared to perfusion fMRI 332–3
    and concurrent arterial spin labeling (ASL)
        perfusion MRI 333–4
    longitudinal studies, test-retest reliability 394
    need for longitudinal studies 393
    for real-time therapy and neurofeedback 397
    technological advances 391
    test-retest reliability 393–4
    see also functional magnetic resonance
        imaging (fMRI)
brain development
    cortical expansion 10–12
    models of normal function 394
    results of neuroimaging studies 149–50
    schizophrenia as a disorder of 148–9
    studies of atypical 395–6
brain functioning, differential effect of
        pharmacotherapy vs. psychotherapy 397
BrainVISA, software platform 401–2
BrainVoyager, software 402
Brownian motion 315, 408, 411
bulimia nervosa (BN)
    functional neuroimaging studies
        receptor imaging studies 252–4
        task activation studies 252
    phenotype of 245–6
    risk factors
        congenital risks 248
        early sexual maturation 249
        genetic influences 247–8
        serotonin (5HT) 247
        trait behaviors 248
    structural brain imaging studies 249

capillary water permeability 408
carbon-13 (^{13}C) MRS 301–2, 310, 408
CASL *see* continuous arterial spin labeling (CASL)
        perfusion imaging ASL
catechol-*O*-methyltransferase (COMT)
    COMT gene and inherited variations 369–70
    effects on functional neuroimaging measures

catechol-*O*-methyltransferase (COMT) (cont.)
    COMT haplotype predicts cortical efficiency 373
    Val158Met predicts measures of frontal
        function 373
    Val158Met and frontal cognition
      prediction of cognitive performance in varied
        groups 371–2
      regulation of dopamine level in prefrontal
        cortex (PFC) 371
    Val158Met and stress resiliency 374
      balancing selection: warrior vs. worrier 373
      stressful stimuli and brain activation 374–5
caudate nucleus (CN)
    activation
      increased activation in obsessive-compulsive
        disorder (OCD) 208–9
      increased during suppression of tics 207
      increased by methylphenidate medication in
        attention deficit/hyperactivity disorder
        (ADHD) 351
      underactivation in ADHD 116
    volumetric reduction
      in ADHD subjects 114, 389
      in fetal alcohol syndrome/fetal alcohol spectrum
        disorders (FAS/FASD) subjects 230
      in Tourette syndrome 203
    see also striatum
CBT *see* cognitive behavioral therapy (CBT)
Center for Computational Biology (CCB) 392, 404
cerebellum
    activation
      in autistic subjects 138
      in fragile X-associated tremor/ataxia syndrome
        (FXTAS) subjects 224
    prenatal development, effects of alcohol on 229
    vermian hypoplasia in autism 131
    volumetric reduction
      in ADHD 351
      in adolescent alcohol use disorder (AUD) 235
      in fragile X syndrome and FXTAS 222, 230
      in schizophrenia 155
cerebral blood flow (CBF) 326, 409
    activation studies, autism 137–42
    after methylphenidate treatment 350
    measurement of using arterial spin labeling (ASL)
      perfusion MRI 326–7
    in adolescents exposed to cocaine in utero 337–8
    in different age groups 331–2
    in sickle cell disease 338–9

    and longitudinal stability 332
      normal development 334–7
    normal development 334–7
    quantification in age groups and longitudinal
      stability 331–2
    see also regional cerebral blood flow (rCBF)
cerebral metabolic rate of glucose (CMRglu) 121, 136, 409
chemical shift 409
chemical shift imaging (CSI) 302, 391, 409
child assent (informed consent) 270–1
chromosome 429
chromosome 22q11 deletion syndrome 372, 429
chronometry 285, 409
clearinghouse 392, 403–4
cocaine, effects of in utero exposure to 337–8
co-causality 429
co-dominant alleles 369–70, 375, 429
codon 429
cognitive ability
    deficits in
      attention deficit/hyperactivity disorder (ADHD) 235
      schizophrenia 147–8, 371
      substance use disorders (SUDs) 235
    differences in, pediatric research 288–9
    effects of *COMT Val158* genotype 371
    link to morphology 16–17
    see also intelligence
cognitive behavioral therapy (CBT)
    effects on brain function 347
    for OCD and effect of selective serotonin reuptake
      inhibitors (SSRIs) 357–61
cognitive control 22
    future directions 33–4
    in schizophrenia 152
    neural substrates of 23–4
    neurodevelopmental changes
      error- and feedback-based performance
        adjustments 29–31
      interference suppression 24–7
      manipulation 27
      relational reasoning 31–3
      response inhibition 27–8
      response selection 28–9
      working memory 24
    neuroimaging studies 149–50
    processes 22–3
    and study of neurodevelopmental disorders 34
coil loading 410
coils 409–10

Common Rule, regulation on human subjects research 267
complex diseases 367, 369, 430
compulsions, differentiating from complex tics 206
COMT *see* catechol-*O*-methyltransferase (COMT)
concordance 430
concordance ratio 430
consequentialism 274–5
continuous arterial spin labeling (CASL) perfusion
        imaging ASL 328, 332, 410
contrast 410
contrast agents 326, 410
convolution 410
corpus callosum (CC)
    abnormalities in fetal alcohol syndrome (FAS) 230
    reduced volume and fractional anisotropy (FA) in
        autism 131, 133
    reduced volume in Tourette syndrome (TS) 203
    volumetric reduction in attention deficit/hyperactivity
        disorder (ADHD) 114
cortical pattern matching (CPM) 410
    normal morphological maturation 7–9
        and brain surface asymmetry 12–15
        and cortical expansion 10–12
        and gray matter density (GMD) 9–10
        and subcortical growth 15
cortical thickness 410
    in peri-Sylvian region 9, 16
        link to phonological processing skills 17, 104
    and intelligence 17, 87
    see also gray matter density (GMD)
cortisol secretion
    healthy adults and children 46–7
    in depressed individuals 47
creatine/phosphocreatine (Cr) 297–8, 410–11
cytosine (C) 430
cytosolic choline compounds (Cho) 297–8, 411
    changes with age 306

databases
    Biomedical Informatics Research Network (BIRN)
        392, 404
    National Database for Autism Research (NDAR) 392, 403
    NIH Pediatric MRI Data Repository 393, 403
    Single Nucleotide Polymorphism database (dbSNP) 369
decision-making 55–6
    fMRI paradigms of reward-related 60–1
    fMRI studies of adolescent 61
        global analysis of processes involved in 61
        outcome analysis 68–9

preference formation/selection 68
        reward anticipation 61–8
    fractal triadic model 56–60
    neural correlates of 124
deconvolution 411
default networks 285–6, 395, 411
demethylation 430
deontological theories of ethics 274
deoxyribonucleic acid (DNA) 365, 430
dephasing and dephasing spins 411
depression 184
    brain glucose metabolism 347
    neurobiology of 44–6
    neuroendocrinology of 47
    and serotonin transporter gene 375–8
    see also anxiety and depressive disorders;
            bipolar disorder (BD)
depth-resolved surface coil spectroscopy
            (DRESS) 302, 411
design *see* experimental design
developmental clinical neuroscience 387
    atypical development, understanding 395–6
    biomarker and therapeutics development 396–7
    contributions of neuroimaging
        normal development 388–9
        psychiatric disorders 389
    neuroethics 397–8
    normal development, understanding 393–5
    rationale for developmental approach 387–8
    technological advances
        imaging modalities 391–2
        neuroinformatics 392–3
diffusion 411
    MRI measurement of 316–17
diffusion anisotropy 317, 407
diffusion coefficient 411
diffusion tensor 316, 411
diffusion tensor imaging (DTI) 290, 314–15, 411–12
    autism studies 132–3
    conventional DTI studies in adults, limitations of 319
    DTI studies in children 321–3
    fiber tractography 317–18
    MR imaging parameters 319
    MRI measurement of diffusion 316–17
    multimodal combination of DTI with fMRI 320–1
    non-conventional DTI 319–20
    principles of 315
    schizophrenia, neural connectivity 151
    technological advances 391

diffusion-weighted imaging (DWI) 412
  high *b* value for tissue structural characterization 323
diplotype 430
discourse comprehension, neural underpinnings
      of 99–100
disruptive behavior disorders (DBD), comorbid
      with AUD 234
dizygotic twin 430
dopamine (DA)
  and ADHD
    COMT genotype and working memory 372
    genes implicated in 113–14
    response to stimulant medication 347–50
    transmission deficits 123, 125
  amygdala nuclei regulating release of 59
  COMT genotype
    chromosome 22q11 deletion syndrome 372
    and cognitive function in healthy children 371–2
    modulating dopamine transmission in prefrontal
      cortex 371
  risk factor in eating disorders 247
  and striatum
    abnormal dopamine transmission in Tourette
      syndrome and obsessive-compulsive
      disorder 201–3
    reward processing 186, 206
Doppler ultrasound 330, 412
dorsal and ventral systems, role in emotion regulation 45
dorsolateral prefrontal cortex (DLPFC)
  emotion regulation 40
  and feedback-based learning 30–1
  increased gray matter volume following paroxetine
      treatment 361
  interference suppression 27
  and limbic inhibition 45
  and manipulation 27
  and regional cerebral blood flow in adults with
      obsessive-compulsive disorder 209
  and relational reasoning 31–3
  response selection 115
  working memory in schizophrenia 152
drugs *see* illicit drug use; substance use disorder (SUD)

eating disorders (EDs) 245
  brain imaging studies
    functional neuroimaging 250–4
    structural and resting-state metabolic
      neuroimaging 249
  phenotypes of 245

  risk factors 246
    childhood developmental factors 248–9
    congenital risks 248
    dopamine (DA) 247
    genotype 247–8
    hormonal and neuropeptide influences 247
    serotonin (5HT) 247
    trait behaviors 248
echo 412
echo time (TE) 412
echo-planar imaging (EPI) 412
eddy currents 412
educational resources 401
eigenvalues (of a diffusion tensor) 317, 412
eigenvector (of a diffusion tensor) 317–18, 412–13
electrophysiology, studies in anxiety and major
      depressive disorder 194–5
emission computed tomography 413
emotion
  contextual influences 193–4
  perception, development of 78–9
  and serotonin transporter gene 375–8
  understanding of, development of brain circuitry for 82–4
emotion regulation 38–9
  future research directions 49
  neurobiological correlates
    depression 44–6
    healthy adults 39–43
    healthy children 43–4
    in pediatric anxiety and depression 185–6
  neuroendocrine aspects
    depression 47
    healthy adults and children 46–7
  and stress reactivity 39
endogenous tracer 428
endophenotypes 126, 155–6, 367–8, 430
enhancer 430
epistasis 430
epoch 413
Ernst angle 413
error-based learning 29–30
ethical issues
  neuroethics 397–8, 405–6
  theoretical ethics 274–5
  use of PET and SPECT with children 396
  see also legal and ethical issues
event-related fMRI (erfMRI) 284–5, 413
  advantages over block designs 103
  anxiety and MDD studies in adolescents 191–3

event-related potentials (ERPs) 355, 413
  early language development 91–2
    phonological processing 92–4
    syntactic processing 94
evolutionary theory of adolescence 53–4
executive attention 121, 122–3
executive function see cognitive control
exon 430
experimental design 283–4
  age-related performance differences, controlling for 394
  block vs. event-related 103, 284–5
  for clinical trials 345–6
  event-related fMRI 191–3, 284–5, 413
  imaging genetic studies 369
  subtractive and additive logic 285
expressivity 430
eye contact, development of 78
eye gaze perception see gaze perception

face perception/processing
  autism studies 138–9
    and theory of mind 140
  brain regions involved in 81–2
  deficits in bipolar disorder (BD) 172
  development of 79
  role of amygdala 172
    in anxiety and depressive disorders 189, 191–2
    in response to fear 82–4
  see also gaze perception
false belief task, theory of mind 80
fear 183
  amygdala (AMY)
    activation in response to 46, 82–4, 191
    and HTTLPR genotypes 377
    role in conditioning of 185–6
  response of infants 79
feedback-based learning 23, 30–1
ferromagnetic 413
fetal alcohol spectrum disorders (FASD) 229–32
fetal alcohol syndrome (FAS) 229–32
fiber assignment by continuous tracking (FACT) 317–18, 413
fiber tractography 317–18
field of view (FOV) 413
field strength 282–3, 297, 413
filtering 413
flip angle 413
fluorine-19 (^{19}F) MRS 302, 310, 413
fluorodeoxyglucose (FDG) 327, 414
fluoxetine treatment 347, 357

FMR1 gene 217, 430
FMR1 protein (FMRP) 430
  and arithmetic processing 223
  and brain development 218–19
  effect on brain activation 223, 390
  lack of causing fragile X syndrome (FXS) 217
  levels of affecting severity of FXS 218
  and overall cognitive-behavioral problems 219–20
  and white matter changes 221–2
fMRI relaxometry 124, 414
FMRIB Software Library (FSL) 402
Food and Drug Administration (FDA), federal regulations
        on conducting research 264, 266–7
Fourier analysis 323, 414
fractal triadic model 56–7
  structural basis of
    amygdala 58–9
    prefrontal cortex (PFC) 59–60
    striatum 57–8
fractional anisotropy (FA) 24, 414
  calculation of 317
  reduced white matter tracts
    in autism 132–3
    in fetal alcohol syndrome 230
    in fragile X syndrome 221
    in schizophrenia 151
fragile X-associated tremor/ataxia syndrome (FXTAS) 218
  gene-behavior studies 221
  gene-brain-behavior studies 224
  gene-brain studies 222
fragile X syndrome (FXS) 217
  association with autism 218, 225
  cause of 217
  FMR1 protein and brain development 218–19
  gene-behavior studies 221
    full mutation findings 219–20
    premutation findings 220–1
  gene-brain-behavior studies
    full mutation findings 222–3
    premutation findings 223–4
  gene-brain studies
    full mutation findings 221–2
    premutation findings 222
  molecular research 218
  phenotypical features 217–18
  as a spectrum condition 218
  therapeutic advances 224–5
  see also fragile X-associated tremor/ataxia
        syndrome (FXTAS)

free induction decay (FID) 296, 414, 422
  repeated free induction decay 423
FreeSurfer, software 402
frequency encoding 414
frontal cortex
  activation deficits in attention deficit/hyperactivity
    disorder (ADHD) subjects 115–16
  changes in gray matter density (GMD) with
    maturation 9–10
  effects of S allele of HTTLPR 377–8
  glutamatergic neurotoxicity and decreased
    N-acetylaspartate (NAA) in adolescent
    bipolar disorder 176
  increased activity during tic suppression 207–8
  and language processing in infants 96
  smaller in ADHD 114, 125
  structural abnormalities in obsessive-compulsive
    disorder (OCD) and Tourette syndrome
    (TS) 205–6
  see also prefrontal cortex (PFC)
full mutation 430–1
  see also fragile X syndrome (FXS)
functional magnetic resonance imaging (fMRI) 414
  advantages of perfusion MRI 332–3
  analytic strategies 286
  experimental design 283–4
    block and event-related 284–5
    resting state baselines 285–6
    subtractive and additive logic 285
  language development 96–100
  normal developmental processes 389
  pediatric imaging, special considerations for 287
    differences in cognitive ability 288–9
    fMRI reproducibility issues 289–90
    longitudinal vs. cross-sectional studies 289
    physiological differences between adults
      and children 287–8
    scanning procedure 287
    spatial normalization 288
    principles of MRI 281–3
    recent advances and future directions 290–1
functional polymorphism (functional allele) 431
Funk-Radon transform 320, 414
fusiform face area (FFA), activation of during face
    processing 80–1, 138
fusiform gyrus (FFG), face processing 80–1, 81–2
FXS see fragile X syndrome (FXS)
FXTAS see fragile X-associated tremor/ataxia
    syndrome (FXTAS)

gain of function 378, 431
gamma-aminobutyric acid (GABA) 300, 415
gaze perception
  autism studies 139, 140
  development of 78, 79
  incongruent vs. congruent paradigm 82
gene 365–6, 431
gene dose–response curve 431
gene expression 431
general linear model (GLM) 286, 415
genetic association and linkage 367
genetic heterogeneity 366, 431
genetic loci 431
genetic markers 431
  ancestry-informative markers (AIMs) 369
genetically complex diseases 367, 369, 430
genetics 365
  catechol-O-methyltransferase (COMT)
    COMT gene and its inherited variations 369–70
    COMT Val158Met and frontal cognition
      COMT regulates dopamine level in prefrontal
        cortex (PFC) 371
      prediction of frontal cognitive performance
        in varied groups 371–2
    COMT Val158Met and stress resiliency
      association with stress resiliency 374
      balancing selection: warrior vs. worrier 373
      stressful stimuli and brain activation 374–5
    effects of COMT on functional neuroimaging
      measures
      COMT haplotype predicts cortical efficiency 373
      Val158Met predicts measures of frontal
        function 373
  concepts of
    alleles 366
    functional alleles 366
    genes 365–6
    haplotypes 366
  fragile X syndrome 217–25
  future directions 378–9
  gene identification approaches 367
  genetic models of psychiatric diseases 366–7
  imaging genetics and its principles 369
  intermediate phenotypes and endophenotypes 367–9
  of eating disorders (ED) 247–8
  serotonin transporter gene and emotional circuits
    functional 5HTT promoter alleles 375–7
    functional imaging of HTTLPR 377–8
    serotonin (5HT) pathway and emotion 375

genome 367, 431
genome-wide scan 367, 431
genomics 431
genotype, definition 432
glutamate (Glu) 299–300, 415
glutamatergic neurotoxicity, adolescent BD 176
Glx resonance 300, 415
glycerol phosphocholine (GPC) 300, 415
glycerophosphoethanolamine (GPE) 300, 415
goal-directed behavior 53
  approach and avoidance behavior 54–5
  decision-making 55–6
    fMRI paradigms 60–1
    fMRI studies 61–9
  definition of 54
  evolutionary theory 53–4
  fractal triadic model 56–60
  innate vs. learned behaviors 54
gradient 415
gradient coils 283, 410
gradient field 283, 415
gradient pulses 316, 415
GRASE (gradient and spin echo) sequence 330, 415
gray matter density (GMD) 415
  decrease in FXS premutation carriers 222
  normal maturation 149
    asymmetry in 13–15
    brain growth studies 10–12
    cortical pattern matching (CPM) 9–10
    summary of studies 15–16
    volumetric image analysis 5–6
    voxel-based morphometry (VBM) 6–7
Grimes case, ethical opinion on pediatric research 272–3
guanine 432
gyromagnetic ratio (γ) 316, 415, 417, 418, 419

haplotypes 366, 432
  COMT haplotype and pain threshold 374
  COMT haplotype predicts measures of frontal
      function 373
  COMT haplotype predicts cortical efficiency 373
harm avoidance in anorexia nervosa 252
Health Insurance Portability and Accountability (HIPAA)
      Act of 1996, privacy regulations 271–2
hemispheric specialization, language processing 100
hemodynamic response (HR) 415–16
hemodynamic response function (HRF) 282, 415–16
heritability 432
  intermediate phenotypes 367–8

of attention deficit/hyperactivity disorder 126
of eating disorders 247–8
of psychiatric disorders 365
see also genetics
heterozygous/heterozygote 432
high angular resolution diffusion imaging (HARDI)
      320, 416
high b value DWI 323, 412
high field 300, 328–30, 330–1, 416
hippocampus
  link to schizophrenia 148
  normal development of 15
  role in threat response 186
  volumetric abnormalities
    in adolescent alcohol use disorder (AUD) 235–6
    in bipolar disorder 174
    in major depressive disorder and post-traumatic
        stress disorder 187, 188
histone 432
homogeneity (of a magnetic field) 416
homozygotes/homozygous 371, 374, 432
hormonal and neuropeptide influences on eating
      disorders 247
Human Brain Project, ASD children 336–7, 339
hybridization 432
hypothalamic-pituitary-adrenocortical (HPA) axis
  and emotion regulation 39
    depression due to HPA dysfunction 47
    in healthy adults and children 46–7

illicit drug use
  adolescent AUD 232–4, 235–8
  prenatal exposure 230, 337–8
  see also substance use disorder (SUD)
image registration 422
image selected in vivo spectroscopy (ISIS) 416
ImageJ, image processing program 405
imaging genetics 179, 369, 390, 416, 432
in utero cocaine exposure (IUCE) 337–8
in vivo MRS 150, 297–302, 310
"incidental" findings in neuroimaging research 269, 273–4
indels, genetic markers 431
informed consent 267, 273–4
inhomogeneity effects 416
innate vs. learned behaviors 54
inorganic phosphate (Pi) 300, 416
intelligence
  cognitive deficits in adolescent alcohol use
      disorder (AUD) 235

intelligence (cont.)
    decline in fragile X syndrome (FXS) 220, 221
    link to morphology 16–17
    relational reasoning 31–3
    see also cognitive ability
intentions of others
    autism studies 140
    biological motion 73, 74, 81, 82
    development of perception of 74
    linking to eye gaze 78
interference suppression 24–7
intermediate phenotypes 367–9, 390, 432
interpulse time 416
intron (intervening sequence) 432
inversion 416
inversion pulse 421
inversion recovery/inversion recovery
            sequence 416
inversion time (TI) 416, 427
isotropic diffusion 416–17
isotropic voxel 417

justice, ethics of research 274

k-space 283, 417
kilobase (kb) 432
knockout mice 432

labeling efficiency 328, 332, 417
lactate (lactic acid), metabolic marker 299, 417
lamotrigine treatment for bipolar disorder 356, 357
language 91, 389
    evidence from event-related potentials 91–2, 94
        early phonological processing 92–4
        early syntactic processing 94
    evidence from fMRI 96–7
        phonological processing 97–8
    evidence from near-infrared spectroscopy 94–5
        higher-level linguistic processing 99–100
        semantic processing 98–9
        syntactic processing 99
    future research directions 103–4
        learning a new language 104
        relating functional and structural brain
            development 103–4
    language skills, link to morphology 17, 103–4
    language stimulation studies, autism 137–8
    methodological issues
        block vs. event-related designs 103

    head motion 103
    interference from scanner noise 103
    summary of imaging studies 100
Larmor equation 417
Larmor frequency 296, 417
lattice 417
lawsuits 263–4
learning
    error-based learning 29–30
    feedback-based learning 23, 30–1
    learning effects during task performance 60
    a novel language 104
    stimulus-response (S-R) learning 210–11
legal and ethical issues 263
    federal regulation of human subjects research
        assent and permission 270–1
        other laws and regulations 271–2
        subpart D, Code of Federal Regulations 264–8
            component analysis 270
            ethical considerations 268–70
    federal regulatory agencies 264
    lawsuits and regulations 263–4
    paradigm case studies and applied ethics 275
    standards of care 273–4
    state law
        state court cases and the Grimes opinion 272–3
        state statutes 272
    theoretical ethics 274–5
liability, genetics 432
Linear Combination of Model (LCModel) spectra 303–5
linkage, genetic 367, 432
linkage disequilibrium (LD) 366, 432
lithium-7 (^{7}Li) MRS 302, 310, 417
lithium treatment for bipolar disorder 177, 355–6
locus/loci 367, 432
longitudinal relaxation 417
longitudinal relaxation time (T_1) 417, 423, 426
longitudinal studies, critical need for 393–5
low birth weight, risk factor in eating disorders 248
lymphoblasts 432

magnet coils 410
magnetic field gradient 417
magnetic field inhomogeneity effects 416
magnetic field strength 282–3, 328–30, 417
magnetic moment 417
magnetic resonance 417
magnetic resonance imaging (MRI) 417–18
    limitations of 314–15

principles of 281
   blood-oxygen-level-dependent (BOLD) contrast 281–2
   magnetic strength 282–3
   see also structural MRI (sMRI)
magnetic resonance spectroscopy (MRS) 296, 418
   advances and future directions 310–11
   anatomical localization of 302
   anxiety and depression 194
   application of
      healthy brain development 305
         phosphorus MRS 306–9
         proton MRS 306
      pediatric psychiatric disorders 309
         lithium and carbon MRS 310
         phosphorus MRS 310
         proton MRS 309–10
   autism studies 133–6
   in vivo studies of normal brain development 150
   limitations of 305
   methods 297
      other MR visible nuclei 300–2
      phosphorus (^{31}P) MRS 300
      proton (^{1}H) MRS 297–300
   principles of 296–7
   schizophrenia studies 153–4
   spectral analysis and quantification 303–5
   technological advances 391
magnetic susceptibility 333, 418, 419
magnetization transfer 418
major depressive disorder (MDD) *see* depression
malpractice 263–4
manipulation, working memory 27
marijuana use, alcohol use disorder subjects 235–8
marker 432
mean diffusivity 418, 428
medial prefrontal cortex (mPFC), implicated in theory
      of mind and self-reflection 81, 84
Medical Imaging Processing Analysis and
      Visualization (MIPAV) 405
meiosis 432
memory
   role in mood/emotion regulation 42–3
   Sternberg memory search task paradigm 285
   see also working memory (WM)
mental chronometry 285, 409
mental retardation, inheritance of *see* fragile
      X syndrome (FXS)
mentalization *see* theory of mind (ToM)
messenger RNA (mRNA) 218, 365, 370, 432–3, 434

metabolite ratios, measuring 305, 311
metabotropic glutamate receptor (mGluR) theory of
      fragile X syndrome (FXS) 433, 407
methylation 433
methylphenidate (MPH)
   and cerebral blood flow (CBF) 350
   effect on synaptic dopamine 350
   effects on children with attention deficit/hyperactivity
      disorder 351–5
microarray 433
microRNAs (miRNAs) 433
microsatellites, genetic markers 431
minisatellites, genetic markers 431
mirror neuron system (MNS)
   autism studies 140–2
   and perception of biological motion 81
mismatch negativity (MMN) response,
      phonological processing 93
missense variant (missense mutation) 433
monozygotic 433
mood-stabilizing medication for bipolar disorder,
      effects of 176–7
mood states, cognitive strategies for regulating 40–3
morphological development of the brain 5
   anxiety and depressive disorders 187
      data from juveniles 188
      findings from adult studies 187–8
   and cognition 16–17
   normal morphological maturation
      cortical pattern matching (CPM) 7–9
         brain surface asymmetry 12–15
         cortical expansion 10–12
         gray matter density (GMD) 9–10
         subcortical growth 15
      summary of findings 15–16
      volumetric image analysis 5–6
      voxel-based morphometry (VBM) 6–7
mosaic 433
motion compliance training 287
motion correction 287
MR sensitivity/visibility 418
MRIcro, brain image tool 402
multi-compartment perfusion models 332, 418
mutation 433
myelination, and cortical expansion 11–12
myo-inositol (mI) 298, 306, 309–10, 418

N-acetylaspartate (NAA) 418
   in vivo proton ^{1}H spectroscopy 150, 297–8

N-acetylaspartate (NAA) (cont.)
  increased levels with age 306
  reduced levels in pediatric psychotic disorders
        306, 309
    bipolar disorder (BD) 174, 176
    schizophrenia 153, 309
  MRS studies of autism 136
narrative comprehension, neural underpinnings
        of 99–100
National Alliance for Medical Image Computing
        (NA-MIC) 392, 404
National Centers for Biomedical Computing 404
National Commission 263
  Belmont Report, deontological theories of ethics 274–5
National Database for Autism Research (NDAR) 392, 403
National Library of Medicine Insight Segmentation and
        Registration Toolkit (ITK) 405
Near-infrared spectroscopy (NIRS) 392, 418
  early language development 94–5
negligence 263, 273–4
neuroethics 397–8, 405–6
Neuroimaging Informatics Tools and Resources
        Clearinghouse (NITRC) 392, 403–4
neuroinformatics 392–3, 403
NeuroLens, analysis tool 402
neuropeptides, and eating disorders 247, 249
neurotransmitter systems
  measurement of 299–300, 391
    limitation in children 396
  receptor imaging studies, eating disorders 250–2
  risk factor in eating disorders 248–9
  role in attention deficit/hyperactivity disorder
        (ADHD) 125
  schizophrenia studies 156
  see also dopamine (DA); serotonin (5HT)
neutral polymorphisms 433
NIH Pediatric MRI Data Repository 403
NIRS see near-infrared spectroscopy (NIRS)
non-synonymous polymorphism 433
nuclear magnetic resonance imaging see magnetic
        resonance imaging (MRI)
nuclear magnetic resonance (NMR) 418–19
nuclear magnetic resonance spectroscopy (NMRS)
        see magnetic resonance spectroscopy (MRS)
nuclear spin 419
nucleolar RNAs 433
nucleolus 433
nucleoside triphosphate (NTP) 300, 309, 419
nucleosome protein cores 366, 433

nucleotides (bases) 433
nucleus, definition and role in MRS 296–7, 433

obsessive-compulsive disorder (OCD) 199
  anatomical disturbances 205–6
  comorbidity with other disorders 200
  functional disturbances 208–10
  neural basis of 200–3
  paroxetine treatment 357–61
  relationship to Tourette syndrome 200
  role of fronto-striatal systems 212
  selective serotonin reuptake inhibitors (SSRIs) and
        cognitive behavioral therapy 357–61
  stimulus-response (S-R) learning 211
  see also Tourette syndrome (TS)
odds ratio 433
Office for Human Research Protections (OHRP),
        responsibility of 264
olanzapine treatment for bipolar disorder (BD)
        177, 356, 357, 390
orbitofrontal cortex (OFC)
  and compulsions in obsessive-compulsive disorder
        (OCD) 206
  and decision-making 61
  and learned responses 54
  role in emotion regulation 44–5
  role in reward anticipation 61
orientation distribution function (ODF) 320, 419
outer volume suppression (OVS) 419

pain perception
  effect of COMT haplotype on pain threshold 374
  effects of placebo analgesia 346–7
panic disorder, treatment of 357
parallel imaging 330, 419
paramagnetic 419
parcellation 419
parental permission (informed consent) 270–1
paroxetine treatment for OCD 357–61
partial volume effect 419
pediatric imaging issues 287
  cross-site and longitudinal studies, fMRI
        reproducibility 289–90
  differences in cognitive ability 288–9
  ethical constraints on use of PET and SPECT 396
  longitudinal vs. cross-sectional studies 289
  physiological differences between children and
        adults 287–8
  scanning procedures 287, 322

spatial normalization 288
use of sedation 397
see also legal and ethical issues
penetrance 430, 433–4
performance monitoring 29–31
perfusion fMRI, advantages over BOLD fMRI 332–3
permission *see* parental permission
        (informed consent)
phantom (coil loading) 305
pharmacologic MRI (phMRI) 396
phase encoding 419
phased array coils 330, 409
phenotypes 434
    anorexia nervosa 245
    anxiety disorders 183–4
    brain phenotypes 86–7
    bulimia nervosa 245–6
    depressive disorders 184
    see also endophenotypes; intermediate phenotypes
philosophy of ethics 274–5
phobias
    and amygdala dysfunction 189
    treatment of 357
    see also social phobia (SOPH)
phonological processing
    and cortical thickening 17, 389
    event-related potential (ERP) studies 92–4
    fMRI evidence 97–8
phosphocreatine (PCr) 287, 298, 309, 419
phosphodiesters (PDE) 300, 306–9, 420
phosphomonoesters (PME) 300, 420
    changes in ratio of PME/PDE 306–9
    reduced levels in schizophrenia 153–4, 310
phosphorus (^{31}P) MRS 150, 300, 420
    healthy brain development 306–9
    pediatric psychiatric disorders 310
physiological noise 287–8
placebo effect 346–7
planes of the brain 420
pleiotropy, definition 434
point-resolved spectroscopy (PRESS) 302, 420
polygenicity 366–7, 434
polymorphisms 247–8, 365, 434
positron emission tomography (PET) 420
    attention deficit/hyperactivity disorder (ADHD)
        studies 119–20, 121, 123, 124
        effects of stimulants and non-stimulants 347–50
    autism studies 136–7, 137–8, 139–40, 142
    brain glucose metabolism in depression 347

cognitive control studies 149–50
eating disorder studies 249
ethics preventing use on children 345, 397–8
obsessive-compulsive disorder studies 208–9
and serotonin transporter binding 375–6
precession 420
preference formation/selection 68
prefrontal cortex (PFC)
    adults with anxiety and depression 188, 191
    COMT regulating dopamine level in 371
    and development of emotion regulation 43–4
    and goal-directed behaviors 59–60
    modulating stimulus-reinforcement associations 186
    role in cognitive control functions 23–4
    role in reappraisal 40
    see also dorsolateral prefrontal cortex; frontal cortex;
        orbitofrontal cortex ventral prefrontal cortex
premutation 434
    in fragile X syndrome
        gene-behavior studies 220–1
        gene-brain-behavior studies 223–4
        gene-brain studies 222
    see also fragile X-associated tremor/ataxia
        syndrome (FXTAS)
prenatal alcohol exposure
    brain structural abnormalities 230–2
    developmental effects 229–30
probabilistic brain atlas 154, 420
probabilistic classification learning (PCL) 210–11
probability density functions (PDFs) 320, 420
probability distribution 420–1
probability in decision-making 60
proband 434
prognostication 155
promoters 366, 375, 434
prosody processing, brain lateralization 100
protected health information (PHI), privacy
        regulations 271–2
protein 434
proton (^{1}H) MRS 150, 297–300, 421
    and pediatric psychiatric disorders 309–10
    studies of healthy brain development 306
proton decoupling 300, 421
proton density (ρ) 421
pseudo-continuous ASL (pseudo-CASL) 328, 332, 421
psychotherapy, effects on the brain 347, 357–61
pulse 90° 421, 423
pulse 180° 416, 421
pulse sequence 421

pulse sequence programming 421
pulsed arterial spin labeling (PASL) perfusion
       imaging 328, 332, 421

q-ball imaging 421
q-space imaging 323, 422
quantitative electroencephalography (qEEG) 195, 422

raclopride 349–50, 422
radiofrequency (RF) 422
radiofrequency (RF) coil 409
radiofrequency (RF) pulse 281, 421, 422
radioisotope 422
radioligands 250–2, 422
reasoning ability 31–3
recombination 434
recovery (relaxation) 296, 422
region of interest (ROI) 286, 422
regional cerebral blood flow (rCBF)
    autism studies 136–7, 339
    normal development 335–6
    see also cerebral blood flow (CBF)
registration 422
relational reasoning 31–3
relative anisotropy (RA) 422
relaxation 422
relaxation rates 422
relaxation times 422
repeat number polymorphism 434
repeated free induction decay 423
repetition time (RT) 423
rephase/rephasing pulse 423
rephasing gradient 423
research legislation
    federal laws and regulations 264–72
    state law 272–3
resonances 423
resources
    educational 401
    neuroethics 405–6
    neuroinformatics 403
    software 401–2
response inhibition 27–8
    in attention deficit/hyperactivity disorder (ADHD) 115–19
    in fragile X syndrome (FXS) 223
    in Tourette syndrome and obsessive-compulsive
           disorder, role of fronto-striatal systems
           211–12
    neural correlates of 172

response selection 28–9, 34
resting state networks 285–6, 395
resting state studies
    attention deficit/hyperactivity disorder (ADHD)
           124, 350
    autistic spectrum disorders (ASDs) 136–7
    obsessive-compulsive disorder (OCD) 208–9
reward processing
    adolescent responses to outcome 68–9
    anticipation of monetary reward 61–8
    eating disorders
        anorexia and 5HT2A-1438A/A allele 247
        bulimia and anterior cingulate response 252
        role of dopamine 247
    preference formation/selection phase 68
    primary vs. secondary reward 60
    role of striatum 61, 124, 174, 186, 206
    stimulus-response (S-R) learning 186
    see also decision-making
ribonucleic acid (RNA) 366, 434
ribosome 434
RNA polymerase enzyme 366, 434
RNA toxicity 221, 434

saturation 423
saturation pulse 421
saturation recovery (SR) 423
schizophrenia 147–8
    brain development 148–9
    cognitive control 152
    cortical inefficiency of Val allele carriers 372–3
    diagnostic value of neuroimaging 155
    neurobiology of 148–9
    neuroimaging research 155
        etiopathology 155–6
        methodologic issues 154–5
        treatment implications 156
    neuroimaging studies 390
        brain function, changes in 152–3
        brain structure, changes in 150–1
        connectivity, changes in 151
        neurochemical brain alterations 153–4, 309, 310
    pathophysiological models of 148
    prognostic value of neuroimaging 155
    and working memory 152, 371
Secretary's Advisory Committee on Human Research
       Protections (SACHRP) 268–9
seeding region of interest (seeding ROI) 317, 320–1, 423–4
selective attention 22, 120, 121–2

selective serotonin reuptake inhibitors (SSRIs)
    and cognitive behavioral therapy in OCD 357–61
    and serotonin transporter function 375
    treatment of bulimia nervosa (BN) 254
self-reflection, role of medial prefrontal cortex 81, 84
semantic processing
    event-related potential studies 93–4
    evidence from near-infrared spectroscopy 98–9
sequence delay time 424
serotonin (5HT)
    and emotion 375
    risk factor for eating disorders 247
serotonin transporter gene (*5HTT*) 375
    functional 5HTT promoter alleles
        functional promoter polymorphism (5HTTLPR) 375
        HTTLPR predicts anxiety and dysphoria 376–7
        a third common, functional *HTTLPR* allele 375–6
    functional imaging of *HTTLPR*
        genotypes alter amygdala/frontal cortex coupling 377
        effect of *S* allele on amygdala-prefrontal cortex
            coupling 377–8
        stronger amygdala activation/volume 377
sex differences
    amygdala development 83
    brain activation in adolescent alcohol use disorder 236
    brain metabolites 309
    hippocampal volumes in bipolar disorder 174
    in cerebral metabolic rate of glucose (CMRGlc) in
        attention deficit/hyperactivity disorder
        (ADHD) 121
shim coils 424
shimming 424
sickle cell disease (SCD) 338–9
signal attenuation, diffusion tensor imaging 424
signal averaging 424
signal loss 316, 424
signal-to-noise ratio (SNR) 424
    arterial spin labeling (ASL) imaging 328–30
single nucleotide polymorphisms (SNPs) 369–70, 431, 435
single photon emission computed tomography
        (SPECT) 424
    abnormal dopaminergic transmission in TS 201–3
    autism studies 136–7
    cerebral blood flow after methylphenidate (MPH)
        treatment 350
    ethical constraints on use with children 345, 396
    fronto-striatal hyperactivity in obsessive-compulsive
        disorder (OCD) 208–9
    task activation studies in bulimia nervosa 252

single-shot radiofrequency excitation 330, 424
single-voxel magnetic resonance spectroscopy 302,
        310, 391, 424
slice selection 424
smoothing 424–5
social cognition 73–4
    brain mechanisms 80
        functional connectivity 85–6
        social perception 80–1
        theory of mind and self cognition 81
    development of
        perceiving emotions 78–9
        perceiving eye gaze 78
        theory of mind (ToM) 79–80
        understanding others' goals and intentions 74
    functional neuroimaging
        emotion understanding 82–4
        social perception 81–2
        theory of mind (ToM) 84–5
    value of developmental approach 86–7
social perception
    brain regions involved in 80–1
        functional neuroimaging of 81–2
    early development of 74
social phobia (SOPH)
    imaging studies showing activation 189–91
        enhanced amygdala activation 189
sodium-23 (^{23}Na) MRS 301, 425
software tools for image analysis 286, 303–4, 401–2, 405
spatial frequency filtering 425
spatial normalization 288, 425
spatial resolution 23, 302, 389, 425
spatial smoothing *see* smoothing
specific absorption rate (SAR) 425
spectroscopic imaging (SI) 302, 409, 425
speech processing 95, 96, 137–8
spin 425
spin axes 425
spin coupling 425
spin density 421, 425
spin echo 425–6
spin labeling schemes 328
spin lattice relaxation 417
spin-lattice relaxation time 423, 426, 428
spin-quantum number 426
spin-spin broadening 426
spin-spin relaxation time 423, 426
spin-spin splitting 426
spiral sequences/imaging 283, 426

standards of care 273–4
Statistical Parametric Mapping (SPM), software 402
stereotaxic/stereotaxic space 132, 426
stimulated echo acquisition (STEAM) 302, 426
stimulus-response learning in TS and OCD 210–11
stream-tubes 426
stress reactivity
   COMT Val158Met genotype and stress
      resiliency 373–5
   effects of *HTTLPR S* allele on stress exposure
      376–8, 377–8
   and emotion regulation 39
   role of hypothalamic-pituitary-adrenocortical
      (HPA) axis
    in depression 47
    in healthy adults and children 46–7
striatum
   abnormalities in bipolar disorder 174–5
   decreased blood flow in boys with attention
      deficit/hyperactivity disorder (ADHD) 124
   fractal triadic model 57–8
   implicated in Tourette syndrome and obsessive
      compulsive disorder circuitry 201–3, 211–12
   role in reward processing 124, 174
   role in S-R learning 210, 211
   structural abnormalities in ADHD 125
   see also caudate nucleus; ventral striatum
structural MRI (sMRI)
   ADHD studies 114, 351
   adolescent alcohol use disorder 236
   autism studies 131–2
   bipolar disorder studies 173
   eating disorders (EDs) 249
   fragile X syndrome 221
   legal implications 398
   morphometry comparisons 187–8, 418
   normal brain development 149, 288, 389
    NIH Pediatric MRI Data Repository 403
   obsessive-compulsive disorder studies 205–6
   prenatal alcohol exposure 230–2
   schizophrenia studies 150–1
   see also magnetic resonance imaging (MRI)
substance use disorders (SUD) 232
   cognitive deficits 235
   comorbidity with behavior disorders 234
   models of development of 232–4
   see also adolescent alcohol use disorder (AUD);
      illicit drug use

subtraction technique 427
subtractive logic 285
superior temporal sulcus (STS) 81
   abnormalities in autism 132, 137
    eye gaze processing 140
     in voice-selective areas of STS 139
     mentalization/theory of mind 140
    biological motion 82
    face sensitivity region 82
    intentions of actions 82
surface coil 409
susceptibility artifacts 427
susceptibility gene 435
sustained attention 120, 121
syntactic processing
   event-related potential studies 94
   fMRI studies 99

T_1 relaxation time 423
T_1-weighted MRI 427
T_2 relaxation time 423
T_2-weighted MRI 427
T_2^* 408, 423, 426
tagging efficiency *see* labeling efficiency
Talairach coordinates 427
TE (time to echo) 427
technological advances
   imaging modalities 391–2
   neuroinformatics 392–3
temporal parietal junction (TPJ), implicated in theory
      of mind 81, 85
temporal resolution 92, 195, 283–4, 427
terminology
   genetic terms 428–35
   imaging terms 407–28
Tesla 282–3, 427
test-retest reliability, longitudinal studies 394
theory of mind (ToM) 73, 79–80
   autism studies 139–40
   and eye contact 78
   functional neuroimaging 84
    behavioral vs. imaging studies, age-related
      differences 84–5
   and self cognition 81
threats, brain response to 185–6
thymine 435
TI (time to inversion) 427
tics *see* Tourette syndrome (TS)

tissue segmentation 6, 305, 427
Tourette syndrome (TS) 199
    anatomical disturbances 203–5
    functional disturbances 207–8
    neural circuitry 200–3, 390
    phenomenology 199–200
    stimulus-response learning 210–11
    see also obsessive-compulsive disorder (OCD)
TR (time to repetition) 427
trace (D) 316–17, 428
tracer half-life, blood T1 327, 328, 331
tracers 326, 428
tractography 317–18, 319, 428
trait behaviors, risk factor for eating disorders 248
transcranial Doppler sonography (TCD) 338, 428
transcranial magnetic stimulation (TMS) 290–1, 428
transcription 366, 435
transcription factors 435
transfer RNA (tRNA) 435
translation 435
transmission disequilibrium test 435
transverse magnetization 428
transverse relaxation time 423, 426, 428
treatment effects 344
    effects of placebo and cognitive behavioral therapy (CBT) on brain function 346–7
    future directions 361
    methodological issues 345–6
treatment responses
    mood stabilizers and antipsychotics in bipolar disorder
      functional and anatomic MRI 356–7
      magnetic resonance spectroscopy (MRS) 355–6
      normalizing neural circuitry 176–7
      potential type 1 biomarkers 357
    selective serotonin reuptake inhibitors (SSRIs) and CBT in obsessive-compulsive disorder 357–61
      potential type 1 biomarkers 361
    stimulant and non-stimulant effects in ADHD
      anatomic MRI 351
      event-related potentials (ERPs) 355
      functional MRI 351–5
      PET studies 351
      potential type 1 biomarkers 355
    see also treatment effects
trinucleotide repeat expansion 217, 435
triplet repeat expansion 435

utilitarianism, ethical issues 274–5

variance, attempts to reduce inter-site sources of 395
velocardiofacial syndrome (VCFS) 372, 429
ventral prefrontal cortex (VPFC)
    changes during adolescence 171–2
    dysfunction in bipolar disorder 175–6, 178
    effects of olanzapine monotherapy 177
    effects of psychotropic medication 356
    increased activation with lithium therapy 177
ventral striatum 57, 59
    abnormalities in bipolar disorder 174–5
    and reward expectancy 206
visuospatial working memory (VSWM) 24–7
    females with FXS 223
voice perception, autism studies 139–40
volumetric image analysis 5–6
voxel 428
voxel-based morphometry (VBM) 428
    autism studies 132
    bipolar disorder studies 174–5
    schizophrenia studies 150–1
    studies of normal development 6–7
voxelwise 428

white matter
    autism studies 132–3, 314
    density increase with age 7, 11
    DTI fiber tractography 317–18
    enlargement in FXS 222
    see also diffusion tensor imaging (DTI); fractional anisotropy (FA)
Williams syndrome 55
working memory (WM)
    developmental studies of 24
    effect of Val158Met in healthy adults 371
    effects of chronic alcohol exposure 236–8
    enhanced by dopamine 371
    in ADHD studies 119–20, 372
    and manipulation 27
    schizophrenia studies 152, 371
    visuospatial working memory 24–7, 223

X-chromosome 435
X inactivation 435
X-inactivation ratio 220, 435